BIOTECHNOLOGY AND PHARMACY

Biotechnology and Pharmacy

Editors

John M. Pezzuto, Ph.D.
Professor and Interim Head
Department of Medicinal Chemistry and Pharmacognosy
College of Pharmacy, University of Illinois at Chicago
Associate Director
Specialized Cancer Center, University of Illinois College of Medicine at
  Chicago
833 South Wood Street
Chicago, Illinois 60612

Michael E. Johnson, Ph.D.
Professor and Interim Director
Center for Pharmaceutical Biotechnology
College of Pharmacy
University of Illinois at Chicago
833 South Wood Street
Chicago, Illinois 60612

Henri R. Manasse, Jr., Ph.D.
Professor, College of Pharmacy,
Interim Vice Chancellor for Health Services
University of Illinois at Chicago
1737 West Taylor Street, Suite 203
Chicago, Illinois 60612

BIOTECHNOLOGY AND PHARMACY

Edited by JOHN M. PEZZUTO, MICHAEL E. JOHNSON, and HENRI R. MANASSE, Jr.

CHAPMAN & HALL

New York • London

First published in 1993 by
Chapman and Hall
29 West 35th Street
New York, NY 10001-2299

Published in Great Britain by
Chapman and Hall
2–6 Boundary Row
London SE1 8HN

Library of Congress Cataloging in Publication Data

Biotechnology and pharmacy / editors, John M. Pezzuto, Michael E. Johnson, Henri R. Manasse, Jr.
   p. cm.
  Includes index.
  ISBN 0-412-03861-7 (CL). — ISBN 0-412-03871-4 (PB)
  1. Pharmaceutical biotechnology. 2. Biotechnology. I. Pezzuto, John M. (John Michael) II. Johnson, Michael E., 1945– .
III. Manasse, Henri R. (Henri Richard)
  [DNLM: 1. Biotechnology—congresses. 2. Pharmacy—congresses. 3. Technology, Pharmaceutical—congresses. QV 778 B6167]
RS380.B59′1993
615′.1—dc20
DNLM/DLC
for Library of Congress   92-49685
                                           CIP

British Library of Congress Cataloguing in Publication Data also available

Contents

Contributors   vii

Preface   xi

Section I   *Basic Elements of Biotechnology*   *1*

1. Background to Recombinant DNA Technology   3
   Dr. Leonard G. Davis, Du Pont Merck Pharmaceutical Co.
2. Background to Monoclonal Antibodies   39
   Dr. Melvin E. Klegerman, UIC
3. Lymphokines and Monokines   53
   Drs. Melvin E. Klegerman and Nicholas P. Plotnikoff, UIC
4. Analytical Methods in Biotechnology   71
   Drs. R. J. Prankerd and S. G. Schulman, University of Florida
5. The Impact of Biotechnology on Analytical Methodology   97
   Dr. John F. Fitzloff, UIC
6. Drug Delivery Aspects of Biotechnology Products   116
   Dr. Diane J. Burgess, UIC

Section II   *Applications of Biotechnology in the Pharmaceutical Sciences*   *153*

7. Applications of Recombinant DNA Technology to the Diagnosis of Genetic Disease: Molecular Methods for Detecting the Genetic Basis of Diseases   155
   Drs. Donna R. Maglott and William C. Nierman, American Type Culture Collection
8. Human Genome Mapping and Sequencing: Applications in Pharmaceutical Science   193
   Drs. C. E. Hildebrand, R. L. Stallings, D. C. Torney, J. W. Fickett, N. A. Doggett, D. A. Nelson, A. A. Ford, and R. K. Moyzis, Los Alamos National Laboratory
9. Clinical Use of Monoclonal Antibodies   227
   Dr. John M. Brown, Centocor

10. Anti-AIDS Drug Development   250
    Dr. Prem Mohan, UIC
11. Oral Adenoviruses as the Carriers for Human Immunodeficiency Virus   275
    or Hepatitis B Virus Surface Antigen Genes
    Drs. Michael D. Lubeck, Satoshi Mizutani, Alan R. Davis, and Paul
    P. Hung, Wyeth-Ayerst Research
12. The Use of Nonclassical Techniques in the Production of Secondary   290
    Metabolites by Plant Tissue Cultures
    Drs. B. O'Keefe and C. Wm. W. Beecher, UIC
13. Applications of Biotechnology in Drug Discovery and Evaluation   312
    Drs. Cindy K. Angerhofer and John M. Pezzuto, UIC
14. Peptide Turn Mimetics   366
    Drs. Michael E. Johnson and Michael Kahn, UIC

Section III   *Biotechnology and the Practice of Pharmacy*   *379*

15. Biotechnology Products: An Overview   381
    Dr. Diana Brixner, SmithKline Beecham Pharmaceutical Corporation
16. The Pharmacist Practitioner's Role in Biotechnology: Clinical   402
    Application of Biotechnology Products
    Drs. Janet P. Engle, Donna M. Kraus, Louise S. Parent, and Mary
    Dean-Holland, UIC
17. Human Trials of Biotechnology Products: A Perspective   435
    Dr. W. Leigh Thompson, Eli Lilly and Company

Index   447

Contributors

Cindy K. Angerhofer, Department of Medicinal Chemistry and Pharmacognosy and Program for Collaborative Research in the Pharmaceutical Sciences, College of Pharmacy, University of Illinois at Chicago, 833 S. Wood St., Chicago, IL 60612

C. W. W. Beecher, Department of Medicinal Chemistry and Pharmacognosy and Program for Collaborative Research in the Pharmaceutical Sciences, College of Pharmacy, University of Illinois at Chicago, 833 S. Wood St., Chicago, IL 60612

Diana Brixner, SmithKline Beecham Pharmaceutical Corporation, 15375 S.E. 30th Place, Bellevue, WA 98007

John M. Brown, Centocor, Inc., 244 Great Valley Parkway, Malvern, PA

Diane J. Burgess, Department of Pharmaceutics, College of Pharmacy, University of Illinois at Chicago, 833 S. Wood St., Chicago, IL 60612

Alan R. Davis, Biotechnology and Microbiology Division, Wyeth-Ayerst Research, P.O. Box 8299, Philadelphia, PA 19101

Leonard G. Davis, CNS Diseases Group, Du Pont Merck Pharmaceutical, Experimental Station (E328/146A), Wilmington, DE 19880-0328

Mary Dean-Holland, Department of Pharmacy Practice and Clerkship Coordinator, College of Pharmacy, University of Illinois at Chicago, 833 S. Wood St., Chicago, IL 60612

N. A. Doggett, Life Sciences Division, Los Alamos National Laboratory, Los Alamos, NM

Janet P. Engle, Department of Pharmacy Practice and Office of Student Affairs, College of Pharmacy, University of Illinois at Chicago, 833 S. Wood St., Chicago, IL 60612

J. W. Fickett, Theoretical Division, Los Alamos National Laboratory, Los Alamos, NM 87545

John F. Fitzloff, Department of Medicinal Chemistry and Pharmacognosy, College of Pharmacy, University of Illinois at Chicago, 833 S. Wood St., Chicago, IL 60612

A. A. Ford, Life Sciences Division, Los Alamos National Laboratory, Los Alamos, NM 87545

C. E. Hildebrand, Life Sciences Division, Los Alamos National Laboratory, Los Alamos, NM 87545

Paul P. Hung, Biotechnology and Microbiology Division, Wyeth-Ayerst Research, P.O. Box 8299, Philadelphia, PA 19101

Michael E. Johnson, Department of Medicinal Chemistry and Pharmacognosy and Center for Pharmaceutical Biotechnology, College of Pharmacy, University of Illinois at Chicago, 833 S. Wood St., Chicago, IL 60612

Michael Kahn, Department of Chemistry, University of Illinois at Chicago, P.O. Box 6998, Chicago, IL 60680

Melvin E. Klegerman, Department of Pharmaceutics and Institute for Tuberculosis Research, College of Pharmacy, University of Illinois at Chicago, 833 S. Wood St., Chicago, IL 60612

Donna M. Kraus, Department of Pharmacy Practice, College of Pharmacy, University of Illinois at Chicago, 833 S. Wood St., Chicago, IL 60612; Department of Pediatrics, College of Medicine, University of Illinois at Chicago; Pediatric Intensive Care, University of Illinois Hospital and Clinics, Chicago, IL

Michael D. Lubeck, Biotechnology and Microbiology Division, Wyeth-Ayerst Research, P.O. Box 8299, Philadelphia, PA 19101

Donna R. Maglott, American Type Culture Collection, Rockville, MD 20852-1776

Satoshi Mizutani, Biotechnology and Microbiology Division, Wyeth-Ayerst Research, P.O. Box 8299, Philadelphia, PA 19101

Prem Mohan, Department of Medicinal Chemistry and Pharmacognosy, College of Pharmacy, University of Illinois at Chicago, 833 S. Wood St., Chicago, IL 60612

R. K. Moyzis, Center for Human Genome Studies, Los Alamos National Laboratory, Los Alamos, NM 87545

D. A. Nelson, Theoretical Division, Los Alamos National Laboratory, Los Alamos, NM 87545

William C. Nierman, American Type Culture Collection, Rockville, MD 20852-1776

B. O'Keefe, Department of Medicinal Chemistry and Pharmacognosy and Program for Collaborative Research in the Pharmaceutical Sciences, College of Pharmacy, University of Illinois at Chicago, 833 S. Wood St., Chicago, IL 60612

Louise S. Parent, Department of Pharmacy Practice, College of Pharmacy, University of Illinois at Chicago, 833 S. Wood St., Chicago, IL 60612; Family Practice/Rheumatology, University of Illinois Hospital and Clinics, Chicago, IL

John M. Pezzuto, Department of Medicinal Chemistry and Pharmacognosy, College of Pharmacy, University of Illinois at Chicago; Cancer Center, University of Illinois College of Medicine at Chicago, 833 S. Wood St., Chicago, IL 60612

Nicholas P. Plotnikoff, Department of Pharmacodynamics, College of Pharmacy, University of Illinois at Chicago, 833 S. Wood St., Chicago, IL 60612

R. J. Prankerd, College of Pharmacy, University of Florida, J. Hillis Miller Health Center, Gainesville, FL 32610

S. G. Schulman, College of Pharmacy, University of Florida, J. Hillis Miller Health Center, Gainesville, FL 32610

R. L. Stallings, Life Sciences Division, Los Alamos National Laboratory, Los Alamos, NM 87545

W. Leigh Thompson, Lilly Research Laboratories, Eli Lilly and Company, Indianapolis, IN 46220

D. C. Torney, Theoretical Division, Los Alamos National Laboratory, Los Alamos, NM 87545

Preface

During the 10 years that have elapsed since FDA approval of recombinant human insulin, over 100 products derived from biotechnological processes have progressed to the stage of clinical use or human trials. Over the next 10 years, global sales of biotechnology products are expected to approximate $30 billion per annum. It is therefore abundantly clear that the practice of pharmacy has been, and will continue to be, strongly influenced by the field of biotechnology and that the practicing pharmacist should have a firm understanding of the area. Because it is a relatively new field of scientific endeavor, however, traditional pharmacy curricula have not included comprehensive coverage of basic biotechnology or the resulting products and processes. Bearing these factors in mind, we have assembled *Biotechnology and Pharmacy*. It is anticipated to be of key interest to pharmacy students and practicing pharmacists who have not had extensive exposure to the field of biotechnology but who have received basic training in biological sciences (e.g., biochemistry, genetics) at the undergraduate professional level. The book should also be of value to other health professionals such as physicians, dentists, and nurses, who are interested in an overview of this important field of science. In addition, graduate students in the biological sciences should benefit by its use.

Biotechnology and Pharmacy has been organized into three parts: "Basic Elements of Biotechnology," "Applications of Biotechnology in the Pharmaceutical Sciences," and "Biotechnology and the Practice of Pharmacy." The chapters in these sections tend logically to fall into these groupings, but it is not necessary to peruse the parts or chapters in a sequential manner. Each chapter was prepared in a stand-alone manner, employing cross-references when appropriate to avoid redundancy. The text in its entirety is intended to provide a reasonably comprehensive overview of the history of many key elements of biotechnology, as well as a comprehensive description of the current state of the field and an indication of future trends. The text should also provide a solid foundation for more advanced studies in a number of related areas.

This volume developed through a faculty retreat and symposium on biotechnology supported through the Grant Awards for Pharmacy Schools program conducted by the American Association of Colleges of Pharmacy and funded by SmithKline Beecham. We gratefully acknowledge the financial support that made this work possible. We are also grateful to Carol Lewandowski and Martha Hoskins for assistance in many areas during the production of this volume.

Chicago   J.M.P.
January 1993   M.E.J.
    H.R.M., Jr.

SECTION I

Basic Elements
of Biotechnology

1

Background to Recombinant DNA Technology
Leonard G. Davis

1.1. Introduction

This chapter serves as a basic primer for those not familiar with the principles and techniques used in molecular biology and is not designed for those already aware of its value. It will emphasize terminology and provide general descriptions of the procedures, with a primary focus on why molecular biology is having such an impact (some call it a revolution) on biological studies. The chapter will also include a few scientific examples, since it is worth emphasizing that these techniques are best utilized as a complement of current approaches designed to answer research questions. Until a few years ago these techniques were employed mostly in academic investigations of the basic mechanisms of cell function. In a few cases they were applied by small biotechnology companies that sought to utilize the procedures for the production of bioactive proteins for use as potential drugs, rather than the typical heterocyclic organic chemicals. More recently this technology has been recognized as a powerful tool for facilitating more classical pharmaceutical and drug development efforts. The reader is, however, referred to the other chapters for more detailed examples, especially of pharmaceutical applications. For descriptions beyond the concise presentation necessitated herein, the reader is referred to some excellent books on the subject (for textbook descriptions, see references 1 to 3; for methodology, see references 4 to 7).

A convenient starting point is to recall the Central Dogma of Biology,[8] which dictates that information for the development, organization, and function of living systems is stored in discrete units (genes) within the linear deoxyribonucleic acid (DNA) molecules (chromosomes) of each cell.[9] At appropriate times, portions of the DNA information are transferred by transcription from the gene into linear ribonucleic acid (RNA) molecules for translation into functionally active proteins (see Figure 1.1). The basic building blocks used for DNA are four nucleotides (2-deoxyriboses linked through the 1 position to one of the nucleic acid bases, guanine [G], thymine [T], cytosine [C], or adenine [A]), which are linked to each

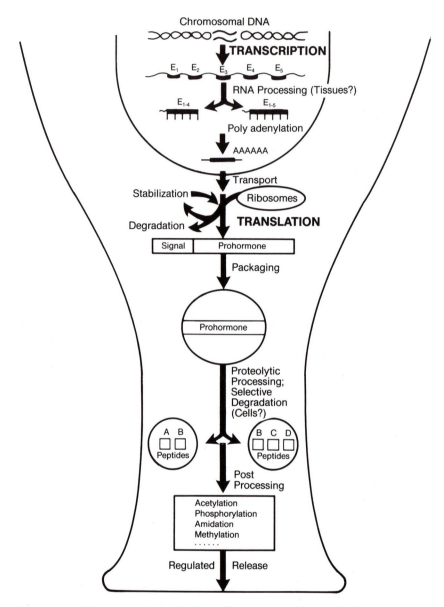

Figure 1.1. Schematic depicting the flow of information within a cell. The example used is a neuron that is generating a peptide neurotransmitter.

other in the linear DNA molecule through a phosphate ester bond at the 5' position of the deoxyribose of one nucleotide to the 3' position of the deoxyribose moiety on the adjacent nucleotide. DNA exists as a double helix with one linear strand proceeding in the 5' to 3' direction and the other aligned parallel to it but in the opposite orientation (i.e., 3' to 5'). The two strands are held together by hydrogen bonding, in which Cs pair with Gs and Ts pair with As. The resulting DNA forms two complementary strands such that the sequential order of the nucleotides of one of the strands (a gene) can be exactly copied into a messenger RNA (mRNA) molecule (copying of both strands occurs in cell duplication).

RNA also uses four nucleotides (uridine is used in place of thymidine) that are attached to ribose (no deoxy positions) and the same 5' to 3' covalent phosphoester linkage between adjacent riboses is used as the backbone of the polymer. These RNA units are linear and single stranded, since their synthesis depends upon the opening of the double-stranded DNA where RNA polymerase copies one strand through the use of the base pairing principles. This ensures that an exact complementary copy of the DNA exists as RNA. In most lower organisms (prokaryotes) the exact copy of the DNA information is represented in the RNA and directly codes for proteins; in higher organisms (eukaryotes) this linear RNA copy of the DNA contains regions that code for amino acids (called exons, since they are excised for use; see Figure 1.2) and intervening sequences of DNA that apparently do not code for the protein (called introns).

The RNA (after processing to remove introns in eukaryotes) then serves as the template (as mRNA) for translation into a protein molecule. The mechanism of converting linear nucleotide sequence information into a linear array of amino acids (proteins) depends upon the recognition of a nucleotide triplet (usually AUG for methionine) near the 5' end of the mRNA, which encodes the first amino acid (N-terminal). Each amino acid is carried by small transfer RNA molecules, each possessing a specific recognition site triplet (again complementary base pairing is used). The transfer RNA triplet complementary for the next three nucleotides in the mRNA binds and transfers its amino acid to the growing polypeptide chain. Consecutive triplets of nucleotides determine the orderly sequence of the amino acids until a stop codon (e.g., UAG) is found in the reading frame. The complete translation of mRNA occurs in the ribosomes (a RNA and protein complex) and results in a cellular protein. These proteins can then be posttranslationally modified (i.e., through phosphorylation, glycosylation, etc., of specific amino acids) and folded into three-dimensional topological units. The proteins serve many functions within the cell, ranging from the enzymatic machinery of energy metabolism and structural components of the membrane to the very specialized functions of different cell types (e.g., neurotransmitters for neurons, hormones for endocrine cells, and myosin for muscle cells). What is important here is that the Central Dogma and the triplet coding system are consistent from bacteria to humans; thus, the tools developed over many years of research by microbiologists can also be utilized.

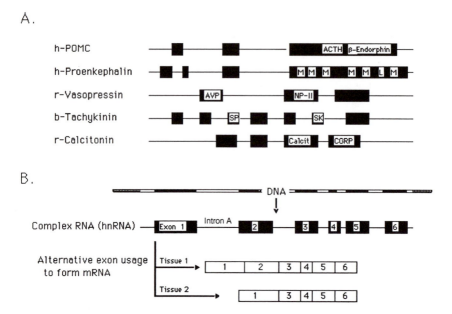

Figure 1.2. (**A**) Representation of eukaryotic gene organization using five different neuropeptide genes to demonstrate various spatial relationships between exons (boxes) and introns (lines); h, human; r, rat; b, bovine; ACTH, adrenocorticotrophin hormone; M, methionine enkephalin; L, leucine enkephalin; AVP, arginine vasopressin; NP-I, neurophysin; SP, substance P; SK, substance K; calcit, calcitonin; CGRP, calcitonin gene-related peptide. (**B**) Example of alternate processing of transcribed RNA (hnRNA) in eukaryotes; hn, heteronuclear (containing both exons and introns); the boxed numbers indicate exons within the intervening DNA sequences (introns) that are spliced out to form the different mRNAs for tissue-specific translation.

1.2. Bacteriological Impact

An apparently simple but major impact from the techniques of molecular biology is the ability to insert (using recombinant techniques) a single gene of interest, even from a different species, into bacteria.[10] These transformed bacteria can then serve as small "biofactories," since bacteria double in number at a fast rate (every 30 minutes or so) and have active machinery for biosynthesis. Thus, one can use them as vehicles to amplify an individual product thousands of times, especially one that occurs endogenously in a minute quantity of the original starting material, in a very short time (a few days). This product can be the original gene, complementary DNA (cDNA representing the mRNA), or its protein product (if a cDNA was used); fortunately, any of these can be rather easily isolated from the bacteria. The second bacteriological procedure that is commonly used is the ability to isolate an individual bacterium (containing the recombinant product of interest, called a clone) from a population of bacteria by a physical separation (spreading out on bacterial agar plates). This is a very powerful procedure when compared with classical protein isolation procedures for isolating and identifying the product of interest from a mixture. To evaluate how important these bacterial procedures are, one needs to understand how mammalian genes are converted into products to allow their manipulation in the bacterial systems.

1.3. Recombinant DNA Impact

The conversion of the normal cellular DNA or its RNA constituents into components that can be utilized in the bacteria is called cloning (see Figure 1.3). Two basic types exist: genomic clones (from cellular DNA) and cDNA clones (from mRNA).

With genomic clones bacterial viruses (bacteriophage[11]) are utilized as the "carrier vehicle" (vector), and these have been modified so that they can accept large pieces [up to 20 kilobase pairs (kb) in length] of exogenous DNA without altering their ability to infect and replicate within bacteria. Each chromosome contains millions of nucleotide base pairs (thousands of genes), which, prior to the discovery of restriction enzymes,[12] were largely unmanageable in a reproducible fashion; however, due to its large size and stability DNA was easy to isolate to study its overall structure.[13] Restriction enzymes, which are bacterial enzymes that have evolved with the ability to recognize unique stretches in double-stranded DNA (a very specific sequence of nucleotides), have had a revolutionary impact, since they are able to reliably hydrolyze (cut) the phosphoester bonds between nucleotides (see Figure 1.4).

The use of a specific restriction enzyme results in reproducible pieces of DNA, typically of manageable size, that still maintain the linear order of the nucleotide sequence, yet each now has at its ends (termini) the residual nucleotides after cleavage by the specific restriction enzyme. If the digested DNA is separated for analysis on an agarose gel (called a Southern blot[14]), DNA pieces of various sizes can be observed with a general stain (e.g., ethidium bromide). All these cut pieces (of various lengths) together encompass all the components of the chromosomes where this restriction enzyme recognized its stretch of DNA base pairs. About 100 such restriction enzymes are known to date. Thus, for example, if the restriction enzyme *Eco*RI (named as restriction enzyme 1 isolated from *Escherichia coli*[15]) is used to cut DNA, it will cut at every location within the chromo-

somal DNA sequence where a $\begin{smallmatrix} GAATTC \\ | \; | \; | \; | \; | \; | \\ CTTAAC \end{smallmatrix}$ exists. This results in DNA fragments

varying in size but each having the residual of the cleaved $\begin{smallmatrix} G & AATTC \\ | & | \\ CTTAA & G \end{smallmatrix}$

General CLONING Principles

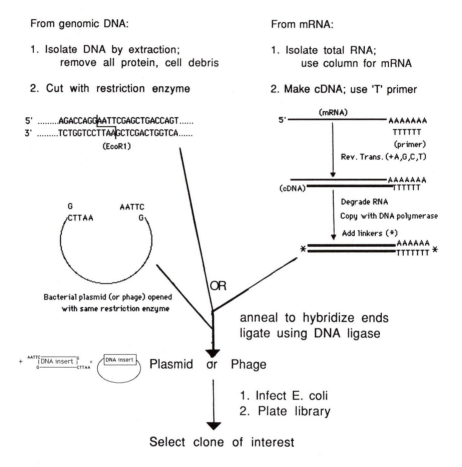

From genomic DNA:

1. Isolate DNA by extraction;
   remove all protein, cell debris

2. Cut with restriction enzyme

5' ........AGACCAGGAATTCGAGCTGACCAGT......
3' ........TCTGGTCCTTAAGCTCGACTGGTCA......
(EcoR1)

G   AATTC
CTTAA   G

Bacterial plasmid (or phage) opened
with same restriction enzyme

+ DNA insert = DNA insert   Plasmid or Phage

From mRNA:

1. Isolate total RNA;
   use column for mRNA

2. Make cDNA; use 'T' primer

(mRNA)
5'————————AAAAAAA
           TTTTTT
(primer)
Rev. Trans. (+A,G,C,T)

————————AAAAAAA
(cDNA)———————TTTTTT

Degrade RNA
Copy with DNA polymerase
Add linkers (*)
*============AAAAAA *
           TTTTTTT

OR

anneal to hybridize ends
ligate using DNA ligase

1. Infect E. coli
2. Plate library

Select clone of interest

1. With synthetic oligonucleotides that code for known peptide sequence
2. With cDNA clone from related gene-family or expected species homolog
3. With antibody if specific for target & available (use expression vector)

Figure 1.3.   Generalized procedure used to clone either genomic DNA or cDNA (from mRNA).

A.

Example of Restriction Enzyme Cleavage:

(5')..... AGTTGACAGGAATTCAGGCGACTGTAGGAC .....(3')

(3')..... TCAACTGTCCTTAAGTCCGCTGACATCCTG .....(5')
DNA + EcoR1 (R.E.)

YIELDS 2 DNA FRAGMENTS:

(5')..... AGTTGACAGG   AATTCAGGCGACTGTAGGAC .....(3')
(3')..... TCAACTGTCCTTAA   GTCCGCTGACATCCTG .....(5')

B.

Representative Restriction Enzymes and Cleavage Sites:

BamH1:	GGATCC CCTAGG	Cla1:	ATCGAT TAGCTA
Hind3:	AAGCTT TTCGAA	Kpn1:	GGTACC CCATGG
Pst1:	CTGCAG GACGTC	Xho1:	CTCGAG GAGCTC

Figure 1.4. (**A**) Schematic of how restriction enzymes cleave DNA to yield fragments with sequence-specific termini. (**B**) Representative examples of common restriction enzymes; arrows indicate location in sequence where that restriction enzyme cleaves the double-stranded DNA.

sequence at both its ends (specifically, AATTC exists on the 5′ termini and G on the 3′ ends, since restriction enzymes only cleave double-stranded DNA).

The resulting termini contain the appropriate complementary base pairs (since the sites are palindromes), and, naturally, if they were repositioned adjacent to each other, they would rehybridize (due to the hydrogen bonds) and could be ligated back together (using the enzyme DNA ligase) with the original covalent bonds reconstructed. Most powerful is the ligation of dissimilar DNA pieces (from different sources) by using the similar termini from restriction enzyme cutting (called recombinant molecular genetics[16]). For example, if a phage is also cut with *Eco*RI, a mammalian DNA piece can be inserted into the phage at that restriction cut location. This results in a recombinant molecule. Interestingly, the bacteriophage (lambda) has been engineered to accept pieces of foreign DNA (between about 7 and 20 kb in length) without affecting its function.[17] If this is done quantitatively (with a statistical representation of the DNA pieces from an entire genome) with an appropriate amount of viral DNA[5], one can insert at least a single representative of each DNA piece into a different phage particle. The sum total of all these insert-containing phages is called a genomic library. In a complete library every piece of chromosomal DNA (all genes) is represented, and within this population it ought to be possible to locate a sought-after gene. The potential exclusion of many small pieces (less than 7 kb) is avoided by performing a partial restriction digest of the original DNA; this is accomplished by monitoring and stopping the DNA cutting reaction (not allowing it to go to completion) when the average size of DNA pieces is between 7 and 20 kb. These insert-containing phage particles are used to infect bacteria [note that phage particles without inserts are defective, since they lack a DNA piece (insert), which does not allow the modified phage to replicate properly]. Before describing additional bacterial procedures, which are similar for genomic and cDNA libraries, it is worthwhile to describe cDNA cloning procedures.

Rather than obtain DNA copies of the entire genome, which is present in all cells, it is sometimes of interest (and easier) to determine which RNA molecules are being expressed in a given tissue (or cell) during certain conditions that interest the investigator, for example, developmental stages. A cDNA (copied or complementary DNA) is generated from the existing RNA molecules present in the tissue. The RNA is easily isolated by centrifugation through a cesium chloride cushion (which forms a gradient during the centrifugation);[18] an important step is homogenization in a buffer that will inhibit the commonly found and very active enzymes that destroy RNA, RNases. The pellet of RNA is separated from the genomic DNA, which only partially penetrates the gradient, and other cell constituents, such as proteins, which are too buoyant to enter the cesium chloride solution. The pellet, called total RNA, consists of the three RNA forms: ribosomal and transfer, which are part of the machinery used to translate the mRNA into proteins, and messenger (mRNA). The mRNA species are the transcripts that encode for proteins and can be purified from the other two forms. Purification is accomplished by taking advantage of the observation that most mRNA molecules have a long chain of adenosines [poly(A) tail] on the 3' end.[19] These can be used to hybridize (base pair) with a synthetic string of thymidines [oligo(dT)s] attached to an insoluble resin by using a standard column separation procedure.[20] The other RNA species (ribosomal and transfer) will run through the column while the mRNA molecules with the poly(A) tail will be retained and can be eluted by disrupting the A-T hydrogen bonds with a change in the ion (salt) concentration. The isolated mRNAs can then be copied (this means that a newly synthesized strand will be created as a complementary DNA strand from base pairing to this original RNA molecule[21]) with reverse transcriptase, a viral enzyme normally used by RNA viruses to accurately convert the viral strands into DNA.

To accomplish this copying, a poly(T) oligonucleotide is hybridized to the poly(A) tail of the mRNA as a starting point (primer) and then the reverse transcriptase adds, in the correct order, nucleotides complementary to each mRNA nucleotide. Once this strand has been generated, the original RNA template is destroyed so that the cDNA can be copied to generate a double-stranded cDNA molecule. This time a modified DNA polymerase is used; this enzyme is naturally used to copy DNA molecules, especially in repair and duplication of cells. The resulting double-stranded piece of DNA exactly reflects the original RNA in one strand (the other strand is the complement), which can now be manipulated as described previously for genomic DNA pieces. One important difference is that ends of the cDNA fragment need to have sequences added (linkers) that can be recognized by restriction enzymes for proper ligation into the phage vector. The addition of the unique linkers avoids further reducing the integrity and size of the cDNA. The typical cDNA is 2 to 3 kb long owing to limitations in the ability of the enzymes to copy for long stretches and because many mRNA species are in this range. The vectors described previously (lambda phages) are not appropriate for these small pieces; however, other modified phages (called λgt10 and λgt11[22]) can accommodate these small pieces and are commonly in use today.

Historically, however, plasmids were used for the cloning of cDNAs.[23] Plasmids are small pieces of DNA that exist autonomously in bacteria, independent of their chromosomes, and can replicate and, for our purposes, possess DNA sites recognized by some restriction enzymes. A common plasmid in recombinant use is called pBR322,[24] which has the ability to accept small DNA pieces (e.g., cDNA). An interesting feature of the plasmids is their ability to confer drug resistance to those cells that possess the plasmid;[11] this allows selection (killing) against those bacteria that do not contain a plasmid and an insert.

The insertion of all cDNAs generated from a given tissue by ligation to a restriction-enzyme–cut plasmid results in a cDNA library. In theory it represents all mRNAs made in that tissue at the time of the experiment. These insert-bearing plasmids do not have the ability, as phages do, to infect bacteria (they are typically passed in the wild from one bacteria to another during bacterial conjugation).

Instead, the plasmids can be added mechanically to the bacteria to allow uptake (transformation) in a procedure that uses a temperature shock of calcium-treated bacteria (called competent cells).[25] At this time whether one has a genomic or cDNA library the next procedure that one needs to perform is the selection of the DNA sequence of interest from the hundreds of thousands of library members.

The key to selection is having a tool that can identify the clone of interest. Bacteriological techniques allow one to physically separate individual bacterial cells from one another by pouring a dilute suspension of cells onto an agar plate (plating). Exponential growth of one cell overnight into hundreds of identical cells results in a colony; if the bacterium contained a plasmid, the colony now consists of identical cells, all of which contain that plasmid. If a bacterium contains a bacteriophage, the originally infected cell releases phage particles to adjacent bacterial cells, and the adjacent cells are continuously lysed after infection by the phage so that a "clear" circular area containing phage particles (called a plaque) exists among the remaining uninfected cells of the "lawn." If one considers that a single cDNA or gene of interest exists as one of 100,000 clones, it should be represented in the plating of 50 plates with 2,000 colonies (or plaques) each (an easy one-day task). The only limitations to obtaining the pure clone of interest is the statistical chance that it is within the population plated and the possession of a tool to identify the one wanted. Recall that in addition to a numerical increase in bacteria during culture growth, a given bacteria will also make multiple copies of the plasmid or phage, allowing an amplification (making for easier detection) of the sought-after product.

A variety of procedures are in use for "finding unknowns." The two most common selection procedures are (1) using an oligonucleotide (made synthetically so that it represents the codons for an amino acid sequence determined from, for example, an isolated peptide fragment) as a probe to hybridize to a specific stretch of DNA, or hybridizing with a clone for a related molecule or even a clone from another species if some DNA homology is expected to exist; and (2) for cDNA libraries detection of the protein product can be performed, but this approach requires that an antibody be available for the protein of interest and that an expression cloning vector be used (e.g., pUC,[26] a modified plasmid, or the λgt11 vector[21]). The radioactive pieces (probes) are made through procedures (e.g., "nick" translation, end labeling, etc.[4-7]) in which the nucleotide fragment is incubated with radioactive nucleotides and enzymes that will incorporate the radiolabel; the antibody probe finds a protein epitope and is typically located with a second "detector" antibody that is directed at the constant region of the primary antibody and carries an enzyme to serve as the reporter.

In either case the library is plated onto an agar substratum plate for growth, after which it is transferred to a solid support such as nitrocellulose or nylon membranes (filters). This transfer results in a replica of the plate existing on the filter, which can be probed for the molecule of interest (note that, fortunately, only a portion of the colonies or plaques is effectively removed). These filters are permanently marked in unique positions so that any positively identified signals can be relocated on the original agar plate by realignment of them. The support membrane is incubated under conditions that will minimize nonspecific binding. The conditions used depend upon whether a radioactively labeled DNA piece (hybridization) or antibody is used and on the expected specificity or signal intensity from the probe. The incubation allows an equilibration of the probe with the target clone. These filters are then washed to remove the background, and the signal is identified autoradiographically for radioactive probes or enzymatically for antibody probes. Once a positive signal is identified, the filter is realigned with the plate to allow isolation (picking) of the original clone or plaque. Picking this putative positive bacterial clone or plaque allows regrowth, plating, and so forth. Note, the signal intensity reflects the bacterial efficiency while the number of positives reflects the frequency of the cDNA in the starting material. The foregoing identification procedure is repeated for the putative positive to confirm its purity. Thus, in a matter of a few weeks a pure clone of a DNA (or protein) of interest can be obtained. The preceding case utilized known information (related DNA clone or antibody) and is expected to proceed smoothly. When little information is known or few tools are available various heroic and clever efforts have been made to obtain a clone of interest without ever purifying or characterizing the protein of interest.[27] The diversity of these various methods is not within the scope of this chapter. Regardless, these procedures are significantly faster than most protein purification schemes and, more important, will routinely result in obtaining significant information on the DNA sequence, which can be converted to protein sequence information directly. Sequencing these clones is the next major advantage from recombinant molecular biology.

1.4. Structural Information Impact

The ability to determine the sequence of the isolated genomic or cDNA clones has revolutionized the rate and ease with which someone can obtain very specific primary structural information. The nucleotide sequence obtained can be easily converted into an amino acid sequence. Predictions on the three-dimensional structure can then be made based on the order of hydrophobic, hydrophilic (positive and negative charges), and unique amino acids when compared with patterns learned from previously studied proteins. A typical amino acid analyzer might be able to determine up to 40 amino acids a day, but then other purified peptide fragments are required to continue the sequence; various methods have been successfully devised to accomplish this. However, since the genetic code of triplet nucleotides that is used to encode specific amino acids is conserved, the ability to sequence a clone allows direct prediction of the amino acid sequence by scanning for an open reading frame (an initiation and termination codon within a continuous frame of triplets). Peptide sequences can also be identified if they are known to occur in the peptide of interest. Furthermore, DNA sequencing allows for a continuous determination until the sequence of the entire insert within the clone is completed; significantly more information per day can be obtained, since a typical laboratory can easily read about 240 nucleotides (i.e., about 80 amino acids) in a single sequencing run. Various strategies have been devised so that thousands of nucleotides can be determined in a week or so. Certainly, some pitfalls do exist in this approach (e.g., posttranslationally modified amino acids are not identified, and care is required to avoid misreads); however, the procedure is very effective in obtaining the complete sequence of a protein, even if the initial sequence was obtained following purification and partial amino acid sequencing.

The procedure commonly used is one devised by Sanger,[28] which takes advantage of the linkage structure of DNA and enzymes that efficiently and accurately copy DNA (see Figure 1.5) as was described previously. Recall that each nucleotide is covalently attached to the next through a phosphate ester linkage of the 5' hydroxyl of one sugar moiety to the 3' hydroxyl of the next (DNA uses 2'-deoxyriboses). When a strand is copied, the enzyme (e.g., Klenow fragment of DNA polymerase) uses the 5'-triphosphate of one nucleotide and adds it to the free 3' hydroxyl on the existing chain of nucleotides. The principle is simple in that if the last nucleotide of the chain does not have a 3' hydroxyl (3' deoxy; the ribose is now called dideoxy), another nucleotide *cannot* be added and the chain elongation procedure terminates. Since the DNA of interest is contained in a vector (typically a special single-stranded vector called M13,[29] but more recently performed by denaturing double-stranded plasmids into single-stranded pieces[30]), the vector and insert serve as the template for copying; an oligonucleotide complementary to a known sequence in the vector adjacent to the location of the insert serves as a primer (since it will hybridize through hydrogen bonding of base pairs) to start the chain for elongation. The chain will now elongate (copy) if the four nucleotide triphosphate substrates and a copying enzyme are present. If the reaction mixture contains a few nucleotides that are the dideoxy species of a specific nucleotide, some of the chain elongations will terminate each time that particular dideoxynucleotide is incorporated. The remaining molecules (which have the normal 2' deoxy, 3' hydroxyl) will continue to serve as a substrate to further the elongation reaction by the continued additions of subsequent nucleotides, until a dideoxynucleotide is incorporated in each. As expected, this results in a percentage of the total population of molecules terminated at each position. Adjusting the ratio of dideoxy- and deoxyriboses for one nucleotide at a time results in a percentage of chain terminations at each position where that nucleotide occurs; repeating the reaction (actually performed simultaneously) for the other three nucleotides in separate reaction vials results in a chain termination at each possible base position in the clone.

Finally, if a radioactive (^{32}P or ^{35}S) tracer nucleotide has been included in each reaction the termination site can be detected by autoradiography after separation on a polyacrylamide sequencing gel. The questions are, just where (at what

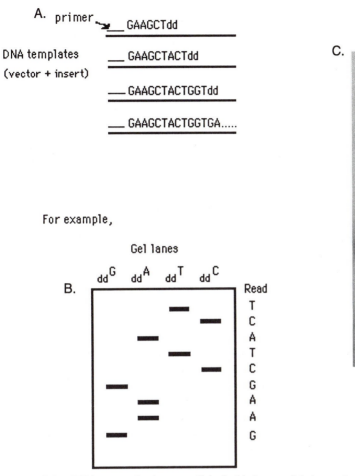

Figure 1.5. DNA sequencing is accomplished with the use of chain terminators (dideoxy-nucleotides). (**A**) How a typical reaction might proceed. (**B**) Drawing of an idealized autoradiogram after separation on a gel. (**C**) An actual gel.

position) and in which tube (determines nucleotide) did the "termination" occur, since that will determine which nucleotide exists at which position. This sequencing gel system can resolve single nucleotide differences in the length of the molecules. If the preceding four reactions (one for each dideoxynucleotide) are run adjacent to each other on the gel, the DNA sequence is read by starting at the bottom of the gel [shortest strands (earliest terminated positions) and thus closest to the starting primer] and then "walking up" the ladder of the autoradiographic banding pattern of the four reactions. If the length of the fragment of interest exceeds the reaction or gel's ability to separate and identify each termination, continued reading can be accomplished by creating a new oligonucleotide primer, based on the just read sequence, and performing another set of reactions. This can be continued until all the insert is sequenced; other approaches have been devised to speed the determination of the complete sequence.

1.5. Informational Impact

During the last few years the introduction of molecular genetic techniques to biological questions has opened new vistas and provided the tools to obtain additional information on how cells and systems operate. For example, in the study of cellular changes induced by ligand–receptor interactions, it has long been known that second messengers (e.g., cyclic AMP, ion permeability, and phosphtidylinositols) mediate the internalization of information and assist the cell in determining its response to the external change in status quo. The application of these molecular techniques has demonstrated that not only do these transient changes occur but changes in gene expression with a longer-term implication also occur. This section will attempt to concisely highlight how the use of the previously described molecular genetic technology has had an impact in one scientific discipline, neuroscience.

The initial impact of molecular genetic techniques in neurobiology was the ability to use recombinant cloning and cDNA sequencing to determine the primary structure of the neuropeptide transmitters and their precursor proteins. The initial studies focused on somatostatin,[31] proopiomelanocortin (POMC),[32] arginine vasopressin (AVP),[33] oxytocin,[34] and methionine enkephalin (ENK).[35] More recently, many more neuropeptide precursor structures have been determined through cloning and DNA sequencing with this technology, as well as most neurotransmitter synthesizing and degrading enzymes,[36–38] many receptor proteins,[39–43] a few ion channel proteins,[44, 45] and some structural proteins. This work has allowed the prediction of the primary structure for these precursor molecules. Furthermore, the results from these studies have clarified some issues in peptide neurobiology. For example, the localization of vasopressin in hypothalamic–pituitary cells was always with another peptide, neurophysin II, which occurred within the same cellular vesicles, whereas the hormone oxytocin was always localized with neurophysin I. The reason for this cellular separation became apparent immediately after determining the sequence of the precursor for these two hormones.[46] Each neurophysin is specifically encoded within the gene along with the respective hormone with which it is colocalized. When the hormone precursor is proteolytically processed into the hormone peptide within the vesicle during transport, the appropriate neurophysin is generated. Similarly, several different opioid peptides (those that bind to the endogenous morphine receptor) had been identified and each was known to have a common stretch of amino acid sequence, but the biosynthetic relationship between methionine enkephalin, leucine enkephalin, β-endorphin, and dynorphin was far from clear. Sequence information directly showed that each peptide was derived primarily from one of three distinct prohormones (both enkephalins are contained in the same precursor) and that each prohormone was encoded by a separate gene;[47–49] thus, their expression and interaction with the receptor were individualized.

Sequence information has also led to the discovery of new neurotransmitters. For example, upon sequencing the calcitonin gene, R. Evans' group determined from the complete cDNA sequence that in some tissues a modified mRNA exists that upon translation might result in an additional peptide fragment.[50] This altered mRNA does not occur in the parathyroid and was a result of a different utilization of exons (the coding regions of the genes) in other tissues, especially brain. This alternate utilization is achieved by a complex process that occurs in the cell nucleus called alternate splicing of heteronuclear RNA (direct copy of the gene) to generate the final mRNA. Recent studies indicate that the peptide, called calcitonin gene-related peptide (CGRP), is primarily located in nerve terminals and has specific binding sites in the brain, suggesting that CGRP may have a role as a neurotransmitter in the neuronal control of cardiovascular function.[51] Molecular genetic studies of brain function have demonstrated that this selective utilization of genetic information through tissue-specific processing of the RNA to yield different combinations of exons from a single gene is commonly used to make mRNA species encoding different peptide combinations in various cells. Another example of this is the generation of the substance-P and -K peptides, in which the coding sequence for each occurs on different exons but within the same gene; each exon is differentially used depending on the tissue and cell type.[52]

Another area where the determination of the primary structure of a prohormone clarified a question on whether peptide fragments in nearby adjacent cells were the result of different genes, alternative splicing, uptake from a nearby cell, or differential posttranslational proteolytic processing of a single gene product is exemplified by understanding the POMC gene.[53, 54] In this case a single gene is transcribed to the same mRNA species in all cells and it is translated into the identical prohormone in each of these cells. However, different cells process different peptide products of the prohormone, so that each cell type maintains and uses the appropriate, but different, peptides. For example, intermediate lobe cells of the rodent pituitary primarily make β-endorphin and α-MSH, whereas anterior lobe cells make adrenocorticotropin (ACTH), even though all these proteins are derived from the same gene, same message, and same prohormone. The key is the posttranslational processing of the prohormone into peptides that are cell specific by the selective destruction of some peptide fragments, which results in different gene products in adjacent cells.

In addition to providing the structural information on peptide neurotransmitters, the cloning of these genes has provided powerful tools to be used as probes to detect and measure the amount of specific mRNA species in a tissue through a procedure called Northern blot hybridization.[55] Northern blots are similar to Southern blots[14] except that RNA is separated and transferred rather than restric-

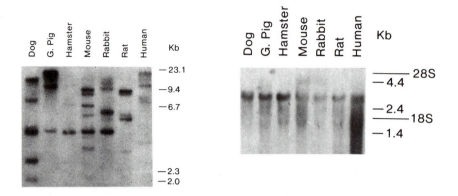

Figure 1.6. (**Left**) A Southern blot autoradiogram; DNA from different species were cut with a restriction enzyme and separated on a gel before transfer to a nitrocellulose filter. The filter was hybridized with a radioactive probe and the darkened areas (bands) were determined by exposure to X-ray film and indicate where the DNA pieces that bind the radiolabeled probe are located. (**Right**) A Northern blot; similar electrophoretic separation, transfer, and hybridization procedures were performed as described for DNA except that RNA was used. Note that the gene detected here has homologous forms in these different species.

tion-enzyme–cut DNA (see Figure 1.6). The use of radiolabeled probes (e.g., "nick translation" of DNA inserts that are isolated from amplified clones by restriction enzyme cleavage of vectors and gel separation) in hybridization experiments has allowed the detection of changes in levels of specific mRNA species after numerous and various experimental manipulations. These hybridization experiments are most often conducted by Northern gel analysis. In this method RNA is isolated from a homogenized tissue sample, denatured and separated on an agarose gel, transferred to a nitrocellulose or nylon based support membrane sheet, and hybridized with a radiolabeled probe that is complementary to the mRNA under investigation.

One of the earliest studies demonstrating the regulation and expression of a neuropeptide gene utilized a radioactively labeled cDNA probe to measure specific changes in the mRNA levels.[56] Long-term treatment with glucocorticoids was known to cause a decrease in the POMC level in the anterior lobe of the rat pituitary without affecting POMC in the intermediate lobe. It was unclear whether this was a result of increased secretion, decreased translation, or a suppression of the transcription of the POMC gene. Steroid treatment induced a decrease in the mRNA level that slightly preceded but paralleled (as expected if it was a decreased mRNA synthesis) the decrease demonstrated for the peptide. Thus, molecular biology techniques add to the types of information that can be obtained in an experiment.

In similar experiments performed in vivo,[57] adrenalectomy (removal of glucocorticoids) is known to produce a prolonged increase in POMC, suggesting that the expression of the POMC mRNA is bidirectionally influenced by the levels of circulating glucocorticoids. Presumably, the glucocorticoid-activated receptor, which is internalized once the steroid hormone binds to it, is transported to the nucleus, where it causes a set of cell specific changes in gene expression as a consequence of the receptor complex binding to unique regulatory DNA sequences associated with certain genes, and is typically located "upstream" (5') from the coding region. The translocation of information to the chromosomes is usually mediated by protein factors that bind to very specific stretches of DNA positioned near different genes; this results in the differential activation in the transcription of specific genes as needed. These DNA sequences are referred to as regulatory regions (promoters, enhancers, silencers, etc.), and regulatory factors bind to them to turn genes on or off under specified conditions;[58] a detailed description is beyond the scope of this chapter (see Figure 1.7).

The dynamic up- and down-regulation of the expression levels of several other hypophyseal hormone genes by peptide releasing factors and steroids has been addressed by several laboratories using radioactive cDNA probes.[59] The explanted adrenal gland has been used in elegant experiments to demonstrate that apparently short-term membrane-mediated activation or inhibition can also influence mRNA levels.[61, 62] For example, quantitative increases in ENK mRNA levels were induced by an elevation of cAMP. These changes were blocked by an opposing depolarization induced by potassium or veratridine. Moreover, the effects of veratridine could be blocked by tetrodotoxin to demonstrate an ion-channel-specific effect, suggesting that the expression and regulation of some cellular genes is under fine control and can be influenced by changes in neuronal activity. As impressive as the aforementioned in vitro studies are in describing the biochemical and molecular events that result from membrane or receptor activation leading to changes in the regulation of the recipient cell's genes, they lack the anatomical orientation so important for the study of brain function. A critical question in neuroscience is how neuronal activity and in vivo development regulates the production of cell specific products (e.g., neurotransmitters) in the highly defined connection pathways in the brain.

A method has been developed that allows identification of mRNA molecules within individual cells after histological sectioning by using hybridization procedures similar to those used in Northern blots coupled with microscopic analysis at the cellular level (see Figures 1.8 and 1.9). Following the pioneering description by Gee and Roberts of a method called in situ hybridization histochemistry,[62] many labs, including our own, have utilized and refined this procedure for detecting the presence of, and changes in, specific mRNA levels occurring in different brain region cell types after experimental manipulation. Conceptually in situ hybridization is analogous to immunocytochemistry or receptor autoradiography except that mRNA molecules are being identified by the "reporter" molecule. The reporter molecules typically are radiolabeled DNA probes that are

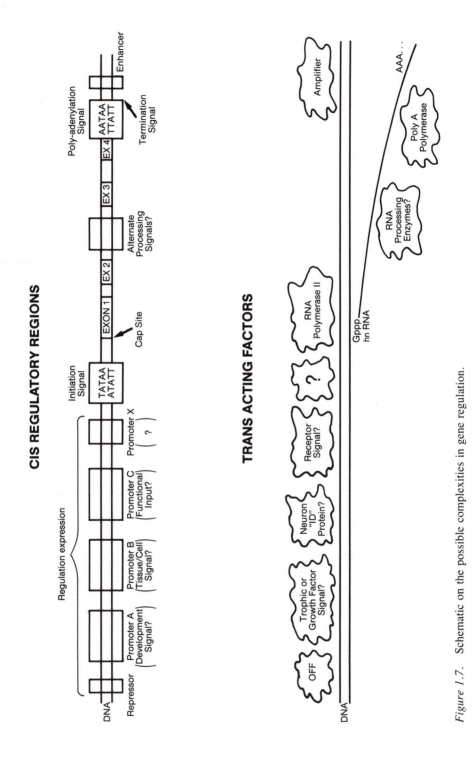

Figure 1.7. Schematic on the possible complexities in gene regulation.

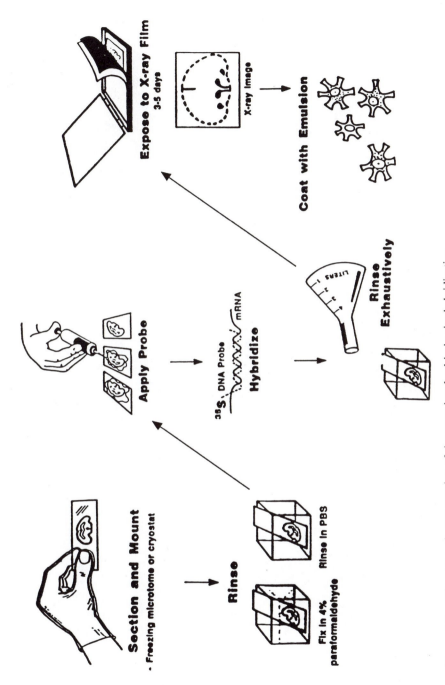

Figure 1.8. Diagrammatic presentation of the steps involved in in situ hybridization.

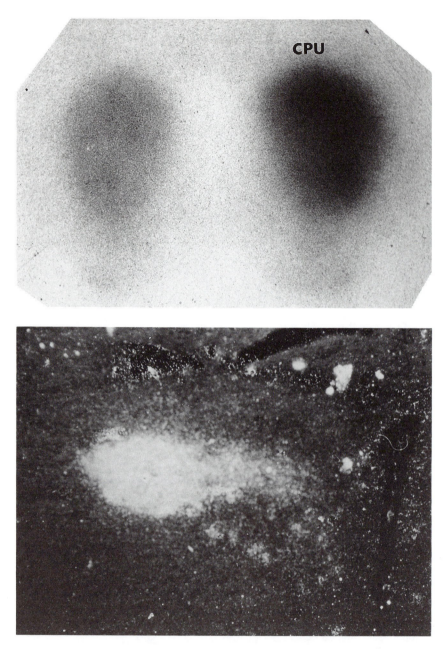

Figure 1.9. Examples of data obtained from in situ hybridization experiments. (**Top**) X-ray film data of mRNA changes following an ipsilateral lesioning experiment (see text). (**Bottom**) The type of signal obtained in discrete brain regions after emulsion coating and autoradiography.

complementary to the sequence of interest. Although longer cDNA (and cRNA) probes have been used successfully, we have used synthetic oligodeoxyribo-nucleotide (oligonucleotide) probes,[63-65] since they offer a number of advantages over cDNA probes which include the following: (1) they allow the use of published sequences, which alleviates the need to clone the mRNA of interest as well as the task of implementing all the requisite microbiological techniques; (2) they can be designed to distinguish two homologous mRNAs from the same gene family; and (3) they can be designed to be complementary to specific regions of a given mRNA to study differential utilization of exons in various cell types. Many laboratories have utilized in situ hybridization histochemistry to investigate the anatomical localization and expression of many neuropeptide mRNAs (a few examples include somatostatin,[66] cholecystokinin,[67] calmodulin,[68] vasopressin,[63] oxytocin,[69] ENK,[64] *Aplysia* egg-laying hormone,[70] and POMC.[65])

One example will be presented here on the tachykinin system to demonstrate how molecular genetic techniques can still provide additional detailed information on a system that has been investigated since the early 1970s.[71] The substance-P peptide (SP) is derived by proteolytic processing of the tachykinin prohormones, which in bovine brain are encoded by two related mRNA species (α and β).[52] It has been reported that the two forms of the tachykinin prohormone found in bovine tissues are derived from the same gene by alternative splicing.[52] In particular, the α form results from a deletion of exon 6 (54 nucleotides) compared with the β form, which utilizes all seven exons. The missing exon in the α form, upon translation, results in a prohormone that encodes SP but lacks substance-K (SK), another tachykinin peptide more recently discovered.[72] The β-tachykinin mRNA encodes this SK-peptide as well as the SP-peptide. We were interested in determining whether a similar molecular genetic scheme existed in the rat as described previously for bovine.[73]

Based on the highly homologous nucleotide sequence (evolutionarily conserved between species), cloning and sequencing the rodent tachykinin cDNAs, and some experiments to determine the absolute presence of a mRNA species by stringent hybridization experiments (S1-nuclease and RNase protection assays), rat striatum was shown to contain another form (γ) of tachykinin mRNA that is different from the two forms previously reported to occur in bovine.[73,74] The difference is the apparent exclusion of a different exon (4) in rodent brain that occurs between the SP and SK coding regions. It is the predominant form found in rat striatum; the other two forms have also been reported to exist in reduced amounts and in other tissues.[74] Based on our determination of the rat tachykinin mRNA sequence, we used a synthetic oligonucleotide probe that is complementary to this tachykinin mRNA. The probe specifically identified a mRNA of the correct size on Northern blots of striatum.[75] Using in situ hybridization, a labeled mRNA was detected by autoradiographic localization in the striatum, the central nucleus of the amygdala, the habenula, and the hypothalamus, all brain structures that have been shown to contain SP-peptide by immunocytochemical methods.

This anatomical information confirms the identity of this sequence as encoding a tachykinin precursor and allows a regional analysis of the changes that occur as a result of pharmacological manipulation.

The antipsychotic efficacy of haloperidol, as well as the development of side effects during chronic treatment, are most likely due to its primary action as a dopamine (DA) receptor antagonist in the central nervous system.[76] DA-containing perikarya in the substantia nigra project to the striatum and impinge upon, among others, SP- and ENK-containing neurons.[77] SP-containing neurons in the striatum project, in turn, back to the substantia nigra. This reciprocal anatomical arrangement suggests an involvement of both SP and dopamine in regulating basal ganglia function. ENK-containing neurons in the striatum project primarily to the globus pallidus. We sought to extend previous studies[78] that determined specific changes in neuropeptide levels after pharmacological manipulation of the DA receptor. Of particular interest was striatal tachykinin and ENK mRNAs levels after blockade of the DA receptor with haloperidol. Chronic administration of haloperidol to rats causes a 40% reduction in the concentration of SP-peptide in the substantia nigra and a 200% increase in ENK in the globus pallidus. These changes develop gradually during the first week of haloperidol administration.[79] The hypothesis tested here is whether the DA-receptor–mediated, haloperidol-induced reduction of SP-peptide levels and the elevation of ENK-peptide levels are the result of a bidirectional change in the mRNA levels from the same pharmacological input. We found an mRNA reduction of tachykinin mRNA and an elevation of ENK mRNA in the striatum, which indicates that these two neuropeptide genes are under a reciprocal control that is responsive to blockade by DA receptor antagonists. Interestingly, the changes found in the mRNA levels are reversible upon removal of the drug. Moreover, we utilized in situ hybridization histochemistry to evaluate the anatomical distribution of the haloperidol-induced changes in these neuropeptide mRNA levels.

The decrease in tachykinin mRNA was uniformly distributed in the striatum when the autoradiographic signal from the in situ hybridization results were analyzed. Although the increase in ENK mRNA was uniformly increased in the striatum, it was more pronounced in the dorsal portion. The effect of chronic haloperidol treatment was regionally specific, since no effect on ENK or tachykinin mRNA levels in other brain regions was observed.[73] Although haloperidol induced the changes in both mRNA levels, it could potentially have been due to an action of haloperidol that did not involve dopamine receptors; coadministration of apomorphine, a DA receptor agonist, prevented the decrease in tachykinin mRNA. Another experimental manipulation that chronically reduces DA neurotransmission in the striatum is the unilateral destruction of substantia nigra DA cells with 6-hydroxydopamine. Such lesions result in a significant elevation of striatal ENK mRNA on the ipsilateral side when compared with the contralateral side, since no DA input is being received on the ipsilateral side (Figure 1.9A). Moreover, these results focus attention on the sequelae of molecular events secondary to chronic interruption of DA-mediated neurotransmission to understand mechanisms of how neurotransmission influences postsynaptic neuropeptide gene expression. These results could have a significant effect in understanding the long-term usage of antipsychotic drugs in the treatment of schizophrenia and are examples of using molecular biological tools to increase our ability to obtain information on biological systems.

1.6. Unraveling the Molecular Biology of Disease

Molecular genetic procedures are having an impact on our understanding of human diseases. Already the molecular genetic defect of a number of disorders has been identified. These include oncogenes in cancer,[80] Huntington's chorea,[81] Duchenne muscular dystrophy,[82] cystic fibrosis,[83] atherosclerotic factors,[84] neurofibromatosis,[85] and possibly even affective disorders.[86] This new information on the precise defect in these problems will, it is hoped, lead to effective treatments (or prevention) in the near future where no hope realistically existed before. One example is Alzheimer's disease, which may affect as much as 20 percent of the population over 80 years of age, yet only a pathological description of the neuronal abnormalities has existed since it was first described in 1907.[87]

Alzheimer's disease is a form of dementia marked by progressive intellectual deterioration without focal, motor, or sensory signs.[88] Definitive diagnosis of the disease is often difficult, since many other disorders and neurological deficits are also accompanied by cognitive loss; however, three characteristic histopathological structures are present in the brains of Alzheimer's patients: neurofibrillary tangles, neuritic plaques, and cerebrovascular plaques. Neuritic and cerebrovascular plaques are found in highest concentration in the hippocampus and the neocortex and result from a pathological deposition of a fragment of the recently discovered amyloid precursor protein(s) (APP). The hallmark pathology of the disease is currently identified with standard stains; however, recent studies have shown that plaque formation is one of the earliest events in the progression of the disease.[89, 90] Moreover, a significant correlation exists between the number of neuritic plaques and neurofibrillary tangles and the clinical severity of dementia;[91, 92] the most serious cognitive deficits are correlated with the greatest number of plaques. These observations underscore the importance of recent efforts to define plaque composition and the mechanism(s) of plaque generation.

Although the etiology of the disease is unknown, researchers have made major advances in the last few years in understanding the disease by using a molecular biological approach, particularly in understanding the composition of the characteristic amyloid plaques. It is worth noting that in some cases the disease appears inherited (familial), whereas in others no family linkage is known. Studies of the familial Alzheimer's disease with molecular genetic techniques has identified an apparent genetic defect (on chromosome 21) through linkage analysis of chromosomes from these family members (those with and without the symptoms).[93] Linkage studies use specific DNA markers for various regions of each of the chromosomes, and these are then hybridized to restriction-enzyme–digested DNA from each patient. If a defect is "linked," it can be identified by studying the DNA fragment pattern after cutting with specific restriction enzymes. If a certain pattern shows up with a unique probe and restriction enzyme (called a restriction fragment length polymorphism, RFLP) in the afflicted group and not in the "normal," a linkage to the disease is found. Note that the chromosome markers, even though the RFLP might be associated with the "linkage" through transmission, might be millions of nucleotides away from the actual defect; thus, it is necessary to "walk" the chromosome to find the specific mutation, a potentially difficult task. Moreover, studies utilizing human pedigrees are difficult owing to the limited generations available for sampling. For example, further studies performed with DNA from additional Alzheimer's disease families indicate that the pathology might not be localized to a single, homogeneous genetic defect.[94]

While these genetic studies were being performed, a parallel approach in some laboratories was to try to analyze the pathological marker: amyloid plaques which are the molecular components characteristic of the disease. After the initial isolation and purification of amyloid plaques isolated from an Alzheimer's-diseased brain, a peptide fragment (called β-amyloid) has been shown to be the major constituent of the pathological plaques.[95–98] Glenner and Wong isolated, from the insoluble birefringent core of cerebrovascular plaques, a 4,200 Da peptide that was sequenced and shown to be a peptide of about 36 amino acids, a previously unknown sequence.[95] It was obvious that this insoluble peptide was derived as a fragment of a larger protein. Analysis of additional peptides isolated from neuritic plaques from Alzheimer's and Down's syndrome patients was performed (older Down's syndrome patients also develop progressive dementia and plaque formation quite similar to that associated with Alzheimer's disease[99]), and these additionally isolated peptides were found to contain a similar amino acid sequence.[96]

The amino acid sequence of the β-amyloid peptide has enabled researchers to use standard recombinant DNA procedures to identify a larger protein that contains this peptide sequence by isolating cDNA clones that encode the mRNA for a precursor protein (APP).[93, 100–102] It became obvious in the analysis of the mRNA encoding APP that it was not unique to Alzheimer's disease (see Figure 1.10) but was a protein that played a role in normal cellular function, since it had a wide tissue distribution and was conserved across species.[101, 103] Thus, in the disease state it is believed that APP (or a peptide fragment) is inappropriately processed and deposited as part of plaques in neuronal tissue. Recent studies have supported this proposal in that the β-peptide is apparently not derived by normal processing.[104] The availability of cDNA clones and antibodies raised against the predicted amino acid sequence have facilitated major advances in describing the

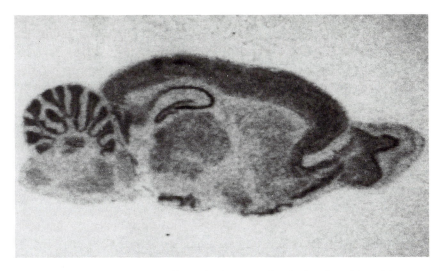

Figure 1.10. X-Ray film autoradiography demonstrating the location of the APP precursor mRNA in normal rat brain.

structure of the APP molecule and its possible function(s). For example, it has been recently demonstrated that more than one form of mRNA is expressed for APP, and these additional transcripts can be derived through alternate processing of the precursor mRNA.[105-107] Based on these cloned cDNA sequences for APP, the amino acid sequences of each of the protein molecules can be predicted from the coding region of the mRNA. While all three forms contain sequences identical to the β-peptide, it is interesting that two of the forms use an additional, common exon to encode an amino acid sequence of Kunitz-type protease inhibitors. It has been shown that APP is glycosylated in a stepwise fashion,[108] can be phosphorylated,[109] and is highly susceptible to certain proteases.[110] Most exciting was the report that the APP gene was located on chromosome 21,[101] the same one reported from the linkage analysis of familial Alzheimer's disease. Further research indicated that the locations within the chromosome were not the same for the APP gene and the inherited form. Interestingly, a mutation has been found in the APP gene for a few inherited forms of the disease.[111]

Using molecular genetic techniques to obtain the cDNA for the precursor protein for the amyloid plaques found in Alzheimer's disease has facilitated studies aimed at understanding how this protein might function normally and malfunction in the disease state. To study APP(s), antibodies have been developed based on the predicted amino acid sequence that can detect APP on Western blots of tissue extracts or in tissue sections with immunocytochemical procedures. For example, Card et al., using an antibody that would detect all three forms of APP, were able to map the overall distribution of APP in normal rat brain,[112] and Siman et al. were able to detect the "induction" of APP in reactive astroctyes following neuronal damage.[113] The ability to localize and detect APP in neuronal function has increased dramatically with the use of antibodies directed at specific but different portions of the APP molecule and by in situ hybridization techniques in order to understand its function in normal tissue, as well as the aberrant processing that might lead to plaque deposition. Thus, the evidence is accumulating on the complex nature of the APP molecules,[114, 115] and the opportunity exists now to fully understand the biosynthetic and posttranslational events as well as the proteolytic processing that leads to various bioactive fragments. Although the function of APP is not known, it appears that APP or specific fragments of the APP molecule may play various roles in cellular growth and neuronal development such as occurs for a homologue identified in *Drosophila*.[116] Characterizations in some cases have led to an identification of APP fragments that may be related to (or even identical to) other previously known molecules such as nexin-II[117] or coagulation factor XIa inhibitor[118] or a heparin binding protein.[119]

Thus, the identification from cDNA cloning of the predicted amino acid sequence corresponding to the three forms of APP allows the creation of specific cDNA and antibody tools for studying the genomic location and its functional regulation and disposition in the context of both normal cellular function and under experimental conditions. Moreover, it is hoped that its role in the disease process, whether it be a direct or indirect effect, will be forthcoming, since the suspected abnormal breakdown of APP into the characteristic plaques may accompany (or even cause) the disease process. The major interest is in determining a cause and effect relationship, if one exists, for Alzheimer's disease, and after 80 years, molecular biology may have opened the door, albeit only a little, to understanding this disease.

1.7. Summary and Future Prospects

The impact of molecular genetic approaches on biological research has been tremendous. In neurobiology, benefits from this technology range from the basic information of gene families and protein structure, which allows both a prediction of the amino acid sequence for understanding function and for use in the generation of specific antibodies, to the design of complementary probes for mRNA analysis to study events in gene regulation. The development of in situ hybridization histochemistry has had a particularly profound influence in neurobiology because of the anatomical complexity and chemical diversity of the nervous system. These tools allow detection of changes in specific mRNA levels in discrete brain regions or single neurons following experimental manipulation. Although significant progress and insight has already been achieved in understanding the regulation of neuropeptide mRNAs, many studies have been aimed at confirming and extending information previously suspected from immunocytochemical data. New procedures are expected to provide further insights into the specific mechanisms of how neurotransmitter regulation is affected by synaptic activity. For example, a very exciting recent report demonstrated the ability to detect by in situ hybridization the unprocessed complex RNA (containing both exons and introns) in the nucleus with an intron specific probe;[120] these RNA molecules exist immediately after transcription of the gene. This finding suggests that transcriptional activity can be detected in individual cells with in situ hybridization histochemistry, and serves as the next logical extension of the previous studies of steady-state cytoplasmic RNA levels. Moreover, these in situ hybridization histochemistry procedures will lend themselves directly to studying receptor and enzymatic regulation at the cellular level as well.

The continual infusion of molecular genetic technologies into all scientific disciplines promises to have a significant impact on answering fundamental biological questions. Some of these approaches have been initiated and include nuclear run-on assays to study actual gene transcriptional rates,[121] DNA hypersensitivity sites to determine potential regulatory regions on genes,[122] determining domains (cis regions) within each gene that are "receptors" for DNA binding proteins that control gene function (transacting factors),[123] the powerful DNA amplification procedures,[124] and in the diagnosis of genetically based disorders. Particular attention should be paid to the gene amplification procedures (PCR)[124] since these powerful techniques are improving many of the basic cloning methods. Their application in the context of development, neurotransmitter actions, aging, disease, and control of specific phenotypic expression at the individual neuron-cell will lead to a better understanding of the system complexities and, certainly, pose new fundamental questions on its plasticity.

One of the newer and more powerful techniques aimed at understanding the function of specific proteins in the context of biology is the creation of transgenic animals.[125] In these animals a gene of interest is inserted into a fertilized oocyte and the resultant offspring possess the new gene, the expression of which is under experimental control. These studies should allow determinations of gene function and malfunction in certain experimental conditions or diseases, since it is possible to observe the effects of a gene in a controlled whole-animal setting. The addition of a whole new array of techniques to the traditional interdisciplinary study of biological problems promises a bright future in our quest for understanding cellular and system function.

Acknowledgments

The comments and suggestions offered during the preparation of this manuscript by Bob Manning, Leslie Logue, Chris Reid, and Dan Tyrrell are greatly appreciated. I would also like to further acknowledge my collaborators during many of the referenced experiments: J. Angulo, R. Arentzen, F. Baldino, B. Burkhart, J. Card, G. Christoph, R. Krause, R. Lampe, B. Lampe, M. Lewis, R. Manning, R. Meade, R. Nelson, and R. Siman.

References

1. Watson, J.D., Hopkins, N.H., Roberts, J.W., Steitz, J.A., and Weiner, A.M. 1987. *Molecular biology of the gene*. 4th ed. 2 vols. Menlo Park, Calif: Benjamin Cummings.

2. Darnell, J., Lodish, H., and Baltimore, D. 1986. *Molecular cell biology*. New York: Scientific American Books.

3. Alberts, B., Bray, D., Lewis, J., Raff, M., Roberts, K., and Watson, J.D. 1989. *Molecular biology of the cell*. New York: Garland.

4. Davis, L.G., Dibner, M.D., and Battey, J.F. 1986. *Basic methods in molecular biology*. New York: Elsevier.

5. Maniatis, T., Fritsch, E.F., and Sambrook, J. 1982. *Molecular cloning: A laboratory manual*. Cold Spring Harbor, N.Y.: Cold Spring Harbor Laboratory Press.

6. Sambrook, J., Fritsch, E.F., and Maniatis, T. 1989. *Molecular cloning: A laboratory manual. 2d ed*. Cold Spring Harbor, N.Y.: Cold Spring Harbor Laboratory Press.

7. Ausubel, F.M., Brent, R., Kingston, R.E., Moore, D.D., Smith, J.A., Seidman, J.G., and Struhl, K., eds. 1988. *Current protocols in molecular biology*. New York: Greene, Wiley-Interscience.

8. Crick, F.H.C. 1958. On protein synthesis. *Symp. Soc. Exp. Biol.* 12:138–163.

9. Watson, J.D. 1970. *Molecular biology of the gene,* New York: W.A. Benjamin.

10. Cohen, S.N., Chang, A.C.Y., Boyer, H.W., and Helling, R.B. 1973. Construction of biologically functional bacterial plasmids in vitro. *Proc Natl. Acad. Sci.* U.S.A. 70:3240–3244.

11. Ingraham, J.L., Low, K.B., Magasanik, B., Schaechter, M., and Umbarger, H.E., eds. 1987. *Escherichia coli and Salmonella typhimurium: Cellular and molecular biology*. In particular, pp. 1169–1178. Washington, D.C.: American Society for Microbiology.

12. Smith, H.O., and Wilcox, K.W. 1970. A restriction enzyme from *Hemophilus influenzae*. 1. Purification and general properties. *J. Mol. Biol.* 51:379–391.

13. Watson, J.D., and Crick, F.H. C. 1953. Molecular structure of nucleic acids: A structure for deoxyribonucleic acid. *Nature (London)* 171:737–738.

14. Southern, E.M. 1975. Detection of specific sequences among DNA fragments separated by gel electrophoresis. *J. Mol. Biol.* 98:503–517.

15. Mertz, J.E., and Davis, R.W. 1972. Cleavage of DNA by RI restriction endonuclease generates cohesive ends. *Proc. Natl. Acad. Sci. U.S.A.* 69:3370–3374.

16. Jackson, D.A., Symons, R.H., and Berg, P. 1972. Biochemical method for inserting new genetic information into DNA of simian virus 40: Circular SV40 DNA molecules containing lambda phage genes and the galactose operon of *E. coli. Proc. Natl. Acad. Sci. U.S.A.* 69:2904–2909.

17. Blattner, F.R., Williams, B.G., Blechl, A.E., Denniston-Thompson, K., Faber, H.E., Furlong, L.A., et al. 1977. Charon phages: Safer derivatives of bacteriophage lambda for DNA cloning. *Science* 196:161–169.

18. Chirgwin, J.M., Przybyla, A.E., MacDonald, R.J., and Rutter, W.J. 1979. Isolation of biologically active ribonucleic acid from sources rich in ribonuclease. *Biochemistry* 18:5294–5299.

19. Darnell, J.E., Wall, R., and Tushinski, R.J. 1971. An adenylic acid-rich sequence in messenger RNA of HeLa cells and its possible relationship to reiterated sites in DNA. *Proc. Natl. Acad. Sci. U.S.A.* 68:1321–1325.

20. Aviv, H., and Leder, P. 1972. Purification of biologically active globin messenger RNA by chromatography on oligothymidylic acid-cellulose. *Proc. Natl. Acad. Sci. U.S.A.* 69:1408–1412.

21. Okayama, H., and Berg, P. 1982. High efficiency cloning of full length cDNA. *Mol. Cell Biol.* 2:161–170.

22. Young, R.A., and Davis, R.W. 1983. Efficient isolation of genes by using antibody probes. *Proc. Natl. Acad. Sci. U.S.A.* 80:1194–1198.

23. Bolivar, F., Rodriques, R.L., Greene, P.J., Betlach, M.C., Heyneker, H.L., Boyer, H.W., Crosa, J., and Falkow, S. 1977. Construction and characterization of new cloning vehicles. II. A multi-purpose cloning system. *Gene* 2:95–113.

24. Sutcliffe, J.G. 1978. Nucleotide sequence of the ampicillin resistance gene of *Escherichia coli* plasmid pBR322. *Proc. Natl. Acad. Sci. U.S.A.* 75:3737–3741.

25. Mandel, M., and Higa, A. 1970. Calcium-dependent bacteriophage DNA infection. *J. Mol. Biol.* 53:159–162.

26. Vieira, J., and Messing, J. 1982. The pUC plasmids, an M13mp7-derived system for insertion mutagenesis and sequencing with synthetic primers. *Gene* 19:259–265.

27. Barnard, E.A. 1986. Application of molecular biology to receptors. In *Innovative approaches in drug research*. A.F. Harms, ed. Amsterdam: Elsevier.

28. Sanger, F., Nicklen, S., and Coulson, A.R. 1977. DNA sequencing using chain-terminating inhibitors. *Proc. Natl. Acad. Sci. U.S.A.* 74:5463–5467.

29. Sanger, F., Coulson, A.R., Barrell, B.G., Smith, A.J.H., and Roe, B.A. 1980. Cloning in single-stranded bacteriophage as an aid to rapid DNA sequencing. *J. Mol. Biol.* 143:161–178.

30. Zagursky, R.J., Baumeister, K., Lomax, N., and Berman, M.L. 1985. Rapid and easy sequencing of large linear double-stranded DNA and supercoiled plasmid DNA. *Gene Anal. Technol.* 2:89–94.

31. Shen, L.P., Pictet, R.L., and Rutter, W.J. 1982. Human somatostatin I: Sequence of the cDNA. *Proc. Natl. Acad. Sci. U.S.A.* 79:4575–4579.

32. Whitfeld, P.L., Seeburg, P.H., and Shine, J. 1982. The human pro-opiomelanocortin gene: Organization, sequence, and interspersion with repetitive DNA. *DNA* 1:133–143.

33. Land, H., Schutz, G., Schmale, H., and Richter, D. 1982. Nucleotide sequence of cloned cDNA encoding bovine arginine vasopressin-neurophysin II precursor. *Nature (London)* 295:298–303.

34. Land, H., Grez, M., Ruppert, S., Schmale, H., Rehbein, M., Richter, D., and

Schutz, G. 1983. Deduced amino acid sequence from the bovine oxytocin-neurophysin I precursor cDNA. *Nature (London)* 302:342–344.

35. Comb, M., Seeburg, P.H., Adelman, J., Eiden, L., and Herbert, E. 1982. Primary structure of the human Met- and Leu-enkephalin precursor and its mRNA. *Nature (London)* 295:663–666.

36. Grima, B., Lamouroux, A., Blanot, F., Biguet, N.F., and Mallet, J. 1985. Complete coding sequence of rat tyrosine hydroxylase mRNA. *Proc. Natl Acad. Sci. U.S.A.* 82:617–621.

37. Joh, E.E., Suh, Y.H., and Joh, T.H. 1986. Complete nucleotide and deduced amino acid sequence of bovine phenlethanolamine N-methyltransferase: Partial amino acid homology with rat tyrosine hydroxylase. *Proc. Natl. Acad. Sci. U.S.A.* 83:5454–5458.

38. Lee, D.C., Carmichael, D.F., Krebs, E.G., and McKnight, G.S. 1983. Isolation of a cDNA clone for the type I regulatory subunit of bovine cAMP-dependent protein kinase. *Proc. Natl. Acad. Sci. U.S.A.* 80:3608–3612.

39. Mishina, M., Kurosaki, T., Tobimatsu, T., Morimoto, Y., Noda, M., Yamamoto, T., et al. 1984. Expression of functional acetylcholine receptor from cloned cDNAs. *Nature (London)* 307:604–608.

40. Kobilka, B.K., Dixon, R.A.F., Frielle, T., Dohlman, H.G., Bolanowski, M.A., Sigal, I.S., et al. 1987. cDNA for the human b_2-adrenergic receptor: A protein with multiple membrane-spanning domains and encoded by a gene whose chromosomal location is shared with that of the receptor for platelet-derived growth factor. *Proc. Natl. Acad. Sci. U.S.A.* 84:46–50.

41. Schofield, P.R., Darlison, M.G., Fujita, N., Burt, D.R., Stephenson, F.A., Rodriquez, H., et al. 1987. Sequence and functional expression of the GABA-a receptor shows a ligand-gated receptor super-family. *Nature (London)* 328:221–227.

42. Masu, Y., Nakayama, K., Tamaki, H., Harada, Y., Kuno, M., and Nakanishi, S. 1987. cDNA cloning of bovine substance-K receptor through oocyte expression system. *Nature (London)* 329:836–838.

43. Sokoloff, P., Giros, B., Martes, M.P., Bouthenet, M.L., and Schwartz, J.C. 1990. Molecular cloning and characterization of a novel dopamine receptor as a target for neuroleptics. *Nature (London)* 347:146–151.

44. Noda, M., Shimizu, S., Tanabe, T., Takai, T., Kayano, T., Ikeda, T., et al. 1984. Primary structure of *Electrophorus electricus* sodium channel deduced from cDNA sequence. *Nature (London)* 312:121–127.

45. Tanabe, T., Takeshima, H., Mikami, A., Flockerzi, V., Takahashi, H., and Kangawa, K., et al. 1987. Primary structure of the receptor for calcium channel blockers from skeletal muscle. *Nature (London)* 328:313–318.

46. Ruppert, S., Scherer, G., and Schutz, G. 1984. Recent conversion involving bovine vasopressin and oxytocin precursor genes suggested by nucleotide sequence. *Nature (London)* 308:554–557.

47. Drouin, J., and Goodman, H.M. 1980. Most of the coding region of rat ACTH/b-LPH precursor gene lacks intervening sequences. *Nature (London)* 288:610–612.

48. Horikawa, S., Takai, T., Toyosato, M., Takahashi, H., Noda, M., Kadidani, H., et al. 1983. Isolation and structural organization of human preproenkephalin B gene. *Nature (London)* 306:611–614.

49. Civelli, O., Douglas, J., Goldstein, A., and Herbert, E. 1985. Sequence and expression of the rat prodynorphin gene. *Proc. Natl. Acad. Sci. U.S.A.* 82:4291–4295.

50. Amara, S.G., Jonas, V., Rosenfeld, M.G., Ong, E.S., and Evans, R.M. 1982. Alternative RNA processing in calcitonin gene expression generates mRNAs encoding different polypeptide products. *Nature (London)* 298:240–244.

51. Rosenfeld, M.G., Mermod, J.J., Amara, S.G., Swanson, L.W., Sawchenko, P.E., Rivier, J., Vale, W.W., and Evans, R.M. 1983. Production of a novel neuropeptide encoded by the calcitonin gene via tissue-specific RNA processing. *Nature (London)* 304:129–135.

52. Nawa, H., Hirose, T., Takashima, H., Inayama, S., and Nakanishi, S. 1983. Nucleotide sequences of cloned cDNAs for two types of bovine brain substance P precursor. *Nature (London)* 306:32–36.

53. Gee, C.E., Chen, C.C., Roberts, J.L., Thompson, R., and Watson, S.J. 1983. Identification of proopiomelanocortin neurones in rat hypothalamus by *in situ* cDNA-mRNA hybridization. *Nature (London)* 306:374–376.

54. Watson, S.J., Albala, A.A., Berger, P., and Akil, H. 1983. Peptides and psychiatry. In *Brain peptides*. D.T. Krieger, M.J. Brownstein, and J.B. Martin, eds. New York: Wiley-Interscience. pp. 359–368.

55. Alwine, J.C., Kemp, D.J., and Stark, G.R. 1977. Method for detection of specific RNAs in agarose gels by transfer to diazobenzyloxymethyl-paper and hybridization with DNA probes. *Proc. Natl. Acad. Sci. U.S.A.* 74:5350–5354.

56. Schachter, B.S., Johnson, L.K., Baxter, J.D., and Roberts, J.L. 1982. Differential regulation of propiomelanocortin mRNA levels in the anterior and intermediate lobes of the rat pituitary. *Endocrinology* 110:1442–1444.

57. Birnberg, N.C., Lissitzky, J.C., Hinman, M., and Herbert, E. 1983. Glucocorticoids regulate proopiomelanocortin gene expression *in vivo* at the levels of transcription and secretion. *Proc. Natl. Acad. Sci. U.S.A.* 80:6982–6986.

58. Briggs, M.R., Kadonaga, J.T., Bell, S.P., and Tjian, R. 1986. Purification and biochemical characterization of the promoter specific transcription factor, Sp1. *Science* 234:47–52.

59. Schmitt, F.O., Bird, S.J., and Bloom, F.E., eds. 1982. *Molecular genetic neuroscience*. New York: Raven.

60. LaGamma, E.F., White, J.D., Adler, J.E., Krause, J.E., McKelvy, J.F., and Black, I.B. 1985. Depolarization regulates adrenal preproenkephalin mRNA. *Proc. Natl. Acad. Sci. U.S.A.* 82:8252–8255.

61. Kley, N., Loeffler, J.P., Pittius, C.W., and Hollt, V. 1986. Proenkephalin A gene expression in bovine adrenal chromaffin cells is regulated by changes in electrical activity. *EMBO J.* 5:967–970.

62. Gee, C.E., and Roberts, J.L. 1983. Laboratory methods: *In situ* hybridization

histochemistry—a technique for the study of gene expression in single cells. *DNA* 2:157–163.

63. Davis, L.G., Arentzen, R., Reid, J.M., Manning, R.W., Wolfson, B., Lawrence, K.L., and Baldino, F. 1986. Glucocorticoid sensitivity of vasopressin mRNA levels in the paraventricular nucleus of the rat. *Proc. Natl. Acad. Sci. U.S.A.* 83:1145–1149.

64. Angulo, J.A., Davis, L.G., Burkhart, B.A., and Christoph, G.R. 1986. Reduction of striatal dopaminergic neurotransmission elevates striatal proenkephalin mRNA. *Eur. J. Pharm.* 130:341–343.

65. Lewis, M.E., Sherman, T.G., Burke, S., Akil, H., Davis, L.G., Arentzen, R., and Watson, S.J. 1986. Detection of proopiomelanocortin mRNA by *in situ* hybridization with an oligonucleotide probe. *Proc. Natl. Acad. Sci. U.S.A.* 83:5419–5423.

66. Uhl, G.R., Evans, J., Parta, M., Walworth, C., Hill, K., Sasek, C., et al. 1986. Vasopressin and somatostatin mRNA *in situ* hybridization. *In situ hybridization in brain.* G. Uhl, ed. New York: Plenum. pp. 21–47.

67. Ingram, S.M., Krause, R.G., Baldino, F., Skeen, L.C., and Lewis, M.E. 1989. Neuronal localization of cholecystokinin mRNA in the rat brain by using *in situ* hybridization histochemistry. *J. Comp. Neurol.* 287:260–272.

68. Roberts-Lewis, J.M., Cimino, M., Krause, R.G., Tyrrell, D.F., Davis, L.G., Weiss, B., and Lewis, M.E. 1990. Anatomical localization of calmodulin mRNA in the rat brain with cloned and synthetic oligonucleotide probes. *Synapse* 5:247–254.

69. McCabe, J.T., Morrell, J.I., and Pfaff, D.W. 1986. *In situ* hybridization as a quantitative autoradiographic method: Vasopressin and oxytocin gene transcription in the Brattleboro rat. In *In situ hybridization in brain.* (G. Uhl, ed. New York: Plenum. Pp. 73–95.

70. McAllister, L.B., Scheller, R.H., Kandel, E.R., and Axel, R. 1983. *In situ* hybridization to study the origin and fate of identified neurons. *Science* 222:800–808.

71. Chang, M.M., Leaman, S.E., and Niall, H.D. 1971. Amino acid sequence of substance P. *Nature New Biol. (London)* 232:86–87.

72. Nawa, H., Kotani, H., and Nakanishi S. 1984. Tissue-specific generation of two preprotachykinin mRNAs from one gene by alternative RNA splicing. *Nature (London)* 312:729–734.

73. Angulo, J.A., Christoph, G.R., Manning, R.W., Burkhart, B.A., and Davis, L.G. 1987. Reduction of dopamine receptor activity differentially alters striatal neuropeptide mRNA levels. In *Molecular mechanisms of neuronal responsiveness.* Adv. in Exp. Med. Series. Y.H. Ehrlich, R.H. Lenox, E. Kornecki, and W.O. Berry, eds. New York: Plenum. Pp. 385–391.

74. Krause, J.E., Chirgwin, J.M., Carter, M.S., Xu, Z.S., and Hershey, A.D. 1987. Three rat preprotachykinin mRNAs encode the neuropeptides substance P and Neurokinin A. *Proc. Natl. Acad. Sci. U.S.A.* 84:881–885.

75. Angulo, J.A., Davis, L.G., Burkhart, B., and Christoph, G.R. 1987. Chronic neuroleptic administration reduces striatal substance P-mRNA and the localization

of SP-mRNA in rat brain by *in situ* hybridization. In *Substance P and neurokinins*. J.L. Henry et al., eds. New York: Springer-Verlag. Pp. 37–39.

76. Kebabian, J.W., and Calne, D.B. 1979. Multiple receptors for dopamine. *Nature (London)* 277:93–96.

77. Kandel, E.R., and Schwartz, J.H. 1988. *Principles of neural science*. 2d ed. New York: Elsevier.

78. Hong, J.S., Yong, H.Y., and Costa, E. 1978. Substance P content of substantia nigra after chronic treatment with antischizophrenic drugs. *Neuropharmacology* 17:83–85.

79. Bannon, M.J., Lee, J.M., Giraud, P., Young, A., Affolter, H.U., and Bonner, T.I. 1986. Dopamine antagonist haloperidol decreases substance P, substance K and preprotachykinin mRNAs in rat striatonigral neurons. *J. Biol. Chem.* 261:6640–6644.

80. Bishop, J.M. 1983. Cellular oncogenes and retroviruses. *Annu. Rev. Biochem.* 52:301–354.

81. Gilliam, T.C., Tanzi, R.E., Haines, J.L., Bonner, T.I., Faryniarz, A.G., Hobbs, W.J., et al. 1987. Localization of the Huntington's disease gene to a small segment of chromosome 4 flanked by D4S10 and the telemere. *Cell* 50:565–571.

82. Monaco, A.P., Neve, R.L., Colletti-Feener, C., Bertelson, C.J., Kurnit, D.M., and Kunkel, L.M. 1986. Isolation of candidate cDNAs for portions of the Duchenne muscular dystrophy gene. *Nature (London)* 323:646–650.

83. Riordan, J.R., Rommens, J.M., Kerem, B.-S., Alon, N., Rozmahel, R., Grzelczak, Z., et al. 1989. Identification of the cystic fibrosis gene: cloning and characterization of the cDNA. *Science* 245:1066–1073.

84. Young, S.G., Hubl, S.T., and Chappell, D.A. 1989. Familial hypobetalipoproteinemia associated with a mutant species of apolipoprotein B. *N. Engl. J. Med.* 320:1604–1610.

85. Seizinger, B.R., Houleau, G.A., Ozelius, L.J., Lane, A.H., Faryniarz, A.G., Chao, M.V., Huson, S. et al. 1987. Genetic linkage of von Recklinghausen neurofibromatosis to the nerve growth factor receptor gene. *Cell* 49:589–594.

86. Hodgkinson, S., Sherrington, R., Gurling, H., Marchbanks, R., Reeders, S., Mallet, J., et al. 1987. Molecular genetic evidence for heterogeneity in manic depression. *Nature (London)* 325:805–806.

87. Alzheimer, A. (1907). Über eine eigenartige Erkrankung der Hirnrinde. *Allg. Z Psychiat.* 64:146–148.

88. Katzman, R. 1986. Alzheimer's disease. *The New England Journal of Medicine* 314:964–973.

89. Terry, R.D., Hansen, L.A., DeTeresa, R., Davies, P., Tobias, H., and Katzman, R. 1987. Senile dementia of the Alzheimer's type without neocortical neurofibrillary tangles. *J. Neuropathol. Exp. Neurol.* 46:262–268.

90. Wisniewski, H.M., Rabe, A., and Wisniewski, K.E. 1988. *Banbury report*. P. Davies and C. Finch, eds. Cold Spring Harbor, N.Y.: Cold Spring Harbor Laboratory. Pp. 1–26.

91. Roth, M., Tomlinson, B.E., and Blessed, G. 1966. Correlation between scores for dementia and counts of "senile plaques" in cerebral grey matter of elderly subjects. *Nature (London)* 50:109–110.

92. Blessed, G., Tomlinson, B.E., and Roth, M. 1968. The association between quantitative measures of dementia and of senila changes in the cerebral grey matter of elderly subjects. *Br. J. Psych.* 114:797–805.

93. St. George-Hyslop, P.H., Tanzi, R.E., Polinsky, R.J., Haines, J.L., Nee, L., Watkins, P.C., et al. 1987. The genetic defect causing familial Alzheimer's disease maps on chromosome 21. *Science* 238:885–890.

94. St. George-Hyslop, P.H., Haines, J.L., Farrer, L.A., Polinsky, R., Van Broeckhoven, C., Goate, A., et al. 1990. Genetic linkage studies suggest that Alzheimer's disease is not a single homogeneous disorder. *Nature (London)* 347:194–197.

95. Glenner, G.G., and Wong, C.W. 1984. Alzheimer's disease: Initial report on the purification and characterization of a novel cerebrovascular amyloid protein. *Biochem. Biophys. Res. Commun.* 122:885–890.

96. Wong, C.W., Quaranta, V., and Glenner, G.G. 1985. Neuritic plaques and cerebrovascular amyloid in Alzheimer's disease are antigenetically related. *Proc. Natl. Acad. Sci. U.S.A.* 82:8729–8732.

97. Masters, C.L., Simms, G., Weinman, N.A., Multhaup, G., MacDonald, B.L., and Beyreuther, K. 1985. Amyloid plaque core protein in Alzheimer's disease and Down syndrome. *Proc. Natl. Acad. of Sci. U.S.A.* 82:4245–4249.

98. Selkoe, D.J., Ihara, Y., and Salazar, F.J. 1985. Alzheimer's disease: Insolubility of partially purified helical filaments in SDS and urea. *Science* 115:1243–1245.

99. Ellis, W.G., McCulloch, J.R., and Corley, C.L. 1974. Presenile dementia in Down's syndrome. *Neurology* 24:101–106.

100. Kang, J., Lemaire, H.-G., Uterbeck, A., Salbaum, J.M., Masters, C.L., Grzeschik, K.-H., et al. 1987. The precursor of Alzheimer's disease amyloid A4 protein resembles a cell-surface receptor. *Nature (London)* 325:733–736.

101. Tanzi, R.E., Gusella, J.F., Watkins, P.C., Bruns, G.A. P., St. George-Hyslop, P., Van Keuren, M.L., et al. 1987. Amyloid b protein gene: cDNA, mRNA distribution, and genetic linkage near the Alzheimer's locus. *Science* 235:880–884.

102. Goldgaber, D., Lerman, M.I., McBride, O.W., Saffiotti, U., and Gajdusek, D.C. 1987. Characterization and chromosomal localization of a cDNA encoding brain amyloid of Alzheimer's disease. *Science* 235:877–880.

103. Manning, R.W., Reid, C.M., Lampe, R.A., and Davis, L.G. 1988. Identification in rodents and other species of an mRNA homologous to the human b-amyloid precursor. *Mol. Brain Res.* 3:293–298.

104. Sisodia, S.S., Koo, E.H., Beyreuther, K., Unterbeck, A., and Price, D.L. 1990. Evidence that b-amyloid protein in Alzheimer's disease is not derived by normal processing. *Science* 248:492–495.

105. Kitaguchi, N., Takahashi, Y., Tokushima, Y., Shiojiri, S., and Ito, H. 1988. Novel precursor of Alzheimer's disease amyloid protein shows protease inhibitory activity. *Nature (London)* 331:530–532.

106. Tanzi, R.E., McClatchey, A.I., Lamperti, E.D., Villa-Komaroff, L., Gusella, J.F., and Neve, R.L. 1988. Protease inhibitor domain encoded by an amyloid protein precursor mRNA associated with Alzheimer's disease. *Nature (London)* 331:528–530.

107. Ponte, P., Gonzalez-DeWhitt, P., Schilling, J., Miller, J., Hsu, D., Greenberg, B., Davis, K., et al. 1988. A new A4 amyloid mRNA contains a domain homologous to serine proteinase inhibitors. *Nature (London)* 331:525–527.

108. Weidemann, A., Konig, G., Bunke, D., Fischer, P., Salbaum, J.M., Masters, C.L., and Beyreuther, K. 1989. Identification, biogenesis, and localization of precursors of Alzheimer's disease A4 amyloid protein. *Cell* 57:115–126.

109. Gandy, S., Czernik, A.J., and Greengard, P. 1988. Phosphorylation of Alzheimer disease amyloid precursor peptide by protein kinase C and Ca^{2+}/calmodulin-dependent protein kinase II. *Proc. Natl. Acad. Sci. U.S.A.* 85:6218–6222.

110. Siman, R., Card, J.P., and Davis, L.G. 1990. Proteolytic processing of b-amyloid precursor by calpain I. *J. Neurosci.* 10:2400–2411.

111. Goate, A., Chartier-Harlin, M., Mullan, M., Brown, J., Crawford, F., Fidani, L., Giuffra, L., Haynes, A., Irving, N., James, L., Mant, R., Newton, P., Rooke, K., Roques, P., Talbot, C., Pericak-Vance, M., Roses, A., Williamson, R., Rossor, M., Owen, M., and Hardy, J. 1991. Segregation of a missense mutation in the amyloid precursor protein gene with familiar Alzheimer's disease. *Nature (London)* 349:704–706.

112. Card, J.P., Meade, R.P., and Davis, L.G. 1988. Immunocytochemical localization of the precursor protein for b-amyloid in the rat central nervous system. *Neuron* 1:835–846.

113. Siman, R., Card, J.P., Nelson, R.B., and Davis, L.G. 1989. Expression of b-amyloid precursor protein in reactive astrocytes following neuronal damage. *Neuron* 3:275–285.

114. Abraham, C.R., and Potter, H. 1989. Alzheimer's disease: Recent advances in understanding the brain amyloid deposits. *Bio/Technology* 7:147–153.

115. Palmert, M.R., Podlisny, M.B., Witker, D.S., Oltersdorf, T., Youngkin, L.H., Selkoe, D.J., and Younkin, S.G. 1989. The b-amyloid protein precursor of Alzheimer disease has soluble derivatives found in human brain and cerebrospinal fluid. *Proc. Natl. Acad. Sci. U.S.A.* 86:6338–6342.

116. Rosen, D.R., Martin-Morris, L., Luo, L., and White, K. 1989. A Drosophila gene encoding a protein resembling the human b-amyloid protein precursor. *Proc. Natl. Acad. Sci. U.S.A.* 86:2478–2482.

117. Van Nostrand, W.E., Wagner, S.L., Suzuki, M., Choi, B.H., Farrow, J.S., Geddes, J.W., et al. 1989. Protease nexin-II, a potent antichymotrypsin, shows identity to amyloid b-protein precursor. *Nature (London)* 341:546–549.

118. Smith, R.P., Higuchi, D.A., and Broze, G.J. 1990. Platelet coagulation factor XI_a-inhibitor, a form of Alzheimer amyloid precursor protein, *Science* 248:1126–1128.

119. Schubert, D., LaCorbiere, M., Saitoh, T., and Cole, G. 1989. Characterization of

an amyloid b precursor protein that binds heparin and contains tyrosine sulfate. *Proc. Natl. Acad. Sci. U.S.A.* 86:2066–2069.

120. Fremeau, R.T., and Roberts, J.L. 1986. Use of *in situ* hybridization histochemistry to analyze gene transcription in individual cells. In *In situ hybridization in brain*. G. Uhl, ed. New York: Plenum. Pp. 193–202.

121. Lindholm, D., Heumann, R., Hengerer, B., and Thoenen, H. 1988. Interleukin 1 increases stability and transcription of mRNA encoding nerve growth factor in cultured rat fibroblasts. *J. Biol. Chem.* 263:16348–16351.

122. Crowder, C.M., and Merlie, J.P. 1986. DNase I-hypersensitive sites surround the mouse acetylcholine receptor d-subunit gene. *Proc. Natl. Acad. Sci. U.S.A.* 83:8405–8409.

123. Von der Ahe, D., Renoir, J.-M., Buchou, T., Baulieu, E.-E., and Beato, M. 1986. Receptors for glucocorticosteroid and progesterone recognize distinct features of a DNA regulatory element. *Proc. Natl. Acad. Sci. U.S.A.* 83:2817–2821.

124. Erlich, H.A., ed. 1989. *PCR technology: Principles and application for DNA amplification*. New York: Stockton.

125. Brinster, R.L., Chen, H.Y., Trumbauer, M.E., Yagle, M.K., and Palmiter, R.D. 1985. Factors affecting the efficiency of introducing foreign DNA into mice by microinjecting eggs. *Proc. Natl. Acad. Sci. U.S.A.* 82:4438–4442.

2

Background to Monoclonal Antibodies

Melvin E. Klegerman

2.1. What Are Antibodies?

Antibodies are the most important elements of what is known as the *humoral immune system,* which comprises soluble proteins that help to defend against foreign entities such as microorganisms and transformed (malignant) cells. The other major branch of the immune system is termed *cell-mediated immunity* (CMI) and denotes the actions of such cells as lymphocytes, macrophages, and granulocytes against foreign entities. The distinction between humoral and cellular immunity is somewhat artificial, however, since the two aspects are intimately interrelated. For instance, antibodies often function to trigger the destruction of target organisms by cells, and soluble proteins known as *cytokines* act as messengers between cells during the immune response.

Antibodies belong to a group of serum proteins known as γ-*globulins,* which is a designation based on the electrophoretic separation of these proteins. The γ-globulins migrate the farthest toward the cathode in an electric field and, therefore, are among the most positively charged of the serum proteins at physiologic pH. Of the electrophoretically separated serum protein fractions, γ-globulins are second in abundance only to albumin, comprising about 25% of the total protein. Most γ-globulins are *immunoglobulins*; an antibody, in turn, is an immunoglobulin that binds specifically to a particular structural component (*epitope*) of an *antigen,* which can be almost any foreign compound. The antibody is generated naturally under conditions such as infection, or by immunizing an animal with a high-molecular-weight substance (macromolecule) that contains the epitope.

2.2. Immunoglobulin Structure

Immunoglobulins are glycoproteins containing 4–18% carbohydrate by weight. The structure of one kind of immunoglobulin, IgG, is shown schematically in

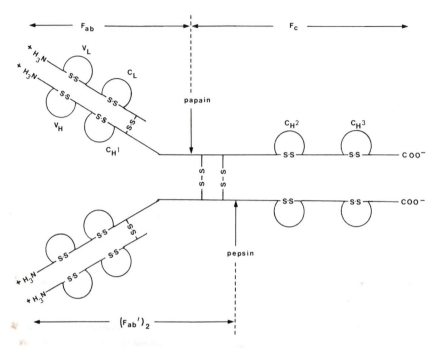

Figure 2.1. Schematic structure of an immunoglobulin G (IgG) molecule, showing the location of domains. C, constant region; H, heavy chain; L, light chain; V, variable region.

Figure 2.1. An immunoglobulin subunit consists of a short polypeptide chain, the *light chain* (approximately 25,000 molecular weight), and a longer polypeptide chain, the *heavy chain* (approximately 50,000 molecular weight), covalently linked together by a single cystine disulfide bond. The immunoglobulin monomer consists of two subunits linked by two disulfide bonds. Three dimensionally, each chain consists of a series of globular *domains*, like a string of beads, each containing a single intrachain disulfide bond. The heavy chain usually consists of four domains and the light chain has two.

The amino acid sequences of immunoglobulin molecules produced in animals vary greatly. The amino acid sequences of all immunoglobulin molecules produced by a single B-lymphocyte are identical. For immunoglobulins synthesized by different B cells, the amino acid sequences of the first (amino-terminal) domain in each chain are usually very different; thus, these are known as *variable domains* and are responsible for the antibody's ability to bind to epitopes. Although the remaining portions of the immunoglobulin chains are termed *constant*, their structure also varies according to *class*, *subclass*, and *allotype*.

There are five classes of immunoglobulins: G, A, M, E, and D. Each is determined by the amino acid sequence of the constant portions of the heavy chains, which are termed γ (IgG), α (IgA), μ (IgM), ε (IgE) and δ (IgD).

There are also some characteristic differences in the molecular structures of the immunoglobulin classes. For instance, IgG is a single monomer, while IgM consists of five monomers linked together by peptide chains known as *J-chains*. The μ-chain also contains five domains instead of four.

Based on the amino acid sequences of their constant regions, there are two types of light chains, κ and λ, either of which can be found in any of the immunoglobulin classes. Further differences in constant region amino acid sequences within an immunoglobulin class fall into subclasses that are found in every individual or define allotype *alleles* that are inherited by different individuals. In addition, characteristic sequence differences exist among species. Any of these amino acid sequence differences can be exploited in producing antibodies specific for particular immunoglobulin molecules. For instance, antibodies that bind only to rabbit IgG (via epitopes consisting of unique sequences on the γ-chain) can be produced in goats, which respond to rabbit-specific sequences as foreign antigens.

Any domain can be specified by chain type, region and distance from the amino-terminus. For example, C_H2 refers to the second constant domain of the heavy chain. Another way of viewing immunoglobulin structure is through "enzyme cleavage fragments." The proteolytic enzyme *papain* cleaves heavy chains at the amino-terminal side of the intersubunit disulfide bonds, forming two amino-terminal Fab fragments and one carboxy-terminal Fc fragment, while the protease *pepsin* cleaves on the other side of the disulfide bonds, forming one $F(ab')_2$ fragment and two heavy chain carboxy termini. This structural classification is also shown in Figure 1.

The Fab or $F(ab')_2$ fragments, therefore, retain the antigen-binding properties of the antibody. The Fc fragment, however, possesses the functional attributes (explained below) of the molecule, since the heavy chain carboxy-terminal domains contain structures that are recognized by immune cell-surface receptors. Fab fragments are often used to utilize the specific binding properties of anitbodies without activating undesired immune processes.

2.3. Development of the Humoral Immune Response

Epitope-containing macromolecules that elicit an immune response are known as *immunogens*. Many factors determine whether an immunogen will produce an effective immune response. These include the size of the immunogen molecule, the route by which it is administered, and nonspecific immune stimulants known as *adjuvants* with which it may be associated. *Hyperimmunization* of an animal with an immunogen requires the use of strong adjuvants, such as Freund's complete adjuvant, resulting in the production of antibodies specific for many epitopes. The serum from a hyperimmunized animal, therefore, is termed *polyclonal*, since it contains antibodies produced by the progeny of many different activated B cells.

The molecular mechanisms of antibody synthesis have been elucidated during the past 30 years. It is now known that each immunoglobulin domain and the amino acid sequences that link them are coded for by separate genes. The genes coding for the variable domains are *hypermutable*, and, in stem cells of the B-lymphocyte lineage during fetal development, mutations and recombinations of these genes result in millions of different randomly produced antigen-binding sites. The variable region genes then become "locked into" each mature B cell so that each of these cells can synthesize immunoglobulins of only one specificity. The progeny of each B cell or *clone* produce antibodies of the same specificity (but not necessarily the same Ig class; see below); therefore, the immunoglobulin product of these cells is *monoclonal*, at least in terms of specificity.

B cells utilize IgM and IgD as cell-surface antigen receptors. During fetal development, B cells that bind to antigens (which, at that point, are almost all self-antigens) with a critical affinity are eliminated in order to prevent autoimmune reactions later. After birth, however, relatively high-affinity antigen binding triggers differentiation of the B-lymphocyte into a *plasma cell*, which migrates to the spleen or a lymph node, proliferates, and secretes IgM antibodies into the circulation as part of a *primary antibody response* to antigen. Alternatively, some of the stimulated B cell's clonal progeny differentiate into circulating *memory cells*. When one of these cells encounters antigen, it triggers a *secondary antibody response*, becoming an IgG-secreting plasma cell that produces much more antibody of higher affinity for antigen for a longer period of time than the primary-response plasma cells. The immunoglobulin "class switch" triggered in the secondary response is sometimes to IgA or IgE. Thus, the progeny of a plasma cell produce antibodies that are monoclonal in terms of immunoglobulin class, as well as specificity.

Antigen alone is sufficient to activate B cells in the case of carbohydrate (especially bacterial polysaccharide) antigens. Protein antigens, however, require the intercession of two other types of cells, T-lymphocytes and *antigen-presenting cells* (APC). A *T-dependent* antigen must first be ingested by an APC such as a macrophage or B cell. Peptides resulting from digestion of the protein antigen are displayed on the surface of the cell by Class II *major histocompatibility complex* (MHC) proteins. Helper T cells recognizing these peptides and the MHC proteins associated with them become activated and secrete *lymphokines*. In addition to antigen, B cells require the lymphokines IL-2, IL-4, and IL-5 for a T-dependent response.

2.4. Antibody Functions

As previously stated, IgM is the immunoglobulin class of the primary antibody response to antigen. Its polymeric structure makes it very effective in *agglutinating* particulate antigens such as bacteria and even eukaryotic cells. The resultant large clumps of particles are then rapidly and easily cleared from the body by both mobile and fixed phagocytes such as macrophages and Kupffer cells in the liver.

IgG, the primary product of the secondary immune response, is most versatile in causing the destruction of foreign cells. High-affinity binding of IgG to microbial cell-surface antigens promotes ingestion (phagocytosis) and killing of the microbes by macrophages and neutrophils. It also activates the *classical complement cascade*, which results in the lysis of eukaryotic cells and Gram-negative bacteria.

Complement is another major component of the humoral immune system. Consisting of nine proteins that sequentially activate each other, the classical complement pathway culminates in a five-protein "attack complex" that punches holes in the lipid bilayer of membranes. The first complement protein, C1, is activated by binding to the Fc portions of IgG antibodies bound to closely spaced antigens on a surface. IgM also efficiently activates this pathway. Antigen-bound IgG promotes phagocytosis directly by binding to cell-surface Fc receptors on macrophages and neutrophils and indirectly, by fixing the activated third complement protein fragment, C3b, to bacteria, which in turn is bound by phagocyte cell-surface C3b receptors.

IgG also promotes the destruction of virally infected and malignantly transformed cells through a process known as *antibody-dependent, cell-mediated cytotoxicity* (ADCC). Virus infection or malignant transformation of normal cells often generate new cell-surface antigens that are recognized as foreign by the host's immune system. The association of natural killer cells, which also possess Fc receptors, with antibodies bound to these antigens, facilitates killing of the altered cells by these lymphocytes.

IgA is the first line of defense against many microorganisms and noxious agents. The major immunoglobulin component of bodily secretions such as mucus, tears, saliva, and milk, IgA is often polymerized (1–3 monomers) by J-chains. This class of antibody functions mainly in two ways: (1) immobilization of microbes, facilitating their clearance by physiologic mechanisms such as mechanical movement of mucus out of body cavities, and (2) neutralization, including inhibition of microbial adherence to membranes, or viral penetration of cells and the effects of toxins.

As mentioned, IgD acts as B cell receptor for antigens. IgE mediates immediate hypersensitivity, or allergic, reactions. IgE associates with ε chain-specific Fc receptors on basophils and mast cells. Binding of antigen by the Fab portion of the antibody stimulates these cells to release various effectors that orchestrate a physiologic response designed to expel foreign material.

2.5. Monoclonal Antibodies

2.5.1. History

The first production of monoclonal antibodies, reported by Köhler and Milstein in 1975,[1] represented the convergence of three areas of scientific research: immunochemistry, the in vitro cultivation of cancer cells, and the molecular biology of malignant transformation.

A modern understanding of the development of the humoral immune response, summarized above, is the result of the *clonal selection theory* advanced during the 1950's by Macfarland Burnet and others.[2] This theory, largely corroborated since its inception, made possible the entire concept of monoclonal antibodies since it postulated "one B cell, one antibody." Thus, immunization of an animal with an immunogen could be expected to produce many plasma cells or B cell clones producing antibodies specific for a number of different epitopes of that immunogen. Unfortunately, normal antibody-producing cells could not be maintained in culture, prohibiting the production of monoclonal antibodies with desired specificities for particular epitopes. Köhler and Milstein's solution of this problem provided the most outstanding proof of the clonal selection theory.

Malignant cells had been grown from tumors in vitro almost from the beginning of this century, but until the 1960s cell culture could be performed only with difficulty by highly skilled investigators. At that time, uniform quality culture media, animal sera, and specially treated disposable tissue culture flasks and other plasticware became commercially available, enabling many biomedical scientists to take advantage of this technique. Since then, innovations such as multiwell culture plates and methods for cell cloning and serum-free (and even protein-free) media have been introduced, allowing optimization of monoclonal antibody production.

Among the many kinds of cancer cells that could be established and maintained indefinitely in culture after the introduction of better tissue culture materials and techniques were several mouse myeloma or plasmacytoma lines. Plasmacytoma, as the name implies, is a cancer composed of endlessly proliferating plasma cells. Establishment of such cells in culture overcame the problem of plasma cells dying in culture, but an inability to control which of these cells become malignant meant that the immunoglobulins that they produce are generally useless. If the specificity of normal plasma cells produced after immunization of an animal could be combined with the proliferative capacity of plasmacytoma cells in culture, a powerful immunochemical technology would be created.

The missing element of this technology was provided by research into the molecular genetics of cancer during the decade preceding Köhler and Milstein's breakthrough. This research involved the production of *heterokaryons* or cell hybrids formed by the fusion of normal and malignant cells in order to determine whether the genes governing the transformation process are dominant or recessive. The fusion was initially accomplished with Sendai virus,[3,4] but polyethylene glycol (PEG) was later found to be a much more efficient fusogen.[5]

The most important aspect of heterokaryon research in terms of monoclonal antibody technology was the ability to select hybrids of the two parent cell lines. Littlefield[6] devised an ingenious method of ensuring that all other cells would die by exploiting the redundant nature of nucleic acid biosynthesis. The de novo synthesis of nucleotides requires tetrahydrofolate derivatives in three places. Two of the carbons in the purine ring, which is the central structure of adenine and guanine, are provided by N^{10}-formyltetrahydrofolate, while the methylation of deoxyuridine monophosphate (dUMP) to form thymidine monophosphate (TMP) requires N^5,N^{10}-methylenetetrahydrofolate. The folic acid analog aminopterin inhibits the enzyme dihydrofolate reductase, which is needed to provide tetrahydrofolate.

Blockage of nucleotide biosynthesis by aminopterin can be overcome, however, by taking advantage of *salvage pathways* designed to utilize bases and nucleosides produced by degradation of nucleic acids. The enzyme hypoxanthine-guanine phosphoribosyltransferase (HGPRT) catalyzes the conversion of guanine to guanosine monophosphate (GMP) and hypoxanthine to inosine monophosphate (IMP), which is converted to both adenosine monophosphate (AMP) and GMP. The inhibition of TMP synthesis by aminopterin is overcome by providing thymidine, which is converted to TMP by the enzyme thymidine kinase.

Littlefield selected cells of a given line that were deficient in HGPRT by adding 8-azaguanine or 6-thioguanine to the culture medium. Conversion of these bases to their corresponding nucleotides by HGPRT caused the incorporation of guanine analogs into nucleic acids, resulting in cell death. Only cells that did not possess a functional HGPRT gene could avoid this poisoning and survive. Culture of these HGPRT-deficient cells in medium containing hypoxanthine, aminopterin, and thymidine (HAT medium), however, would kill them since they could not overcome inhibition of nucleotide synthesis by the aminopterin. Only hybrids formed by fusion of HGPRT⁻ cells with cells possessing a functional HGPRT gene would produce the enzyme and proliferate in HAT medium.

If the HGPRT⁺ cells are likely to persist in culture for extended periods of time, a strategy to kill these cells as well would also have to be employed. For instance, Littlefield selected his second cell line for thymidine kinase deficiency and others[3, 4] selected their first cell line for resistance to a toxic compound, ouabain, as well as HGPRT deficiency. Thus, cells from each line would provide the other with genes needed for survival and only fusion products of the two could survive and proliferate. Köhler and Milstein knew that normal plasma cells could survive in culture for only about a week. Use of HGPRT⁻ plasmacytomas alone, therefore, would be sufficient to guarantee selection of *hybridomas* in HAT medium.

2.5.2. Hybridoma Production

The establishment of hybridoma cell lines that consistently secrete useful monoclonal antibodies requires several steps: immunization, cell fusion, selection, screening, cloning, characterization and storage. This process is summarized in Figure 2.2.

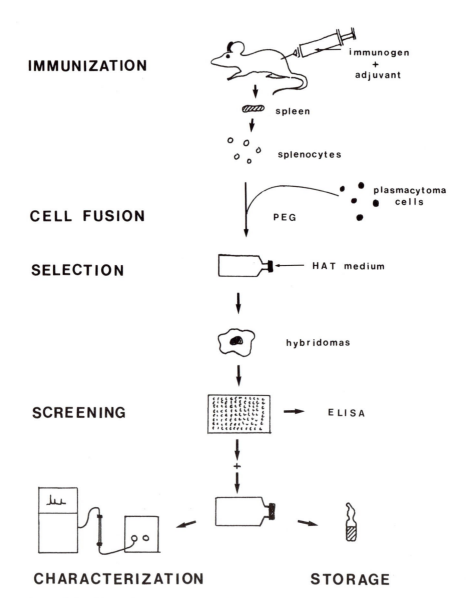

IMMUNIZATION

immunogen
+
adjuvant

spleen

splenocytes

CELL FUSION

PEG

plasmacytoma
cells

SELECTION

HAT medium

hybridomas

SCREENING

ELISA

+

CHARACTERIZATION   STORAGE

Figure 2.2. The major stages of monoclonal antibody production. ELISA, enzyme-linked immunosorbent assay; HAT, hypoxanthine–aminopterin–thymidine; PEG, polyethylene glycol.

2.5.2.1. Immunization

Immunization of animals with immunogens is performed in much the same way as it is for production of polyclonal antisera. Microgram–milligram quantities of the immunogen mixed with an adjuvant (aluminum salts, Freund's complete or incomplete adjuvant, or one of several recently developed commercial formulations) are injected intradermally or subcutaneously at multiple sites repeatedly at different times. Often, the animal is bled at various times and the serum is tested using a convenient assay to determine the relative concentration of antibodies of the desired specificity. When the response is judged to be nearly optimal, the animal is sacrificed and the spleen, which contains a large number of plasma cells, is dissociated into single *splenocytes* by mechanical and/or enzymic methods.

2.5.2.2. Cell Fusion

Splenocytes are then mixed with plasmacytoma cells in an appropriate medium, exposed briefly to a high concentration (e.g., 50%) of PEG, and fusion is allowed to proceed over a period of time. The method of producing monoclonal antibody-secreting hybridomas with splenocytes has made the mouse the animal of choice for this technology since these animals are very inexpensive and several billion cells can be obtained from a mouse spleen. In contrast, mice are rarely used for the preparation of polyclonal antisera, since they have a very low blood volume (about 1 ml); guinea pigs, rabbits, goats, sheep, and horses are most commonly used for this purpose.

The first mouse plasmacytoma lines used to produce hybridomas were immunoglobulin-secreting cells. Thus, only a portion of the resultant hybridomas secreted antibodies having the structure of the splenocyte immunoglobulin.[7] Since then, this problem has been solved by using nonsecreting lines.[8] Hybridomas from many different species have been prepared. Of greatest interest, however, is the production of human monoclonal antibodies for therapeutic use, since immunoglobulins from other species are perceived as foreign antigens by the immune system. The immune response greatly reduces the effectiveness of the monoclonal antibody and may cause adverse effects in the patient.

Obviously, humans cannot be experimentally immunized and splenectomized. The method most commonly used for preparation of human monoclonal antibody-secreting cells is to isolate circulating B cells and transform them with an oncogenic virus. Cell lines prepared in this way, however, tend to be unstable and complete removal of virus must be assured before the antibodies can be used clinically. Recently, mouse–human hybrids prepared by genetic engineering techniques, in which the variable immunoglobulin domains are of mouse origin and the immunogenic constant domains are of human origin, have shown much promise.[9, 10]

2.5.2.3. Selection and Screening

Selection of hybridoma cells in HAT medium is usually followed directly by screening of hybridomas for secretion of antibodies of the desired specificity. After fusion, the cells may be transferred to HAT medium in tissue culture flasks. After an appropriate incubation time, all viable cells will be hybridomas. They are removed from the flasks, transferred to regular culture medium and aliquots are distributed among the wells of 96-well plastic culture plates. Alternatively, the initial suspension of cells in HAT medium may be apportioned into wells. In either case, the medium in each well is periodically tested for the desired antibody reactivity.

The most commonly used screening assay is the enzyme-linked immunosorbent assay (ELISA). The antigen is adsorbed to the bottom of 96-well plates. Samples to be tested are then incubated in the wells for an appropriate period of time. If a sample contains the desired antibody, it will bind to the antigen and remain associated with the well as unbound material is washed off. This antibody is then detected by an *immunoconjugate* consisting of two components covalently linked to each other. One component is an antibody specific for an epitope in the constant domain of the first antibody (e.g., mouse Fc or mouse γ-chain). The second component is an enzyme such as alkaline phosphatase or horseradish peroxidase. If immunoglobulin from the first incubation is immobilized in the well, the immunoconjugate will be retained also.

After another washing step, a colorless substrate that is converted to a colored product by the enzyme of the immunoconjugate is added. For instance, alkaline phosphatase converts paranitrophenyl phosphate to the yellow compound paranitrophenol, which has an absorbance maximum near 405 nm. After an appropriate incubation, the enzymic reaction is stopped and the optical density of each well at the product's characteristic absorbance wavelength is determined with a specialized colorimeter known as a "plate reader." The ELISA procedure is easily automated, and motorized, computerized plate readers are commercially available, making it possible to screen thousands of culture wells routinely.

2.5.2.4. Cloning

Single cells secreting the desired antibody must then be isolated from positive cultures and propagated into cell lines. Two cloning techniques are most widely used: (1) limiting dilution and (2) soft agar. In limiting dilution, cells in the culture are enumerated, diluted, and aliquoted into new wells so that it is statistically unlikely that more than one cell will be found in any well. Cells are allowed to regrow and the procedure is repeated several times to increase the probability that all the cells in a given well are monoclonal. The second method exploits the fact that many malignant cells will proliferate, forming spherical colonies, in a semisolid medium containing low amounts of agar. If the culture can be reliably dispersed into single cells and the cell concentration is such that the colonies will be well spaced, then visible colonies picked out of the agar are likely to be monoclonal.

It is often advisable to combine both methods, especially since clones require rescreening. For instance, positive cultures after initial screening can be subjected to several rounds of limiting dilution. The medium from each well is then screened with ELISA and positive cultures are dispersed in soft agar. Individual colonies are cultured in new wells and transferred to successively larger vessels until seed stocks can be frozen and enough material is available for characterization.

2.5.2.5. Characterization and Storage

The final determination of monoclonality requires biochemical and biophysical characterization of the immunoglobulin. Spectroscopic, electrophoretic, and chromatographic methods, which can detect multiple molecular populations, are frequently used for this purpose. At this time, the antibody is tested for suitability in the context of its intended application (e.g., diagnostic or therapeutic; coupled to a drug, enzyme, or radionuclide; associated with a target cell population after injection into an animal). It is also characterized immunochemically to determine its affinity for antigen, its immunoglobulin subclass (assuming that its class was determined during initial screening), the epitope for which it is specific, and the effective number of binding sites that it possesses.

Knowledge regarding stability, not only of the antibody but also of the cell line, is important. Genetic stability of the cell line can be determined by monitoring several of its properties during serial passage of the cells in culture. These include the number, shape, and size of the chromosomes (karyology), biochemical properties of the immunoglobulin product and other proteins, and various metabolic and growth characteristics. The ability of the cells to withstand freezing, storage for various periods of time, thawing, and reculture must be ascertained. Finally, the physical and chemical stability of the antibody itself during different conditions of storage and use, especially for its intended purpose, must be determined. In short, characterization should be aimed at answering vital questions concerning applications, stability of both cells and antibody in conditions of handling and storage, and large-scale cell culture and antibody purification.

Samples of incipient cell lines secreting antibodies of the desired specificity should be frozen in liquid nitrogen at several stages of cloning and culture so that unforeseen mishaps will not destroy useful clones. Seed stocks should also contain a maximum number of clones secreting antibodies of the desired specificity so that alternatives are available as intended applications of the antibody change or clones or their immunoglobulin products prove to be unstable.

2.5.3. Uses of Monoclonal Antibodies

The applications of monoclonal antibodies are governed by their advantages and disadvantages relative to polyclonal antibodies. Advantages include (1) unlimited supply without change, making unnecessary the extensive recharacterization previously required when producing new antisera with the same immunogen; (2) no need to store antibody preparations for long periods of time; (3) defined specificity, making unnecessary the isolation of antibodies of a defined specificity once the initial selection has been performed; (4) homogeneous properties, including affinity for antigen, interaction with other immunologic components, and other characteristics; (5) ease of antibody purification, often requiring only a salt precipitation or adsorption to protein A; and (6) ability to use a crude, poorly characterized mixture as an immunogen, with selection and characterization of epitope specificity at the hybridoma stage. The last advantage makes monoclonal antibody technology a particularly powerful research tool.

Disadvantages include (1) lack of usefulness in investigating the physiologic role of a particular antibody, unless a considerable amount of information is already known; (2) tendency to be nonprecipitating, because antigen–antibody network formation is precluded by only one epitope per protein molecule; (3) tendency toward instability, especially in vivo; and (4) inability to discriminate among different antigens possessing the same epitope.

The major applications of monoclonal antibodies are investigational, diagnostic, therapeutic, and preparative. An example of the research applications of monoclonal antibodies is the delineation of lymphocyte subpopulations by phenotypic markers. Previously, cells (both prokaryotic and eukaryotic) were *serotyped* by injecting animals with the whole cells or extracts prepared from them. The resultant antisera were methodically and tediously absorbed with cell clones to produce panels of antisera that reacted with particular antigens found on one or more groups or types of cells and not others. Now, the same kind of immunogens can be used to produce monoclonal antibodies that may be directly tested for their ability to bind to subpopulations of cells.

This approach has been used to catalog the cell-surface proteins of lymphocytes, monocytes, and other cells according to the *cluster determinant* (CD) system. At least 78 proteins have thus far been identified by monoclonal antibodies that bind to epitopes specific to each of them and their functions are gradually being determined. For instance, CD16 is the B cell Fc receptor.[11] Much of the recent explosion of knowledge concerning immunologic mechanisms is due to monoclonal antibody technology.

Medical diagnostics were the first, and continue to be the main, commercial application of monoclonal antibody technology. The major in vitro use of monoclonal antibodies is immunoassay, which utilizes antibodies for their specificity, and is based on the principle of "labeling" the antibody or the antigen so that it may be quantitated in a reaction mixture. Measurement of antigen concentration is exemplified by radioimmunoassay (RIA), in which the antigen contains or is coupled to a radionuclide such as ^{125}I (an emitter of γ-radiation), ^3H, or ^{14}C (emitters of β-radiation). A small quantity of this "tracer" is allowed to bind to an even smaller number of antibodies. As an increasing quantity of unlabeled antigen (as in an unknown sample) is added, it competes with labeled antigen for antibody binding sites and a progressively smaller proportion of the tracer is bound to antibody.

Antibody-bound radioactivity may be separated from unbound tracer by a number of different methods and the radioactivity quantitated in a suitable detector such as a gamma-ray spectrophotometer or a liquid scintillation counter. The antigen concentration in the sample is determined from the proportion of tracer bound to antibody by reference to a dose–response curve determined from assay of antigen standards of known concentration. Antigens measured by RIA include hormones, drugs, cytokines, metabolites, and tumor-associated antigens. Use of a labeled antibody to quantitate another antibody is exemplified by ELISA, described above.

Scientists are extensively investigating the possibility of using monoclonal antibodies to target radionuclides and contrast agents to pathologic structures such as tumors and atherosclerotic plaques in vivo. Scanning of the patient with a gamma-ray detector, X-ray machine, or ultrasound, for instance, would detect and locate the diseased tissues. Major problems that have been encountered in these attempts are antibody instability and nonspecific localization of the immunoconjugates in organs such as the liver and spleen. The use of monoclonal antibodies for in vivo diagnosis and therapy is described in Chapter 9 of this volume.

A particularly valuable application of monoclonal antibodies is antigen purification by affinity chromatography, for which the extreme specificity of monoclonal antibodies is especially advantageous. Antibody affinity can be selected to optimize adsorption and elution of antigen, but clones also should be selected for their ability to produce antibodies that can withstand covalent linkage to an inert support and resist degradation under elution conditions.

References

1. Köhler, G., and Milstein, C. 1975. Continuous cultures of fused cells secreting antibody of predefined specificity. *Nature (London)* 256:495–497.

2. Burnet, F.M.A. 1957. A modification of Jerne's theory of antibody production using the concept of clonal selection. *Aust. J. Sci.* 20:67–69.

3. Croce, C.M. 1976. Loss of mouse chromosomes in somatic cell hybrids between HT-1080 human fibrosarcoma cells and mouse peritoneal macrophages. *Proc. Natl. Acad. Sci. U.S.A.* 73:3248–3252.

4. Jha, K.K., and Ozer, H.L. 1976. Expression of transformation in cell hybrids. I.

Isolation and application of density-inhibited Balb/3T3 cells deficient in hypoxanthine phosphoribosyltransferase and resistant to ouabain. *Somat. Cell Gen.* 2:215–223.

5. Davidson, R.L., and Gerald, P.S. 1976. Improved techniques for the induction of mammalian cell hybridization by polyethylene glycol. Somat. Cell Gen. 2:165–176.

6. Littlefield, J.W. 1964. Selection of hybrids from matings of fibroblasts in vitro and their presumed recombinants. *Science* 145:709–710.

7. Waldmann, M.C.H. 1986. Production of murine monoclonal antibodies. In *Monoclonal antibodies*. P.C.L. Beverley, ed. New York: Churchill Livingstone.

8. Kearney, J.F., Radbruch, A., Leisegang, B., and Rajewsky, K. 1979. A new mouse myeloma cell line that has lost immunoglobulin expression but permits the construction of antibody-secreting hybrid cell line. *J. Immunol.* 123:1548–1550.

9. Thompson, K.M. 1988. Human monoclonal antibodies. *Immunol. Today* 9:113–117.

10. Morrison, S.L., Wims, L., Wallick, S., Tan, L., and Oi, V.T. 1987. Genetically engineered antibody molecules and their application. *Ann. N.Y. Acad. Sci.* 507:187–198.

11. Benjamini, E., and Leskowitz, S. 1991. *Immunology: A short course,* 2nd ed. New York: Wiley-Liss. Pp. 154–158.

Additional Reading

Abbas, A.K., Lichtman, A.H., and Pober, J.S. 1991. *Cellular and molecular immunology*. Philadelphia: W.B. Saunders.

Benjamini, E., and Leskowitz, S. 1991. *Immunology: A short course*, 2nd ed. New York: Wiley-Liss.

Beverley, P.C.L 1986. *Monoclonal antibodies*. New York: Churchill Livingstone.

Campbell, A.M. 1984. *Monoclonal antibody technology*. Amsterdam: Elsevier.

Goding, J.W. 1983. *Monoclonal antibodies: Principles and practice*. London: Academic Press.

Groves, D.J. 1992. Monoclonal antibodies. In M.E. Klegerman, and M.J. Groves, eds. *Pharmaceutical biotechnology: Fundamentals and essentials*. Buffalo Grove, IL: Interpharm Press.

Maggio, E.T. 1980. *Enzyme immunoassay*. Boca Raton, FL: CRC Press,

Seaver, S.S. 1987. *Commercial production of monoclonal antibodies*. New York: Marcel Dekker.

3

Lymphokines and Monokines

Melvin E. Klegerman and Nicholas P. Plotnikoff

3.1. Introduction

The cellular parts of the immune system (macrophages, T cells, and B cells) are regulated in their functions, to a large extent, by factors produced by these cells. These factors produced by the cells of the immune system are referred to as cytokines in the generic sense. More specifically, factors produced by lymphocytes are known as lymphokines, whereas those produced by macrophages are known as monokines. Recent progress in biotechnology is based on the production of large amounts of these factors, allowing for clinical evaluation in cancer and AIDS patients as well as other conditions involving the immune system (Figure 3.1).

3.2. Gamma Interferon (IFN-γ; Immune Interferon)

The interferons were discovered in 1957, when it was found that heat-inactivated influenza virus induced the formation of substances in chick chorioallantoic membranes that inhibited the growth of the live virus.[1] In 1965 Wheelock discovered that an interferon-like protein was produced by leukocytes exposed to the T-lymphocyte mitogen phytohemagglutinin (PHA).[2] Subsequently, this T cell factor was found to mediate several important immune mechanisms and was termed immune, nonviral, type II, and finally, gamma interferon (IFN-γ).

The virally induced, nonimmune, type I interferons consist of leukocyte interferon (IFN-α), produced by various lymphocytes and macrophages, and fibroblast interferon (IFN-β), produced by fibroblasts, epithelial cells and macrophages.[3] All the interferons are glycoproteins; IFN-α and IFN-β exhibit about 30% homology in their amino acid sequences, but the type I interferons and IFN-γ show no structural similarities and bind to different cell-surface receptors. Yet they exhibit

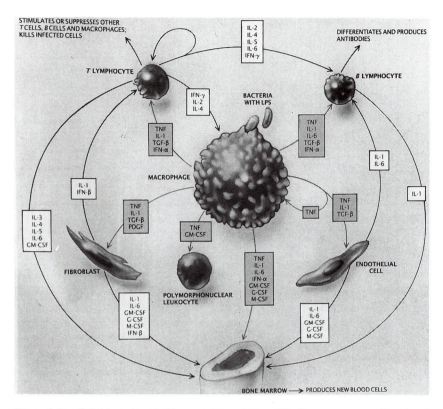

Figure 3.1. Cytokines, lymphokines, and monokines of the immune system. (Reprinted by permission of Dr. Lloyd J. Old from *Scientific American*, May 1988.)

many of the same effects, with type I interferons being more potent in antiviral activity and IFN-γ possessing a primary physiologic function as a lymphokine.[4]

3.2.1. Endogenous Production

Human IFN-γ consists of a single protein chain of 134 amino acids exhibiting a molecular weight of 15,500, which is differentially glycosylated to produce two subunits of molecular weight 20,000 and 25,000. The subunits associate noncovalently, probably by hydrophobic interactions, to produce a dimer, which is the main form of the lymphokine occurring in vivo. After the human IFN-γ gene was successfully cloned and expressed in the bacterium *Escherichia coli* in 1982, the unglycosylated protein was found to possess full biological activity, indicating that the carbohydrate portions of the molecule are not involved in cell-surface receptor binding.[3, 5–7]

IFN-γ is produced by T4 and T8 lymphocytes and natural killer (NK) lymphocytes that have been stimulated both by antigen binding to their T cell receptors and by interleukin-2 (IL-2), which is secreted by activated helper T4 lymphocytes. Antigen binding can be mimicked by mitogens such as PHA, concanavalin A, and staphylococcal enterotoxin A. Binding of IL-2 to the lymphocyte IL-2 receptor activates a cell membrane lipoxygenase, leading to increased production of leukotrienes. Leukotrienes activate guanylate cyclase, which causes an increase of intracellular cyclic guanosine monophosphate (cGMP) levels, resulting in an influx of calcium ion (Ca^{2+}). Increased cytoplasmic Ca^{2+}, with antigen or mitogen binding, triggers production and secretion of IFN-γ and also activates a phospholipase, which causes release of arachidonic acid, the substrate for leukotriene synthesis, from membrane phospholipids. IFN-γ secretion is terminated by development of T8 suppressor lymphocytes, which sequester IL-2. Macrophages control the release of IFN-γ in three ways: (1) by presenting antigen (in association with histocompatibility antigens) to T4 helper and T8 lymphocytes, (2) by secreting IL-1, which induces IL-2 production by antigen-stimulated helper T cells, and (3) by providing leukotrienes to lymphocytes directly, thereby bypassing the IL-2 requirement.[3] The molecular mechanisms of IFN-γ production are summarized in Figure 3.2.

3.2.2. Effects and Functions of IFN-γ

In response to viruses and oligoribonucleotides, both type I interferons and IFN-γ induce an antiviral state in various cells by activating 2′, 5′-oligoadenylate synthetase and a protein kinase.[8] All interferons also inhibit cell proliferation and promote antiviral immunity by activating NK and cytotoxic T lymphocytes, stimulating B lymphocyte differentiation (resulting in antiviral antibody production), and inducing increased expression of class I histocompatibility antigens on the surface of various cells[4]. Recognition of these histocompatibility antigens (when associated with viral antigens) by cytotoxic T8 lymphocytes leads to destruction of virally infected cells[9].

Major immunomodulatory activities exhibited by IFN-γ, but not type I interferons, are induction of cell-surface class II histocompatibility antigens on lymphoid cells and monocytes,[10–12] induction of cell-surface immunoglobulin Fc receptors on monocytes and macrophages,[13] and activation of macrophages. Presentation of antigens with class II histocompatibility proteins to helper T4 lymphocytes results in activation of these cells, which secrete IL-2, amplifying the immune response, and cause production of antibodies instrumental in the clearance and killing of microorganisms, as well as the destruction of malignantly transformed and virally infected cells.[9] The expression of Fc receptors promotes phagocytosis of microorganisms to which antibody has bound.[14] The role of IFN-γ in activation of macrophages to microbicidal and tumoricidal states is discussed in the section on TNF-α production.

IFN-γ also induces differentiation of stem cells into monocytes in bone marrow and inhibits growth of intracellular protozoal pathogens.[4]

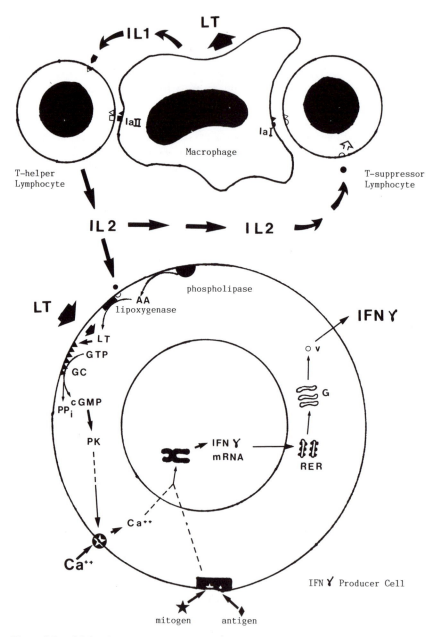

Figure 3.2. Molecular mechanisms involved in INF-γ production. AA, arachidonic acid; cGMP, cyclic guanosine monophosphate; G, Golgi complex; GC, guanylate cyclase; GTP, guanosine triphosphate; Ia-I, class I histocompatibility antigen; Ia-II, class II histocompatibility antigen; IFN-γ, gamma interferon; IL-1, interleukin-1; IL-2, interleukin-2; LT, leukotrienes; mRNA, messenger ribonucleic acid; PK, protein kinase, PP$_i$, pyrophosphate; RER, rough endoplasmic reticulum; v, vesicle. Dashed lines signify unknown second messengers.

3.2.3. Applications

Cloning of the IFN-γ gene by scientists at Genentech after the type I interferon genes had been cloned in 1979–1980 is an excellent example of how the ingenious and powerful methods of recombinant DNA technology can be used to solve both applied and basic science problems. Because of IFN-γ heterogeneity in vivo, investigators lacked a pure preparation and an amino acid sequence of the lymphokine. They first isolated messenger RNA (mRNA) from cultures of human peripheral blood lymphocytes in which the synthesis of IFN-γ had been induced by staphylococcal enterotoxin B and 1-α-desacetylthymosin. The mRNA fraction was then injected into *Xenopus laevis* oocytes, which are highly active in synthesizing proteins; mRNA associated with ribosomal RNA (rRNA, meaning that they were actively being translated on ribosomes) was then converted to complementary DNA (cDNA) with the enzyme reverse transcriptase. Insertion of these cDNAs into bacterial plasmids created a cDNA library of 8,200 clones. When this library was hybridized with ^{32}P-labeled cDNA probes prepared from induced and uninduced lymphocytes, about 20 of the clones contained DNA that bound to the induced, but not the uninduced, probes, which is the behavior expected for the IFN-γ gene. Testing of the proteins produced by prokaryotic and eukaryotic cells transfected with these last clones for the antigenic and biochemical properties of IFN-γ resulted in identification of the IFN-γ gene. The cloned gene proved invaluable for elucidation of the lymphokine's structure and functions, as well as for its manufacture from *E. coli* by bacterial disruption and immunoaffinity chromatography methods.[4]

All the interferons are now manufactured industrially using recombinant DNA technology. The type I interferons show great potential for treatment of viral diseases, including certain cancers that appear to have a viral origin, but the effectiveness of IFN-γ in treating other cancers is more speculative.[15] The U.S. Food and Drug Administration has issued licenses to two firms, allowing the sale of IFN-α for treatment of hairy-cell leukemia, for which the agent is 90% effective. Awaiting licensure is the use of IFN-α for treatment of genital warts, genital and oral herpes, acquired immunodeficiency syndrome (AIDS) and AIDS-related Kaposi's sarcoma, various cancers, upper respiratory infections, and severe papillomavirus-induced infections (condylomata acuminata). Undergoing clinical trials are IFN-α for viral hepatitis, IFN-β for AIDS-related complex (ARC), multiple sclerosis, and cancer, and IFN-γ for various cancers, rheumatoid arthritis, venereal warts, and scleroderma.[15, 16]

3.3. Interleukins

3.3.1. Interleukin-1 (Lymphocyte-Activating Factor)

Interleukin-1 (IL-1) is produced by activated macrophages (see Section 3.4.1.1), signaling the onset of disease by activating T- and B-lymphocytes and natural killer cells.[17, 18] However, IL-1 also has multiple hormonal effects in the body, producing neurological, hematologic, and endocrinologic changes.

There are two forms of IL-1, IL-1α and IL-1β. There is a large prohormone, of molecular weight 31,000 that is cleaved by protease to result in a 17,500 form. Both the α and β forms are recognized by the IL-1 receptor. IL-1 can also be produced by other cells such as Langerhans cells of the skin, B lymphocytes, natural killer cells, and smooth muscle cells, as well as by the thymus. However, the monocyte–macrophage cells are the principal source of IL-1.

3.3.2. IL-1 Effects and Applications

One of the primary roles of IL-1 is to activate T cells by inducing in them the transcription, synthesis, and secretion of IL-2 plus the expression of IL-2 receptors on the T cell surface. This effect has been described as a constimulator activity, making IL-1 a growth factor for T cells. IL-1 also induces production of IFN-γ and other lymphokines by T-lymphocytes and activates B cells, natural killer cells (in synergy with IL-2 and interferons), and macrophages. Its effects on bone marrow cells are unique in that it synergizes with bone marrow growth factors.

IL-1 is pyrogenic by stimulating prostaglandin production in the hypothalamus. It also produces anorexia and somnolence, and diminishes pain. IL-1 stimulates pancreatic β cells, resulting in hyperinsulinemia, causes release of ACTH, and elevates plasma corticosteroids. The monokine induces release of hematopoietic growth factors by bone marrow stromal cells as well as by endothelial cells and fibroblasts. IL-1 produces hypotension by decreasing venous pressure and vascular resistance and increasing heart rate and cardiac output. It stimulates production of prostaglandins, which are potent vasodilators, and induces inflammatory effects on connective tissue and epithelial cells, activates osteoclasts that resorb bone, and chondrocytes that increase cartilage breakdown. Research is in process to find IL-1 antagonists that may be helpful in treatment of rheumatoid or osteoarthritis. Other approaches include treatment of immunodeficiency states that accompany metastatic cancer, nutritional deficiencies, certain autoimmune diseases, irradiation, chemotherapy, wounds, and burns.

3.3.3. Interleukin-2

Interleukin-2 (IL-2) has been identified as a T cell growth factor.[19–21] It is a glycoprotein with a molecular weight of 15,500. The IL-2 receptor is found in activated T cells. The IL-2 receptor meets all the requirements for hormone–receptor interactions, including high affinity, saturability, ligand specificity, and target cell specificity.

The receptor-binding concentration of IL-2 is the same concentration of the lymphokine that promotes T cell proliferation. There is a low- and a high-affinity IL-2 receptor. The high-affinity receptor consists of a large binding protein (75 kDα), the α-chain, and a smaller protein (55 kDα), the β-chain.

In the presence of specific antigens, T cell clones proliferate and differentiate, producing T helper cells as well as cytotoxic T cells, and B cell clones become antibody-producing plasma cells. In the quiescent stage, resting T cells do not respond to IL-2. Normally, 3 days are required for all the T cells to express IL-2 receptors. As the antigen is cleared, IL-2 receptors decline in number (the high-affinity receptor diminishes). The IL-2 receptor binding of the lymphokine has a half-life of 15 minutes as a result of accelerated ligand-mediated internalization.

3.3.4. IL-2 Therapeutic Implications

3.3.4.1. Transplants

Immunosuppressants such as the glucocorticoids and cyclosporine selectively inhibit IL-2 production and thereby limit T cell numbers. Antibodies have been developed that specifically react with the p55 β-chain, specifically preventing cardiac allograft rejection.

3.3.4.2. Autoimmune Diseases

IL-2 has been studied as a suppressant of experimental autoimmune diabetes mellitus as well as systemic lupus erythematosus. Genetic engineering of bacterial toxin and IL-2 conjugates is being studied as immunosuppressants of allograft rejection.

3.3.4.3. Therapy of Cancer in Humans Utilizing LAK Cells and IL-2

IL-2 is being studied for treatment of cancer by adoptive immunotherapy that stimulates NK (natural killer) as well as tumor-reactive T cells. In addition, IL-2 is being studied as an adjuvant for stimulating the immune response to vaccines against tumors.

More recently, lymphokine-activated killer (LAK) cells have been found to lyse tumor cells after activation with the lectin mitogen phytohemagglutinin in combination with recombinant IL-2. Leukopheresis (repeated infusion of LAK cells) is being used to treat cancer patients. Furthermore, the combination of LAK cells plus IL-2 was found to result in varied clinical improvements in colon, renal, and lung adenocarcinoma and melanoma patients. The limited clinical benefits in cancer patients were accompanied by major side effects, such as marked fluid retention (the capillary leak syndrome), resulting in hypotension, tachycardia, oliguria, acute respiratory distress syndrome (hypoxia), and renal dysfunction. Other side effects, such as chills and fever, were found to be reduced by acetaminophen.[19]

3.3.5. IL-3 and GM-CSF

Activated T cells produce multi-colony-stimulating factors (CSF), which collectively have been grouped as IL-3, a glycoprotein of 28 kDa. A major factor in this group is granulocyte macrophage–colony-stimulating factor (GM-CSF), a glycoprotein of 144 amino acids. GM-CSF stimulates the formation of neutrophils, macrophages, and eosinophils. Thus, GM-CSF increases the white blood cell count in neutropenic patients. GM-CSF is also produced by macrophages and endothelial cells, the production being stimulated by IL-1.[21–23]

3.3.6. GM-CSF Effects and Applications

GM-CSF stimulates the production of neutrophils and potentiates antigen processing by macrophages. The latter produce G-CSF (granulocyte CSF), and T lymphocytes produce GM-CSF. Complementary DNA clones (as well as purification of the protein) encoding human GM-CSF have been isolated. Thus, it became possible to produce biosynthetic (recombinant) GM-CSF protein. GM-CSF has multiple effects. It enhances chemotaxis, phagocytosis, oxidative metabolism, and cytotoxicity of eosinophils, macrophages, and neutrophils by increasing production of arachidonic acid and leukotriene B4.[22]

In the bone marrow, GM-CSF increases production of neutrophils, monocytes, and eosinophils. Cell-surface receptor protein for GM-CSF has been isolated from cDNA clones. Receptors on neutrophils, eosinophils, and monocytes have been characterized with ^{125}I-labeled GM-CSF. Approximately 500 to 800 binding sites per cell have been identified, with maximal activity at 100 pM GM-CSF (K_d=30 pM). The molecular weight of human GM-CSF is 84,000. Of special interest is the finding that there are GM-CSF receptors on small-cell carcinoma cells, which may be correlated with inhibition of growth. Thus, GM-CSF has been proposed as adjunctive therapy for patients undergoing radiation or chemotherapy. Recombinant glycosylated GM-CSF has been produced in Chinese hamster ovary cells, and found to increase neutrophils, monocytes, and eosinophils in cynomolgus monkeys. Nonglycosylated recombinant human GM-CSF has also been produced from E. coli and found to increase leukocyte numbers.

In bone marrow ablation by X-ray and replacement with autologous bone marrow infusion, GM-CSF increases leukocytes, platelets, neutrophils, monocytes, and eosinophils.

Neutropenia in AIDS patients has been successfully treated with GM-CSF in a dose range of 0.5 to 8 μg/kg. Side effects observed included fever, facial flushing, and maculopapular rash. Many believe that GM-CSF can increase host defenses and reduce myelosuppression induced by chemotherapy, antibiotic, or antiviral therapy. Reversal of leukopenia in sarcoma patients and neutropenia in aplastic anemia have been reported. Toxicity at high doses (32 mg/kg) includes myalgias, pulmonary toxicity, fever, and inflammatory symptoms. In summary, the therapeutic applications of GM-CSF can improve the host defenses of immunocompromised patients, facilitate recovery of bone marrow transplants, treat infectious parasitic diseases, prevent infections in hospital patients with debilitating burns and complex surgery, treat preleukemia and aplastic anemia, and induce antitumor activity.

3.3.7. IL-4, IL-5, and IL-6

The last three interleukins, IL-4, IL-5, and IL-6, have primary effects in augmenting B cell functions.[21, 23] The human IL-4 consists of 129 amino acids, IL-5 has 123 amino acids, and IL-6 is a polypeptide with 184 amino acids. In addition, IL-4 has T cell growth factor (TCGF) activity (distinct from IL-2) and M-CSF (monocyte CSF) activity separate from IL-3. IL-5 has also been found to exhibit E-CSF (eosinophil CSF) activity. Finally, IL-4 and IL-5 stimulate lymphocytes and hematopoietic cells.

Currently, basic studies are in progress to determine possible clinical applications in the areas of autoimmune disease, allergy, and cancer. For example, the hematological restorative effects of GM-CSF are being widely explored.

3.4. Tumor Necrosis Factor (TNF)

3.4.1. TNF-α (Cachectin)

The discovery of TNF-α resulted from three separate lines of investigation into the causes of (1) septic shock, (2) the wasting syndrome known as cachexia that accompanies cancer, tuberculosis, and other diseases, and (3) hemorrhagic necrosis of tumors sometimes caused by bacterial infections.

Septic shock is an often lethal condition characterized by hypotension, disseminated intravascular coagulation, and renal, hepatic, and cerebral injury occurring in mammals infected with gram-negative bacteria.[24] During the 1940s the ultimate cause of septic shock was shown to be an endotoxin in the cell walls of these bacteria and was identified chemically as a lipopolysaccharide (LPS).[25–27]

Thirty years later, a marked elevation of plasma triglycerides was observed in rabbits infected with *Trypanosoma brucei*, despite a condition of starvation and wasting, and was shown to result from a systemic deficiency of the enzyme lipoprotein lipase (LPL), which enables utilization of ingested fatty acids by tissues. LPL suppression was found to be caused by a factor—termed cachectin—produced by the animal in response to bacterial LPS.[28, 29]

The discovery of TNF-α ultimately had its origin in anecdotal reports during the nineteenth century of sometimes dramatic regressions of malignant tumors after patients contracted bacterial infections, especially if the infections occurred in the tumor.[30] These observations led the U.S. surgeon William B. Coley during 1892–1931 to attempt to treat cancers with injections of heat-killed bacterial preparations known as Coley's toxins.[31] Although extreme variability of results and toxicity (including fever and symptoms of septic shock) of these preparations resulted in abandonment of this form of therapy, Coley's work stimulated interest in the mechanisms underlying bacterially induced tumor regression.[32] In fact, it was an investigation into the phenomenon of hemorrhagic tumor necrosis associated with gram-negative bacterial infections that resulted in identification of LPS as the causative agent of this effect, as well as of septic shock.[26]

At about the same time that cachectin was discovered, Old et al. demonstrated that LPS induced hemorrhagic necrosis of tumors via an endogenous mediator (produced internally in response to LPS and directly causing tumor destruction) termed tumor necrosis factor (TNF).[33-35] Subsequently, this factor, which is a protein, was purified and analyzed. Isolation and replication (cloning) of the gene coding for the protein made possible complete determination of its amino acid sequence, and it was then found that TNF-α (to distinguish it from TNF-β, a similiar protein produced by lymphocytes) and cachectin were the same substance.[36] Recently TNF-α has also been shown to help mediate endotoxic shock.[24, 36, 37]

3.4.1.1. Endogenous Production

TNF-α is one of the major products of activated macrophages, which are the primary source of this monokine.[32] In response to chemical signals (cytokines) released by lymphoreticuloendothelial system (LRES) cells in the vicinity of microorganisms or growing tumors, circulating monocytes migrate to the site and become more metabolically active macrophages. Macrophages are then "primed" by certain lymphokines (INF-γ in mice and, perhaps, one or more other lymphokines in humans). Finally, primed macrophages may be "activated" by substances present on the surfaces of certain bacterial cells, especially LPS and muramyl dipeptide (MDP), the basic repeating subunit of the mycobacterial cell wall peptidoglycan. Alternatively, macrophages may be temporarily activated in the absence of microbial substances by high concentrations of lymphokines. Activated macrophages exhibit a much enhanced ability to engulf (by a process known as phagocytosis) and kill microorganisms, and to kill on contact many different kinds of cancer cells. The microbicidal and tumoricidal capacities of activated macrophages vary according to the circumstances of activation.[38]

Macrophage activation is accompanied by a threefold acceleration of transcription of the gene coding for the TNF-α protein into messenger RNA (mRNA). Cellular levels of this mRNA, however, rise 50 to 100 times, indicating that activation also triggers mechanisms that prevent enzymic degradation of the mRNA. The mRNA is translated in the cytoplasm, by ribosomes of the rough endoplastic reticulum, into a protein containing 233 amino acids (in humans) and exhibiting a molecular weight of about 26,000. High levels of glucocorticoid hormones block TNF-α production by inhibiting both gene transcription and translation of mRNA into protein[32]. Recent research indicates that the 26,000 MW "pro-TNF-α" is integrated into the macrophage plasma membrane by the first (N-terminal) 76 to 79 amino acids, many of which bear hydrophobic side chains. The remainder of the protein, which causes the molecule's biological effects, projects from the outer cell membrane. This cell-bound form of TNF-α apparently is responsible for activated macrophage killing of cancer cells on contact. The biologically active portion of the molecule, consisting of 157 amino acids (in humans) and exhibiting a molecular weight of 17,000, is then cleaved from the cell surface by proteolytic enzymes released from lysosomes subsequent to macrophage activation, associating to form a trimer and resulting in the local cytokinetic and tumor necrotic effects of the monokine. Systemic macrophage activation, caused by gram-negative bacteremic endotoxemia results in widespread release of TNF-α, which produces the symptoms of septic shock.[39]

3.4.1.2. Effects and Functions of TNF-α

TNF-α, along with interleukin-I (IL-I), is a major mediator of the inflammatory response, mobilizing LRES cells and systemic resources to meet exogenous and endogenous challenges. Together, these two monokines are mainly responsible for the macrophage's central role in resisting microbial invasion and the proliferation of malignant cells.

Locally, TNF-α acts as a cytokine to kill or inhibit the growth of most cancer cells, to cause hemorrhagic necrosis of malignant tumors by destroying the blood vessels that feed them, and to recruit and activate cells that mediate the immune response. Malignant cell death or growth inhibition follows binding of TNF-α to specific cell-surface receptors and appears to involve fragmentation of chromosomal DNA, but the exact mechanisms of this effect are not yet known. TNF-α may destroy blood vessels in tumors by essentially the same mechanism that mediates the hemodynamic aspects of septic shock, but at lower doses of the monokine. TNF-α causes neutrophils to adhere to endothelial cells in the walls of blood vessels. Secretion of potent catabolic enzymes by these inflammatory phagocytic cells damages the vessels, which then become leaky (hemorrhage), depriving cancer cells of nutrients and leading to their death (necrosis). Low doses of TNF-α, however, actually cause proliferation of endothelial cells and formation of new blood vessels in normal tissues. The factor also decreases levels of thrombomodulin and plasminogen activator while increasing plasminogen activator inhibitor in endothelial cells, promoting blood coagulation.

In addition, TNF-α increases the quantity of circulating neutrophils and monocytes by causing macrophages, fibroblasts, endothelial cells, and T-lymphocytes to secrete the colony-stimulating factors (G-CSF, M-CSF, and GM-CSF) that induce maturation and release of the former cells from bone marrow. The monokine also promotes complement-mediated phagocytosis and intracellular killing of microorganisms by neutrophils, entrance of monocytes into tissues, and secretion of IL-I and prostaglandin E_2 (PGE_2) from macrophages. TNF-α synergizes with IFN-γ to activate macrophages and may synergize with IL-2 in the absence of IFN-γ to activate macrophages in humans. It helps modulate the immune response by increasing numbers of cell-surface IL-2 receptors on T-lymphocytes, facilitating the activation of these cells, stimulates secretion of IFN-γ by T-lymphocytes, and inhibits conversion of B-lymphocytes to antibody-secreting plasma cells.

Systemically, TNF-α acts as a hormone, affecting the functions of target tissues. It decreases levels of lipoprotein lipase, acetyl CoA carboxylase, and fatty acid synthetase in adipocytes and otherwise inhibits fat storage by preventing differentiation of these cells. The factor diminishes muscle function by lowering the transmembrane potential of myocytes. In liver, it inhibits cytochrome P-450 drug metabolism and albumin synthesis, but stimulates production of acute phase proteins. It causes bowel necrosis, particularly in the cecum, and delays gastric emptying. In the central nervous system (CNS), TNF-α, like IL-1, acts as a pyrogen, inducing the hypothalmus to secrete PGE_2 and resulting in elevation of body temperature. Other CNS effects include anorexia, slow-wave sleep, and stimulation of prolactin and α-melanocyte stimulating hormone secretion by the anterior pituitary gland. Since the latter hormones are known to inhibit the immune response, TNF-α may also terminate the inflammatory response through a negative feedback loop, like IL-1 (via stimulation of corticotropin secretion). TNF-α also causes inflammatory responses of skin, stimulating production of collagenase and PGE_2 by fibroblasts. It promotes wound healing by inducing fibroblast proliferation. The monokine reduces plasma iron and zinc levels, and inhibits cartilage proteoglycan and bone formation by promoting resorption and preventing synthesis of these structures.[31, 32, 40, 41] The effects and functions of TNF-α are summarized in Figure 3.1.

TNF-α plays a major role in producing the familiar symptoms of infection (fever, gastritis, constipation, sleep, lethargy, and loss of appetite) and underlying orchestrated efforts to impede, kill, and eliminate microbial invaders by diverting the body's energy and immune resources to the task. Systemic or chronic release of TNF-α by such disease states as septicemia, cancer, tuberculosis, or AIDS, however, produces the deleterious, even lethal, conditions of cachexia, bowel ulceration, and shock.

3.4.1.3. Applications

Because of its varied local and systemic effects, TNF-α is a good example of why cytokine therapy is problematic. The plasma half-life of TNF-α in humans is only 15 to 17 minutes.[42] As a protein, the monokine would be destroyed by proteases in the gastrointestinal tract and must be administered parenterally. Thus, it is unlikely that an optimal dose of TNF-α, providing an effective concentration at the site of a tumor without producing unacceptably adverse reactions after parenteral administration, can be found. In support of this contention, numerous preliminary clinical trials of TNF-α administered to cancer patients in this way have thus far produced very few positive results. More promising responses have been seen where the factor can be injected directly into the tumor.[31]

More knowledge regarding synergy between TNF-α and other cytokines, as well as other forms of cancer therapy, may make possible effective combination therapy using nontoxic doses of TNF-α. Targeted delivery modalities may also overcome the problems associated with TNF-α administration. For instance, incorporation of 26 kDa TNF-α into the surfaces of liposomes covalently coupled to monoclonal antibodies specific for unique proteins (antigens) found on the surfaces of malignant cells may produce targeted artificial macrophages for cancer therapy. Recent evidence that additional microbial or tumor and host factors are required to produce TNF-α–induced hemorrhagic necrosis and shock [37] suggests that ways can be found to administer therapeutic doses of pure recombinant TNF-α without producing deleterious side effects.

3.4.2. TNF-β (Lymphotoxin)

TNF-β, a protein closely related to TNF-α, is produced by activated T-lympho-cytes. The process of T-lymphocyte activation is much different from that of macrophage activation, involving presentation of protein antigen fragments by certain cells. The fragments must be associated with cell-surface proteins known as histocompatibility antigens. In addition, lymphokines such as IL-2 are required for activation. (For a review of how T-lymphocytes recognize antigen, see reference 9.)

Although TNF-β shows only a 30% homology with TNF-α, meaning that three-tenths of the amino acid sequences in both molecules are identical, TNF-α sequences that are highly conserved among species (implying that these regions are responsible for the protein's biological activities) are also present in TNF-β. The physiological effects of the two proteins are largely the same, but the dose–response differs. This functional similarity derives from the fact that both factors bind to the same cell-surface receptor, but with different affinities.[32, 43–45]

3.5. Future Prospects

The therapeutic applications of lymphokines and monokines illustrate all the problems inherent in the pharmaceutical uses of protein biological response modifiers. Virtually impossible without recombinant DNA technology, such a course provides unprecedented challenges to human ingenuity. Cytokines are short-range effectors that are often rapidly destroyed within millimeters of their origin. Because of this spatial limitation, finely modulated effects involving complex cellular responses to low doses of multiple cytokines have evolved in higher animals. Some effectors, such as IL-I and TNF-α, possess distinct func-tions as cytokines and hormones. Thus, the pharmacology of the lymphokines and monokines is daunting and will require expansion of the boundaries of pharmaceutical science, involving especially questions of stability, delivery, and pharmacokinetics.

Reflecting these realities, IFN-α is the most successful of these agents so far, mainly because its antiviral effects predominate at pharmacological doses. The colony-stimulating factors also appear promising because their functions are long-range, involving communication between sites of infection and bone marrow. IL-I, IL-2, and TNF-α are the most problematic, requiring imaginative approaches to targeting. The problems and their possible solutions are exemplified by the severe adverse effects caused by a need to infuse large quantities of IL-2 into cancer patients to overcome the extremely short half-life of the lymphokine in vivo and genetically engineering tumor-infiltrating lymphocytes (TILs) to secrete large quantities of TNF-α.

3.6. Perspectives

It is becoming apparent that new cytokines, lymphokines, and monokines will be identified in the near future. For example, a pentapeptide, methionine enkephalin (MEK, derived from its prohormone, proenkephalin-A, found in T helper cells and macrophages), has recently been identified to have lymphokine–monokine actions. These include the activation of T cell subsets, NK-K-LAK cells, as well as IL-2 and INF-γ. Preliminary clinical studies indicate that MEK may have beneficial effects in AIDS and cancer patients.[46] Future clinical studies will probably include combinations of the lymphokines and monokines. Already combinations with chemotherapeutic agents in the treatment of cancer patients are in progress.[47] As our knowledge base increases, it may be possible to consider combinations with antiinfective agents. Clinical trials are currently in progress testing combinations with antiviral agents in the treatment of AIDS patients.[48] From the preceding discussion, the identification and isolation of new lymphokines and monokines with more specific effects on the cells of the immune system is envisioned.

References

1. Isaacs, A., and Lindenmann, J. 1957. Virus interference. I. The interferon. *Proc. Roy. Soc. London Ser. B* 147:258–273.

2. Wheelock, E.F., 1965. Interferon-like virus inhibitor induced in human leukocytes by phytohemagglutinin. *Science* 149:310–311.

3. Johnson, H.M. 1985. Mechanism of interferon-gamma production and assessment of immunoregulatory properties. *Lymphokines* 11:33–46.

4. Vilcek, J., Gray, P.W., Rinderknecht, E., and Sevastopoulos, C.G. 1985. Interferon gamma: A lymphokine for all seasons. *Lymphokines* 11:1–32.

5. Torres, B.A., Farrar, W.L., and Johnson, H.M. 1982. Interleukin 2 regulates immune

interferon (IFN gamma) production by normal and suppressor cell cultures. *J. Immunol.* 128:2217–2219.

6. Torres, B.A., Yamamoto, J.K., and Johnson, H.M. 1982. Cellular regulation of gamma interferon production: Lyt phenotype of the suppressor cell. *Infect. Immun.* 35:770–776.

7. Handra, K., Suzuki, R., Matsui, H., Shimizu, Y., and Kumagai, K. 1983. Natural killer (NK) cells as a responder to interleukin 2 (IL2). II. IL2-induced interferon production. *J. Immunol.* 130:988–992.

8. Lengyel, P. 1982. Biochemistry of interferons and their actions. *Annu. Rev. Biochem.* 51:251–282.

9. Grey, H.M., Sette, A.S., and Buus, S. 1989. How T cells see antigen. *Sci. Am.* 261(5):56–64.

10. Sonnenfeld, G., Meruelo, D., McDevitt, H.O., and Merigan, T.C. 1981. Effect of type I and type II interferons on murine thymocyte surface antigen expression: Induction or selection? *Cell. Immunol.* 57:427–439.

11. Steeg, P.G., Moore, R.N., Johnson, H.M., and Oppenheim, J.J. 1982. Regulation of murine macrophage Ia antigen expression by a lymphokine with immune interferon activity. *J. Exp. Med.* 156:1780–1793.

12. King, D.P., and Jones, P.P. 1983. Induction of Ia and H-2 antigens on a macrophage cell line by murine interferon. *J. Immunol.* 131:315–318.

13. Vogel, S.N., and Rosenstreich, D.L. 1979. Defective Fc receptor-mediated phagocytosis in C3H/HeJ macrophages. I. Correction by lymphokine induced stimulation. *J. Immunol.* 123:2842–2850.

14. Vogel, S.N., and Friedman, R.M. 1984. Interferon and macrophages. In *Interferons and the immune system.* J. Vilcek and E. De Mayer, eds. Pp. 323–329. Amsterdam: Elsevier.

15. Merigan, T.C. 1988. Human interferon as a therapeutic agent. *N. Engel. J. Med.* 318:1458–1460.

16. *Biotechnology products in development.* 1988. Washington, D.C: Pharm. Mnfrs. Asso.

17. Dinarello, C.A. 1988. Biology of interleukin 1. *FASEB J.* 2:108–115.

18. Neta, R., and Oppenheim, J.J. 1988. Why should internists be interested in interleukin-1? *Ann. Int. Med.* 109:1–2.

19. Rosenberg, S.A. 1988. The development of new immunotherapies for the treatment of cancer using interleukin-2. *Ann. Surg.* 208(2): 121–135.

20. Smith, K.A. 1988. Interleukin-2: Inception, impact, and implications. *Science* 240:1169—1176.

21. Miyajima, A., Miyatake, S., Schreurs, J., DeVries, J., Arai, N., Yokota, T., and Arai, K.-I. 1988. Coordinate regulation of immune and inflammatory responses by T cell-derived lymphokines. *FASEB J.* 2:2462–2473.

22. Weisbart, R.H., Gasson, J.C., and Golde, D.W. 1989. Colony-stimulating factors and host defense. *Ann. Int. Med.* 110(4):297–303.

23. Kirkpatrick, C.H. 1989. Biological response modifiers, interferons, interleukins, and transfer factor. *Ann. Allergy* 62:170–176.

24. Beutler, B., Milsark, I.W., and Cerami, A.C. 1985. Passive immunization against cachectin/tumor necrosis factor protects mice from lethal effect of endotoxin. *Science* 229:869–871.

25. Favorite, G.O., and Morgan, H.R. 1942. Effects produced by the intravenous injection in man of a toxic antigenic material derived from *Eberthella typhosa*:Clinical, hematological, chemical and serological studies. *J. Clin. Invest.* 21:589–599.

26. Shear, M.G., and Turner, F.C. 1943. Chemical treatment of tumors. Reaction of mice with primary subcutaneous tumors to injections of hemorrhagic-producing bacterial polysaccharide. *J. Natl. Cancer Inst.* 4:461–476.

27. Beeson, P.B. 1947. Tolerance to bacterial pyrogens. II. Role of the reticuloendothelial system. *J. Exp. Med.* 86:39–44.

28. Rouzer, C.Z., and Cerami, A. 1980. Hypertriglyceridemia associated with *Trypanosoma brucei* infection in rabbits: Role of defective triglyceride removal. *Mol. Biochem. Parasitol.* 2:31–38.

29. Kawakami, M., and Cerami, A. 1981. Studies of endotoxin-induced decrease in lipoprotein lipase activity. *J. Exp. Med.* 154:631–639.

30. Terry, W.D. 1975. BCG in the treatment of human cancer. *CA* 25:198–203.

31. Old, L.J. 1988. Tumor necrosis factor. *Sci. Am.* 258(5):59–75.

32. Beutler, B., and Cerami, A. 1989. The biology of cachectin/TNF—A primary mediator of the host response. *Annu. Rev. Immunol.* 7:625–655.

33. Carswell, E.A, Old, L.J., Kassel, R.L., Green, S., Fiore, N., and Williamson, B. 1975. An endotoxin-induced serum factor that causes necrosis of tumors. *Proc. Natl. Acad. Sci. U.S.A.* 72:3666–3670.

34. Helson, L., Green, S., Carswell, E., and Old, L.J. 1975. Effect of tumour necrosis factor on cultured human melanoma cells. *Nature (London)* 258:731–732.

35. Green, S., Dobrjansky, A., Carswell, E.A., Kassel, R.L., Old, L.J., Fiore, N., and Schwartz, M.K. 1976. Partial purification of a serum factor that causes necrosis of tumors. *Proc. Natl. Acad. Sci. U.S.A.*73:381–385.

36. Beutler, B., and Cerami, A. 1986. Cachectin and tumor necrosis factor: Two sides of the same biological coin. *Nature (London)* 320:584–588.

37. Rothstein, J.L., and Schreiber, H. 1988. Synergy between tumor necrosis factor and bacterial products causes hemorrhagic necrosis and lethal shock in normal mice. *Proc. Natl. Acad. Sci. U.S.A.* 85:607–611.

38. Fidler, I.J. 1985. Macrophages and metastasis—A biological approach to cancer therapy: Presidential address. *Cancer Res.*45:4714–4726.

39. Kriegler, M., Perez, C., DeFay, K., Albert, I., and Lu, S.D. 1988. A novel form of TNF/cachectin is a cell surface cytotoxic transmembrane protein: Ramifications for the complex physiology of TNF. *Cell* 53:45–53.

40. Dinarello, C.A. 1988. Cytokines: Interleukin-I and tumor necrosis factor (cachectin).

In *Inflammation: Basic principles and clinical correlates*. J.I. Gallin, I.M. Goldstein, and R. Synderman, eds. New York: Raven. Pp. 195–208.

41. Bermudez, L.E., and Young, L.S. 1988. Tumor necrosis factor, alone or in combination with IL-2, but not IFN-gamma, is associated with macrophage killing of Mycobacterium avium complex. *J. Immunol.* 140:3006–3013.

42. Michie, H.R., Manogue, K.R., Spriggs, D.R., Revhaug, A., O'Dwyer, S., Dinarello, C.A., Cerami, A., Wolff, S.M., and Wilmore, D.W. 1988. Detection of circulating tumor necrosis factor after endotoxin administration. *N. Engl. J. Med.* 318:1481–1486.

43. Aggarwal, B., Eessalu, E., and Hass, E. 1985. Characterization of receptors for human tumour necrosis factor and their regulation by gamma-interferon. *Nature (London)* 318:665–669.

44. Granger, G.A., Masunaka, I., Averbook, B., Kobayashi, M., Fitzgerald, M., and Yamamoto, R. 1988. Differences in the bioactivity of recombinant human TNF, LT, and T-cell-derived LT-3 on transformed cells in vitro and the Meth A tumor growing in BALB/c mice. *J. Biol. Response Mod.* 7:488–497.

45. Paya, C.V., Kenmotsu, N., Schoon, R.A., and Leibson, P.J. 1988. Tumor necrosis factor and lymphotoxin secretion by human natural killer cells leads to antiviral cytotoxicity. *J. Immunol.* 141:1989–1995.

46. Wybran, J., Schandene, L., VanVooren, J.P., Vandermoten, G., Latinne, D., Sonnet, J., DeBruyere, M., Taelman, H., and Plotnikoff, N.P. 1987. Immunologic properties of methionine-enkephalin, and therapeutic implications in AIDS, ARC and cancer. *Ann. N.Y. Acad. Sci* 496:108–114.

47. McLeod, G.R. 1988. Alpha interferons in malignant melanoma. *Br. J. Clin. Pract. Suppl.* 62:22–26.

48. Bartlett, J.A., Blankenship, K., Waskin, H., Sebastian, M., Shipp, K., and Weinhold, K. 1990. Zidovudine and Interleukin-2 in WR2 HIV infected patients: Evidence for stimulated iommunologic reactivity against HIV. *Sixth Int. Conf. AIDS* (San Francisco, 3, S.B.) 421:191

Additional Reading

Arai, K.I., and de Vries, J. 1989. *Lymphokines: The molecular biology of regulators of immune and inflammatory responses*. New York: Stockton.

Bonavida, B., Gifford, G.E., Kirchner, H. and Old, L.J., eds. 1988. *Tumor necrosis factor/cachectin and related cytokines*. New York: Karger.

Fradelizi, D., and Bertoglio, J. 1989. *Lymphokine receptor interactions*. Paris: Inst. Natl. Sante Recherche Med.

Groopman, J.E., Golde, D.W., and Evans, C.H. 1989. *Mechanisms of action and therapeutic applications of biologicals in cancer and immune deficiency disorders*. New York: Alan R. Liss.

Mannel, D.N., Northoff, H., Bauss, F., and Falk, W. 1987. Tumor necrosis factor: A cytokine involved in toxic effects of endotoxin. *Rev. Infect. Dis.* 9 suppl. 5:S602–S606.

Medunitsyn, N.V., Litvinov, V.I., and Moroz, A.M. 1987. *Mediators of the immune response: The biology and biochemistry of lymphokines, monokines and cytokines.* New York: Harwood.

Oldham, R.K. ed. 1987. *Principles of cancer biotherapy.* New York: Raven.

Sorg, C., ed. 1989. *Macrophage-derived cell regulatory factors.* New York: Karger.

Urbaschek, R., and Urbaschek, B. 1987. Tumor necrosis factor and interleukin 1 as mediators of endotoxin-induced beneficial effects. *Rev. Infect. Dis.* 9, suppl. 5: S607–S615.

Waldmann, T.A. 1986. The structure, function, and expression of interleukin-2 receptors on normal and malignant lymphocytes. *Science* 232:727–732.

Webb, D.R., Pierce, C.W., and Cohen, S. 1987. *Molecular basis of lymphokine action.* Clifton, N.J.: Humana.

4

Analytical Methods in Biotechnology

R. J. Prankerd and S. G. Schulman

4.1. Introduction

Advanced biotechnology is inextricably linked to analytical chemistry. Not only is it necessary to verify the identities and purities of the products of recombinant DNA (rDNA) technology, but it is also essential to be able to detect and quantify these substances, their decomposition products, and their metabolites (when they are used as pharmacons), at trace concentrations. In this chapter we shall consider some of the modern analytical methodology used for the detection and determination of the products of the "new" biotechnology.

Because the pharmaceutically significant products of rDNA technology are almost invariably polypeptides, the chemistry and instrumentation discussed herein will, in the interests of brevity, be confined to those analytical tools particularly relevant to peptide and protein chemistry. There is no intention of focusing on the polynucleotides and nucleic acids, which are the machinery of biotechnology. However, the analytical methods described here are often applicable to them, because their polymeric nature is in common with peptides and proteins. Much of the pertinent analytical methodology is relevant only to the analysis of large thermo- and hydrolytically labile molecules. Space does not permit a review of all analytical techniques of possible interest. Some, such as direct spectrophotometry or fluorometry, which by themselves are of very limited use in the complex mixtures that must be dealt with in biotechnology and the applications of its products, are not considered here except in so far as they are linked to the techniques that are of considerable utility in this context. Rather, the focus will be on separation, immunoanalytical, and chiral methods which account for the bulk of the analytical methodology in current use in advanced biotechnology. The reader seeking a more extensive description of these, as well as the peripheral analytical methods, is referred to the recent excellent texts by Willard et al.,[1] Snyder and Kirkland,[2] Christian and O'Reilly,[3] and Skoog.[4]

Early quantitative analysis of biologicals, biopolymers, endotoxins and other pharmaceuticals (such as antibiotics) relied on biological assays using cell cultures, isolated organ baths, or whole animals. These bioassays are expensive, labor intensive, and cumbersome and give results with considerable variability. However, they have the undeniable advantage of measuring the one property that the product is intended to have: a biological or pharmacological effect. This is especially important in view of the fact that the product may be either a heterogeneous mixture of several components with differing activities, or else a very potent active species, inseparably associated to a slightly greater or lesser extent with a large amount of inert material. Hence, bioassays are still of considerable importance and potency is conventionally reported in terms of units of activity per mass or amount of the product. Conversely, quantitative instrumental analytical techniques measure properties that are directly related to the molar amount of the product and which are hoped to be proportional to the biological potency of the product. This is not necessarily the case, particularly for proteins which have delicate tertiary or quaternary structure and can unfold or aggregate to an extent that may not be readily controlled. The requirements for the validation of analytical methodology need to be as rigorous as possible so that biological potency can be inferred with confidence from analyses based on physicochemical properties. In vitro methods based on the physiological processes that involve the biopolymers of interest are expected, a priori, to be more likely to reflect their biological potency. These include affinity chromatography and immunological methods. The correlation of pharmacological effect with the physicochemical properties of biopolymers is the major challenge facing practical instrumental analysis of rDNA biotechnology products.

4.2. Chromatography

A variety of chromatographic methods have been applied to the qualitative and quantitative analysis of peptides and proteins. Because of the low volatility of protein and peptide analytes, gas chromatographic methods are of limited use unless additional derivatization steps are performed. In principle, supercritical fluid chromatography could be applied to the analysis of peptides and proteins. The vast majority of chromatographic applications for peptides and proteins use either high- or low-pressure liquids as mobile phases. The present discussion will be mainly confined to high-performance liquid chromatography (HPLC). Liquid chromatographic separation methods are usually specific for small organic drug molecules and elution conditions can be manipulated to give good resolution of one analyte from other compounds of similar structure. Peptides comprising up to 10 to 15 amino acids can be regarded as behaving similarly to small drug molecules. However, it has been hypothesized that the surface contact area between a protein and the HPLC stationary phase is a very small part of the total surface.[5] If proteins are modified (either in the primary amino acid sequence or conformationally) at sites remote from such a specific "adsorption site," the modified versions might be expected to coelute with the native protein. For example, the catalytically active form of an enzyme may coelute with an inactive form if the active site is remote from the adsorption site. Thus, HPLC analysis should not be expected a priori to be a reliable means of quantifying the biological or pharmacological activity of proteins that have pharmaceutical applications. It should, however, be useful for quantifying the activity of smaller peptides (< 20 amino acids), in the same way as for traditional small drug molecules. HPLC methods can be applied to studies of the identity, purity, and stability of pharmaceutical peptides and proteins.[6]

4.2.1. Reversed-Phase HPLC

A description of conventional reversed-phase (RP) HPLC is not given here, as there are many textbook accounts.[2, 7] Reversed-phase chromatographic columns used in the separation of conventional drugs can also be useful for separating a wide range of peptides and small proteins.[8] However, problems are normally encountered with large proteins, especially those that aggregate in aqueous solution, either through complexation with metal cations (e.g., zinc insulin) or as a result of their hydrophobic exterior surfaces.[9] There are major difficulties of poor resolution, surface adsorption and unfolding or denaturation of such hydrophobic proteins when hydroorganic cosolvent mobile phase mixtures are used. A variety of mobile phases have been used to address this problem, for example, water and propanol, water and butanol, aqueous acids, aqueous urea or guanidine-HCl, detergent solutions or a combination of these. Also, hydrophobic proteins may cause band-broadening problems on silica support phases, for example, from mixed retention mechanisms (the secondary organic modifiers used in HPLC, such as tetrahydrofuran, were not found to be useful). These problems have been addressed by the use of column packings based on polystyrene or polymethacrylate resins and also by the use of microbore columns. Resin support phases are more stable than silica to the acidic elution conditions commonly used (e.g., dilute trifluoroacetic acid). However, commercially available resin supports are less rigid than conventional silica supports and are limited in the back pressures to which they may be subjected. Improved chemical stability of silica supports to acidic eluants has been achieved by the use of more bulky groups in the bonded stationary phase.[10] Changes in mobile phase composition during gradient runs can cause protein molecules to unfold or swell and then become trapped within porous support phases, to be released and eluted in a subsequent gradient run. Large pore (200–300 Å) packing materials have been found to be necessary for protein HPLC, especially for gradient elution. The use of very small silica particles (<3 μm) allowed separation of peptide (e.g., enkephalins) and protein (MW 6,500–100,000) mixtures in a few minutes.[11] However, the advent of highly efficient separations using very small (<1 μm) nonporous particles is expected to reduce the current reliance on highly porous particles which may trap large analyte molecules.[10] A reduction in internal column diameter is also expected to lead to improved separations of proteins. However, the reduced flow-rates required impose increasingly greater demands on the accuracy and stability of the pumping systems needed for gradient elution.

A further technique for solving the problems of RP separation of large proteins is hydrophobic interaction chromatography (HIC). This approach is based on the use of ionic strength control in the salting out of proteins from aqueous phases onto weakly hydrophobic stationary phases. A salt gradient of decreasing concentration elutes the analytes in order of increasing hydrophobicity. Commonly used silica stationary phase supports are too hard for many biopolymers leading to denaturation and irreversible binding. Modified hydrophilic silicas (e.g., dihydroxypropoxypropyl-silica) with bonded polyether or polyvinyl alcohol stationary phases in combination with aqueous salt gradients and additives such as Carbowax 4000 or PEG have demonstrated promise in overcoming these difficulties for the separation of proteins and also for polynucleotides.[12]

4.2.2. Size Exclusion Chromatography

High-performance size exclusion chromatography (SEC), or gel permeation chromatography (GPC), has been developed from the low-pressure (gravity feed) chromatographic methods first introduced in the 1960s for the purification of proteins, especially enzymes. This technique separates dissolved solute molecules based on differences in their sizes, and assuming that the solutes have similar shapes, based on their molecular weights (MW). The column packing material is in the form of polysaccharide, silica, or resin beads (usually 10 μm diameter) that are very porous. The pores are of controlled size-range. Depending on the porosity and tortuosity of the packing, larger molecules enter the pores less readily than smaller molecules; hence, a smaller column bed volume is available to large molecules. Consequently, the largest molecules are excluded from the pores altogether and elute with the solvent front. Medium size molecules migrate more slowly through the column and smaller molecules migrate very slowly. Several ranges of pore size are available, from 50 Å up to 10,000Å. These exclude solutes with molecular weights ranging from <1,000 g/mol for the smallest pore sizes, up to ~5 × 10^6 g/mol for the largest. Detection of eluates is typically by UV absorption at 220 or 280 nm although other methods are also used.

In addition to preparative applications, SEC columns can be used for the identification and quantitative analysis of proteins if an authentic sample of the protein is available. They may also be used for MW estimation of an unidentified protein if the column is calibrated with a series of molecular weight standards. For such columns, there will be a MW range for which there is a linear relationship between log MW and elution volume. Recently, a combination of low-angle laser light scattering (LALLS) detection in conjunction with a concentration-sensitive detector (e.g., refractive index, ultraviolet absorption, or fluorescence) was suggested for use with a SEC column to directly estimate the molecular weights of eluted solutes.[13, 14] This takes advantage of the capability of light-scattering methods in determining MW, provided that the molar concentration of the solute is known. Consequently, precalibration of the SEC column can be avoided. The determination of molecular size as well as MW from multiangle laser light scattering has also been reported,[15] using a commercially available detector.

Silica can also be used as a SEC packing support phase if the surface is modified with a hydrophilic bonded phase [e.g., the TSK SW series (Toyo Soda)]. These have been used for the characterization of a sustained release formulation of nafarelin (a decapeptide) incorporated into lactide–glycolide copolymer microspheres.[16] A TSK 3000SW column was able to retain nafarelin by electrostatic (coulombic) attraction to residual silanol groups, as well as to determine the MW distribution of the copolymer microspheres (or their hydrolytic degradation products) by SEC, using a single mobile phase. The results obtained were equivalent to those obtained by conventional RP-HPLC (for nafarelin) and SEC on μStyragel columns (for the copolymer). Other commercially available hydrophilic silica supports (GPC, SOTAPhase) have been used for the separation of plasma proteins.

4.2.3. Ion-Exchange Chromatography

Ion-exchange (IE) resins based on the modified (sulfonated) cross-linked styrene-divinylbenzene polymers used in conventional low-pressure column chromatography of inorganic and small organic ions have been adapted for use in high-performance applications by increasing pore sizes, reducing the resin bead diameter, and operating under higher pressures. These may be used for the separation of peptides and smaller proteins. The analytes are applied to the column in acidic solution, and may then be eluted isocratically or by using temperature, cosolvent, pH, or ionic strength gradients.

4.2.4. Affinity Chromatography

This is a separation method that depends on specific interactions between the analyte and a complementary group which is covalently bonded to the column packing. Initially, affinity chromatography was developed as a preparative method for purification of enzymes and other proteins in the 1970s. More recently, it has been adapted for quantitative analytical purposes by fusion with HPLC techniques.[17] The method uses solid support phases similar to those used for SEC. The support phase should have a large specific surface area, large pores, and hydrophilic properties, as well as being stable and insoluble. The surface of the support phase is modified with a chemically reactive group, which is then covalently bonded to a ligand. The ligand, in turn, can interact with a greater or lesser degree of specificity with the protein or peptide of interest. The interaction is thermodynamically controlled and is characterized by an equilibrium binding constant. A spacer group between the support surface and the ligand may also be necessary to optimize interaction with bulky analytes. After application of a solution containing the analyte and then washing off any contaminants, the retained analyte is eluted, either isocratically (as in conventional HPLC) or by changing the mobile phase composition so that the specific interaction is weakened or abolished. Compositional changes include alterations in pH or ionic strength and the addition of organic cosolvents, chaotropes, or a second substance that binds even more tightly to the ligand thus, displacing the analyte. These changes may be made as step or continuous gradients. The ligand may be specific for a group of related analytes (low specificity) or it may be specific for a single substance (high specificity). The ligand group may be a protein such as protein A (reversibly binds to immunoglobulins) or it may be a small group such as *m*-aminophenylboronic acid (binds at pH 8 to 10 to a variety of diols such as catechols and sugars, but not at lower pH values).[18, 19]

4.2.5. Two-Dimensional Chromatography

Early two-dimensional chromatographic methods included thin-layer chromatography (TLC) separations, in which a complex mixture of analytes was spotted in one corner of a square TLC plate and developed using one mobile phase. Then the plate was rotated through 90° and redeveloped using a second mobile phase. The method was adapted for use in conventional HPLC, in which the term *two dimensional* refers to the consecutive use of two different kinds of HPLC column. These methods are often automated and employ an initial separation of the mixture into groups of analytes using one type of column [e.g., cation exchange (separation of cationic from anionic and neutral species)], then separation of the group of interest into individual analytes using a second column of a different type (e.g., reversed-phase separation of several catecholamines). The method is often used for separation of exogenous or endogenous analytes from plasma samples. Two-dimensional HPLC was recently applied to the peptide mapping technique.[20] A mixture of peptides obtained from tryptic digestion of cellobiohydrolase I (*Trichoderma reesei*) was initially separated using SEC, then nine eluted fractions were further separated using a RP column. Neither column alone could give satisfactory resolution of the components of the mixture, but the combination gave considerably improved resolution. However, analysis by SDS-PAGE showed that some of the fractions eluted from the two-dimensional separation system contained more than one peptide.

4.2.6. Primary Amino Acid Sequence of Proteins

The percentage content of each of the amino acids that make up proteins and peptides is usually determined by complete acid hydrolysis of the protein then analysis of the hydrolysate with a dedicated autoanalyzer. The analysis uses ion exchange separation of the acidic amino acid solution followed by postcolumn color formation with ninhydrin and absorptiometric detection (440 and 570 nm). Quantitation is by comparison of peak heights or areas from the sample with those from a standard hydrolysate of known amino acid composition obtained under the same experimental conditions. Analyses with greater speed and sensitivity have been recently achieved using precolumn derivatization with reagents such as 2,3-naphthalenedialdehyde-cyanide (NDA), *o*-phthaldialdehyde (OPA), phenylisothiocyanate (PITC), and 9-fluorenylmethyloxycarbonyl chloride (FMOC), followed by RP-HPLC and fluorescence detection.[21, 22] However, the amino acid content is a crude measure of protein identity and purity. A change of one residue in the amino acid content of a protein is likely to go undetected by this type of analysis. Of far greater importance to the quality control and purity analysis of a recombinant protein is the full determination of its primary amino acid sequence. The methods required for this task are considerably more involved. A major tool for this purpose is automated Edman degradation[23–25] of the N-terminal end of the protein, Scheme 1.

Phenylisothiocyanate (PITC) couples with the free amino group to form a phenylthiocarbamoyl derivative. The first peptide bond in the protein is then cleaved with anhydrous acid to form a new N-terminus on the protein and a substituted anilinothiazolinone. The anilinothiazolinone is extracted and treated with aqueous acid, causing rearrangement to the phenylthiohydantoin (PTH) derivative of the first amino acid of the original protein sequence. The PTH derivative can be identified by its retention volume on HPLC or by chemical ionization mass spectrometry (CIMS).[26] The remaining protein can then be subjected to further cycles of degradation, leading to identification of the linear sequence of the first 5 to 15 amino acids. The degradation is not quantitative (30–70%), and some PTH amino acid derivatives are not stable to the reaction conditions.

The C-terminal amino acid sequence may also be determined by chemical or enzymatic degradation procedures.[27] Chemical degradation uses reagents such as ammonium thiocyanate and acetic anhydride, which result in the formation of thiohydantoin derivatives.[28, 29] These procedures are not as well established as the Edman procedure for the N-terminus. Enzymatic methods involve the use of carboxypeptidases.[30–32]

4.2.7. Peptide Mapping

Edman degradation allows identification of the sequence of the first 5 to 30 amino acids at the N-terminus of the protein and some identification is also possible at the C-terminus. However, the bulk of the protein sequence remains unknown and the protein may not be distinguishable from a corresponding one in which a single amino acid residue is changed. Digestion of the protein with enzymes such as

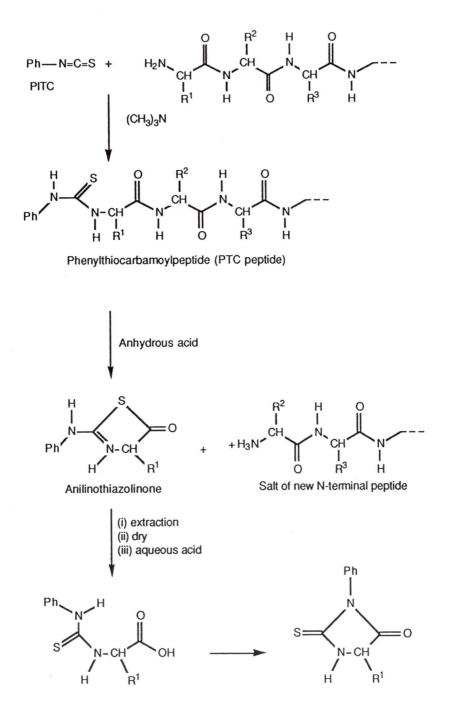

Phenylthiocarbamoylpeptide (PTC peptide)

Anilinothiazolinone

Salt of new N-terminal peptide

Substituted phenylthiohydantoin

Scheme 1. Edman degradation of proteins.

trypsin or V8 protease (*Staphylococcus aureus*) produces a characteristic set of polypeptides that may be separated by HPLC, gel electrophoresis or two-dimensional TLC. This powerful technique is called peptide mapping.[33, 34] The resulting peptides are small enough that careful gradient HPLC analysis will create a pattern of eluted peaks that is unique to the protein of interest. If a single amino acid is changed in the protein, the resultant single altered peptide will have a different elution volume. The peptide maps of the original and altered proteins will usually be clearly distinguished by the appearance of one new peak and the loss of a peak from the original map.[6] The separated peptides may also be isolated by small scale preparative HPLC in sufficient quantity to be sequenced by Edman degradation or by "soft ionization" mass spectrometric methods such as liquid secondary ion mass spectrometry (LSIMS) or plasma desorption mass spectrometry (PDMS), especially in conjunction with a tandem double-focusing ion analyzer.[26, 35] Trypsin digestion cleaves peptide bonds on the C-terminal side of arginine and lysine residues, and V8 protease cleaves on the C-terminal side of glutamic acid residues. Other enzymes may also be used. Chemical or radiochemical labeling of the N- and C-terminal amino acids (e.g., by maleylation and esterification respectively) followed by tryptic digestion allows isolation and identification of the N- and C-terminal peptides. Arginine-containing peptides can be selectively identified by base-catalyzed derivatization with benzoin. Peptides containing free thiol groups can be located by S-carboxymethylation of the intact protein with iodoacetic acid (and can be tritiated for scintillation counting), followed by tryptic digestion. Similarly, the location of disulfide bonds can be demonstrated by comparing a peptide map of the intact protein with that obtained after the protein has been treated first with a reducing agent and then trapping the liberated free thiols by S-carboxymethylation. Reducing agents include 2-mercaptoethanol and dithiothreitol. Methionine-containing peptides may be identified by S-oxidation of the protein before mapping.

The separation of the peptides from tryptic or other enzymatic digestion of proteins is usually performed using gradient elution HPLC. Very accurate pumping systems are required for precise control of the HPLC gradients needed to obtain reproducible peptide maps. Good column temperature control is also important. The value of ion-pairing agents in RP-HPLC of amino acid and peptide mixtures has also been stressed.[36]

4.2.8. Glycoproteins

Naturally occurring proteins are often glycosylated with oligosaccharide chains of 5 to 20 sugars. Glycosylation occurs at amino acid residues such as asparagine, serine, or threonine. Thus, it is necessary to characterize the types, extent and sites of glycosylation and to determine if glycosylation is necessary for biological activity of the protein. Usually, there is a mixture of closely related materials even if the glycoprotein has only one site of glycosylation [e.g., human interleukin-2 (microheterogeneous)]. This is due to the fact that although the primary amino acid sequence of the protein is under genetic control, glycosylation of the protein is dictated by the type of host cell and the conditions of fermentation. Bacterial host cells do not produce glycosylated recombinant proteins, whereas mammalian host cells do. In the case of recombinant tissue plasminogen activator (TPA), it was necessary to use a mammalian host cell (Chinese hamster ovary) for commercial production as glycosylation is necessary for its biological properties.

Mono- and oligosaccharide characterization by HPLC has always been more difficult than for many other analytes owing to the following conditions: (1) normal phase HPLC cannot always resolve α- from β-linked isomers; (2) RP-HPLC has difficulties with the anionic oligosaccharides that occur naturally in glycoproteins; (3) separation of anionic oligosaccharides can be performed with anion-exchange resins, but attempted simultaneous separation of neutral oligosaccharides gives poor resolution; and (4) UV detection is difficult due to the lack of a good chromophore.[37] Approaches used to resolve these difficulties have included (1) strong anion-exchange chromatography (SAX) in very basic solution, in which deprotonation of probably one hydroxyl group occurs per sugar molecule (this allows selective retention on the SAX column as well as detection by UV absorption); (2) chemical precolumn derivatization of the sugars to permit separation by RP-HPLC and UV or fluorescence detection; and (3) a combination of high-pH SAX separation followed by pulsed amperometric detection (Dionex).

Chemical methods for oligosaccharide cleavage (e.g., hydrazinolysis) from the protein are vigorous, difficult to reproduce and produce chain fragments as well as intact chains. The use of enzymatic methods such as endoglycosidase H or glycopeptidase F[37] is preferred. These methods are more highly reproducible than chemical degradation methods and generally give cleavage of complete chains. Owing to coulombic repulsion, monosaccharides generally have one ionized site per molecule at pH $\sim$ 13, whereas oligosaccharides have multiple sites of ionization and are retained more strongly by a SAX column. Therefore, separation of monosaccharides from oligosaccharides can be achieved with an increasing sodium acetate gradient on a base eluant of 0.1 to 0.2 M NaOH. In addition, total monosaccharide analysis can be achieved by SAX chromatography of the products of acid hydrolysis of the intact protein and of the enzymatically liberated oligosaccharides. Identification of oligosaccharides has also been performed using "soft ionization" mass spectrometry techniques.[26, 35]

4.3. Gel Electrophoresis

Electrophoretic methods involve the migration of charged particles on a support such as polyacrylamide gel (in the form of a sheet, slab, or tube) under the influence of a strong electric field, typically several kilovolts. Proteins can be identified by comparison of their electrophoretic mobility with those of standards.

The purity of the protein can be qualitatively monitored by determining the number of bands produced on analysis. Quantitative analysis is also possible by visually or optically scanning the density of each band and comparing with standards, if the band identities are known. Common impurities in proteins manufactured by recombinant methods include endotoxins, DNA and other proteins from the host cell organism used for the preparation, point mutant proteins from genetically altered cells after gene-splicing has been performed and degradation products of the desired product (e.g., from deamidation reactions or proteolysis). The separated bands can be isolated and subjected to further analysis,[38] for example, by immunological methods or amino acid sequencing by Edman degradation or mass spectrometry.

4.3.1. Electrophoresis Techniques

Sodium dodecyl sulfate-polyacrylamide gel electrophoresis (SDS-PAGE) is a very commonly used method for protein characterization. SDS is a strongly active surfactant that dissociates noncovalently linked protein aggregates and binds to protein molecules at an approximately constant ratio (1.4 g SDS/g protein). The resulting complex has an overall negative charge. Thus, separation in an electric field is solely a function of the molecular size. Assuming similarity of shape, the size of a protein is proportional to its molecular weight (MW). SDS-PAGE is thus widely used in the determination of molecular weight when the unknown protein is concurrently run with MW standards. Disulfide linkages in the protein can be demonstrated by reduction with mercaptoethanol or dithiothreitol to give either (1) unfolded proteins that have an apparently higher MW than the unreduced protein, owing to the change in shape of a protein that has a single polypeptide chain and intramolecular S—S linkages; or (2) lower MW fragments that are characteristic of a protein comprising two or more polypeptide chains held together by disulfide bonds.

The bands separated by SDS-PAGE are visualized by a variety of staining methods, of which the two most common are Coomassie brilliant blue and silver stains. Coomassie stain has a 100 to 200 ng absolute sensitivity limit which is quantifiable by laser densitometry. The silver stain is much more sensitive (0.2 to 2 ng absolute sensitivity limit) than Coomassie stain but not generally quantifiable. Hence, it is more useful for the detection of very low levels of impurities by simply examining for the absence of any contaminating bands.

Isoelectric focusing (IEF) is an electrophoretic separation method that is based on the overall charge balance on a protein. The charge balance results from the effects of solution pH on amino acid side chains that contain ionizable groups. The groups may be negatively charged (aspartic acid and glutamic acid) or positively charged (histidine, lysine, and arginine) when ionized at or near physiological pH. An electric field across a gel containing a series of immobilized organic ampholyte buffers produces migration of the buffers until they form a pH gradient. The same electric field causes slower migration of applied proteins until each protein reaches the pH on the gradient that corresponds to its isoelectric point (pH = pl). At the isoelectric point, the net charge on the protein is zero and it is no longer moved by the electric field. IEF gives rise to very sharp bands (compared with SDS-PAGE) as thermal diffusion of the analytes during IEF brings the analyte molecules into regions where pH ≠ pl and the electric field will again induce mobility. Hence, smaller amounts of protein are detected using Coomassie or silver stains with IEF than with SDS-PAGE. Deamidation of asparagine and glutamine residues is a common degradation reaction that results in the formation of proteins with lower isoelectric points. IEF is a useful quality control method for the detection of this type of degradation. Deamidation results in a small change in MW that is not detected by SDS-PAGE. For glycosylated proteins, the relative amounts of cationic (amino) and anionic (sialic acid) sugars contributes to the isoelectric point. As the exact composition of the oligosaccharides comprising the glycosyl residues is not under genetic control, some variation is always to be expected. This variation must be controlled as much as possible during manufacture with recombinant techniques so that other sources of isoelectric point variation such as different amino acid composition from point source mutation, are not masked. Just as two-dimensional techniques have evolved for TLC and HPLC, similar combinations have appeared for gel electrophoresis. The method typically involves initial separation by IEF then a second separation by SDS-PAGE.[39]

Gel electrophoretic methods are invaluable for the qualitative and quantitative characterization of large peptides and proteins of pharmaceutical interest. However, they are labor intensive and slow (2 hours to 2 days analysis time). This has led to the recent development of several capillary electrophoretic techniques.

4.4. Capillary Electrophoresis

A variety of electrophoretic methods have been developed for use in open capillary tubes rather than on traditional gel slabs or tubes.[40–42] Capillary electrophoresis (CE) methods are much faster than gel electrophoresis as the narrow bore of the capillary tube (25–100 μm) allows greater heat dissipation: hence, larger voltage differences (up to 100 kV with forced capillary cooling) and currents (300 μA) may be used. CE methods can be similar in speed to HPLC and often more efficient, owing to the very large surface-area-to-volume ratio of the capillary tube. The efficiency of CE methods can be as high as 500,000 plates/m or more. They are expected to complement HPLC methods, rather than replace them. The absolute mass sensitivity of CE is very good (typically picograms, compared with nanograms for HPLC and micrograms for slab gel electrophoresis) but high concentrations are needed for normal UV detection as the system volume is very small. The applied sample volume may be 1 to 50 nl in a typical capillary volume of 2 to 5 μ. Such low sample volumes have provided an incentive to the development of methods for analysis of the contents of single cells.[43] Methods that have been used for sample introduction involve dipping the capillary into an autosampler vial containing the analyte solution and (1) the use of a hermetic seal on the vial and time-controlled head pressure application to pump the analyte solution into the capillary; (2) time-controlled vacuum application to the detector end of the capillary; and (3) direct electromigration of the analyte (electrokinesis) under the influence of a low potential difference for a controlled time interval.

In capillary zone electrophoresis (CZE) or free-solution capillary electrophoresis (FSCE), ionized solutes migrate as narrow bands or zones in a buffer solution.[43] The migration of solutes results from the separation of ions that occurs at the mobile diffuse layer and the stagnant double layer that exist adjacent to glass or silica surfaces (usually negatively charged from ionized silanol groups). Net migration of the solvent by electroosmosis is a result of drag by solvated solutes and the flow direction is normally toward the cathode. Separation of the solutes is based on the differences in their electrophoretic mobilities. Chiral additives can be used to resolve racemic mixtures by diastereomer formation. Isotachophoresis is performed on aqueous solutions of analytes with like charges (e.g., a mixture of protonated amines) and has been traditionally performed in capillary tubes. The solution of charged analytes and their counterions is confined between two electrolyte solutions, one (the leading electrolyte) containing an ion of the same charge type as the analytes but of higher mobility than any of them and the other (the trailing electrolyte) containing an ion of the same charge type but lower mobility than any of the analytes. Application of the electric field then sorts the analytes into sharply defined contiguous bands of differing mobilities.

Micellar electrokinetic capillary chromatography (MECC)[44] is a particularly interesting technique, as it permits electrophoretic separation of neutral as well as ionized molecules. The electrolyte is a micellar detergent solution (e.g., of SDS) in which the analytes partition to different extents into the charged micelles, thus, causing differences in their electrophoretic mobilities. Capillary gel electrophoresis (separation by a combination of molecular sieving and electrophoretic mobilities) and capillary isoelectric focusing (separation by differences in isoelectric points) are microscale adaptations of the more traditional slab or tube gel methods.

Although the developments in CE methods offer considerable potential for the future, some significant problems remain. Quantitative analysis by CE methods is imprecise as sample matrix problems cause variations in response. Precise temperature control of the capillary is important for reproducibility of migration times.[41] Hydrophobic proteins adsorb strongly onto silica capillary walls. There have been some attempts to minimize this problem in the separation of peptides and proteins by wall surface modification.[45] Improvements are also needed in the methods for precise sample introduction[46, 47] and in detector sensitivity.[48] Developments in detection methods include laser induced fluorescence,[44] thermal lensing and electrochemical and mass spectrometric detection.

Although CE methods are not yet established for widespread routine use, there are already several commercial instruments available that combine at least some of the aforementioned techniques. They include the P/ACE System 2000 (Beckman Instruments), Applied Biosystems 270A and the HPE 100 (Bio-Rad) as well as commercially available power supply and detector modules.

4.5. Immunoanalytical Chemistry

Immunoanalysis has become an integral part of the tools employed in the biotechnological revolution.[49, 50] Its ability to distinguish, detect, and quantitate quite similar substances (e.g., peptides and proteins), in situ, has proven invaluable in many aspects of biomedical analysis. Analyses of subnanomole quantities of small and large biopolymers present in biological fluids and fermentation media which are often unassayable by other methods, have provided scientists and engineers with much useful information. The relatively low cost and adaptability to automation of immunoassay, coupled with its sensitivity and specificity, have made it competitive with chromatographic methods in the analysis of complex biological samples.

Immunoanalytical methods are based upon the competition that occurs when a labeled and an unlabeled ligand compete for highly specific binding sites on antibodies. The evaluation of this competitive binding, accomplished by measuring a physical or chemical property associated with the label, allows the construction of a calibration curve representing a measured physical signal that is altered by changes in the concentration of the bound labeled ligand as a function of the concentration of the unlabeled ligand, the latter being the analyte. Unknown analyte concentrations are determined from this calibration curve.

Distinction of the signal corresponding to either the bound or free labeled ligand from that of the total labeled ligand population can be made in two ways. In most cases it is necessary to carry out a physical separation of the antibody-bound labeled material from the free labeled analyte. This can be accomplished by precipitation of the antibody-containing material with a concentrated salt solution such as ammonium sulfate or a hydrateable organic substance such as polyethylene glycol. Alternatively (and of great popularity, recently), solid phase extraction is possible. In the latter approach, analyte, labeled ligand, or antibody is covalently attached to a solid surface such as glass or plastic beads, a tube wall, or a dip stick and then reaction between substances in the liquid phase with those immobilized on the solid phase provides the basis for physical separation. These methods are known as heterogeneous immunoassays and are required when there is no way to distinguish, in situ, between the physical signals derived from bound and free labeled ligand, as in the case of radioactivity, which is unaffected by the chemical environment and yet provides the measurable physical signal in radioimmunoassay. Also, many immunoassay methods in which the label is an enzyme, such as enzyme-linked immunosorbent assay (ELISA), are heterogeneous.

Homogeneous immunoassays make up the second category in which physical separation of the bound from the free labeled ligand is not required. Signals that are different for the bound and free labeled ligands are obtained in situ. The signal producing species may be derived enzymatically when the label is an enzyme whose rate of reaction with a labeled substrate is reduced by complexation of the labeled ligand with an antibody or the label can be a fluorophore, a chromophore or a chromophore subject to induced chirality upon complexation with a protein. Here, environmental effects can lead to signal modification. For example, induced circular dichroism resulting from the electronic interaction associated with labeled ligand binding to immunoglobulin protein can be monitored with a spectropolarimeter. Alternatively, the complexation of a fluorescent label resulting in its incorporation into an antibody binding site may quench or enhance its fluorescence relative to that it demonstrated in aqueous solution. Fluorescence immunoassays account for the majority of homogeneous immunoassays in use today. Homogeneous immunoassay offers several advantages over heterogeneous analysis. Since no separation is involved in homogeneous immunoanalysis, this approach is more economical of time and materials. Additionally, errors due to sample loss in the separation aspect of the heterogeneous approach are minimized.

4.5.1. Analytical Aspects of the Immune Response

Analytes that can function as ligands in immunochemical reactions fall into that class of substances known as antigens. An antigen is any substance capable of reacting with an antibody. However, molecules of MW $< 10,000$ are not capable of inducing antibody formation (i.e., functioning as immunogens). Therefore, all immunogens are antigens, although antigens may or may not be immunogens. Antigen–antibody reactions are so specific that an antibody will usually react only with the immunogen that caused its formation or with molecules that are very similar structurally. Reaction with the latter is called cross reactivity and is a potential source of interference in immunoanalysis.

Antibodies are immunoglobulin-type proteins produced in mammalian blood in response to invasion by foreign immunogens. Immunogens are usually naturally occurring macromolecules (e.g., proteins, polysaccharides, nucleic acids) or microorganisms containing such molecules on their outer surfaces. The antibodies developed in response to those molecules or particles will recognize and bind only a small portion of the antigen. Antibody specificity usually involves not more than six to seven amino acid residues of the large protein structure.[51] The failure of antigens with molecular weights of less than 1×10^4 to elicit antibody production presents problems for the acquisition of antibodies directed against drugs that are to be analyzed by immunoassay procedures since drugs, including most peptides derived from biotechnology, usually have molecular weights much lower than 1×10^4.

Investigations performed by Landsteiner determined that an immunoresponse could be elicited from a mammal by small molecules (haptens) if they were coupled to macromolecules of MW $> 1 \times 10^4$.[52] The antibodies produced may react with the hapten even when it is not coupled to a carrier macromolecule. It is often more difficult to induce antibody production with hapten–carrier conjugates than with naturally immunogenic macromolecules and the conjugation site must be carefully chosen to maximize exposure of the hapten molecule. Moreover, haptens such as drugs that undergo biotransformation to inactive metabolites should be coupled to carrier molecules in such a way that those portions of the molecules subject to metabolic change are spatially available for the initial immunochemical antigenic recognition process. This reduces the cross-reactivity to the metabolites of the antibodies produced. In modern immunochemistry human and bovine serum albumins, synthetic peptides such as polylysine, and a number of polysaccharides have been used as immunogenic carriers for small drug molecules.

4.5.2. Hapten–Carrier Coupling

Coupling of the drug of interest to the macromolecular carrier is usually effected through derivatized carboxy-, amino-, hydroxy- or sulfhydryl groups on the hapten or the carrier. Occasionally a bridging group such as a carbodiimide is used to join the drug and the carrier. This often improves the specificity for the hapten of the antibodies produced. Once produced, different members of the heterogeneous population of antibodies may recognize parts of the carrier and the bridging molecule along with sites present on the hapten itself. This is a consequence of a lack of antibody uniformity and gives rise to various degrees of specificity for the hapten among the antibody population. Since antibodies produced in vivo are not isolated as chemically distinct species and their compositions and distributions are generally unknown, quantitative descriptions of immunoreactions in terms of simple competitive equilibria are not rigorous but have been dealt with in the literature as if this were the case.

4.5.3. Detection of Immunoreaction

Most immunoassays currently employed in the biomedical field are either radioimmunoassays, enzyme multiplied immunoassays, or luminescence immunoassays (including fluorescence immunoassay and chemiluminescence immunoassay). Radioimmunoassay, which is based on the scintillation counting of radioisotopically labeled ligands displaced from antibodies by the unlabeled material of analytical interest is historically the first kind of immunoassay and is currently the most sensitive. Concentrations of 10^{-12} to 10^{-15} M are often detectable. Because of the problems inherent to dealing with radioactive materials, such as licensing, radiation hazard, short shelf life of expensive radioisotopes, the expense of the counting equipment and the tedium associated with heterogeneous immunoassay, it has fallen in popularity behind the nonisotopic methods of analysis.

In the enzyme multiplied immunoassay techniques (EMIT), the antigen or antibody is labeled with an enzyme (e.g., lysozyme, alkaline phosphatase, horseradish peroxidase, or glucose-6-phosphate dehydrogenase) instead of a radioisotope. For example, an alkaline phosphatase–labeled drug can be made to compete with an unlabeled drug for binding sites on a drug-directed antibody. Owing to steric inhibition of the enzyme by the antibodies, when the enzyme-labeled drug is bound to the antibody the enzyme will lose its activity (ability to hydrolyze phosphates). However, the free enzyme–labeled drug retains its enzymatic activity. If a potentially absorbing or fluorescing organic phosphate whose optical properties are altered by the condition of esterification by phosphate is put into the solution containing the enzyme-labeled drug, drug-directed antibody, and unlabeled drug, only that fraction of the labeled drug population not bound to the antibody will be capable of hydrolyzing the phosphate ester and thereby generating the absorption or fluorescence spectrum of the hydrolysis products. Moreover, the amount of absorptiometrically or fluorimetrically measured, hydrolyzed material will depend upon the concentration of unlabeled drug added to the test solution, as it is this concentration that ultimately determines how much enzymic activity is released to the solution. Commercial EMIT kits based upon absorptiometric (colorimetric) estimation of enzymatically oxidized NADH to quantitate a variety of drugs have been popular for over a decade. The sensitivity is not especially high, drug concentrations down to about 0.5 μM being detectable. The measurement of the light emitted by a fluorescent or chemiluminescent enzyme substrate is capable of extending the limits of detection of the analyte down to 10^{-12} to 10^{-9} M with conventional arc lamp excitation and down to perhaps 10^{-15} M with laser excitation.[53]

Fluorescence immunoassay (FIA) involves the measurement of the fluorescence of a luminescent label that is part of a competitive immunochemical binding system and whose spectral position, intensity, and polarization vary with differences in the concentrations of the analyte. Fluorescent labels are used in homogeneous and heterogeneous immunoassay systems and may be bound to antigens, antibodies, or solid phases, or they may exist free in solution as enzyme substrates.

A fluorescent label to be used in homogeneous immunoanalysis should fulfill several requirements. Since the fluorescence signal must be measured in a serum matrix, the probe should have a high fluorescence quantum yield and the excitation and emission maxima of the probe should occur at wavelengths longer than those of the serum. Excitation spectra of dilute serum show excitation maxima at 280 and 340 nm. Serum fluoresces at 348 nm and weakly at 520 nm. A fluorescent label useful for in situ measurements in serum should be excited at wavelengths longer than about 400 nm and should fluoresce at wavelengths longer than about 450 nm.

To date the most popular fluorescent labels for FIA have been those derived from the long-wavelength, strongly emitting xanthene dyes fluorescein and rhodamine B. The isothiocyanates or isocyanates of these fluorophores can be used to label primary and secondary aliphatic amines in aqueous solutions by simple chemistry under ambient conditions. Consequently, they can be used to label antibiotics or polypeptides. Even those drugs that do not have indigenous alkylamino groups can often be labeled by introducing bridging groups (e.g., aminoethyl) that are amenable to coupling with the isothiocyanato or isocyanato functions.

Heterogeneous fluorescence immunoassays can be carried out with the aid of the same separation procedures used in radioimmunoassay. The more expedient homogeneous fluorescence immunoassays usually require quenching, enhancement, or shifting of the fluorescence of the label on binding of the labeled drug to its antibody. Occasionally, a second antibody, directed at the drug-directed antibody, will be used in a "double-antibody" method to precipitate the bound labeled and unlabeled drug or to alter the optical properties of the label in such a way as to make the analysis more sensitive.[54]

Homogeneous fluorescence immunoassays can often be effected even when there is no obvious change in the intensity or spectral position of fluorescence of the label upon binding to the antibody. If light-polarizing polymer films or crystals are used to polarize the exciting light and analyze the fluorescence of the sample excited by polarized light, it will generally be observed that the amount of fluorescence reaching the detector will be considerably smaller in those samples where a greater amount of antibody binding occurs. This is a result of the higher degree of polarized fluorescence emitted from the labels affixed to the large, slowly rotating antibody. The polarized emission is more efficiently attenuated by the analyzer polarizing film than the unpolarized light emitted by the rapidly rotating, relatively small labeled drug molecules that are not bound to macromolecules. This phenomenon forms the basis of fluorescence polarization immunoassay. The increase in fluorescence intensity measured with increasing unlabeled drug (analyte) added occurs as a result of the liberation of the labeled drug from the antibody with attendant fluorescence depolarization and can be used to construct a calibration curve from which the concentrations of unknown drug samples can be determined when their polarized fluorescences are measured. Fluorescence polarization immunoassay probably accounts for a majority of fluorescent immunoassays currently performed.

Chemiluminescence immunoassay,[55] a technique that is rapidly gaining popularity because its sensitivity is comparable to that of radioimmunoassay is, in a sense, a variation of fluorescence immunoassay. Chemiluminescence (also called bioluminescence when it occurs in fireflies, some dinoflagellates, coelenterates, and fungi) is fluorescence. However, what is usually termed fluorescence is light emission that is activated by prior light absorption by the emitting molecule. This is properly termed photofluorescence or photoluminescence. Chemiluminescence occurs in the oxidation products of some highly strained, highly reduced molecules (e.g., peroxyoxalate esters and amino-substituted cyclic hydrazides of phthalic acid). Oxidation results in intermediates possessing such great quantities of thermal energy from the reaction that they spontaneously (without photoexcitation) become electronically excited and subsequently fluoresce as a means of achieving the state of lowest energy. The oxidations are frequently catalyzed by metal ions and occur at appreciable rates only when the precursors of the chemiluminescent species (the labels) are freely diffusible in solution (i.e., they will not generate light rapidly when bound to antibodies or perhaps prior to release in an enzymic reaction). In this sense they are analogous to many of the fluorescent labels. However, photofluorescence analysis entails the measurement of light emitted by a fraction of an even smaller portion of molecules absorbing light for an instant. In chemiluminescence it is possible to gather and integrate the light output associated with the chemiluminescent reaction over the entire course of the reaction. Consequently, very low detection limits can be attained if the luminescence efficiency of the chemiluminescent reaction is reasonably good (say, > 0.1) Unfortunately, very few chemiluminescent reactions have high luminescence efficiencies, so that the choice of labels is much more restricted than in fluorescent immunoassay. However, chemiluminescence immunoassay is a field whose popularity is increasing, and this may inspire the discovery of new labels.

4.6. Optical Rotatory Dispersion and Circular Dichroism

The electronic structures of peptides and proteins give rise to UV absorption spectra. These spectra result from absorption of radiation by the peptide bonds (far UV; 190–240 nm) forming the protein backbone, and by the aromatic side chains of tyrosine, phenylalanine, and tryptophan (near UV; 250–295 nm). These absorption bands can be used for both qualitative and quantitative analysis. As most other methods of quantitative analysis tend to be more sensitive than UV methods, the qualitative aspects of absorption spectra tend to be more important. In particular, structural aspects of protein molecules can be studied by a variety of spectral methods (e.g., second order and difference spectra). The use of these methods has been facilitated by the advent of computer-controlled diode array spectrophotometers, which can obtain spectral data rapidly, store it on magnetic media and subsequently, perform numerical manipulation of the data.

Optical rotatory dispersion (ORD) and circular dichroism (CD) are spectroscopic methods of analysis that are derived from the ability of certain asymmetric molecules to alter the plane of polarization of incident light. These phenomena are related to light absorption (which gives rise to ultraviolet and visible spectrophotometry) and light scattering in that they arise from the alteration of molecular electronic motion. However, they are rather more complicated than absorption or simple dispersion phenomena. On the molecular level, ORD and CD derive from the weak abilities of the magnetic field associated with light to induce electronic polarization and the electric field of light to induce magnetic polarization in the electronic structures of molecules that lack a plane of symmetry. These subjects are of interest here because virtually all peptides and proteins lack a plane of symmetry and therefore demonstrate optical activity (i.e., give rise to ORD or CD spectra or both). Furthermore, small molecules that do have a plane of symmetry often will exhibit ORD or CD upon binding to a peptide or protein because the electronic structures of the ligand and ligate become intermingled.

Optical rotatory dispersion arises from the scattering or instantaneous absorption and reemission of polarized light by asymmetric molecules.[56] It is manifested as a rotation of the plane of the incident polarized light by an angle (α) that varies with the wavelength (λ) of the polarized light. The intensity of optical rotation for a dissolved asymmetric solute is expressed as a function $[\alpha]$ (specific rotation) of the observed rotation α and is given by

$$[\alpha]_\lambda^t = \frac{100\alpha}{lC}$$

where C is the concentration of the optically active solute in grams per 100 ml of solvent, l is the optical path in decimeters, and the indices t and λ refer to Celsius temperature and the wavelength of the incident light, both of which influence $[\alpha]$. A plot of $[\alpha]$ against λ is called an ORD spectrum. Alternatively, rotatory power for different compounds may be expressed as the molecular rotation $[\phi]$:

$$[\phi] = \frac{\alpha \, MW}{lC}$$

where MW is the molecular weight of the compound of interest. If the asymmetric molecule demonstrating ORD does not absorb light in the spectral range over which the ORD experiment is carried out, the ORD spectrum will demonstrate a gradual and continuous increase or decrease (depending on the absolute molecular configuration) of $[\alpha]$ with λ. This kind of spectrum is said to be a "plain curve." On the other hand, if the asymmetric analyte contains a chromophore that is responsible for light absorption in the spectral region of interest, the ORD spectrum will show a maximum and a minimum in the spectral region of each absorption band. The order of the maxima and minima (whether the maximum or minimum occurs at longer wavelengths) is a function of the absolute configuration of the solute, and the ORD spectrum is then said to exhibit a "Cotton effect."

Compounds that absorb light and are optically active may also demonstrate circular dichroism, which is differential absorptivity for enantiomers of opposite absolute configuration. Because the light emerging from a sample demonstrating circular dichroism is elliptically polarized rather than linearly polarized,[57] the measured differential absorbance of the absorbing sample is called the molecular ellipticity, $[\Theta]$. A plot of $[\Theta]$ against λ is called a CD spectrum and consists of peaks (positive bands) or troughs (negative bands) at the same wavelengths as bands in the absorption spectrum and no features at all in spectral regions where no absorption occurs. Whether a peak or trough is observed in the vicinity of an optically active absorption band depends upon the absolute configuration of the analyte, or if a mixture, on which isomer is predominant.

ORD and CD can both be used for the detection and quantitation of peptides and proteins with sensitivity limits approximating those of conventional absorption spectrophotometry (i.e., down to about 10^{-6}–10^{-5} M). However, they are used far more often as the best means of establishing the absolute configuration of chiral compounds than for their quantitative analysis. The secondary and tertiary structure of proteins have effects on absorption spectra that are particularly apparent in CD spectra. The backbone of the protein comprises three structural types: α-helix, β-pleated sheet, and random coil. Each of these types exhibits CD spectra with different features. Analysis of the wavelength maxima and the intensities of the positive and negative bands in a CD spectrum is used to estimate the percentage of the protein that exists in each of the three configurations. This method can demonstrate the close similarity of configuration for the same protein from different origins, for example, for human and porcine insulin,[58] pituitary and recombinant human growth hormone[59] and human, equine and porcine prolactin.[60] ORD and CD spectra are also very useful in studies of the interactions in solution of small molecules with biopolymers such as proteins and polynucleotides.[61]

4.7. Mass Spectrometry

The high sensitivity and selectivity of mass spectrometric (MS) techniques have been exploited to a great extent in the qualitative and quantitative analysis of peptides and proteins in general.[26, 35] Such exploitation will continue in the study of genetically engineered products with the increasing availability and sophistication of commercial instrumentation. Early MS studies on peptides and proteins were limited by the low volatility of the analytes and the vigorous electron impact and chemical ionization methods available. Low analyte volatility was addressed with numerous methods for chemical derivatization, especially in the use of coupled gas chromatography-MS (GC–MS). The value of gas chromatographic separation of analytes prior to MS detection led to similar attempts at combining liquid chromatography with MS (LC–MS). These have not been so successful, although the development of novel moving belt, thermospray, and microbore continuous-flow interfaces holds promise for future studies of proteins and peptides.[62]

The advent of sophisticated ionization methods such as plasma desorption MS (PDMS) and liquid secondary ion MS (LSIMS) and of new ion analyzers (e.g., tandem double-focusing mass spectrometry [MS–MS])[63] has been responsible for major advances in the detection and characterization of native proteins and peptides. These methods are also invaluable for the characterization of the products of chemical modification of peptides, e.g., enkephalins[64] and endorphins.[65] In LSIMS, the analyte is dissolved in a polar liquid such as glycerol, thioglycerol, aminoglycerol, diethanolamine, or nitrobenzene[66] and bombarded with an energetic beam of either neutral or charged particles. A high yield of molecular ion clusters is formed in the first few layers of the liquid matrix ("sputtering"). Critical factors in LSIMS include the nature of the liquid matrix and the incident beam energy (typically in the keV range). PDMS is similar to LSIMS except that the analyte is deposited in solid form on an inert probe and the incident beam energy is very high (typically in the MeV range). These advanced methods have been responsible for a continued increase in the upper limit of molecular ion mass range such that the MW of small proteins (4,000–16,000) can be readily determined to an accuracy of ±0.3 with LSIMS–MSMS combinations. Larger proteins (MW 16,000–25,000) have been determined by similar methods, giving MW data with an accuracy of several fold. For polypeptides (<4,000), the accuracy and versatility of these techniques are able to provide direct evidence for both the amino acid composition and the amino acid sequence. This can permit independent verification of the results of Edman degradation studies of the products of proteolytic digests. The location of sites of glycosylation and the identification of the oligosaccharide chains have been performed with LSIMS, for example, for recombinant human interferon-β expressed by Chinese hamster ovary cells.[67]

The chief limitation of LSIMS is its relative insensitivity, owing to "chemical noise" from the liquid matrices. Molecular ion clusters may not be observed at all, depending on the matrix. The combination of LSIMS with MS–MS has been vital in reducing this limitation although at the cost of substantially increased capital outlay. Further interdisciplinary research aimed at understanding the factors controlling secondary ion formation at liquid surfaces was recognized as essential for obtaining sequence data from picomole amounts of biopolymers.[26] A further limitation of secondary ion methods is that some peptides are not readily ejected as ion clusters from the liquid or solid matrix surface, compared with others. Expected significant developments in techniques for biotechnologically related applications include the increased use of Fourier transform ion cyclotron resonance analyzers and improved thermospray coupling of HPLC with MS–MS.[62]

References

1. Willard, H.H., Merritt, L.L., Dean, J.A., and Settle, F.A., Jr. 1988. *Instrumental methods of analysis*, 7th ed. Belmont, Calif: Wadsworth.

2. Snyder, L.R., and Kirkland, J.J. 1979. *Introduction to modern liquid chromatography*, 2nd ed. New York: Wiley-Interscience.

3. Christian, G.D., and O'Reilly, J.E. 1986. *Instrumental analysis*, 2nd. ed. Boston: Allyn and Bacon.

4. Skoog, D.A. 1987. *Principles of instrumental analysis*, 2nd ed. Philadelphia: Saunders.

5. Regnier, F. 1987. Liquid chromatography as a tool for differentiating among similar structural forms of proteins. *LC-GC* 5(1): 230–234.

6. Garnick, R.L., Ross, M.J., and du Mée, C.P. 1988. Analysis of recombinant biologicals. In *Encyclopedia of pharmaceutical technology*, Vol. 1. J. Swarbrick and J.C. Boylan, eds. New York: Dekker.

7. Willard, H.H., Merritt, L.L., Dean, J.A., and Settle, F.A., Jr. 1988. *Instrumental methods of analysis*, 7th ed., Chaps. 19, 20. Belmont, Calif.: Wadsworth.

8. Rubenstein, M., Rubenstein, S., Familletti, P.C., Miller, R.S., Waldman, A.A., and Pestka, S. 1979. Human leucocyte interferon: Production, purification to homgeneity and initial characterization. *Proc. Natl. Acad. Sci. U. S. A.* 76:640–644.

9. Wehr, C.T., Lundgard, R.P., and Nugent, K.D. 1989. Hydrophobic proteins: A challenge for reversed-phase HPLC. *LC-GC.* 7(1): 32–37.

10. Shively, J.E. 1989. Protein chemistry in the 1990's: Part 1. *Pharm. Technol.* February: 32–44.

11. Danielson, N.D., and Kirkland, J.J. 1987. Synthesis and characterization of 2-μm wide-pore silica microspheres as column packing for the reversed-phase liquid chromatography of peptides and proteins. *Anal. Chem.* 59:2501–2506.

12. El Rassi, Z., and Horvath, C. 1986. Hydrophobic interaction chromatography of t-RNA's and proteins. *J. Liquid Chromatogr.* 9(15):3245–3268.

13. Kaye, W., and Havlik, A.J. 1973. Low angle laser light scattering—absolute calibration. *Appl. Opt.* 12:541–550.

14. Stuting, H.H., Krull, I.S., Mhatre, R., Krzysko, S.C., and Barth, H.G. 1989. High performance liquid chromatography of biopolymers using on-line laser light scattering photometry. *LC-GC.* 7(5):402–417.

15. Wyatt, P.J., Jackson, C., and Wyatt, G.K. 1988. Absolute GPC determinations of molecular weights and sizes from light scattering. *Am. Lab.* May:86–91.

16. Kenley, R.A., Hamme, K.J., Lee, M.O., and Tom, J. 1987. Silica-based size exclusion chromatography to characterize the decapeptide nafarelin in a controlled release pharmaceutical formulation. *Anal. Chem.* 59:2050–2054.

17. Larsson, P.-O., Glad, M., Hansson, L., Månsson, M.-O., Ohlson, S., and Mosbach, K. 1983. High-performance liquid affinity chromatography. In *Advances in chroma-*

tography, Vol. 21, ch. 2. J.C. Giddings, E. Grushka, J. Cazes, and P.R. Brown, eds. New York: Dekker.

18. Hjertén, S., and Yang, D.J. 1984. High-performance liquid chromatographic separations on dihydroxyboryl-agarose. *Chromatography* 316:301–309.

19. Glad, M., Ohlson, S., Hansson, L., Månsson, M.-O., and Mosbach, K. 1980. High-performance liquid affinity chromatography of nucleosides, nucleotides and carbohydrates with boronic acid-substituted microparticulate silica. *J. Chromatogr.* 200:254–260.

20. Bhikhabhai, R., Lindblom, H., Källman, I., and Fägerstam, L. 1989. Automated 2-D peptide mapping: Gel filtration and reversed-phase chromatography. *Am. Lab.* May: 76–81.

21. Ogden, G., and Földi, P. 1987. Amino acid analysis: An overview of current methods. *LC-GC* 5(1):28–40.

22. Betnér. I., and Földi, P. 1988. The FMOC-ADAM approach to amino acid analysis. *LC-GC* 6(9):832–840.

23. Edman, P. 1950. Method for the determination of the amino acid sequence in peptides. *Acta Chem. Scand.* 4:283–293.

24. Edman, P., and Begg, G. 1967. A protein sequenator. *Eur. J. Biochem.* 1:80–91.

25. Hunkapiller, M.W., Hewick, R.M., Dreyer, W.J., and Hood, L.E. 1983. High sensitivity sequencing with a gas phase sequenator. *Methods Enzymol.* 91(Part 1):399–413.

26. Burlingame, A.L., Maltby, D., Russell, D.H., and Holland, P.T. 1988. Mass spectrometry. *Anal. Chem.* 60:294R–342R.

27. Croft, L.R. 1980. *Introduction to protein sequence analysis.* New York: Wiley.

28. Stark, G.R. 1972. Sequential degradation of peptides and proteins from their COOH termini with ammonium thiocyanate and acetic anhydride. *Methods Enzymol.* XXV:369–384.

29. Schlesinger, D.H., Weiss, J., and Audhya, T.K. 1979. Isocratic resolution of amino acid thiohydantoins by high performance liquid chromatography. *Anal. Biochem.* 95:494–496.

30. Hayashi, R. 1976. Carboxypeptidase Y. *Methods Enzymol.* XLV:568–587.

31. Tsugita, A., Ataka, T., and Uchida, T. 1987. Approaches for submicrosequencing. *J. Prot. Chem.* 6:121–130.

32. Pilosof, D., Kim, H.-Y., Vestal, M.L., and Dyckes, D.F. 1984. Direct monitoring of sequential enzymatic hydrolysis of peptides by thermospray mass spectrometry. *Biomed. Mass Spectrom.* 11:403–407.

33. Fontana, A., and Gross, E. 1986. Fragmentation of peptides by chemical methods. In *Practical protein chemistry: A handbook,* Pp. 67–120. A. Dabre, ed. New York: Wiley.

34. Hartman, P.A., Stodola, J.D., Harbour, G.C., and Hoogerheide, J.G. 1986. Reversed phase high performance liquid chromatography mapping of bovine somatotropin. *J. Chromatogr.* 360:385–395.

35. Burlingame, A.L., Baillie, T.A., and Derrick, P.J. 1986. Mass spectrometry. *Anal. Chem.* 58:165R–211R.

36. Hearn, M.T.W. 1985. Ion-pair chromatography of amino acids, peptides and proteins. In *Ion-pair chromatography*, M.T.W. Hearn, ed, chap. 5. New York: Dekker.

37. Hardy, M.R. 1989. Liquid chromatographic analysis of the carbohydrates of glycoproteins. *LC-GC* 7(3):242–246.

38. Hunkapiller, M.W., Lujan, E., Ostrander, F., and Hood, L.E. 1983. Isolation of microgram quantities of proteins from polyacrylamide gels for amino acid analysis. *Methods Enzymol.* 91:227–236.

39. Jorgenson, J.W. 1986. Electrophoresis. *Anal. Chem.* 58(7):743A–760A.

40. Gordon, M., Huang, X., Pentoney, S.L., Jr., and Zare, R.N. 1988. Capillary electrophoresis. *Science* 242:224–228.

41. Burolla, V.P., Pentoney, S.L., Jr., and Zare, R.N. 1989. High performance capillary electrophoresis. *Am. Biotechnol. Lab.* November/December:20–25.

42. Tehrani, J., and Day, L. 1989. High performance capillary electrophoresis using a modular system. *Am. Biotechnol. Lab.* November/December:32–40.

43. Ewing, A.G., Wallingford, R.A., and Olefirowicz, T.M. 1989. Capillary electrophoresis. *Anal. Chem.* 61:292A–300A.

44. Burton, D.E., Sepaniak, M.J., and Maskarinec, M.P. 1986. Analysis of B6 vitamers by micellar electrokinetic capillary chromatography with laser-excited fluorescence detection. *J. Chromatogr. Sci.* 24:347–351.

45. McCormick, R.M. 1988. Capillary zone electrophoretic separation of peptides and proteins using low pH buffers in modified silica capillaries. *Anal. Chem.* 60:2322–2328.

46. Huang, X., Gordon, M.J., and Zare, R.N. 1988. Bias in quantitative capillary zone electrophoresis caused by electrokinetic sample injection. *Anal. Chem.* 60:375–377.

47. Rose, D.J., Jr., and Jorgenson, J.W. 1988. Characterization and automation of sample introduction methods for capillary zone electrophoresis. *Anal. Chem.* 60:642–648.

48. Pickering, M.V. 1989. Capillary electrophoresis: Proteins, problems and promises. *LC-GC* 7(9):752–756.

49. Thomas, A.H. 1988. Quality control in genetic engineering. *Chem. Br.* 24(10):1031–1035.

50. Schulman, S.G., Hochhaus, G., and Karnes, H.T. 1991. In *Luminescence techniques in chemical and biochemical analysis*, W.R.G. Baeyens, D. DeKeukelaire, and K. Korkidis, eds., Chap. 11. New York: Dekker.

51. Lerner, R.A., et al. 1989. Cloning of the immunological repertoire in *Escherichia coli* for generation of monoclonal catalytic antibodies: Construction of a heavy chain variable region-specific DNA library. *Proc. Natl. Acad. Sci. U. S. A.* 86:5728,–5732.

52. Landsteiner, K. 1962. *The specificity of serological reactions*. Boston:Harvard University Press.

53. Van den Beld, C.M.B., et al. 1988. Laser-induced fluorescence detection in liquid chromatography. *Chimi Oggi,* November:33–37.

54. Fullmann, E., Langer, J., and Clapp, J.J. eds. 1981. *Liquid assay: Analysis of international development on isotopic and non-isotopic immunoassay*. P. 113. New York: Masson.

55. Karnes, H.T., O'Neal, J.S., and Schulman, S.G. 1985. In *Molecular luminescence spectroscopy: Methods and applications*, S.G. Schulman, ed., Pt. 1, Chap. 8. New York: Wiley-Interscience.

56. Djerassi, C. 1968. *Optical rotatory dispersion*, New York: McGraw-Hill.

57. Connors, K. 1982. *A textbook of pharmaceutical analysis*, 3rd ed. New York: Wiley-Interscience.

58. Johnson, I.S. 1982. Authenticity and purity of human insulin (recombinant DNA). *Diabetes Care* 5(Suppl. 2):4–12.

59. Jones, A.J.S., and O'Connor, J.V. 1982. Chemical characterization of methionyl human growth hormone. In *Hormone drugs*. Rockville, Md: USP Convention. Pp. 335–351.

60. Bewley, T.A., and Li, C.H. 1983. Studies on prolactin: Conformation comparison of human, equine and porcine pituitary prolactins. *Arch. Biochem. Biophys.* 227:618–625.

61. Perrin, J.H., and Hart, P.A. 1970. Small molecule-macromolecule interactions as studied by optical rotatory dispersion-circular dichroism. *J. Pharm. Sci.* 59(4):431–448.

62. Caprioli, R.M., DaGue, B., Fan, T., and Moore, W.T. 1987. Microbore HPLC / mass spectrometry for the analysis of peptide mixtures using a continuous flow interface. *Biochem. Biophys. Res. Commun.* 146:291–299.

63. Johnson, J.V., and Yost, R.A. 1985. Tandem mass spectrometry for trace analysis. *Anal. Chem.* 57:758A–788A.

64. Fales, H.M., McNeal, C.J., Macfarlane, R.D., and Shimohigashi, Y. 1985. Californium-252 plasma-desorption mass spectrometry of polymethylenediamine linked enkephalin peptides. *Anal. Chem.* 57:1616–1621.

65. Hochhaus, G., Gibson, B.W., and Sadée, W. 1988. Biotinylated human β-endorphins as probes for the opioid receptor. *J. Biol. Chem.* 263(1):92–97.

66. Gower, J.L. 1985. Matrix compounds for fast atom bombardment mass spectrometry. *Biomed. Mass Spectrom.* 12:191–196.

67. Conradt, H.S., Egge, H., Peter-Katalinic, J., Reiser, W., Siklosi, T., and Schaper, K. 1987. Structure of the carbohydrate moiety of human interferon-β secreted by a recombinant Chinese hamster ovary cell line. *J. Biol. Chem.* 262(30):14600–14605.

5

The Impact of Biotechnology on Analytical Methodology

John F. Fitzloff

5.1. Introduction

What is encompassed by the term *biotechnology products?* In terms of numbers, proteins and peptides predominate. This includes enzymes, antibodies, receptors, structural proteins, hormones (including an increasing number of cell growth factors), and carrier or transport proteins. Second are the poly- and oligonucleotides, the primary information carriers of molecular biology. From a pharmaceutical point of view they are the "reagents" needed to produce the proteins and peptides for therapy, but they also can be considered therapeutic agents, being polymeric extensions of the nucleotides and nucleosides already used in therapy. Third are the poly- and oligosaccharides, either alone or as part of glycoproteins. Usually thought of as playing only a structural role, polysaccharides are now recognized to have additional functions. Last, there are cells and organisms that have been genetically altered. As with all new pharmaceutical products, new products derived through biotechnology require characterization.

Like most areas of science, in the pharmaceutical industry there was previous experience with similar types of products and processes. Fermentation to obtain antibiotics, the preparation of bacterial and viral vaccines, and the isolation and purification of vitamins and industrial enzymes are examples of biotechnology that predated the common use of the term. Although fermentation was used earlier in the food industry, the pharmaceutical industry was first in using genetic manipulation for improving the yield of its products. This occurred because of the relative ease with which microbial organisms could be modified genetically and the research support for health-related products.

The impact of biotechnology on analysis can be subdivided into two parts: (1) the use of biotechnology-derived products as analytical tools and components in analytical systems, and (2) the improvement of existing methods and the invention of new analytical systems to provide the necessary specificity and sensitivity for quality control of biotechnology products and their production processes.[1] This chapter will briefly discuss these two areas and give some examples of the current status of analytical biotechnology with regard to peptides and proteins, poly- and oligonucleotides, and poly- and oligosaccharides.

5.2. Biotechnology Products as Analytical Tools

5.2.1. Evolution of Immunoassays

The beginnings of biotechnology products as analytical reagents can be attributed to the success of Berson and Yalow in developing the radioimmunoassay of insulin, for which Rosalind Yalow shared the Nobel Prize in medicine in 1977. This technique was expanded to include smaller molecules, by covalently linking them to antigenic macromolecules and producing polyclonal antibodies that would bind only the small molecule. The major drawback of this methodology was cross-reactivity with structurally related compounds. This was particularly troublesome in the quantitation of therapeutic drugs, where inactive or less active metabolites interfered. This problem has essentially been eliminated by the production of monoclonal antibodies (MAbs; see Chapter 2 for a discussion of the production and characterization of antibodies). The number of methods based on MAbs are continuing to increase and include such innovations as fluorescence polarization immunoassays (FPIA)—which are most sensitive for low concentrations of analyte (antigen)—and immunochromatography, in which the analyte releases an enzyme-labeled antigen as it moves through the chromatographic matrix, allowing the enzyme to catalyze a reaction producing a chromogen (the height of the "color bar" is proportional to the analyte concentration). This latter method allows direct visual reading of the result.[2] A thorough review of enzyme immunoassays (EIA) has been published that presents good examples of improvements in the technique.[3] For instance, in 1972 a "sandwich" EIA for human chorionic gonadotropin (hCG) required an incubation of 4.5 hours, whereas in 1986 a radial partition EIA for hCG had a total run time of only 8 minutes. Several of the current immunoassay techniques are reviewed in this volume (see Chapter 4).

5.2.2. Polymerase Chain Reaction

The polymerase chain reaction (PCR) is a technique widely used in biotechnology. Although this technique uses biotechnology products and involves several analytical methods, it per se is not an analytical method. The essence of the technique is to incubate DNA polymerase, two oligonucleotide primers, and a small amount of some genomic or cloned DNA. The target DNA is produced rapidly (e.g., up to 10^5-fold increase in 3 hours). Some applications for target DNAs produced include direct sequencing, genomic cloning, and site-directed mutagenesis. Uses that are analytical include DNA typing for prenatal genetic disease or forensic evidence, detection of infectious microorganisms, and interpretation of allelic variations. Once obtained, the first process for the target DNA is purification, usually by gel electrophoresis. This and other methods will be discussed below. Because of its increasing use, PCR has been largely automated and is now linked to anion-exchange HPLC for purification.[4]

5.2.3. DNA Hybridization Assay

The DNA hybridization assay, or dot blot assay, is used primarily to determine trace amounts of rDNA in recombinant protein products. However, it is also useful for detecting DNAs from viruses, mycoplasms, bacteria, and fungi that may be contaminating biotechnology products. It is a very sensitive assay with detection limits at the femtogram (10^{-15}) level. The hybridization, between the sample DNA and ^{32}P-labeled DNA probes, is accomplished on a nitrocellulose matrix. The matrix is exposed to X-ray film and the resultant autoradiograph is compared with one with a concentration series of DNA standards.[5]

5.3. Analytical Demands of Biotechnology Products

5.3.1. Characterization of Biotechnology Products

The use of antibodies has now come full circle as MAbs themselves are being used as therapeutic agents, and this requires that they be rigorously characterized. The analytical problem of biotechnology-derived products such as MAbs is the second issue of this chapter. The specificity engendered by MAbs, posttranslational modified protein, genetically altered stem cells, or other "biotechnology products" demands that during their production, and particularly in their final form, a level of quality control be exerted that exceeds that for most small organic drug products. Proteins in particular represent an analytical challenge. Those resulting from recombinant DNA (rDNA), even though their production may be amplified (20% or more of the total soluble expressed protein from bacterial hosts and 3 to 5% from yeast hosts), require high-efficiency separations. It is necessary at this point to remove proteins that differ by one or two amino acids, as a result of expression errors or chemical changes, such as oxidation or hydrolysis that may occur postexpression during processing.[6] Such separations depend on the mainstays of protein and peptide separation and characterization, chromatography and electrophoresis (see Chapter 4, this volume).

As the genomic DNA is characterized (see Chapter 8, this volume) for sequences whose expression results in diseases, the use of antisense oligonucleotides to prevent expression becomes a possible treatment modality and a molecular tool to understand the pathological state. This approach has been recently reviewed.[7] The necessary minimum size of these oligonucleotides for high specificity is 15 to 25 bases, and one or two incorrect bases may make them inactive.[8]

Although it has recently been determined that the normal phosphodiester antisense oligonucleotides are more stable in cell culture than originally thought, there is still concern about how well they will survive hydrolysis by nucleases in vivo. As a result, analogues to the phosphodiesters are being considered. These include phosphothioates, methylphosphonates, ethylphosphotriesters and phosphomorpholidates to increase stability. As with the other biopolymer products, the same thorough analytical methods must be applied to these synthetic antisense oligonucleotides and analogues before their therapeutic use.

5.3.2. Monitoring of Biotechnology Products

As equally challenging as characterization is the task of monitoring biotechnology products. In vivo, most of these products are present in less than microgram per milliliter concentrations. In many instances, the active agent will be immeasurable, and one will have to rely on measuring a specific biological response as a reflection of the actual administered agent. Fortunately from the regulatory point of view, bioassays have been used before. Nonetheless, it will be interesting to read NDAs with pharmacokinetics based on a response rather than the direct measurement of the drug concentration.

5.4. Pharmaceutical Analyses Using Immunoassays

5.4.1. Therapeutic Drug Monitoring (TDM)

With the advent of MAbs, immunoassays of a number of therapeutic drugs have become routine. Specificity and sensitivity, the major attributes of immunoassays, are particularly important to many of those drugs that most often require TDM, the classic example being digoxin. For TDM of drugs with low therapeutic concentrations and with demand for rapid results, FPIA will be the method of choice for some time. The major competing technique to FPIA for TDM is high-performance liquid chromatography (HPLC). HPLC is more cost-effective in cases where samples contain multiple drugs or drugs that are metabolized extensively with one or more of the metabolites also being active. Improvements in columns, detectors, and sample preparation, coupled with automation, indicate that the use of HPLC will increase again for TDM.[9]

5.4.2. Peptide and Other Hormone Diagnostic Assays

The use of radioimmunoassay (RIA) to assess insulin levels in patients was the first diagnostic immunoassay. It was quickly followed by immunoassays for other peptide hormones for which polyclonal antibodies could be produced. When it became possible to produce antibodies using antigens with a covalently bound hapten, immunoassays for thyroxine and the steroid hormones soon were developed.

As was indicated earlier, MAbs are now being used as therapeutic and diagnostic agents; most of these MAbs are hybrids or chemically modified antibodies. Those that are to be used in vivo require sensitive and selective analytical methods.

The bifunctional or hybrid MAbs are produced by the fusion of two different hybridomas to form a quadradoma producing the two different MAbs and hybrid MAbs. These hybrid MAbs contain one heavy and one light chain of each of the MAb immunoglobulins from the original hybridomas.[10]

These hybrid MAbs offer several diagnostic and therapeutic possibilities. For example, one site can be a tumor cell antigen and the other an antigenic (binding) site for an antitumor agent. It appears that retention of specificity and activity are better in the hybrid MAbs than in the chemically modified MAbs. Improvements in chemical modification, by conjugating anticancer agents to cancer cell MAbs[11–13] or linking two different FAB′ molecules (obtained by breaking disulfide bonds in light chains) through disulfide bonds in the heavy chain,[14] have been reported. Another interesting chemically modified MAb recognizes an antigen found in human granulocytes and is made bifunctional by reaction of ^{99m}Tc with thiols on the MAb. Since granulocytes accumulate where inflammation is occurring, these sites can be radioimaged through administration of this radiolabeled MAb.[15]

5.4.3. Drug Abuse Testing

The number of different immunoassays for drug abuse testing grew rapidly during the Vietnam War era, with homogeneous assays predominating, like the free radical analysis technique (FRAT, also called spin immunoassay). This immunoassay was relatively quick and sensitive, but it required an expensive electron spin resonance (ESR) spectrometer for measuring the free radical spin label. Of more widespread use with time was the enzyme-mediated immunoassay technique (EMIT), which required only a standard UV-VIS spectrophotometer. Both FRAT and EMIT were good for eliminating most of the negative samples. However, from a legal standpoint, both required that positive samples be confirmed by gas chromatography–mass spectrometry (GC–MS), since both used polyclonal antibodies that could show substantial cross-reactivity. For example, codeine could not be distinguished from morphine. This problem has been eliminated with the use of MAbs.

5.5. Methods Characterizing Biotechnology Products

5.5.1. Proteins and Peptides

5.5.1.1. Purity

The substantial improvement in filtering materials and techniques during processing or prior to analyses should be noted. This results in cleaner products and more trouble-free analyses. Also, well-controlled sample volume reduction should not be overlooked as a way to enhance HPLC sensitivity and to take advantage of microbore column technology. Automated equipment providing moderate temperature and pressure control and giving high sample recoveries without degradation is available. In general, the quality control of biologically derived pharmaceutical products has been aided substantially by robotics and the objectivity and error-reduction capability of computers.[16] Automation of the isolation and purification processes is particularly important, since this represents 80 percent of the production costs, most of which is due to labor.[17]

Utilizing microbore HPLC and applying partially purified recombinant protein at 100 micrograms, about 1% is separated out as a variant, that is, 50 pmol in 20 μl. This and an equivalent amount of the major protein are subjected to tryptic digestion, and each digest is injected into the microbore HPLC system. The differing peptide fragment from the tryptic map of the variant is collected and subjected to microsequencing. From knowledge of the error, the process is improved to produce a purer product, typically with less than 0.1% variant being produced.[17] Coupling a diode array detector to the microbore HPLC for peptide mapping, provides considerably more information for comparison of the two proteins. One can identify differences due to point mutations or posttranslational modification. Monitoring at different wavelengths, such as 215 and 280 nm, allows calculation of the aromatic amino acid content in each peptide fragment.[18] A rapid, less than 5 minute, HPLC separation of small amounts of recombinant protein, useful for both at-line and on-line process monitoring, has been reported.[19]

The purification of synthetic peptides is more difficult than for those produced by recombinant techniques. For example, a 55 amino acid peptide synthesized at a 98% coupling efficiency at each step will result in a final product that contains only 33.6% of the target peptide, and 37% will be a mixture of 54 residue peptides differing from the target peptide by only one amino acid.[20] The remaining mixture is of smaller peptides. This is clearly a challenging separation.[21]

Antibodies can be purified by affinity chromatography, using specific ligands or protein G (a natural binder of IgG) bound to stable silica-based supports that have minimal nonspecific binding.[22] Conversely, antibodies to any desired biopolymer can be made, purified, and bound to a solid support to make affinity columns for that biopolymer. A variation of this is so-called "paralog" chromatography, which instead of an antibody uses a small peptide that mimics the antibody's paratope (antigen recognition site). It is hoped that paralog affinity columns will be able to separate variant proteins that differ by one or two amino acids from the expected recombinant target protein.[23] Similarly, other proteins (e.g., enzymes) that have very specific recognition sites can also use affinity chromatography for purification. Affinity columns can be produced with good capacity, up to 10 mg/g of support.[24]

5.5.1.2. Sequence and Structure

Once the purity of a protein has been established, the sequence can be obtained, which helps to confirm the purity and complements the biological activity information. A major analytical area for biotechnology in general, and for pharmaceutical biotechnology in particular, is the sequencing of peptides and proteins derived from recombinant DNA or chemical synthesis. It is necessary to confirm the correct amino acid sequence and to detect errors in sequence translation or posttranslational modifications. This need is the driving force for improvements in the traditional chemical and enzymatic sequencing methods, in separation and detection techniques, and in sequencing by mass spectrometry (several of these improvements are discussed in Chapter 4, this volume). An excellent overview of the traditional sequencing methods, the complementary use of mass spectrometry, and the different ionization methods for peptide sequencing has been presented by Stults.[25] Another source for practical approaches to mass spectrometry of proteins is also available.[26] A two-part review of protein chemistry dealing with isolation and characterization has been published.[27, 28] The method to be used for sequencing is determined by several factors: the molecular weight, preknowledge about the sequence, sample availability, sample purity, and the presence of glycosylation or other modifications. Similar needs can also be noted for oligosaccharides and oligonucleotides, but to a lesser extent. Some additional methods for dealing with these biopolymers will be briefly discussed below.

Complementing sequence determination is amino acid composition analysis. Methods currently in use routinely measure amino acids in the low picomole range. Recently, quantitation of even lower amounts has been reported using the combination of gas phase hydrolysis, derivatization with naphthalene-2,3-dicarboxaldehyde (NDA), separation on a C_{18}-coated open tubular column and electrochemical detection. Quantitation of 4 fmol (4×10^{-15} mol) of bovine chymotrypsinogen was demonstrated, and the current limit of the detector is 1 amol (10^{-18} mol).[29]

Two of the most useful checks on the solution conformation of proteins and peptides are circular dichroism (CD) and optical rotatory dispersion (ORD) spectra (these are discussed in Chapter 4, this volume). For these techniques to be of maximal value, a reference protein or peptide must be available for comparison. The standard analytical tool for the higher-order structure of peptides and proteins is X-ray crystallography, which requires the availability of appropriate crystals. The fact is that the majority of proteins do not crystallize readily. In addition to ORD and CD, modern high-field multidimensional NMR spectroscopy can also provide higher order structural information.[30, 31] NMR-derived structure is more relevant to function, since it can be done in solution. Analysis of proteins in solution also allows NMR studies of the effects of ligand binding, the kinetics of unfolding, and conformational equilibria. Proteins for NMR structural determinations should be soluble at millimolar concentrations, nonaggregating, and stable up to 40°C. Good chemical shift dispersion is seen for β-pleated sheets, but not for α-helices, β-bends, and turns. Therefore, domains with the latter features are difficult to define structurally.[32] The current molecular weight limit of NMR three-dimensional structural determinations of proteins is about 20,000 using 500 or 600 MHz instruments. With isotope labeling and more powerful magnets, this could increase to about 40,000. This is probably the practical limit, as the resulting large line width would reduce the sensitivity of the experiment.

Interesting research has also been done using reverse phase HPLC to predict the presence of amphipathic α-helical structure in peptides. The method is based on differences in retention time between two series of model peptides of known sequence and α-helicity, as measured by CD.[33] A new HPLC detector that is particularly useful for at-line process quality control has been developed. Based on photon correlation spectroscopy, it measures the diffusion coefficient in the flow cell. This can be converted by instrument software to yield the molecular size (radius in nanometers) or molecular weight. In the case of multiple molecular species, measurements of polydispersity can be converted to an estimate of protein purity. This detector is particularly useful for determining aggregation, dimerization, and degradation of products.[34]

5.5.1.3. Biological Activity

Biological activity testing is absolutely essential for biotechnology products. The nature of the testing used is, of course, dependent on the specific product. Enzymes can be well characterized by determining their kinetic parameters with the appropriate substrate. Antibodies and carrier proteins can be evaluated by determining equilibrium binding constants with their respective optimal ligands. Peptide hormone activities can be determined by standardized bioassays.

5.5.2. Poly- and Oligonucleotides

5.5.2.1. Purity

Gel electrophoresis is the standard method for separating and purifying polynucleotides and is capable of excellent resolution.

Denaturing polyacrylamide gel electrophoresis (PAGE) is competitive with HPLC in separating an oligonucleotide from possible error products and can resolve 600 base pair (bp) oligonucleotides from 599 or 601 bp error products.[35] Thicker and longer gel plates have become available that give even more resolution and recovery.[36] However, the interest in oligonucleotides such as single-stranded DNA probes has spurred the development of HPLC to provide rapid isolation of highly purified DNA prior to amplification using the PCR technique. Reverse phase columns can be used if the synthetic oligonucleotide still has dimethoxytrityl protecting groups, but this is capacity limited, up to 1 mg/g of support. Ion-exchange polymer columns can be used for the deprotected or native oligonucleotides with capacities up to 100 mg/g of support.[24] Ion-exchange columns are available to separate and purify DNA fragments such as plasmids, synthetic oligomers, and restriction fragments. These columns can separate fragments from 10 to 50,000 bp with purification of 50 to 150 μg per run and a detection limit of 50 pg.[37]

5.5.2.2. Sequence and Structure

The fundamental DNA sequencing problem is to determine the order of the four bases in the strands. The two main methods that developed in the 1970s were the enzymatic method of Sanger and Coulson and the chemical method of Maxam and Gilbert. Both require the high-resolution denaturing PAGE to separate the cleaved or terminated DNA fragments produced. Fragments in the range of 300 to 600 bases can be resolved. Improved versions of these two methods together have so far sequenced 26 million of the 3 billion bases of the human genome.

Sequencing is essentially divided into three parts: first, the production of fragments of manageable size and the purification of the fragments, followed by performing the necessary chemical or enzymatic methods; second, determination of the actual sequence; and third, the analysis of the results so that the thousand or so 300 to 600 base sequences can be assembled into a single genomic sequence. In order to automate sequencing, an alternative to the use of radiolabel (^{32}P) to identify fragments on the gels was needed. This was done by using different fluorescent tags for each of the bases. This allows sensitive detection of the fragments, while they are being separated. Some variation in this modification exists (e.g., the fluorescent labels can be in DNA primers or on terminal bases on the DNA fragment itself). This automated sequencing should enable 10,000 bases to be sequenced per day. Appropriate computer hardware and software is available to handle all the data to be analyzed.[35]

A practical example of the need for sequencing genomic DNA is as follows: The evaluation of an antiviral agent in addition to the normal pharmacokinetic analysis requires establishing its activity or lack of activity at the molecular level. In the case of acyclovir the mechanism of action is the inhibition of herpes simplex viral (HSV) DNA polymerase by acyclovir triphosphate, but resistance is known to develop. To establish the cause of resistance, the lack of thymidine kinase or its mutation or the alteration of HSV DNA polymerase must be shown. The latter two causes require that the enzymes or the gene coding for the enzyme be sequenced.[38]

Secondary and tertiary solution structures of oligo- and polynucleotides are now quite well understood and can be predicted knowing the primary structure (sequence) and confirmed by spectroscopic methods, such as NMR and CD.[39, 40]

5.5.2.3. Biological Activity

The ultimate determination of rDNA activity is the production of the correct recombinant protein, the analysis of which has been discussed previously.

5.5.3. Poly- and Oligosaccharides

5.5.3.1. Purity

The separation of erythropoietin (EPO) from fermentation media is a good example of the purification of a valuable glycoprotein, derived from recombinant technology, that takes advantage of its high oligosaccharide content (30 to 50%) depending on the culture used. Only 1% of the protein in the culture media is EPO, and initial cleanup takes advantage of the net negative charge of EPO at pH 6.8. It is not retained on a DEAE Sephacel anion exchange column, whereas 98% of the non-EPO protein is. The next step is to suppress the ionic character of the remaining protein using an acetonitrile–trifluoroacetic acid in water gradient and a C4 reverse-phase HPLC column. After this step the EPO is 15% of the isolated protein. The final step uses a concanavalin A-agarose affinity column, which retains only the oligosaccharide containing EPO. EPO is then eluted with α-methyl-D-mannose to give greater than 99% EPO.[24]

5.5.3.2. Sequence and Structure

The separation of glycoproteins can be accomplished on high-pH anion-exchange (HPAE) columns, followed by the treatment of each glycoprotein with endo β-N-acetylglucosaminidase H and the peptide, N-glycosidase F. The released oligosaccharides after desialylation can then be separated on HPAE columns.[41] The remaining proteins can be sequenced by one of the previously described methods and the oligosaccharide analyzed as will be indicated below. The possible complexity of oligosaccharides is illustrated by the fact that a hexasaccharide composed of some combination of the eight most common sugars leads to 4.76 $\times$ 10^9 different structures.[42]

The sequence and structure of oligosaccharides can be studied by the following methods: X-ray crystallography, chiroptical methods, nuclear magnetic resonance, theoretical calculations, and mass spectrometry. X-Ray crystallography requires crystals, and many oligosaccharides cannot be crystallized. However, if obtainable, the X-ray crystallographic structure is a good starting point for the solution structure. Chiroptical methods give considerable information on solution conformation, but it is only readily interpretable for polymers with the same repeating monomer unit. NMR, using high-field instruments (500 MHz and above), is the method of choice as the ^{1}H and ^{13}C parameters contain detailed conformational information. Theoretical calculations, particularly hard-sphere *exo*-anomeric (HSEA) effect calculations, can give good results even on fairly large oligosaccharides.[43] Mass spectrometry is particularly useful for sequence and substructural analysis of polysaccharides, but defined chemical modification is usually required to produce fragments and derivatives suitable for the determination of sequence and linkages.[42]

5.6. Future Analytical Needs for Biotechnology Products

5.6.1. Needs Ancillary to Actual Analysis

All the analytical methods discussed here and by Prankerd and Schulman (Chapter 4, this volume) will continue to be refined as the number and quantity of biotechnology products grow. Particular needs should result in more rapid development as follows: Both research laboratories and production facilities will take advantage of the time saving engendered by robotic and automated devices that can do increasingly complex handling tasks. Through the use of artificial intelligence, the incorporation of decision making can also be done based on increased data from continuous process monitoring. The current bioassays, notably those done in vivo, which take many days to complete yet are required to demonstrate the potency of the product, will be replaced with in vitro or analog assays. These assays must be equivalent and have a short turnaround time. Modifications to analytical systems so that they can be used for in-line process control is an area in which considerable efforts will be made, especially biosensors[44–47] and flowing systems, HPLC, and capillary electrophoresis (CE). Much improved methods for the mapping and sequencing of oligosaccharides in conjunction with the characterization of high-molecular-weight glycoproteins and proteoglycans are under development. More efficient, specific, and sensitive methods for trace DNA (10–100 pg/dose) in recombinant protein are also needed to meet regulatory requirements.

Most of the methods discussed here and by Prankerd and Schulman (Chapter 4, this volume) are applicable to biotechnology-derived pharmaceuticals in delivery systems. The intent, however, is to focus on establishing the integrity of the product in the delivery system matrix. Is potency maintained? Has any degradation occurred? What is the nature of the degradation products? As was indicated earlier for most biotechnology-derived pharmaceuticals, pharmacodynamic studies (bioavailability, absorption, distribution, metabolism, and excretion) are difficult, if not impossible, to accomplish analytically by direct means. Yet many innovative techniques will meet the challenge. For example, single-cell monitoring is possible;[48] why not continuous microsampling? Stable isotope markers in specific parts of the molecule would allow sorting out some of the pharmacokinetics.

Two analytical areas in which the needs of biotechnology are stimulating very rapid developments are CE and MS (both of these methods are also discussed in Chapter 4, this volume).

5.6.2. Capillary Electrophoresis

Research in CE has proceeded rapidly since 1983 in large part owing to the publications of Jorgenson and Lukacs.[49–52] The first separate review on CE by Kuhr[53] has appeared in *Analytical Chemistry*'s annual fundamental reviews. Of the many other recent reviews, the one by Wallingford and Ewing[54] is recommended for comprehensive coverage of the fundamentals of CE. The attractiveness of CE is based on its speed, high efficiency of separation (e.g., 10^6 plates/m), and sensitivity (femtomoles or less). Drawbacks include irreproducible separations (due in part to Joule heating by using high-ionic-strength buffers and poor control of sample introduction), high operating voltages, and a lack of preparative capability. These problems are being addressed (e.g., instrument automation[55]) and will be overcome or circumvented, except for the lack of preparative capability. This latter problem is not as critical for DNA, since it can be amplified by the PCR technique mentioned previously. The uses of CE will center on purity and identification.

Two major types of CE are emerging, free solution CE (FSCE; also known as capillary *zone* electrophoresis or CZE) and capillary *gel* electrophoresis (CGE). The major problem with FSCE is in the control of electroosmotic flow; however, a recent report presents the use of an external electric field for control.[56] The FSCE open tubular capillary columns are much easier to manufacture reproducibly than the gel-filled capillaries for CGE. The components and capabilities of a commercial FSCE system have been well presented.[57] Linking reverse phase HPLC to FSCE takes advantage of the abilities to separate by two different properties. This orthogonal relationship gives a system that provides more capacity and resolution than either alone. An example of this concept is its successful testing, using fluorescently labeled peptide products, of a tryptic digest of ovalbumin.[58] Another new methodology that uses FSCE systems is the application of the technique to isoelectric focusing (IEF). These two would seem incompatible, since at increasing pH an electrically charged layer forms on the inner surface of the silica capillary and supports a strong electroosmotic flow, preventing the necessary pH gradient for focusing proteins. However, by coating the inside of the capillary with a covalently bonded linear hydrophilic polymer, electroosmosis is reduced as well as adsorption. These coated columns are then filled with a mixture of proteins and ampholytes. Finally, a voltage is applied to obtain a gradient and focus the proteins according to their isoelectric points (p*I*). By monitoring current, focusing can be terminated when the current reaches a minimum. Focused protein can then be removed from the column by adding salt to the cathodic or anodic reservoirs. The mobility time to remove a given protein is linearly related to p*I*. This is a very high resolution separation and will be additionally valuable for the characterization of proteins.[59]

The retention variability due to electroosmotic flow in FSCE is obviated when one goes to CGE. CGE systems have been shown to be very useful for separating polynucleotides. The gel-filled capillaries are capable of repeated use, up to 150 runs.[60] Of particular interest is the rapid separation of fluorescently labeled DNA fragments, which can be detected by laser-induced fluorescence at a level of 1 amol.[61]

5.6.3. Mass Spectrometry

The major breakthroughs in using MS in biotechnology were (1) the sample introduction methods; (2) the ionization methods (which produced mass ions of biopolymers, which then were measured with some accuracy by a variety of one or more mass spectrometers); and (3) the understanding of how to interpret MS data to give the sequences of biopolymers. Continuing work is being done to make these methods generally applicable. It is expected that optimal measurements on femtomole quantities, accurate molecular mass measurements (within one Da at mass 100 kDa), and the measurement of variants present below the 1% level will become routine. Burlingame et al. point to matrix-assisted laser ionization (MALI) combined with time-of-flight (TOF) mass spectrometers and electrospray sample introduction-ionization (ESI) as the two most significant developments in this field in the past 2 years.[62]

The optimization of ESI has been summarized by Meng and Fenn[63] along with some notable examples. The spectra of four proteins are presented, which range in molecular weight from 8,560 to 39,830. Each spectra has a distribution of multiple charged peaks, each separated by an m/z of one. A thorough review to date of ESI-MS compares its capabilities to other ionization techniques for biopolymers. ESI-MS can currently determine protein molecular weights accurately over 50,000 and some reported over 100,000, whereas fast atom bombardment (FAB) and plasma desorption (PD) are limited to about 25,000 and 45,000, respectively. ESI of oligonucleotides up to 25,000 (76-mer) has also been done.[64] A related technique, called ion spray, is more amenable to completely aqueous samples and use with atmospheric pressure ionization (API) sources.[65] The problems and advantages of linking high-resolution FSCE to API-MS have been investigated and reviewed.[66] Interest in the MS of biopolymers has stimulated an interest in TOF mass spectrometers, since they are capable of recording masses higher than most magnetic sector mass spectrometers. They operate at a higher scan rate than quadrupole mass spectrometers. Combined with the ionization energy control of a laser, protein molecular ions of masses over 200 kD using matrix-assisted UV laser ionization and TOF-MS have been reported.[67] This method has also shown an accuracy of ±0.01% for as little as 1 pmol of protein in the 10,000 to 20,000 MW range. When the matrices are cinnamic acids, which allow longer laser wavelengths to be used, these measurements appear to be independent of varying concentrations of small organic or inorganic components in the sample.[68] Following this work, the combination of HPLC interfaces with TOF-MS has been investigated.[69]

Infrared laser desorption (IRLD) TOF-MS has been found to be useful for oligosaccharides and glycoconjugates. As fragmentation occurs in the sugar ring rather than at the glycosidic bond, this method should prove valuable for understanding the structure of complex carbohydrates.[70] IRLD Fourier transform (FT) MS provides sensitive analysis of amino acid peptide mixture.[71]

5.7. Conclusion

Biotechnology has contributed greatly to analytical methodology in the form of monoclonal antibodies. They have made a tremendous impact on bioanalysis, from the home pregnancy test to the early diagnosis and treatment of cancer. Virtually any organic compound can be identified and quantitated with high specificity and sensitivity with a monoclonal antibody-based immunoassay produced for the compound in question. The great advantage is that it can do so using a sample that is a very complex mixture; the disadvantage being that it can only measure one compound per sample.

Central to biotechnology is determining the human DNA sequence; it will be some time before it is completely known, but work with fragments is proceeding at a rapid pace. This chapter has highlighted the analytical methodology that has developed and is developing to support the human genome project as well as basic research and applications in biotechnology. The methods center around answering two fundamental questions about the "reagents" and products of biotechnology: how pure they are, and what is their exact structure? For example, Do I have a single DNA fragment before amplifying it by PCR? and Is the interleukin-2 we produced using recombinant DNA identical in structure to the human interleukin-2? Methods of characterizing the three main compound groups that are used as tools in, or are products of, biotechnology—namely peptides and proteins, poly- and oligonucleotides, and poly- and oligosaccharides—have been discussed. The analytical chemists have risen to the challenge of biotechnology, as is illustrated by methods such as CE and ESI–MS. Every indication is that the bioanalytical needs of the future will also be met.

Acknowledgments

The critical review of this chapter by my colleagues Charles P. Woodbury and Bruce L. Currie is gratefully acknowledged.

References

1. Garnick, R.L., Solli, N.J., and Papa, P.A. 1988. The role of quality control in biotechnology: An analytical perspective. *Anal. Chem.* 60:2546–2557.

2. Conboy, C., Ellis, E., Jenne, J., Shaughnessy, T., Szefler, S., Weiner, M., Milavetz, G., Vaughan, L., Weinberger, M., Carrico, J., and Tillson, S. 1985. Evaluation of a whole blood theophylline test requiring no instrument. *J. Allerg. Clin. Immunol.* 75, pt. 2:128.

3. Hubbach, A., Debus, E., Linke, R., and Schrenk, W. J. 1986. Enzyme immunoassay: A review. In *Progress in clinical biochemistry,* vol. 4. Berlin, Heidelberg: Springer-Verlag. Pp. 110–143.

4. Katz, E.D., and Dong, M.W. 1990. Rapid analysis and purification of polymerase chain reaction products by high-performance liquid chromatography. *BioTechniques* 8:546–555.

5. Pepin, R.A., Lucas, D.J., Lang, R.B., Liao, M.-J., and Testa, D. 1990. Detection of picogram amounts of nucleic acids by dot blot hybridization. *BioTechniques* 8:628–632.

6. Tabor, J.M. 1989. In *Genetic engineering technology in industrial pharmacy.* J. M. Tabor, ed. New York: Dekker.

7. Cohen, J.S., ed. 1989. *Oligonucleotides: Antisense inhibitors of gene expression.* Boca Raton, Fla.: CRC Press.

8. Chambers, A.F., and Denhardt, D.T. 1990. Abatement of gene expression using antisense oligonucleotides. *Pharmaceut. Technol.* 5 (2):24–28.

9. Shihabi, Z.K., and McCormick, C.P. 1990. The role of HPLC in therapeutic drug monitoring and drug-of-abuse screening. *BioChromatography* 5(3):121–126.

10. Reading, C.L., Bator, J., and O'Kennedy, R. 1989. Monoclonal antibodies in tumor therapy. In *Monoclonal Antibodies.* J.M. Moulds and S.P. Masouredis, eds. Arlington, Va: American Association of Blood Banks. Pp. 145–179.

11. Li, S., Zhang, X.-Y., Zhang, S.-Y., Chen, X.-T., Chen, L.-J., Shu, Y.-H, Zhang, J.-L., and Fan, D.-M. 1990. Preparation of antigastric cancer monoclonal antibody MGb_2-mitomycin C conjugate with improved antitumor activity. *Bioconjugate Chem.* 1(4):245–250.

12. Hurwitz, E., Stancovski, I., Wilchek, M., Shouval, D., Takahashi, H., Wands, J.R., and Sela, M. 1990. A conjugate of 5-fluorouridine-poly (*L*-lysine) and an antibody reactive with human colon cancer. *Bioconjugate Chem.* 1(4):285–290.

13. Abrams, P.G. 1989. Specific targeting of cancer with monoclonal antibodies. In *The present and future role of monoclonal antibodies in the management of cancer,* vol. 24, *Frontiers of radiation therapy and oncology.* J.M. Vaeth and J.L. Meyer, eds. Pp. 182–193. Basel: Karger.

14. Runge, M.S., Bode, C., Savard, C.E., Matsueda, G.R., and Haber, E. 1990. Antibody-directed fibrinolysis: A bispecific (Fab')$_2$ that binds to fibrin and tissue plasminogen activator. *Bioconjugate Chem.* 1(4):274–277.

15. Rhodes, B.A., and Martinez-Duncker, C. 1990. Direct labeling of antibodies with Tc-99m. *Am. Biotechnol. Lab.* 8(4):50–53.

16. Tsuji, K., Jenkins, K.M., and Price, J.M. 1988. An expert system for assessing the microbiological quality of pharmaceutical products and materials. *Pharmaceut. Technol.* 12(9):154–158.

17. Ransohoff, T.M., Murphy, M.K., and Levine, H.L. 1990. Automation of biopharmaceutical purification processes. *BioPharm* 3(3):20–26.

18. Posluszny, J.V., and Wickham, D. 1990. Diode array detection enhancement to small-bore chromatography. *Am. Lab.* 22(3):28–39.

19. Nugent, K.D. 1990. A technique for rapid protein and peptide analysis. *Am. Biotechnol. Lab.* 8(7):24–32.

20. Horn, M., and Novak, C. 1987. A monitoring and control chemistry for solid phase peptide synthesis. *Am. Biotechnol. Lab.* 5(5):12–21.

21. Benedek, K., and Swadesh, J.K. 1991. HPLC of proteins and peptides in the pharmaceutical industry. In *HPLC in the pharmaceutical industry,* vol. 47, Chap. 11. G.W. Fong and S.K. Lam, eds. *Drugs and the pharmaceutical sciences.* J. Swarbrick, ed. New York: Dekker. Pp. 241–302.

22. Cook, C., Hazen, B., and Pang, D. 1990. Luer tip quick antibody purification by affinity chromatography. *Am. Biotechnol. Lab. News* 8(2):16–18.

23. Kauvar, L.M., Cheung, Y.K., Gomer, R.H., and Fleischer, A.A. 1990. Paralog chromatography. *BioTechniques* 8(2):204–209.

24. Newton, P. 1990. Complex biological matrices: Column capacity and separation strategy. *LC-GC* 8(2):116–122.

25. Stults, J.T. 1990. Peptide sequencing by mass spectrometry. In *Biomedical applications of mass spectrometry,* Vol. 34. C.H. Suelter and J.T. Watson, eds. New York: Wiley. Pp. 145–201.

26. McEwen, C.N., and Larsen, B.S., eds. 1989. In *Mass spectrometry of biological materials.* Vol. 8. Practical spectroscopy series, E.G. Brame, Jr., ed. New York: Dekker.

27. Shively, J.E. 1989. Protein chemistry in the 1990s, part I. *Pharmaceut. Technol.* 13(2):32–44.

28. Shively, J.E. 1989. Protein chemistry in the 1990's, part II: Microstructural analysis. *Pharmaceut. Technol.* 13(3):38–46.

29. Oates, M.D., and Jorgensen, J.J. 1990. Quantitative amino acid analysis of subnanogram levels of protein by open tubular liquid chromatography. *Anal. Chem.* 62:1577–1580.

30. Bax, A. 1989. Two dimensional NMR and protein structure. *Annu. Rev. Biochem.* 58:223–256.

31. Wuethrich, K., Basus, V.J., Billeter, M., Kuntz, I.D., Thomason, J.F., Oshiro, C.M., Sheek, R.M., van Gunsteren, W.F., Kaptein, R., Altman, R.B., Jardetzky, O., Bertini, I., Banci, L., Luchinat, C., Vogel, H.J., Wilde, J.A., Bolton, J.H., Hibler, D.W., Harpold, L., Pourmotabbed, T., Dell'Acqua, M., and Gerlt, J.A. 1989. Section 2, protein structure. *Methods Enzymol.* 177:125–292 (Chaps. 6–14).

32. Groenenborn, A.M., and Clore, G.M. 1990. Protein structure determination in solution by two dimensional and three dimensional nuclear magnetic resonance. *Anal. Chem.* 62:2–15.

33. Zhou, N.E., Mant, C.T., and Hodges, R.S. 1990. Effect of preferred binding domains

on retention behavior in reversed phase chromatography: Amphipathic alpha helices. *Peptide Res.* 3:8–20.

34. Claes, P., Fowell, S., Woollin, C., and Kenny, A. 1990. On-line molecular size detection for protein chromatography. *Am. Lab.* 22(3):58–62.

35. Smith, L. 1989. DNA sequence analysis: Past, present and future. *Am. Biotechnol. Lab.* 7(5):10–25.

36. Bush, C.N., and Walsh, K.D. 1990. S. and S SEQ3545 sequencing system for isolation and purification of oligonucleotides for DNA sequencing, PCR or southern hybridization probes. *Biomed. Prod.* 8(7):52.

37. Anon. 1988. Waters Chromatography Division, Millipore Corporation. *Biomed. Prod.* 6(1):15.

38. Stollar, V. 1989. Drug-resistant viral mutants. *Pharmaceut. Technol.* 13(11):24–29.

39. Kearns, D.R. 1984. NMR studies of conformational states and dynamics of DNA. *CRC Crit. Rev. Biochem.* 15:237–290.

40. Wolk, S., Thurmes, W.N., Ross, W.S., Hardin, C.C., and Tinoco, I., Jr. 1989. Conformational analysis of d(C_3G_3), a β-family duplex in solution. *Biochemistry* 28:2452–2459.

41. Spellman, M.W. 1990. Carbohydrate characterization of recombinant glycoproteins of pharmaceutical interest. *Anal. Chem.* 62:1714–1722.

42. Hellerquist, C.G., and Sweetman, B.J. 1990. Mass spectrometry of carbohydrates. In *Biomedical applications of mass spectrometry*, Vol. 34. C.H. Suelter and J.T. Watson, eds. New York: Wiley. Pp. 91–143.

43. Bock, K. 1988. Molecular recognition of oligonucleotides related to starch studied by NMR spectroscopy and HSEA calculations. In *NMR spectroscopy in drug research*, J.W. Jaroszewski, K. Schaumberg, and H. Kofod, eds. Copenhagen: Munksgaard.

44. Athani, A., and Banakar, U.V. 1990. The development and potential uses of biosensors. *BioPharm* 3(1):23–30.

45. Schramm, W., Yang, T., and Midgley, A.R. 1987. The commercialization of biosensors. *Med. Device Diag. Ind.* 9(11):53–57.

46. Janata, J. 1990. Chemical sensors. *Anal. Chem.* 62:33R–44R.

47. Guilbault, G.G., and Suleiman, A. 1990. Piezoelectric crystal biosensors. *Am. Lab.* 8(4):28–32.

48. Olefirowicz, T.M., and Ewing, A.G. 1990. Capillary electrophoresis in 2 and 5 um diameter capillaries: Application to cytoplasmic analysis. *Anal. Chem.* 62:1872–1876.

49. Jorgenson, J.W., and Lukacs, K.D. 1981. Zone electrophoresis in open tubular glass capillaries. *Anal. Chem.* 53:1298–1302.

50. Jorgenson, J.W., and Lukacs, K.D. 1981. Free zone electrophoresis in glass capillaries. *Clin. Chem.* 27:1551–1553.

51. Jorgenson, J.W., and Lukacs, K.D. 1981. High-resolution separations based on electrophoresis and electroosmosis. *J. Chromatogr.* 218:209–216.

52. Jorgenson, J.W., and Lukacs, K.D. 1983. Capillary zone electrophoresis. *Science* 222:266–272.

53. Kuhr, W.G. 1990. Capillary electrophoresis. *Anal. Chem.* 62:403R–414R.

54. Wallingford, R.A., and Ewing, A.G. 1989. Capillary electrophoresis. In *Advances in chromatography*, Vol. 29. J.C. Giddings, E. Grushka and P. Brown, eds. New York: Dekker. Pp. 1–76.

55. Moring, S.E., Colburn, J.C., Grossman, P.D., and Lauer, H.H. 1990. Analytical aspects of an automated capillary electrophoresis system. *LC-GC 8* (1):34–46.

56. Lee, C.S., Blanchard, W.C., and Wu, C.-T. 1990. Direct control of the electroosmosis in capillary zone electrophoresis by using an external electric field. *Anal. Chem.* 62:1550–1552.

57. Grossman, P.D., Lauer, H.H., Moring, S.E., Mead, D.E., Oldham, M.F., Nickel, J.H., Goudberg, J.R.P., Krever, A., Ransom, J.H., and Colburn, J.C. 1990. A practical introduction to free solution capillary electrophoresis of proteins and peptides. *Am. Biotechnol. Lab.* 8(2):35–43.

58. Bushey, M.M., and Jorgenson, J.W. 1990. Automated instrumentation for comprehensive two-dimensional high-performance liquid chromatography/capillary zone electrophoresis. *Anal. Chem.* 62:978–984.

59. Wehr, T., Zhu, M., Rodriguez, R., Burke, D., and Duncan, K. 1990. High performance isoelectric focusing using capillary electrophoresis instrumentation. *Am. Biotechnol. Lab.* 8(11):22–29.

60. Terabe, S. 1990. HCPE '90. *Anal. Chem.* 62:605A–607A.

61. Drossman, H., Luckey, J.A., Kostichka, A.J., D'Cunha, J., and Smith, L.M. 1990. High-speed separations of DNA sequencing reactions by capillary electrophoresis. *Anal. Chem.* 62:900–903.

62. Burlingame, A.L., Millington, D.S., Norwood, D.L., and Russell, D.H. 1990. Mass spectrometry. *Anal. Chem.* 62:268R–303R.

63. Meng, C.K., and Fenn, J.B. 1990. Analyzing organic molecules with electrospray mass spectrometry. *Am. Lab.* 8(4):54–60.

64. Smith, R.D., Loo, J.A., Edmonds, C.G., Barinaga, C.J., and Hudseth, H.R. 1990. New developments in biochemical mass spectrometry: Electrospray ionization. *Anal. Chem.* 62:882–899.

65. Huang, E.C., Wachs, T., Conboy, J.J., and Henion, J.D. 1990. Atmospheric pressure ionization mass spectrometry: Detection for the separation sciences. *Anal. Chem.* 62:713A–725A.

66. Smith, R.D., Barinaga, C.H., and Hudseth, H.R. 1989. Capillary zone electrophoresis/mass spectrometry: An alternative to LC/MS? *Spectra* 12(1):10–15.

67. Karas, M., Ingedoh, A., Bahr, U., and Hillenkamp, F. 1989. Ultraviolet-laser desorption/ionization mass spectroscopy of femtomolar amounts of large proteins. *Biomed. Environ. Mass Spectrom.* 18:841–843.

68. Beavis, R.C., and Chait, B.T. 1990. High accuracy molecular mass determination

of proteins using matrix-assisted laser desorption mass spectrometry. *Anal. Chem.* 62:1836–1840.

69. Emary, W.B., Lys, I., Cotter, R.J., Simpson, R., and Hoffman, A. 1990. Liquid chromatography/time-of-flight mass spectrometry with high-speed integrated transient recording. *Anal. Chem.* 62:1319–1324.

70. Spengler, B., Dolce, J.W., and Cotter, R.J. 1990. Infrared laser desorption mass spectrometry of oligosaccharides: fragmentation mechanisms and isomer analysis. *Anal. Chem.* 62:1731–1737.

71. Chiarelli, M.P., and Gross, M.G. 1989. Amino acid and tripeptide mixture analysis by laser desorption Fourier transform mass spectrometry. *Anal. Chem.* 61:1895–1899.

6

Drug Delivery Aspects of
Biotechnology Products

Diane J. Burgess

6.1. Introduction

In recent years, there have been enormous advances in the field of protein and peptide engineering and an increased understanding of the way in which biological response modifiers function in the body. It is now possible, through the use of recombinant DNA techniques or by solid phase protein synthesis, to produce on a commercial scale a large variety of regulatory agents that are therapeutically applicable. The list of these response modifiers is continually expanding and includes interferons, interleukins, monoclonal antibodies, colony-stimulating factors, human insulin of recombinant DNA origin, human growth hormones, anticoagulants, and agents that have potential in inflammation and contraception.

Biotechnology-derived human therapeutic agents currently comprise the largest product category of pharmaceuticals in research and development. For example, there are a growing number of cardiovascular proteins under investigation, such as kidney plasminogen activator (k-PA), protein C, atrial natriuretic peptide (ANP), and apolipoproteins. Thrombolytic agents, which dissolve blood clots and are thus important in the treatment of heart attack, include t-PA (approved, Genentech), which binds specifically to protein in the clot; urokinase, which unlike t-PA is nonspecific and can cause bleeding anywhere; k-PA, which is a precursor of urokinase; and protein C, which is an anticoagulant that helps balance clot dissolution and clot formation. Superoxide dismutase, an enzyme that neutralizes oxygen free radicals, is also under investigation. The blood levels of this enzyme rise when a blood clot is dissolved and blood begins to flow through previously blocked tissues. Atrial natriuretic peptide (ANP) has a large market potential as an agent to prevent heart attack, stroke, and other cardiovascular diseases by controlling blood pressure. ANP has an advantage over existing drugs, as it controls blood pressure through a number of different mechanisms. Currently available antihypertensive agents only affect one mechanism and tend to cause harmful side effects. Apolipoproteins, which are involved in the transport

Table 6.1. Currently Approved Biotechnology-Derived Pharmaceutical Products, Indications and Sales

Products	Indications for Use	1990 Sales, Worldwide (million $)
Interferon-α	Leukemia, AIDS, renal cell carcinoma, Kaposi's sarcoma, malignant melanoma, multiple myeloma, ovarian cancer, genital and oral herpes	235
Interferon-β^a	AIDS, ARC, multiple sclerosis, cancer	15
Erythorpoietin	Anemia, chronic renal failure, AIDS	290
Factor VIII:C[b]	Hemophilia	175
Hepatitis B	Hepatitis B vaccine	160
Human growth hormone	Human growth hormone deficiency in children	285
Human insulin	Diabetes	400
Interleukin-2[c]	Metastic renal cell carcinoma, advanced cases of cancer	15
Monoclonal antibody	Cancer, septic shock, graft-vs.-host body based disease, prevention of blood clot imaging	35
Tissue plasminogen activator	Acute myocardial infarction	210
Granulocyte–macrophage colony-stimulating factor	Bone marrow transplantation, chronic neutropenia	25
Interferon-γ	Renal cell carcinoma, rheumatoid arthritis, chronic granulomatous disease	10
	Total	1,855

[a]Approved in Europe and Japan, not yet approved in United States.

[b]Blood-derived factor VIII that has been purified using monoclonal antibodies. (Recombinant factor VIII has just been approved in Germany).

[c]Approved in some European countries.

of cholesterol in the blood, are also under investigation, as they may help prevent deposition of cholesterol in arteries (arteriosclerosis).

World sales of biotechnology-derived human therapeutics reached $1,855 million in 1990, which is up almost 100% since 1988 ($995 million).[1] Since the technology that allows us to produce these drugs is only a little over a decade old, the level of sales is very respectable. Sales from currently approved biotechnology products are expected to more than double in the next 5 years (projected at $4,915 million for 1995), not including new products that will be rapidly entering the marketplace. Table 6.1 lists currently approved therapeutic products, together with their indications and their 1990 sales.[1]

6.2. Stability of Peptides and Proteins

A major difference between the traditional chemical entities used as drugs and the new genetically engineered peptide and protein drugs lies in their stability.

Peptides and proteins are highly susceptible to both chemical and physical instability when compared with traditional drugs. This presents unique difficulties in the purification, separation, formulation, storage, and delivery of these compounds. Chemical instability results in the generation of a new chemical entity, by bond formation or cleavage. Physical instability involves changes in the secondary, tertiary, or quaternary structure of the molecule, which can be manifested as denaturation (structural rearrangement within the molecule), adsorption, aggregation, and precipitation. Therefore, denaturation rarely occurs in small peptides, which do not possess high levels of structural order. Both physical and chemical changes in peptides and proteins can result in a loss of biological activity. A comprehensive review of protein stability is given by Manning et al.[2]

6.2.1. Physical Stability

6.2.1.1. Denaturation

Peptides and proteins are made up of both polar amino acid residues and nonpolar amino acid residues. In an aqueous environment the hydrophobic, nonpolar amino acid residues fold up on themselves to form globular-shaped molecules, with the hydrophilic, polar amino acid residues exposed to the aqueous environment. If these molecules are then placed in a hydrophobic environment such as a nonaqueous solvent for the purposes of purification, separation, or formulation, they will unfold, exposing their hydrophobic residues to the hydrophobic environment. This results in rearrangement and loss of quaternary and tertiary structure (Figure 6.1). On unfolding, hydrophobic and hydrogen bonds are broken. Hydrophobic bonding is a type of bonding that exists between hydrophobic regions. The secondary structures (α-helix and β-pleated sheets), which have a greater number of stabilizing hydrogen bonds, may or may not be lost. Unfolding and denaturation may also occur if the solvent is changed from an aqueous to a mixed solvent such as acetone and water or alcohol and water. Changes in pH will affect the conformation of polypeptides and proteins, since this will alter the ionization of the carboxylic acid and amino groups and hence the charges carried by the molecules. An increase in the number of like charges on an individual molecule will result in charge repulsion within the molecule and it will unfold. Similarly, changes in ionic strength will affect the charge carried by molecules through the effect of counterion screening. Ions of the opposite charge to that of the molecule will gather around the molecule, hence reducing its effective charge. Changes in temperature may also cause unfolding of polypeptides and proteins. An increase in the temperature will increase the thermal energy of the molecules. The increase in energy may be sufficient to break the hydrogen bonds that stabilize the quaternary, tertiary, and secondary structure of these molecules. At very high temperature, chemical bonds may also be broken.

The biological activity of the polypeptides and proteins is often lost on structural

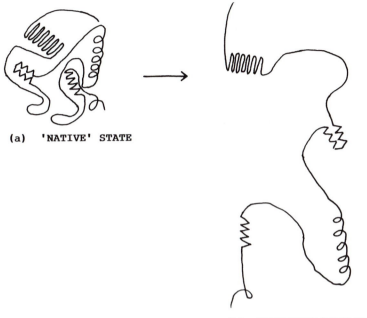

(a)   'NATIVE' STATE

(b)   DENATURED/UNFOLDED STATE

Figure 6.1. Schematic diagram of polypeptide or protein unfolding (denaturation): (a) in the native state; (b) in the denatured state (unfolded).

rearrangement. Particular residues that were previously exposed may be hidden, with a resultant loss of activity, and similarly, new residues may become exposed, which may possess a different activity.

Denaturation can be reversible or irreversible. In cases of reversible denaturation the conformational change in the molecules can be reversed by returning the conditions to their original state. For example, a protein may be denatured by an increase in temperature and then may reform its original structure when the temperature is decreased. Irreversibly denatured proteins cannot reform their native structures by returning the conditions to their original state. These proteins may be simply misfolded, so that the conformation does not allow them to renature properly, or they may have undergone some chemical or physical process that inhibits the original folding pattern.

6.2.1.2. Adsorption

Peptides and proteins are amphiphilic, that is, they possess both polar, hydrophilic and nonpolar, hydrophobic amino acid residues. Consequently, in a similar manner to small surfactant molecules, they have a tendency to be adsorbed at interfaces, such as air–water and air–solid. The hydrophobic, nonpolar, amino

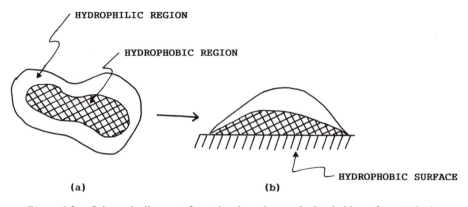

HYDROPHILIC REGION

HYDROPHOBIC REGION

HYDROPHOBIC SURFACE

(a)   (b)

Figure 6.2. Schematic diagram of protein adsorption at a hydrophobic surface: (a) in the native state; (b) adsorbed on a hydrophobic surface (denatured).

acid residues prefer the hydrophobic environment of the air or the surface of a container such as a glass beaker or a plastic bottle, whereas the hydrophilic, polar amino acid residues prefer the aqueous environment. Peptides and proteins may therefore rearrange and denature on adsorption at an interface. Figure 6.2 is a schematic representation of this type of conformational rearrangement. Once adsorbed, the protein molecules will form short-range bonds with the surface, such as van der Waals, hydrophobic, and ion-pair bonds, which may result in further denaturation of the molecules. Peptides and proteins adsorb at interfaces relatively rapidly; however, conformational changes in the protein at the interface are slow and may take hours to complete.[3] Desorption of surface proteins in their "native" state is unlikely. Rather, peptides and proteins will desorb with their hydrophobic residues exposed. Therefore, they will have unfavorable interactions with water and will tend to form aggregates and precipitates.

Biological activity can be lost or changed as a result of the rearrangement of molecular structure on adsorption. For example, an antigenic moiety may become exposed when blood proteins are adsorbed on surfaces. Adsorption of polypeptide and protein drugs at interfaces will also lead to a reduction in the concentration of drug available. The loss of polypeptides and proteins as a result of interfacial adsorption may occur during purification, formulation, storage, or delivery. Difficulties have been encountered with insulin adsorbing to the surfaces of delivery pumps,[4-8] to glass and plastic containers,[9-13] and to plastic intravenous bags and tubing.[14-16]

6.2.1.3. Aggregation and Precipitation

Denatured, unfolded proteins may rearrange to form aggregates in which the hydrophobic amino acid residues of different molecules associate together. Aggregation may be on a macroscopic scale, resulting in precipitation. Aggregation and precipitation can follow interfacial adsorption, as explained previously. The extent of aggregation and precipitation is dependent on the relative hydrophilicity or hydrophobicity of the surfaces with which the polypeptide or protein solution is in contact. Both aggregation and precipitation are accelerated by hydrophobic surfaces. The types of vials and stoppers (as well as their sizes) used to contain polypeptide and protein solutions are therefore of importance. An example is insulin frosting, the formation of finely divided precipitates on the walls of insulin containers. Precipitation is accelerated in the presence of a large air–water interface,[11] indicating that adsorbed protein at the air–water interface is denaturing, aggregating, and finally precipitating. The air interface is very hydrophobic. Agitation of polypeptide and protein solutions increases aggregation and precipitation. This may be a result of the introduction of air bubbles, which increase the exposure of the molecules to the hydrophobic air interface, and may also be a result of the increased thermal motion of the molecules, which results from agitation. The formation of aggregates or precipitates in the insulin reservoir of implanted insulin pumps has been a major problem associated with this delivery system. Aggregates in these pumps form at body temperature, particularly with agitation of the insulin reservoir, and when the insulin is in contact with hydrophobic surfaces such as polyvinyl chlorine and air bubbles.

6.2.1.4. Use of Additives to Improve Polypeptide and Protein Physical Stability

Certain additives such as salts, metal ions, polyalcohols, surfactants, reducing agents, chelating agents, other proteins, amino acids, fatty acids, and phospholipids can improve the stability of polypeptides and proteins. Salts decrease denaturation and increase stability in relation to aggregation and precipitation via nonspecific or specific ion binding to the protein.[17–23] Specific ion binding sites have been introduced into proteins to improve their stability (the reader is referred to Section 6.2.2.6 in this chapter, Methods of Improving the Chemical Stability of Polypeptides and Proteins). When present in aqueous solutions at low concentration, polyalcohols such as glycerol can stabilize proteins by selective solvation.[24–24] Water molecules pack around the protein molecules to exclude the hydrophobic polyalcohol molecules, thereby increasing stability. At high polyalcohol concentrations the opposite effect occurs, as there is insufficient water to protect the protein, and exposure to these hydrophobic solvents denatures the protein.

Surfactants, both nonionic and ionic, can stabilize proteins by preventing adsorption of protein at interfaces.[16, 29] The small surfactant molecule preferentially adsorb at the interfaces. The presence of surfactant molecules also inhibits aggregation and precipitation of polypeptides and proteins.[9, 12, 30, 31] Phospholipids, fatty acids, and other protein molecules are also surface active and therefore act similarly to surfactants.

6.2.2. Chemical Stability

There are nine major reaction mechanisms by which peptides and proteins chemically degrade: hydrolysis, imide formation, deamidation, isomerization, racemization, diketopiperizine formation, oxidation, disulfide exchange, and photodecomposition.

6.2.2.1. Deamidation

Deamidation involves the hydrolysis of the side chain amide linkage of an amino acid residue to form a free carboxylic acid. Glutamine (Gln) and asparagine (Asn) are particularly susceptible to deamidation. This type of hydrolysis of proteins and peptides has been reviewed by Robinson and Rudd.[32] Proteins that undergo deamidation in vitro include insulin, prolactin, human growth hormone, and bovine growth hormone. In vivo deamidation may be enzymatic or nonenzymatic. The rate of deamidation of Asn residues in peptides and proteins is favored by increased pH, temperature, and ionic strength, and is affected by the tertiary structure of the protein. The protein molecule may be folded in such a way that deamidation is sterically hindered. For example, the tertiary structure of trypsin appears to prevent deamidation of the molecule.[33] Deamidation can also alter the activity of peptides and proteins: a decrease in the biological activity of porcine adrenal corticotropic hormone (ACTH) follows deamidation.[34]

6.2.2.2. Oxidation

Oxidation may occur on the side chains of histidine (His), methionine (Met), lysine (Lys), tryptophan (Trp), and tyrosine (Tyr) residues in proteins. Oxidation is very common during isolation, synthesis, and storage of peptides and proteins.[35–38] The thioether group of Met is particularly susceptible to oxidation. Atmospheric oxygen can oxidize Met residues under acidic conditions. Oxidizing agents such as hydrogen peroxide, periodate, iodine, and dimethyl sulfoxide can oxidize Met-to-Met sulfoxides. The thiol group of Cys can be oxidized to sulfonic acid. Again, oxidation occurs in the presence of oxidizing agents such as iodine and hydrogen peroxide, is catalyzed by metal ions, and may occur spontaneously by atmospheric oxygen.

Oxidation of amino acid residues is usually associated with a loss of biological activity. For example, oxidation of the Met residues in corticotropin,[39] calcitonin,[40] and gastrin[41] results in a loss of activity. However, some proteins, such as glucagon, retain their biological activity after oxidation. Lost biological activity may be restored by reduction. Almost full activity is restored to lysozyme[42] and ribonuclease[43] on reduction of Met sulfoxide to Met.

6.2.2.3. Proteolysis

The susceptibility of peptide bonds to cleavage is dependent on the residues involved. Asparagine residues are more unstable than other residues, and the Asp-Pro (proline) bond is particularly labile. In general, at neutral pH and room temperature, peptide bonds are relatively stable in relation to hydrolysis. For example, the half-life of Phe-Phe-Phe-Gly was determined to be of the order of 7 years.[43] Proteolysis occurs in most proteins on heating. For example, lysozyme is inactivated irreversibly by heating at 90–100°C at pH.[44]

6.2.2.4. Disulfide Exchange

Disulfide bonds may break and reform incorrectly, altering the three-dimensional structure and hence the activity of polypeptides and proteins. This reaction can be catalyzed in neutral and alkaline media by thiols, which may arise as a result of hydrolytic cleavage of disulfides. The thiolate ions carry out nucleophilic attack on a sulfur atom of the disulfide. Disulfide exchange may be prevented by the addition of thiol scavengers such as *p*-mercuribenzoate, *N*-ethylmaleimide, and copper ions.

6.2.2.5. Racemization

All of the amino acids, with the exception of Gly, are chiral at the carbon bearing the side chain, and are subject to base-catalyzed racemization. The α-methene hydrogen is removed by the base, forming a carbonion. Racemization of an amino acid residue may render the protein nonmetabolizable, as the racemic peptide bonds are inaccessible to proteolytic enzymes, and may reduce or eliminate biological activity.

6.2.2.6. Methods of Improving the Chemical Stability of Peptides and Proteins

The chemical stability of peptides and proteins can be improved by site-directed mutagenesis, chemical modification of the molecules, and the appropriate choice of conditions.

Site-Directed Mutagenesis. Bioengineering techniques are available that enable specific mutations to occur that alter the primary sequence of the proteins and thus create new chemical entities.[45] Site-directed mutagenesis can be designed to improve both the stability and the intrinsic specificity of proteins. Specific amino acid residues that are chemically reactive can be replaced with others that are not, while retaining the activity of the polypeptide or protein. Ion binding sites can be introduced that improve thermal or other types of stability. Specific sites that are susceptible to chemical degradation, such as deamidation, can be replaced. Physical stability can also be improved. Disulfide bridges can be introduced that stabilize the native conformation and provide stability against reversible and irreversible denaturation. Similarly, residues can be introduced that improve hydrophobic packing and maximize hydrogen bonding.

Chemical Modification of Proteins. Synthetic polymers, such as polyethylene glycol[46–48] and polyoxyethylene,[49, 50] can be chemically conjugated to the surfaces of polypeptide and protein drugs to improve stability and to alter their pharmacodisposition. However, biological activity may be lost on coupling.[51–55] Lipids have been successfully used as coupling agents. Lipids coupled to insulin have improved insulin absorption.[52]

Appropriate Choice of Conditions. Most chemical reactions can be retarded by the appropriate choice of conditions: physical state, pH, ionic strength, temperature, and the addition of preservatives. The presence of water is a most significant factor in both chemical and physical denaturation. The stability of peptides and proteins is usually enhanced when in the dry solid state and at low temperature. Proteins are usually stored and handled as solids. Dosage forms should be formulated in such a way that they can be stored and shipped as solid, preferably freeze-dried products. Most commonly, peptide and protein drugs are given by parenteral injection. These injectable dosage forms are prepared as freeze-dried powders that are reconstituted as solutions prior to use. The stability of peptide and protein drugs is a very important consideration in the selection of an appropriate dosage form and route of administration.

6.2.3. Stability Testing

Most peptides and proteins are extremely stable at low temperature (5°C and below) but are usually extremely thermolabile (especially above body temperature); therefore, accelerated stability testing is not appropriate. Real-time stability studies must therefore be conducted. These studies take much longer to conduct; thus, the development stage of any peptide or protein drug formulation is much longer compared with most conventional drugs.

6.3. Delivery of Protein and Peptide Drugs

6.3.1. Oral Route

The conventional and most convenient route of drug delivery, the oral route, is problematic for protein or peptide drugs. Peptides are degraded by chemical and enzymatic action in the gastrointestinal tract (GIT). The molecular mass of peptides and proteins (approximately 0.5 to over 100 kDa) inhibits their ability to cross the intestinal epithelia. However, limited intestinal absorption of a few small bioactive peptides, such as cyclosporin and thyrotropin-releasing hormone (TRH), has been reported during the past decade.[56] Any peptide or protein drugs that are absorbed intact may undergo first-pass metabolism in the liver.

Proteins and polypeptides are hydrolyzed in the GIT to a mixture of amino acids and small peptides. However, small peptides, di-, and tripeptides, such as TRH, can exhibit metabolic stability in the GIT. TRH has been successfully formulated as a tablet for oral use.[57] Generally, tetra- and larger peptides are hydrolyzed to absorbable products,[58] although there are a few that are absorbed intact by specific carrier transport mechanisms.

Gastric acid contains the important proteolytic enzyme pepsin. Pepsin is specific for peptide linkages to phenylalanine and tyrosine. In the upper small intestine, the breakdown of proteins and polypeptides occurs under the influence of trypsin, chymotrypsin, carboxypeptidases, and aminopeptidases, all of which are excreted by the pancreas. Trypsin is specific for peptide linkages to lysine and to arginine. Chymotrypsin primarily attacks phenylalanine and tyrosine. The carboxypeptidases and aminopeptidases cleave terminal amino acid residues. The pancreatic juice is rich in carboxypeptidases but contains only a small amount of aminopeptidases. The intestinal mucosa is rich in aminopeptidases and dipeptidases.

Although small peptides such as TRH, and a few larger peptides such as cyclosporin (an undecapeptide), may be stable to degradation, they are not absorbed to any significant extent. The bioavailability of TRH is only a few percent.[59, 60] Most drugs are absorbed by a passive diffusion process in which the fraction of the dose absorbed is a function of the dose to solubility ratio, the particle size, the partition coefficient, the pK_a, the stability, and the molecular weight of the drug. Compounds with molecular weights of approximately 500 and above (most peptides) are poorly absorbed from the GIT.[61] This may be, in part, a consequence of the gel-like mucin, which can act as a physical barrier to these large molecules. Absorption of high-molecular-weight compounds, such as peptides, across the intestinal mucosa may only occur by (1) diffusion through the aqueous pores or the intercellular spaces (water-soluble molecules), (2) diffusion through the membrane lipids (lipid-soluble molecules), and (3) uptake into epithelial cells by endocytosis or pinocytosis. Absorption by these mechanisms is very limited, approximately 1 to 5%.[62] For example, the metabolically stable analogues of somatostatin, cyclic hexa- and octapeptides, are absorbed about 2% orally. Some di- and tripeptides, absorbed by a facilitated diffusion transport system, are absorbed to a greater extent.[63]

6.3.1.1. Pharmaceutical Methods to Improve the Oral Bioavailability of Peptides and Proteins

The oral bioavailability of peptides and proteins can be improved by the coadministration of peptidase inhibitors, by saturation of the intestinal peptidases, by avoiding the intestinal peptidases through enteric coating, by the coadministration of penetration enhancers, and by avoiding hepatic first-pass metabolism through promotion of lymphatic uptake.

Use of Peptidase Inhibitors. Pharmaceutically it is possible to improve the stability, absorption, and therefore the bioavailability of peptide and protein drugs. Protection against proteolysis may be achieved by coadministration of peptidase inhibitors. For example, it has been shown that amistatin acts effectively as a peptidase inhibitor, significantly promoting the absorption of enkephalins.[64] Aminopeptidase, phosphoramidon-sensitive endopeptidase-24-II, and captopril-sensitive peptidyl dipeptidase A have all been indicated in the inactivation of [Met[5]]-enkephalin in the rat and mouse vas deferens and the guinea pig ileum.[64]

Saturation of and Bypassing the Intestinal Peptidases. Another approach to improving peptide and protein bioavailability is to saturate the peptidases either by increasing the dosage or by loading with another peptide or protein. The drug must be nontoxic at high dosage if peptidase saturation by increased dosage is to be employed. Alternatively, the drug may be enteric coated so that certain parts of the GIT and hence certain peptidases are bypassed. Acid-resistant acrylic resins will bypass the stomach and release the drug in the small intestine. Polymer coatings, cross-linked with azoaromatic groups, have been developed that form a tough, water-impervious film that remains intact in the stomach and the small intestine. The azo bonds are reduced in the large intestine by indigenous microflora.[65] This methodology avoids the majority of the peptidases; however, the colon is much less permeable than the small intestine, particularly to water-soluble drugs.

Use of Penetration Enhancers. Absorption of peptides and proteins can be improved pharmaceutically by the coadministration of penetration enhancers. Several different groups of penetration enhancers have been studied, including surfactants, nonsurfactants, and chelators. Ionic and nonionic detergents[66–68] and bile salt surfactants[69–72] have all been demonstrated to enhance the absorption of proteins and peptides. Surfactants appear to act by increasing membrane fluidity,[73, 74] which may be linked to surfactant micellar structures. Micelles may interact with the phospholipids of the intestinal membrane, promoting absorption through the membrane; similarly, pericellular permeation may be enhanced by solubilization of drugs within bile salt micelles. Bile salt micelles have been shown to significantly enhance the absorption of insulin.[75]

Ethylenediaminetetraacetic acid (EDTA), trisodium citrate, and salicylates promote absorption of water-soluble drugs by chelation. The mechanism of action is thought to be by depletion of calcium and magnesium ions in the regions of the tight junctions between the epithelial cells, which then forces water through by osmosis, enhancing pericellular absorption of water-soluble drugs. EDTA and trisodium citrate have been shown to increase the absorption of heparin, mannitol, and insulin.[76]

Promotion of Lymphatic Uptake. A third important factor to be taken into consideration, beside enzymatic degradation in the GIT and poor absorption, is hepatic first-pass metabolism. Hepatic first-pass metabolism may be avoided by saturation blockage of specific enzymes or by formulation in a vehicle that promotes lymphatic uptake. Lymphatic uptake of drugs may be increased by the use of surfactants, such as bile salts, which facilitates micellar solubilization; by formulation in lipid vehicles; and by concurrent feeding with lipid. For example, the bioavailability of cyclosporin is increased from approximately 1%, to between 20 and 50% when the drug is given in an olive oil-based formulation.[77] Lipids promote the intestinal absorption of drugs selectively via the lymph system.[78] Digestion of lipids enhances the synthesis of chylomicrons in which drugs may be entrapped and thereby selectively absorbed by the lymphatic system. Surfactants facilitate micellar solubilization of drugs. Salicylates also enhance lymphatic uptake, the more hydrophilic the salicylate the greater the adjuvant activity.[79]

The foregoing methods of improving the oral delivery of peptides are mostly in the experimental stage. Apart from a few exceptions, such as cyclosporin, oral delivery of peptide and protein drugs is not viable at present. However, a significant amount of research is underway, and oral delivery, particularly of the smaller peptides, may indeed become possible in the near future. Formulation aspects of oral delivery are also under investigation[80].

6.3.2. Parenteral Routes

Parenteral delivery has been the sole practical method of administration of proteins, owing to their poor absorption and metabolic instability when given by other routes. Insulin has been delivered parenterally by subcutaneous injection since the early 1930s. The parenteral route also offers the possibility of targeting the drug directly to the site of action in a controlled manner.

6.3.2.1. Parenteral Drug Targeting

Increasing the specificity and effectiveness of drugs by combining the active ingredient with a suitable delivery system is a major research objective in pharmaceutical engineering. As a rule, the active ingredients contained in a conventional dosage form are distributed without sufficient distinction between biological targets and a wide variety of sites, in accordance with their physicochemical properties. Site-specific delivery can improve the therapeutic index of a drug by optimizing the access, amplitude, and nature of interactions with the pharmacological receptor. Drug targeting can also protect the drug and the body from any unwanted and deleterious disposition. Drug release rates can also be controlled using a drug delivery device. The release rate can be reduced by delayed diffusion through a polymer matrix. This allows drugs with very short biological half-lives (most peptide and protein drugs) and very narrow therapeutic indices to be utilized. The advantages of a temporal controlled-release injectable system include maintaining constant levels of the drug in the body and eliminating patient compliance as a variable in therapy.

Genetically engineered peptides and proteins, being very potent and expensive, are ideal candidates for site-specific delivery. Peptide and protein drugs are chemically similar or identical to the endogenous substances they are intended to replace; therefore the body can metabolize them easily. Both plasma and tissue peptidases are present in the body that attack and break down these drugs. As they are subject to enzymatic degradation in the plasma, they may cause immunological reactions and other side effects. For example, asparaginase is not active for long periods in the treatment of acute leukemias.[81] Most peptide and protein drugs have very short half-lives in the bloodstream, of the order of seconds to hours.[82] The vasoactive drug angiotensin has a half-life of 15 seconds. Oxytocin, which stimulates uterine motility, has a half-life of 2 minutes. The antidiuretic vasopressin has a half-life of 4 minutes, and the half-life of insulin is less than 25 minutes. The body normally produces biological response modifiers on demand in close proximity (of the order of a few nanometers) to the site of action; thus, degradation of these substances prior to their action is not a problem. By encapsulating peptide and protein drugs in a targetable delivery device, they will be protected from enzymatic degradation until they reach their target site. Peptide and protein drugs are very potent, with therapeutic ranges in the submicrogram level. However, if these drugs are given conventionally and are distributed throughout the body, large quantities are required to achieve therapeutic levels at the site of action. Peptide and protein drugs, especially if present at high dosage levels, may generate immune responses and other unwanted side effects. Drug targeting can protect the drug and the body from unwanted and deleterious disposition and interaction. The targeting of peptides and proteins may also be necessary, as these molecules may be unable to leave the drug compartment to reach their active sites owing to the large size of these molecules, which restricts their ability to pass through the various membranes and barriers of the body.

6.3.2.2. Methods of Parenteral Targeting

Parenterally administered drug carrier systems are of two types: particulate carriers (either capsular, monolithic, or cellular) and soluble carriers (macromolecular drug conjugates). Particulate carrier systems include liposomes, microspheres, nanoparticles, microemulsions, erythrocytes, and vaccines. Materials used to prepare soluble drug complexes include both natural and synthetic polypeptides, which are covalently linked to the drug. Both soluble and particulate drug carriers are used for drug targeting. These systems may be passively targeted, according to their natural disposition, or they may be actively targeted using a site-specific moiety attached to the carrier system.

The fate of intravenously administered colloidal particles is controlled by particle size and particle surface properties. Particles larger than 7 μm are rapidly entrapped in the fine capillary beds of the lung.[83–85] Particles 100 nm to 7 μm and macromolecules, which are hydrophobic or carry a net negative charge, are rapidly taken up by the cells of the reticuloendoethelial system (RES), if they are recognized as foreign.[86, 87] The rapid and efficient uptake of large particles by the lungs and small particles by the reticuloendoethelial system (RES) provides opportunities for passive drug targeting.[88]

Passive Targeting. Passive targeting according to size is the simplest method of parenteral targeting. Passive targeting to the RES can be utilized to treat diseases of the RES (neoplasms, parasitic, viral, and bacterial infections) and for macrophage stimulation. Macrophage activating factors (such as modified lymphokines and interferons), which activate the macrophage to the tumoricidal state, can be incorporated into colloidal systems such as liposomes that will be taken up at the active site (the macrophage). Fidler has successfully delivered liposomes containing lymphokines that activate macrophages to abrogate the development of micrometastases after surgical removal of a primary tumor.[89] Liposome-encapsulated muramyl dipeptide is currently in clinical trials, with the potential application as a macrophage-stimulating factor (Ciba-Geigy).

Active Targeting. Active targeting includes coupling the drug or drug carrier system to ligands such as erythrocyte membrane glycoproteins, native immunoglobulins, and monoclonal antibodies to impart specificity and alter the natural distribution of the carrier. For example, tumor-reactive antibodies have been used in an attempt to target cytotoxic moieties to neoplastic cells[90] (Figure 6.3). Unfortunately, the use of monoclonal antibodies as targeting agents has met with limited success, owing to difficulties in preserving the recognition ability in vivo. Synthetic polymers have been chemically conjugated to peptides and proteins to alter the pharmacodeposition of these drugs. These conjugates increase the apparent size of the protein or reduce unwanted interactions with blood and tissue components, or both. Conjugation with hydrophilic polymers is particularly successful in reducing the immunogenicity and increasing the duration of action.

Problems Associated with Uptake and Intracellular Transport of Parenteral Delivery Systems. As well as being an opportunity to passively target the RES, the rapid sequestration of particulates by the cells of the RES is a major obstacle to targeting other sites in the body. As a consequence, only a very small fraction of administered drug carriers reach their target sites. A partial answer to this problem is to coat carrier systems with hydrophilic surface active agents. The

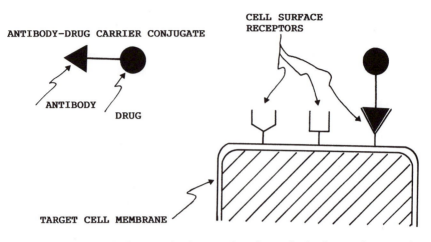

Figure 6.3. Schematic diagram of active targeting of an antibody–drug conjugate carrier system, showing the affinity of the antibody for the target cell.

coating decreases their recognition and subsequent uptake by the cells of the RES.

Another obstacle to particulate drug targeting is the inability of particulates to cross the vascular endothelium. Only a few capillary membranes contain small holes (fenestrae) 50 to 60 nm in size, which allow small particles and large molecules to pass through. The parenchymal cells of the liver have sinusoidal or discontinuous membranes with gaps of the order of 150 nm (sieve plates). However, most particulate targeting systems must release their contents extracellularly, decreasing the potential specificity and subjecting the drugs to possible metabolic degradation prior to cellular uptake.

The intracellular transport of certain soluble drug-carrier complexes may occur via endocytosis.[91] Endocytotic vesicles form at cell surfaces for the purpose of transporting fluids (pinocytic vesicles) and solids (phagocytic vesicles) into the cells, and through the cells to the opposite membrane. Receptor-mediated transcytotic routes exist for certain macromolecules, and these can be exploited for the purpose of drug delivery. A receptor-mediated transcytotic route for albumin is present in the continuous endothelia of the heart, lung, and diaphragm.[92] Drug–albumin complexes could be targeted via this mechanism. Studies are underway to elucidate operational receptors and cell-surface markers that can be used to translocate carriers and drugs through cell membranes.

6.3.2.3. Parenteral Drug Delivery Systems

Parenteral drug delivery systems include those intended for intravenous, intraarteriolar, intramuscular, subcutaneous, intraperitoneal, intraarticular, and intrathecal use. The simplest form of delivery system for peptides and proteins is drug in a buffered aqueous solution. The appropriate dose is then injected intravenously or subcutaneously according to the desired time pattern. However, the therapeutic use of most peptides and proteins is hampered by short half-lives, metabolic lability, immunological reactions, and the inability to achieve sufficient concentrations at the active site. Thus, targetable, prolonged-release carrier systems capable of protecting the drug and the body from unwanted reactions are being developed.

Soluble Carrier Systems. Soluble carrier systems include chemically modified drugs, drug conjugates, and hybrid proteins. Peptide and protein complexes with various polymers and lipids have been successful in increasing drug stability and site specificity, while retaining the biological activity. Specific properties of target cells have been utilized to achieve targeting. For example, antibodies specific for a variant of cell-surface determinants can be complexed to drugs (Figure 6.3). Drug conjugate systems may be active at the cell surface, or active following endocytic capture by cells, or the active drug may be released from the conjugate extracellularly in a controlled manner.[93]

Sophisticated site-specific molecules have been produced by site-directed mutagenesis and hybrid fusion techniques. These molecules have increased half-lives in the body and still retain their biological activity. Hybrid proteins that combine the features of two or more proteins create drugs with the required target-recognition properties and pharmacological activity. Hybrid proteins may also have increased stability and altered immunogenicity. These molecules are produced either by synthetically linking protein fragments or by gene fusion. For example, a portion of the interleukin-2 molecule that interacts with a receptor on T cells is linked to a portion of diphtheria toxin. The resultant site-specific hybrid fusion protein can enter and kill T cells, in the treatment of graft-versus-host disease.[45]

Particulate Carrier Systems. Various types of particulate carrier systems exist, including liposomes, microspheres, nanoparticles, emulsions, red blood cells, and viruses. These systems can either be passively targeted according to their size or can be actively targeted by conjugation with a targetable moiety such as monoclonal antibodies.

Liposomes. Liposomes have attracted a lot of attention, as their structure resembles that of cell membranes. Liposomes are tiny spheres that are formed when phospholipids are combined with water. These are lipid vesicles made up of one or more lipid bilayers alternating with aqueous compartments[94] (Figure 6.4). Liposomes can be categorized as (1) small unilamellar vesicles (SUV), 25 to 70 nm, that consist of a single lipid bilayer; (2) large unilamellar vesicles (LUV), 100 to 400 nm in size, that also consist of a single lipid bilayer; and (3) multilamellar vesicles (MLV), 200 nm to several microns, that consist of two or more concentric bilayers. Liposomes are relatively unstable, have low drug-carrying

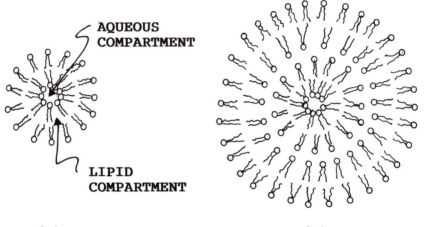

Figure 6.4. Schematic diagram of liposomes: (a) small unilamellar liposome; (b) multilamellar liposome.

capacities, and tend to leak entrapped drug substances. The properties of liposomes can be modified by the incorporation of various molecules into the phospholipid bilayer such as cholesterol, cetylphosphate, and stearylalamine. The presence of cholesterol results in a more stable membrane. Stable, polymerized liposomes composed of phospholipids with polymerizable moieties have been developed. Liposomes passively target the Kupffer cells of the liver,[95] and have been used to target drugs to the liver. SUVs may target the liver parenchymal cells, as they are small enough to pass through the "sieve plates." Immunoglobulins have been attached to the surface of liposomes to actively target them to specific sites. Liposomes are currently being tested as spatial-controlled delivery systems for bioactive peptides and proteins such as the macrophage stimulating agent, muramyl dipeptide, and for enzyme replacement to treat genetic metabolic disorders. For further information the reader is directed to Ostro's tests.[96]

Microspheres. Microspheres are usually solid, approximately spherical particles in the particle size range of 1 to 600 μm, containing dispersed drug in either solution or microcrystalline form. Microspheres are usually formed by emulsification techniques in which the dispersed phase consists of droplets of polymer-drug solution (either aqueous or nonaqueous) or a polymer solution with suspended drug particles. The polymer droplets are cross-linked either chemically

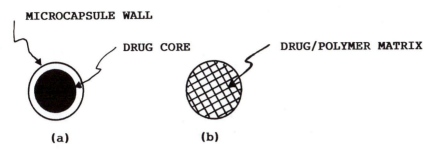

MICROCAPSULE WALL

DRUG CORE

DRUG/POLYMER MATRIX

(a)

(b)

Figure 6.5. Schematic diagram of microencapsulated drug systems: (a) drug encapsulated within a distinct capsule wall; (b) drug dispersed throughout the microcapsule in a polymer matrix.

(using a reactive cross-linking agent), thermally, or by γ-irradiation. The incorporation of drugs can be achieved by physical entrapment during production or by covalent or ionic attachment. Drug-loading capacities are usually fairly high, up to 50%. Microspheres may be prepared from natural polymers such as gelatin and albumin[97, 98] and from synthetic polymers such as poly(lactic acid) and poly(glycolic acid). The drug is either totally encapsulated within a distinct capsule wall or is dispersed throughout the microsphere (Figure 6.5). Drug release is controlled by dissolution and diffusion of the drug through the microsphere matrix or the microcapsule wall.

Microspheres can be passively targeted to the lung capillaries (7–12 μm, particles) or to the RES (1–7 μm, particles) and can be actively targeted by the attachment of specific receptor moieties such as monoclonal antibodies. Magnetic microspheres composed of albumin and iron oxide have been used as a means of targeting specific areas in the body in combination with the use of an external magnet.[99] Polymeric microsphere systems have been reviewed by Davis et al.[83] and by Oppenheim.[100]

Nanoparticles. Nanoparticles are similar to microspheres but have particle sizes in the nanometer range (10–1000 nm). Nanoparticles can be made from nonbiodegradable materials, such as methylmethacrylate[101] or from biodegradable materials such as alkylcyanoacrylate[102] and albumin. Nanoparticles can be used for targeted delivery. Owing to their small size they can pass through the sinusoidal spaces in the spleen and bone marrow. Monoclonal antibodies and other targeting moieties can be attached to nanoparticles to increase their specificity.

Emulsions (Lipid Microspheres). Colloid-sized emulsion droplets are used for parenteral drug delivery to solubilize water-insoluble drug substances, for controlled and sustained release, and drug targeting. The stability and the fate of parenterally administered emulsion droplets is dependent on the nature of the emulsifying agent used. Lecithin is most frequently used to stabilize parenteral emulsions. Emulsion droplets are usually taken up by the cells of the RES. Emulsions of vegetable oils are preferentially deposited in adipose tissue, lactating mammary glands, and in myocardial muscle.[103] Drug release from emulsions is controlled by the droplet size of the dispersed phase and the emulsion viscosity.

Red Blood Cells. Human erythrocytes are under investigation for the delivery of compounds to the body. Peptides and proteins can be introduced into erythrocytes by a hypotonic exchange process. The red blood cells are placed in a hypotonic solution that causes the cell membrane to become permeable. The peptide or protein is added and the membrane is restabilized. The erythrocytes are taken up by the RES and end up in the hepatic lysosomes. This is a useful system for enzyme replacement in genetic metabolic disorders. A limitation to the application of red blood cells is that they can only be used for drugs that are not susceptible to irreversible denaturation under hypotonic conditions.

Viruses. Retroviral gene delivery systems have been developed to assist the entry of genes into cells for the purpose of gene therapy.[104] Retroviral gene delivery systems consist of an RNA copy of a gene package into a viral particle. These provide an excellent alternative to physical methods of gene delivery, such as direct injection into the nucleus. The basic concept of human gene therapy is that functionally active genes are delivered into the somatic cells of a patient with a genetic defect. This can be gene replacement, which involves specific correction of a mutant gene sequence, or gene augmentation, in which a correctly functioning gene is inserted into nonspecific sites in a genome, leaving the malfunctioning gene alone.

A virus consists of an RNA-protein core encapsulated in a lipid envelope. The viral glycoproteins bind with specific receptors on target cells, and the viral envelope then fuses with the cell membrane, resulting in the introduction of the RNA genome into the target cell cytoplasm. Nonreplicating viral vectors have been developed that efficiently infect human cells.[105] An RNA copy of a replacement gene is packaged into a nonreplicating viral particle. When viral gene delivery systems infect target cells, these cells express both the viral and the exogenous genes. This process is not normally harmful to the target cells. Problems that may arise include the deletion of sequences during replication, recombination with endogenous viral sequences to produce infectious recombinant viruses, the activation of cellular oncogenes, the introduction of viral oncogenes, and the inactivation of genes. The interested reader is referred to Wilson.[104]

Preparation of Particulate Carrier Systems Suitable for Peptide and Protein Drugs. The preparation of microcapsules, nanoparticles, liposomes, and microemulsions usually involves at least one of several harsh conditions that may denature proteins and peptide drugs. These are contact with an organic solvent; heating during the process (temperatures of 120°C and above are often em-

ployed[106, 107]); the addition of chemical cross-linking agents (such as glutaralde-
hyde and formaldehyde); adverse pH conditions; the addition of surfactants; and
the application of γ-irradiation.

Systems therefore must be developed that do not involve harsh conditions.
Albumin has been cross-linked under very mild conditions, without the use of
highly reactive chemicals, resulting in a delivery system that showed no loss of
peptide activity for both insulin and urokinase.[108, 109] Cross-linking albumin under
very mild conditions also provides a delivery system that is recognizable to the
body as albumin and therefore should not cause immunological reactions. A
drawback to this method is that the microspheres are very leaky, since they have
few cross-linkages.

An alginate-polylysine-alginate system has been developed that is suitable for
encapsulation of peptides and proteins.[10] The initial step in this process is to
suspend the drug in alginate gel beads, which protects the drug from degradation
on further chemical processing. A sodium alginate–drug solution is sprayed into
a calcium chloride solution. The calcium cross-links the alginate, forming calcium
alginate gel beads (Figure 6.6). The gel beads can be cross-linked with lysine
and then with a second layer of alginate.

Other Types of Parenteral Delivery Systems. The preceding microsphere, nano-
particle, liposome, microemulsion, erythrocyte, and vaccine systems may be
given intravenously, intraarterially, or as intramuscular or subcutaneous implants.
Other implant systems include automatic feedback minipumps, osmotic minipumps, porous polymer fiber systems, and large capsular and tablet-type implants.
These systems are intended to release drugs for periods of up to 1 year. A hollow,
porous, polymer fiber system suitable for implantation has been developed for
Levonorgestrel.[111] The hormone was released at a continuous and stable rate for
approximately 6 months. One must be careful of the potential for denaturation or
aggregation of peptide and protein drugs that remain in the body for a long period
when using these types of systems.

Implantable systems have been designed that release drugs into the body on
demand. For example, small magnetic beads have been embedded in polymer
matrices along with the drug. Drug release rates from these systems can be
enhanced as much as 30 times by an oscillating external magnetic field.[112] Ultrasonic waves can also be used to trigger drug release from polymer matrices.
Ultrasonic waves increase the bioerosion of biodegradable polymers, causing
cavitation effects that enhance drug release. pH-sensitive polymer systems have
been devised, for potential application at tissue sites with unusual pH, for example, cancerous cells or tissues that undergo pH cycling (such as the vagina). An
osmotic minipump system has been developed for an adrenocorticotropic hormone analogue for the regeneration of injured nerves. Osmotic pumps are composed of a semipermeable membrane with a tiny orifice surrounding an osmotic
core reservoir containing the drug. Drug release is controlled by the osmotic

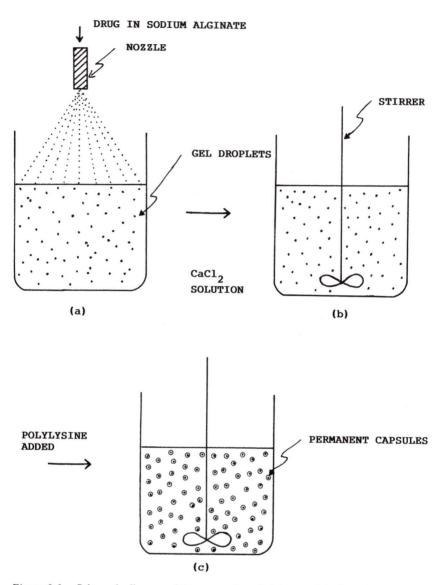

Figure 6.6. Schematic diagram of the preparation of alginate-polylysine microcapsules: (a) spraying of drug in sodium alginate solution into calcium chloride solution; (b) temporary gel microcapsules formed; (c) permanent microcapsules formed.

pressure within the pump. As the osmotic pressure is increased, the drug is forced out. Automatic feedback pumps have been successfully developed for insulin. A system has been designed that increases the release of insulin in the presence of excess glucose.[82] The enzyme glucose oxidase is immobilized within a cross-linked ionizable polymeric membrane. Glucose oxidase catalyzes the conversion of glucose to gluconic acid, which in turn can protonate amine groups within the membrane's network structure. The resultant electrostatic repulsion of the positively charged amino groups causes expansion of the polymeric network structure. Thus, faster delivery of insulin takes place in the presence of glucose, and when the glucose level drops, the rate of release will be decreased.

6.4. Nonconventional Routes of Administration

Nonconventional routes of administration (such as transdermal, nasal, ocular, buccal, rectal, intrauterine, and vaginal) are being investigated as alternatives to the oral and parenteral routes for the delivery of peptides and proteins. These routes avoid both the gastrointestinal enzymes and hepatic first-pass metabolism. Peptidases are present in all of the membranes associated with the aforementioned routes. However, compared with the gastrointestinal tract (GIT), the quantities of enzymes present are much less. Absorption of large-molecular-weight peptides and proteins across the various epithelial barriers associated with these alternative routes are as much of a problem as in the GIT.

6.4.1. Transdermal

Until recently, the skin was used as a route of administration only for treatment of local diseases of the skin. However, with an increased understanding of the anatomy and physiology of the skin, and of percutaneous absorption, this route is now being considered for systemic drug delivery. The transdermal route has several advantages, such as avoidance of hepatic first-pass metabolism, ease of termination of dosing, decreased frequency of dosing, and improved patient compliance. Disadvantages include high intra- and intersubject variability in skin permeation and a low rate of permeation for most drugs.

The stratum corneum is responsible for the low permeability of skin. The stratum corneum is made up of multilayers of horny cells that are compacted, flattened, and keratinized. The water of the stratum corneum is only 20%. Drug transport across the stratum corneum may be transcellular, intercellular, and appendageal. For most compounds the intercellular route is the predominating path.[113] Interference with the packing of the intercellular lipid lamellae can significantly reduce the diffusional resistance of drugs in the stratum corneum.

Peptides and proteins have particular difficulty in penetrating the stratum corneum, owing to their hydrophilicity, ionic character, and large molecular weight. Drug penetration can be improved using penetration enhancers such as azone and fatty acids, and by techniques such as iontophoresis and phonophoresis.

6.4.1.1. Iontophoresis

Iontophoresis is defined as the migration of ions when an electric current is passed through a solution containing ionized species. Drugs in the ionic form can be "phoresed" with a small current and driven into the body. To undergo iontophoresis the peptide and protein molecules must carry a charge. This can be done by controlling the pH and ionic strength of the solution. The drug is contained in a gauze pad that is applied to the skin under an electrode of the same charge as the drug. The other electrode (of opposite charge) is placed on a gauze pad soaked in saline at a distal location of the body. A small current, below the patient's pain threshold, is allowed to flow for a sufficient period of time. It has been shown that insulin can be administered through intact hairless rat skin under iontophoresis using direct current.[114] The rate and extent of delivery of TRH has also been improved by this method.[115]

Disadvantages associated with this method of transdermal delivery include the possibility of burning of the skin (even at low voltages) and denaturation of the peptides and proteins. If pH will need to be changed to achieve a suitable charge for iontophoresis, a loss of quaternary and tertiary structure, and possibly even secondary structure, may result. The process of iontophoresis itself, which can generate a considerable amount of heat, may also result in denaturation of peptide and protein drugs.

6.4.1.2. Phonophoresis

Phonophoresis is also used to enhance drug delivery through the skin. Ultrasound is applied via a coupling-contact agent to the skin. Ultrasonic waves cause thermal effects and temporary alterations in the physical structure of the skin,[116] thereby increasing drug absorption. The heat generated during this process may be sufficient to denature peptide and protein drugs.

6.4.1.3. Penetration Enhancers

Penetration enhancers such as azone and oleic acid have been used to enhance drug absorption through the skin. These molecules fluidize the intercellular lipid lamellae of the stratum corneum.[117] Fatty acids disrupt the packed structure of the lipids in the extracellular spaces. However, skin irritation is a major limiting factor to the use of currently available penetration enhancers.

6.4.1.4. Prodrugs

The use of prodrugs to improve transdermal absorption is a viable option, especially for small peptides. The skin contains a multitude of enzymes that can be utilized to regenerate the active drug. Several peptides and proteins such as TRH, LHRH, neurotensin, and fibrinopeptides possess a terminal pyroglutamyl residue. This residue can be readily derivatized to produce prodrugs with enhanced transdermal permeability, which will undergo spontaneous hydrolysis at physiological pH, regenerating the active drug.

6.4.2. Nasal

The nasal mucosa is an attractive option for the delivery of peptide and protein drugs. As reviewed by Su,[118] the nasal route is convenient; there is a large surface area available for absorption; the mucosa is highly vascularized; hepatic first-pass metabolism is avoided; and self medication is possible. Compared with other routes such as oral and transdermal, the nasal cavity has a relatively high permeability for peptides. Small peptides such as enkephalin analogues,[119] calcitonin,[120] and ACTH,[121] and even polypeptides such as insulin[122] and proteins such as interferon,[123] have been shown to be absorbed efficiently by the nasal mucosa. Pituitary hormones such as oxytocin and vasopressin have been administered by the nasal route for many years.[118] The bioavailability of peptides via the nasal route is of the order of 1 to 20%, depending on the molecular weight and physicochemical properties of the peptide, and can be extremely variable.[122] Absorption enhancers are used to increase the permeation and to decrease the relatively high peptidase activity in the nasal mucosa. Disadvantages of the nasal route include individual variations in mucous secretion and turnover, which affect the extent of absorption; peptidases present in the nasal membrane, which act as an enzymatic barrier to peptide absorption; alteration in absorption, which may occur with disease states such as upper respiratory tract infections and allergic and chronic rhinitis; and the penetration enhancers and preservatives used in nasal formulations may cause severe damage to the mucosal cell membrane and may be ciliotoxic.

The nasal cavity is made up of different types of epithelia. The most anterior portion of the nasal cavity, the vestibule, which is 3 to 4% of the total surface area, is lined with stratified squamous epithelia. The respiratory epithelium comprises ciliated cuboidal and columnar cells, and goblet cells. The olfactory epithelium is pseudostratified neuroepithelium. Clearance of drug formulations from the nasal cavity is a problem. In the nonciliated anterior part of the nasal cavity, clearance is relatively slow; however, in the posterior part of the nasal cavity, where the mucus flow rate is approximately 6 to 10 mm min^{-1} clearance is of the order of 10 to 15 minutes. Thus, to reduce clearance, nasal sprays are designed to deposit mainly in the anterior portion of the nasal cavity.

Absorption of peptides and proteins through the nasal mucosa can be improved by penetration enhancers such as bile salts, other surface active agents, and chelating agents. These penetration enhancers work by increasing the fluidity of the lipid bilayer membrane,[123] and by opening aqueous pores as a result of calcium ion chelation. Peptidase inhibitors can also enhance the absorption of peptides and proteins by decreasing peptidase activity in the mucus itself and within the mucosal cells.[124] However, penetration enhancers can damage the nasal mucosa and may be ciliotoxic. For example, surfactants such as laureth-9 damage membranes by dissolving the membrane lipids or proteins, resulting in cell erosion, cell to cell separation, and the loss of cilia.[124] Promoters such as oleic acid, linoleic acid, salicylic acid, and their salts appear to enhance absorption safely. Microscopic observations do not show the severe damage that occurs with surfactants. The effect of these agents is readily reversible within 15 to 20 minutes after washing. For example, sodium taurodihydrofusidate, a derivative of fusidic acid, is an excellent enhancer of insulin absorption across the nasal mucosa and is nontoxic both locally and systemically. The nasal epithelial layer serves as a defense against infecting organisms and therefore loss or damage of this layer can cause severe problems. Reepithelization may occur but it is a slow process. Ideally, nasal drug formulations should not disturb the ciliary movement or damage the nasal mucosa and therefore should leave the respiratory defense system intact.

For long-term use the nasal route is probably not practical owing to possible mucosa and cilia toxicities, and alterations in absorption due to changes in the nasal environment during disease states such as the common cold. For short-term use the nasal route appears to be advantageous for the delivery of peptides and proteins, particularly small-molecular-weight peptides.

6.4.3. Rectal

The large intestine is drained by the hepatic portal vein and by the lymphatics. In the lower colon the drainage is mostly lymphatic. It has also been shown that the larger the molecular weight of a compound, the greater the lymphatic uptake. Compounds with molecular weights greater than 2,000 principally enter the lymphatic fluid. Thus, if peptides and proteins are delivered rectally, the hepatic first-pass effect can be largely avoided. Rectal delivery also avoids the vast majority of GIT peptidases. The rectal mucosal membrane is a simple columnar epithelium with tight junctions at intercellular contact points. The bioavailability of peptides and proteins is particularly low owing to these tight intercellular junctions, and penetration enhancers are required. The use of surfactants such as sodium lauryl sulfate to enhance drug delivery can result in rectal bleeding and other damage to the rectal mucosa.[125] However, salicylates, which promote rectal absorption by a different mechanism than surfactants, are better tolerated, and lasting changes in the membrane are not observed after the application of salicylates.[126] The enhancing effect of salicylates, but not of surfactants, can be reversed by washing the rectum with buffer. Suppositories can be designed that will release the peptide and the promotion enhancer in close contact with the rectal and colonic mucous membranes.

Peptide and protein drugs may also be delivered to the large intestine via the oral route using an impermeable coating material that will not degrade until it reaches the colon. A polymer material cross-linked with azoaromatic bonds has been designed that is stable throughout the GIT but is susceptible to degradation by microflora normally present in the colon.[65] The thickness and composition of the coating can be adjusted to achieve delivery to the lower colon.

The rectal route appears to be very promising for the delivery of peptides and proteins, particularly if lymphatic uptake can be achieved and hence hepatic first-pass metabolism avoided.

6.4.4. Buccal

The use of the buccal route for absorption of peptides dates back to 1966 when this route was first used for the adsorption of oxytocin.[127] The buccal route has a number of distinct advantages for peptide and protein delivery when compared with other nonparenteral routes. The buccal route may not be as efficient as the nasal route for peptide and protein absorption, but it is much less sensitive, which is important when considering the need for penetration enhancers to aid drug absorption. Drug administration is very easy via this route, and the drug delivery system may be removed at any time. Better patient compliance is expected for this route compared with the rectal, nasal, and vaginal.

The buccal and sublingual membranes are stratified squamous epithelia that are keratinized in some areas. The most common method of delivery to the oral mucosa is via an aqueous solution or a conventional tablet. However, part of the dose may be lost by accidental swallowing of the tablet or solution or by salivary washout. Therefore self-adhesive systems have been designed that are retained in the oral cavity in intimate contact with the mucosa, either buccally, sublingually, or on the gingiva. The adhesion mechanism involves the intercalation and nonspecific or specific interaction between the polymer chains of both the polymer used as the dosage form excipient and the glycoproteins of the mucosal membrane.[128, 129] Both water-soluble and water-insoluble, ionic and nonionic polymers can be used in these adhesive dosage forms. Polyacrylate-type hydrogels are most commonly used.

The adhesive polymer layer may act as the drug carrier itself, or it may act as an adhesive link between a drug-loaded disc and the mucosa. A third possibility is that a drug-loaded disc may be fixed to the mucosa using a self-adhesive shield (Figure 6.7). These systems can be applied for several days, if a slow release rate is desired. However, a maximum buccal residence time of several hours is best, as these devices may interfere with eating, drinking, and possibly even talking. Adhesive polymer delivery systems offer promise as a convenient method of delivery of peptides and proteins via the buccal route.

Although peptide and protein delivery into the body has been successfully achieved for all the preceding nonconventional routes, each has limitations. The

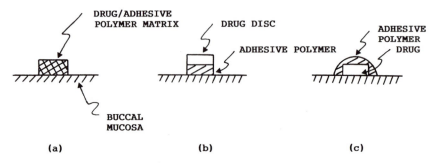

Figure 6.7. Schematic diagram of three different self-adhesive buccal delivery systems: (a) adhesive polymer acts as the drug carrier (drug-and-adhesive polymer matrix); (b) adhesive polymer acts as a link between a drug-loaded disc and the mucosa; (c) a drug-loaded disc is fixed to the mucosa using a self-adhesive polymer shield.

need for penetration enhancers is a major limiting factor, owing to irritation and possible permanent membrane damage. The buccal route appears to be most promising, as it is least susceptible to irritation. Milder, less irritating penetration enhancers are being developed that may increase the attractiveness of these nonconventional routes for both acute and chronic applications.

6.5. Summary

The recent, very rapid advances in biotechnological engineering have resulted in a wide variety of biological response modifiers becoming available for clinical use, such as tissue plasminogen activator and human growth hormone. These peptide and protein drugs are difficult to process, formulate, store, and deliver, and they are highly unstable, both chemically and physically. Chemical degradation may result from hydrolysis, deamidation, oxidation, racemization, disulfide exchange, or photodecomposition. Physical instability is limited to large polypeptides and proteins, which possess high levels of structural order, and involves denaturation (conformation rearrangement), adsorption at interfaces, aggregation, and precipitation. The biological activity of peptide and protein drugs can be lost as a result of chemical degradation or physical structural change. The effective drug concentration may also be reduced as a result of adsorption, aggregation, and precipitation. Care must therefore be taken during the processing, handling, and packaging of these compounds, to eliminate or minimize chemical and physical degradation. Packaging materials must be selected with the aim of reducing drug loss due to interfacial adsorption. Peptides and proteins are generally most stable at low temperature and in the solid state.

The delivery of peptides and proteins into the body is difficult owing to the poor absorption of these large-molecular-weight compounds and their rapid enzymatic degradation in the body. Peptides are present in the gastrointestinal tract and in most mucosal membrane tissue in the body, including rectal, buccal, nasal, dermal, vaginal, and ocular. Thus, peptide administration has been largely confined to the parenteral routes. However, the oral route and alternative routes, such as the buccal, nasal, rectal, vaginal, and transdermal routes, are under investigation. The use of peptidase inhibitors and penetration enhancers is under investigation, and it has been shown, at least in the laboratory situation, that significant bioavailability can be achieved for polypeptide and protein drug delivery via all the foregoing routes. The extent of exposure to peptidase activity is less in these "alternative" routes than in the oral route and appears to be more readily controllable using peptidase inhibitors. The need for penetration enhancers is a limiting factor, however, as these compounds tend to be very irritant and can cause permanent membrane damage. Milder, less irritant enhancers are being developed.

Targeted delivery directly to the site of action is a major objective in the formulation of peptide and protein drugs, owing to their degradative instability, the possibility of unwanted immunological reactions, the specificity of these drugs, and their relatively high cost. The parenteral route offers the possibility of targeted delivery using site-specific delivery systems such as nanoparticles and liposomes coupled with vectoring moieties such as monoclonal antibodies. Recent advances in biotechnology have made possible the production of very sophisticated site-specific peptide and protein drugs by site-directed mutagenesis and hybrid fusion techniques. These techniques can be used to create molecules that are not only targetable but that also possess other desirable properties, such as increased stability and prolonged half-lifes in the body, while retaining the pharmacological activity.

References

1. Shah, H.K., and Rodgers, R.J. 1990. Biopharmaceutical sales and forecasts. In *Spectrum biotechnology overview,* March. Little Decision Resources.

2. Manning, M.C., Patel, K., and Borchardt, R.T. 1989. Stability of protein pharmaceuticals. *Pharm. Res.* 6:903–918.

3. Yoon, J.K., and Burgess, D.J. 1989. Investigation of Interfacial stability of proteins using a surface oscillatory flow technique. *Proc. Int. Sym. Control. Release Biact. Mater.* 16:340–341.

4. Brennan, J.R., Gebhart, S.S.P., and Blackard, W.G. 1985. Pump induced insulin aggregation. A problem with the biostator. *Diabetes* 34:353–359.

5. James, D.E., Jenkins, A.B., Kraegen, E.W., and Chisholm, D.J. 1981. Insulin precipitation in artificial infusion devices. *Diabetologia* 21:554–557.

6. Lougheed, W.D., Woulfe-Flanagan, H., Clement, J.R., and Albisser, A.M. 1980. Insulin aggregation in artificial delivery systems. *Diabetologia* 19:1–9.

7. Peterson, L., Caldwell, J., and Hoffman, J. 1976. Insulin adsorbance to polyvinyl-

chloride surface with implications for constant infusion therapy. *Diabetes* 25:72–74.

8. Sefton, M.V. 1982. Implantable micropump for insulin delivery. Effect of a rate controlling membrane. *ACS Adv. Chem. Ser.* 199:511–522.

9. Chawla, A.S., Hinberg, I., Blais, P., and Johnson, D. 1985. Aggregation of insulin containing surfactants, in contact with different materials. *Diabetes* 34:420–424.

10. Iwamoto, G.K., Van Wagenen, R.A., and Andrade, J.D. 1982. Insulin adsorption. Intrinsic tyrosine interfacial fluorescence. *J. Colloid Interface Sci.* 86:581–585.

11. Massey, E.H., and Sheliga, T.A. 1988. Human insulin (HI) isophane suspension (NPH) with improved physical stability. *Pharm. Res.* 5:S34.

12. Lougheed, W.D., Albisser, A.M., Martindale, H.M., Chow, J.C., and Clement, J.R. 1983. Physical stability of insulin formulations. *Diabetes* 32:424–432.

13. Sato, S., Ebert, C.D., and Kim, S.W. 1983. Prevention of insulin self-association and self-adsorption. *J. Pharm. Sci.* 72:228–232.

14. Twardowski, Z.J., Nolph, K.D., McGary, T.J., and Moore, H.L. 1983. Influence of temperature and time on insulin adsorption to plastic bags. *Am. J. Hosp. Pharm.* 40:583–586.

15. Twardowski, Z.J., Nolph, K.D., McGary, T.J., Moore, H.L., Collin, P., Ausman, R.K., and Slimack, W.S. 1983. Insulin binding to plastic bags: A methodologic study. *Am. J. Hosp. Pharm.* 40:575–579.

16. Twardowski, Z.J., Nolph, K.D., McGary, T.J., and Moore, H.L. 1983. Nature of insulin binding to plastic bags. *Am. J. Hosp. Pharm.* 40:579–581.

17. Arakawa, T., and Timasheff, S.N. 1984. Mechanism of protein salting in and salting out by divalent cation salts: Balance between hydration and salt binding. *Biochemistry* 23:5912–5928.

18. Arakawa, T., and Timasheff, S.N. 1982. Preferential interactions of proteins with salts in concentrated solutions. *Biochemistry* 22:6545–6552.

19. Ahmad, T., and Bigelow, C.C. 1986. Thermodynamic stability of proteins in salt solutions: A comparison of the effectiveness of protein stabilizers. *J. Protein Chem.* 5:355–367.

20. Bhat, R., and Ahluwalia, J.C. 1985. Effect of calcium chloride on the conformation of proteins. *Int. J. Peptide Protein Res.* 30:145–152.

21. Ahmed, F. 1985. Thermodynamic characterization of the partially denatured states of ribonuclease A in calcium chloride and lithium chloride. *Can. J. Biochem. Cell. Biol.* 63:1058–1063.

22. Almog, R. 1983. Effects of neutral salts on the circular dichroism spectra of ribonuclease A. *Biophys. Chem.* 17:111–118.

23. Stellwagen, E., and Babul, J. 1975. Stabilization of the globular structure of ferricytochrome C by chloride in acidic solvents. *Biochemistry* 14:5135–5140.

24. Gekko, K., and Timasheff, S.N. 1981. Mechanism of protein stabilization by glycerol: Preferential hydration in glycerol-water mixtures. *Biochemistry* 20:4667–4676.

25. Gekko, K., and Timasheff, S.N. 1981. Thermodynamic and kinetic examination of protein stabilization by glycerol. *Biochemistry* 20:4677–4686.

26. Lee, J.C., and Timasheff, S.N. 1981. The stabilization of proteins by sucrose. *J. Biol. Chem.* 256:7193–7201.

27. Lee, J.C., and Timasheff, S.N. 1974. Partial specific volumes and interactions with solvent components of proteins in guanidine hydrochloride. *Biochemistry* 13:257–265.

28. Lee, J.C., and Timasheff, S.N. 1975. The reconstitution of microtubules from purified calf brain tubulin. *Biochemistry* 14:5183–5187.

29. Bohnert, J.L., and Horbett, T.A. 1986. Changes in adsorbed fibrinogen and albumin interactions with polymers indicated by decreases in detergent elutability. *J. Colloid Interface Sci.* 3:363–377.

30. Piatigorsky, J., Horwitz, J., and Simpson, R.T. 1977. Partial dissociation and renaturation of embryonic chick S-crystallin. Characterization by ultracentrifugation and circular dichroism. *Biochim. Biophys. Acta* 490:279–289.

31. Tandon, S., and Horowitz, P.M. 1987. Detergent-assisted refolding of guanidinium chloride-denatured rhodanese. The effects of the concentration and type of detergent. *J. Biol. Chem.* 262:4486–4491.

32. Robinson, A.B., and Rudd, C.J. 1974. Deamination of Glutaminyl and Asparanginyl Residues in Peptides and Proteins. In *Current topics in cellular regulations*, Vol. 8. B.L. Horecker and E.R. Stadtman, eds. New York: Academic Press. Pp. 247–295.

33. Kossiakoff, A.A. 1988. Tertiary structure is a principal determinant to protein deamidation. *Science* 240:191–194.

34. Graf, L., Hajos, G., Patthy, A., and Cseh, G. 1973. The influence of deamidation on the biological activity of porcine adrenocorticotropic hormone (ACTH). *Metab. Res.* 5:142–143.

35. Dixon, H.B. F., Moore, S., Stack-Dunne, M.P., and Young, F.G. 1951. Chromatography of adrenotropic hormone on ion-exchange columns. *Nature (London)* 168:1044–1045.

36. Kuehl, F.A., Jr., Meisinger, M.A.P., Brink, N.G., and Folkers, K. 1953. Pituitary hormones. 6. The purification of corticotropin-β by counter-current distribution. *J. Am. Chem. Soc.* 75:1955–1959.

37. Rasmussen, H., and Craig, L.C. 1962. 3. Parathyroid hormone. The parathyroid polypeptide. *Recent Prog. Horm. Res.* 18:269–295.

38. Vale, W., Spiess, J., Rivier, C., and Rivier, J. 1981. Characterization of a 41-residue bovine hypothalamic peptide that stimulates secretion of corticotropic and β-endorphin. *Science* 213:1394–1397.

39. Dedman, M.L., Farmer, T.H., and Morris, C.J.O.R. 1961. Studies on pituitary adrenocorticotrophin. 3. Identification of the oxidation-reduction center. *Biochem. J.* 78:348–352.

40. Riniker, V.B., Neher, R., Maier, R., Kahnt, F.W., Byfield, P.G.H., Gudmundsson, T.V., Galante, L., and MacIntyre, I. 1968. 199. Menschliches calcitonin I. Isolierung und charakterisierung. *Helv. Chem. Acta* 51:1738–1742.

41. Morley, J.S., Tracey, H.J., and Gregory, R.A. 1965. Structure function relationships in the active C-terminal tetrapeptide sequence of gastrin. *Nature (London)* 207:1356–1359.

42. Jori, G., Galiazzo, G., Marzotto, A., and Scoffone, E. 1968. Dye-sensitized selective photooxidation of methionine. *Biochim. Biophys. Acta* 154:1–9.

43. Kahne, D., and Still, W.C. 1988. Hydrolysis of a peptide bond in neutral water. *J. Am. Chem. Soc.* 110:7529–7534.

44. Ahern, T.J., and Klibanov, A.M. 1985. The mechanism of irreversible enzyme inactivation at 100°C. *Science* 228:1280–1284.

45. Tomlinson, E., and Livingston, C. 1989. Therapeutic peptides and proteins. *Pharm. J.* 243:646–648.

46. Abuchowski, A. 1987. Drug delivery/pharmacokinetics using PEG-modified proteins. *J. Cell. Biochem.* 11A:174.

47. Koziej, P., Mutter, M., Gremlich, H.U., and Holzemann, G. 1985. Conformational studies of synthetic polyethylene glycol bound substance P and its lower analogues. *Z. Naturforsch.* 40B:1570–1574.

48. Mutter, M., Mutter, H., Uhlmann, R., and Bayer, E. 1976. Konformation sumtersuchungen and Oligoalanin, substanz P and der myoglobin sequenz (66–73) der zirkulardichroismus von polyathylenglykol-gebundenen peptiden. *Biopolymers* 15:917–927.

49. Rajasekharan, P.V.N., and Mutter, M. 1981. Conformational studies of poly(oxyethylene)-bound peptides and protein sequences. *Acct. Chem. Res.* 14:122–130.

50. Ribiero, A.A., Saltman, R.P., Goodman, M., and Mutter, M. 1982. ^{1}H-NMR studies of polyoxyethylene-bound homo-oligo-L-methionines. *Biopolymers* 21:2225–2239.

51. Hashimoto, M., Takada, K., Kiso, Y., and Muanishi, S. 1989. Synthesis of palmitoyl derivatives of insulin and their biological activities. *Pharm. Res.* 6:171–176.

52. Chow, D.D., and Hwang, K.J. 1987. An investigation of the formulation of an insulin-based drug delivery and drug targeting system. *J. Pharm. Sci.* 76:S49.

53. Towler, D.A., Eubanks, S.R., Towery, D.S., Adams, S.P., and Glaser, L. 1987. Amino-terminal processing of proteins by N-myristoylation. Substrate specificity of N-myristoyl transferase. *J. Biol. Chem.* 262:1030–1036.

54. Ovchinnikov, Y.A., Abdulaev, N.G., and Bogachuk, A.S. 1988. Two adjacent cysteine residues in the C-terminal cytoplasmic fragment of bovine rhodopsin are palmitylated. *FEBS Lett.* 230:1–5.

55. Fojo, A.T., Whitney, P.L., and Awad, M.W., Jr. 1983. Effects of acetylation and guanidination on alkaline conformations of chymotrypsin. *Arch. Biochem. Biophys.* 224:636–642.

56. Gardner, M.L.G. 1984. Intestinal assimilation of intact peptides and proteins from the diet—A neglected field. *Biol. Rev.* 59:289–331.

57. Burger, H.G., and Patel, Y.C. 1977. TSH and TRH: Their physiological regulation

and the clinical applications of TRH. In *Clinical neuroendocrinology,* Chapt. 3. L. Martini and G.M. Besser, eds. New York: Academic Press. Pp. 67–131.

58. Roger, C.S., Heading, C.E., and Wilkinson, S. 1980. Absorption of two tyrosine containing tetra peptides from the ileum of the rat. *IRCS J. Med. Sci.* 8:648.

59. Hichens, M. 1983. A comparison of thyrotropin-releasing hormone with analogues: Influence of disposition upon pharmacology. *Drug Metab. Rev.* 14:77–98.

60. Yokohama, S., Yamashita, K., Toguchi, H., Takeuchi, J., and Kitamori, N. 1984. Absorption of thyrotropin-releasing hormone after oral administration of TRH tartarate monohydrate in the rat, dog and human. *J. Pharm. Dyn.* 7:101–111.

61. Tagesson, C., Anderson, P.A., Anderson, T., Bolin, T., Kallberg, M., and Sjodahl, R. 1983. Passage of molecules through the wall of the gastrointestinal tract; measurement of intestinal permeability to polyethylene glycols in the 634–1338 dalton range (PEG 1000). *Scand. J. Gastroenterol* 18:481–486.

62. Humphrey, M.J., and Ringrose, P.S. 1986. Peptides and related drugs: A review of their absorption, metabolism and excretion. *Dr. Metab. Rev.* 17 (3, 4):283–310.

63. Silk, D.B.A. 1981. Peptide transport. *Clin. Sci.* 60:607–615.

64. Cui, S., Kajiwara, M., Ishii, K., Aoki, K., Sakamoto, J., Matsumiya, T., and Oka, T. 1986. The enhancing effects of amastatin, phosphoramindon and captopril on the potency of [Met^5]-enkephalin in rat vas deferens. *Japan. J. Pharmcol.* 42:43–49.

65. Saffran, M., Kumar, G.S., Savariar, C., Burnham, J.C., Williams, F., and Neckers, C.D. 1986. A new approach to the oral administration of insulin and other peptide drugs. *Science* 233:1081–1084.

66. Engel, R.H., and Riggi, S.J. 1969. Intestinal absorption of heparin facilitated by sulfated or sulfonated surfactants. *J. Pharm. Sci.* 58:706–710.

67. Shichiri, M., Yamasaki, Y., Kawamori, R., Kikuchi, M., Hakui, N., and Abe, H. 1978. Increased intestinal absorption of insulin: An insulin suppository. *J. Pharm. Pharmacol.* 30:806.

68. Kidron, M., Eldor, A., Lichtenberg, D., Touitou, E., Ziv, E., and Bar-On, H. 1979. Enteral administration of heparin. *Tromb. Res.* 16:833–835.

69. Kidron, M., Bar-On, H., Berry, E.M., and Ziv, E. 1982. The absorption of insulin from various regions of the rat intestine. *Life Sci.* 31:2837.

70. Ziv, E., Eldor, A., Kleinman, Y., Bar-On, H., and Kidron, M. 1983. Bile salts facilitate the absorption of heparin from the intestine. *Biochem. Pharmacol.* 32:773–776.

71. Ziv, E., Kidron, M., Berry, E.M., and Bar-On, H. 1981. Bile salts promote the absorption of insulin from the rat colon. *Life Sci.* 29:803–809.

72. Ziv, E., Kleinman, Y., Bar-On, H., and Kidron, M. 1984. In *Lessons from animal diabetes.* E. Shafrier and A.E. Renold, eds. London: Libbey.

73. Muranishi, S. 1985. Modification of intestinal absorption of drugs by lipoidal adjuvants. *Pharm. Res.* 2:108–118.

74. Kajii, H., Horie, T., Hayashi, M., and Awazu, S. 1985. Fluorescence study on the

interaction of salicylate with rat small intestinal epithelial cells: Possible mechanism for promoting effects of salicylate on drug absorption *in vivo*. *Life Sci.* 37:523–530.

75. Hirai, S., Yasiki, T., Matsuzawa, T., and Mima, H. 1981. Absorption of drugs from the nasal mucosa of rat. *Int. J. Pharm.* 7:317–325.

76. Schanker, L.S., and Johnson, J.M. 1961. Increased Intestinal absorption of foreign organic compounds in the presence of EDTA. *Biochem. Pharmacol.* 8:421–422.

77. Wood, A.J., Maurer, G., Niederberger, W., and Beveridge, T. 1983. Cyclosporine: Pharmacokinetics, metabolism and drug interaction. *Transplant. Proc.* 15:2409–2412.

78. Chemma, M., Palin, K.J., and Davis, S.S. 1987. Lipid vehicles for intestinal lymphatic drug absorption. *J. Pharm. Pharmacol.* 39:55–56.

79. Reymond, J., Sucker, H., and Vonderscher, J. 1988. *In vivo* model for cyclosporin intestinal absorption in lipid vehicles. *Pharm. Res.* 5:677–679.

80. Teng, C.D. 1989. Studies on some physical and biological properties of compacted protein matrices. Ph.D. thesis, University of Illinois at Chicago.

81. Killander, D., Dohlwitz, A., Engstedt, L., Franzen, S., Gahrton, G., Gulbring, B., Holm, G., Holmberg, A., Hoglund, S., Killander, A., Lockner, D., Mellstedt, H., Moe, P.J., Palmblad, J., Reizenstein, P., Skarberg, K.O., Swedberg, B., Uden, A.M., Wadman, B., Wide, L., and Ahstrom, L. 1976. Hypersensitive reactions and antibody formation during L-asparaginase treatment of children and adults with acute leukemia. *Cancer* 37:220–228.

82. Langer, R. 1989. Biomaterials in controlled drug delivery: New perspectives from biotechnological advances. *Pharm. Tech.* 13(8):18–30.

83. Davis, S.S., Illum, L., McVie, J.G., and Tomlinson, E. 1984. *Microspheres and drug therapy*. Amsterdam: Elsevier.

84. Tomlinson, E., and McVie, J.G. 1983. New directions in cancer chemotherapy 2. Targeting with microspheres. *Pharm. Ind.* 281–284.

85. Illum, L. and Davis, S.S. 1982. The targeting of drugs parenterally by use of microspheres. *J. Parent. Sci. Technol.* 36:242–248.

86. Ballanti, J.A. 1985. *Immunology III*. Philadelphia: Saunders.

87. Illum, L., Jones, P.D.E., and Davis, S.S. 1984. Drug targeting using monoclonal antibody-coated nanoparticles. In *Microspheres and drug therapy: Pharmaceutical, immunological and medical aspects*. S.S. Davis, L. Illum, J.G. McVie, and E. Tomlinson, eds. Amsterdam: Elsevier. Pp. 353–363.

88. DeDuve, C., DeBarsay, T., Poole, B., Trouet, A., Tulkens, P., and van Hoof, F. 1974. Commentary, Lysosomotropic agents. *Biochem. Pharmacol.* 23:2495–2531.

89. Fidler, I.J. 1980. Therapy of spontaneous metastases by intravenous injection of liposomes containing lymphokines. *Science* 208:1469–1471.

90. Baldwin, R.W., and Garnet, M.C. 1955. *Monoclonal antibodies for cancer detection and therapy*. London: Academic Press.

91. Trouet, A., Deprez-De Compeneere, D., and DeDuve, C. 1972. Chemotherapy

through lysosomes with a DNA-Daunorubicin complex. *Nature* (*New Biol.*) 239:110–112.

92. Ghitescu, L., Fixman, A., Simionescu, M., and Simionescu, N. 1986. Specific binding sites for albumin restricted to plasmalemmal vesicles of continuous capillary endothelium: Receptor-medicated trranscytosis. *J. Biol. Chem.* 102:1304–1311.

93. Kopecek, J., and Duncan, R. 1987. Poly[N-(2-hydroxypropyl)-methacrylamide] macromolecules as drug carrier systems. In *Polymers in controlled drug delivery*. L. Illum and S.S. Davis, eds. Bristol, U.K.: Wright. Pp. 152–187.

94. Szoka, F.J., and Paphadjopoulas, D. 1980. Comparative properties and methods of preparation of lipid vesicles (liposomes). *Annu. Rev. Biophys. Bioeng.* 9:467–508.

95. Godfredsen, C.F., Van Berkel, Th.J.C., Kruijt, J.K., and Goethais, A. 1983. Cellular localization of stable solid liposomes in the liver of rats. *Biochem. Pharmacol.* 32:3389–3396.

96. Ostro, M.J. 1983. *Liposomes*. New York: Dekker.

97. Burgess, D.J., and Carless, J.E. 1985. Manufacture of gelatin/gelatin coacervate microcapsules. *Int. J. Pharm.* 27:61–70.

98. Burgess, D.J., Davis, S.S., and Tomlinson, E. 1987. Potential use of albumin microspheres as a drug delivery system. 1. Preparation and *in vitro* release of steroids. *Int. J. Pharm.* 39:129–136.

99. Widder, K., Flouret, G., and Senyei, A. 1979. Magnetic microspheres: Synthesis of a novel parenteral drug carrier. *J. Pharm. Sci.* 68:79–82.

100. Oppenheim, R.C. 1981. Solid colloidal drug delivery systems: Nanoparticles, *Int. J. Pharm.* 8:217–234.

101. Birrenback, G., and Speiser, F. 1976. Polymerized micelles and their use as adjuvants. *J. Pharm. Sci.* 65:1763–1766.

102. Couvreur, P., Kante, B., Roland, M., Guiot, P., Bauduin, P., and Speiser, P. 1979. Polycyanoacrylate nanocapsules as potential lysosomotropic carriers: Preparation, morphological and sorptive properties. *J. Pharm. Pharmacol.* 31:331–332.

103. Schoeft, G.I., and French J.E. 1968. Vacular permeability to particulate fat: Morphological observations on vessels of lactating mammary gland and of lung. *Proc. Roy. Soc. London* [*Biol.*] 169:153–165.

104. Wilson, G. 1986. Genes therapy: Rationale and realization. In *Site-specific drug delivery*. E. Tomlinson and S.S. Davis, eds. Chichester, U.K.: Wiley. Pp. 149–164.

105. Karlson, S., Humphries, R.K., Gluzman, Y., and Nienhuis, A.W. 1985. Transfer of genes into hematopoietic cells using recombinant DNA viruses. *Proc. Natl. Acad. Sci. U.S.A.* 82:158–162.

106. Gallo, J.M., Hung, C.T., and Perrier, D.G. 1983. Analysis in albumin microsphere preparation. *Int. J. Pharm.* 22:63–74.

107. Waser, P.G., Muller, V., Krueuter, J., Berger, S., Munz, K., Kaiser, E., and Pfluger, B. 1987. Localization of colloidal particles (liposomes, hexyclyanoacrylate

nanoparticles and albumin nanoparticles) by histology and autoradiography in mice. *Int. J. Pharm.* 39:213–227.

108. Royer, G.P., Lee, T.K., and Sokoloski, T.D. 1983. Entrapment of bioactive compounds within native albumin beads. *J. Parenteral Sci. Technol.* 37(2):34–37.

109. Ihler, G.M. 1986. *Methods of drug delivery.* New York: Pergamon.

110. Kwok, K.K., Groves, M.J., and Burgess, D.J. 1989. Sterile microencapsulation of BCG in alginate-poly(1-lysine) by an air spraying technique. *Proc. Int. Sym. Control. Release Bioact. Mater.* 16:242–342.

111. Schakenraad, J., Oosterbaan, J., Nieuwenhuis, P., Molenaar, I., Olojslager, J., Potman, W., Eenink, M., and Feijen, J. 1988. Biodegradable hollow fibers for the controlled release of drugs. *Biomaterials* 9:116–120.

112. Hsieh, D.S.T., Langer, R., and Folkman, J. 1981. Magnetic modulation of release of macromolecules from polymers. *Proc. Natl. Acad. Sci. U.S.A.* 78:1863–1867.

113. Guy, R.H., and Hadgraft, J. 1988. Physicochemical aspects of percutaneous penetration and its enhancement. *Pharm. Res.* 5(12):753–758.

114. Chien, Y.W., Siddiqui, O., Sun, Y., Shi, W.M., and Liu, J.C. 1987. Transdermal inotphoretic delivery of therapeutic peptides/proteins. 1. Insulin. *Ann. N.Y. Acad. Sci.* 507:32–51.

115. Burnette, R.R., and Marrero, D. 1986. Comparison between the iontrophoretic transparent of thyrotropin releasing hormone across excised nude mouse skin. *J. Pharm. Sci.* 75:738–743.

116. Nanavaty, M., Brucks, R., Grimes, H., and Siegel, F.P. 1989. An ATR-FTIR approach to study the effect of ultrasound on human skin. *Proc. Int. Sym. Control. Release Bioact. Mater.* 16:310–311.

117. Golden, G.M., Guzek, D.B., Kennedy, A.H., and Potts, R.P. 1987. Stratum corneum lipid phase transitions and water barrier properties. *Biochemistry* 26:2382–2388.

118. Su, K.S.E. 1986. Intranasal delivery of peptides and proteins. *Pharm. Int.* 7:8–11.

119. Su, K.S.E., Campanale, K.M., Medelsohn, L.G., Kerschner, G.A., and Gries, C.L. 1985. Nasal delivery of polypeptides. 1. Nasal absorption of enkephalins in rats. *J. Pharm. Sci.* 74:394–398.

120. Hanson, M., Gazdick, G., Cahill, J., and Augustine, M. 1986. Intranasal delivery of the peptide salmon calcitonin. In *Delivery systems for peptide drugs.* S.S. Davis, L. Illum, and E. Tomlinson, eds. New York: Plennum. Pp. 233–242.

121. Sandow, J., and Petri, W. 1985. Intranasal administration of peptides. Biological activity and therapeutic efficacy. In *Transnasal systemic medications.* Y.W. Chien, Ed. Amsterdam: Elsevier. Pp. 183–199.

122. Gordon, G.S., Moses, A.C., Silver, R.D., Flier, J.S., and Carey, M.C. 1985. Nasal absorption of insulin: Enhancement by hydrophobic slats. *Proc. Natl. Acad. Sci. U.S.A.* 82:7419–7423.

123. Davies, H.W., Scott, G.M., Robinson, J.A., Higgins, P.G., Wootton, R., and

Tyrrell, D.A.J. 1983. Comparative intranasal pharmacokinetics of interferon using 2 spray systems. *J. Interferon Res.* 3:443–449.

124. Stratford, R.E., Jr., and Lee, V.H. L. 1986. Aminopeptidase activity in homogenates of various absorptive mucosae in the albino rabbit: Implications of peptide delivery. *Int. J. Pharm.* 30:73–82.

125. Nishihata, T., Rytting, J.H., Higuchi, T., and Caldwell, L. 1981. Enhanced rectal absorption of insulin and heparin in rats in the presence of nonsurfactant adjuvants. *J. Pharm. Sci.* 33:334–335.

126. Wespi, H.J., and Rehsteiner, H.P. 1966. Erfahrungen mit Syntocinon und ODA-Buccaltabletten. *Gynaecologia* 162:414–418.

127. Chang, H.S., Park, H., Kelly, P., and Robinson, J.R. 1985. Bioadhesive polymers as platforms for oral controlled drug delivery. 2. Synthesis and evaluation of some swelling water-insoluble bioadhesive polymers. *J. Pharm. Sci.* 74:399–405.

128. Park, H., and Robinson, J.R. 1986. Physico-chemical properties of water insoluble polymers important to mucin/epithelial adhesion. In *Advances in drug delivery systems*. J.M. Anderson and S.W. Kim, eds. Amsterdam: Elsevier. Pp. 47–60.

129. Peppas, N.A., and Buri, P.A. 1986. Surface interfacial and molecular aspects of polymer bioadhesion on soft tissues. In *Advances in drug delivery systems*. J.M. Anderson and S.W. Kim, eds. Amsterdam: Elsevier. Pp. 257–265.

SECTION II

Applications of Biotechnology in the Pharmaceutical Sciences

7

Applications of Recombinant DNA Technology to the Diagnosis of Genetic Disease: Molecular Methods for Detecting the Genetic Basis of Diseases

Donna R. Maglott and William C. Nierman

7.1. Introduction

Disease in humans can be caused by mutations that alter the DNA sequence of the coding or control regions of a gene. These mutations may result in the inappropriate expression or the nonexpression of a functional gene product, or in the production of a nonfunctional gene product. The mutations can be of various categories, ranging from point mutations or small insertions or deletions that are not cytologically detectable, to grosser abnormalities, including rearrangements and aneuploidy. The mutations can be in the germ line and heritable, and thus be studied by analysis of family members, or can arise somatically and be studied by comparing affected cells with nonmutated ones. Diseases have been identified that exemplify each of these genetic alterations. In addition, normally occurring somatic mutations, such as the rearrangements characteristic of immunoglobulin or T cell receptor genes, can be indicative of the differentiation and clonal status of various hematological malignancies. The diagnosis of these diseases, or the determination of carrier status for recessive heritable disorders, can thus be approached by analyzing DNA.

Diagnosis at the DNA level requires methodology that permits specific analysis of the potential mutant sites or regions against the entire background of genomic sequences. There must be sufficient DNA for analysis, and there must be tools to characterize the sequence of interest. DNA can be extracted from tissue for analysis directly, or specific segments can be amplified from smaller tissue samples by the polymerase chain reaction,[1] as will be discussed in more detail later. Methods of analysis include digestion of DNA with restriction endonucleases, separation of DNA fragments by size using agarose or polyacrylamide gel electrophoresis, and detection of specific fragments containing complementary sequences by hybridization with a nucleic acid probe for that sequence.[2]

These procedures can be used to detect a disease-causing sequence directly or indirectly. The word "direct" is used when the exact sequence alteration causing a disease is known, and the presence or absence of that sequence is being determined. The term "indirect" is used when the precise mutation is not known but has at least been mapped to a single chromosomal location. For hereditary diseases, the inheritance of that chromosomal region is determined in family studies by analysis of the linkage of the disease to DNA markers for that chromosomal region. For diseases that may include somatic changes, such as some tumors, the genetic structure of the tumor tissue is compared with that of normal somatic cells. Compilations of diseases with known genetic components and therefore amenable to molecular diagnosis are maintained by the Human Gene Mapping Workshops.[3, 4] Some diseases that have been diagnosed using DNA analysis are listed in Table 7.1.

The following sections summarize the basic aspects of diagnosing diseases with genetic components by recombinant DNA techniques. After a brief review of methods, applications to particular types of disease-associated mutations are discussed. Discussion has been limited to diseases with single genetic components, such as sickle cell disease or cystic fibrosis, and to single genetic components of multifactorial diseases, such as some cancers. Many complexities that must be considered to achieve a diagnosis and to determine probabilities of error have been omitted in this general overview.

7.2. Methodologies

7.2.1. Hybridization of Nucleic Acids

The ability of single-stranded nucleic acids to form stable duplexes with complementary sequences is the basis for many of the diagnostic procedures discussed in this chapter. The specificity of this recognition permits analysis of that small fraction of the genome that may contain a disease-causing mutation, within the complexity of the complete genome for which the haploid size is approximately 3×10^9 base pairs of DNA.

To analyze a particular sequence within the genome, it must be identified. Hybridization, or the formation of a specific duplex from two complementary single strands, is the process that permits this identification. Nucleic acid hybridization has two important features: (1) the two sequences involved in duplex formation must be complementary, and (2) the stability of the resultant double-stranded structure is determined by the extent of that complementarity.

Most diagnostic procedures using molecular biological techniques involve the use of nucleic acid probes that detect single copy or unique DNA sequences that occur once per haploid genome. (Repeated DNA sequences, in contrast, occur more than once and are frequently dispersed among the chromosomes.) The probe is a short DNA or RNA sequence that is labeled with a radioisotope, a fluorescent tag, or some other reporter group. The labeled probe binds its complementary sequence in the DNA being analyzed (the target) by the formation of sequence-

Table 7.1.   Some Disorders for Which Diagnosis Has Been Aided by DNA Analysis

Disease	Reference
α_1-Antitrypsin deficiency	Kidd et al., 1983[66]
Cystic fibrosis	Kerem et al., 1989[27]
Duchenne muscular dystrophy	Darras and Francke, 1988[34]
Fragile-X syndrome	Suthers et al., 1989[67]
Hemoglobinopathies	
Sickle cell anemia	Geever et al., 1981[68]
Thalassemias	Ottolenghi et al., 1974[69]
Hemophilia A	Antonarakis et al., 1985[70]
Hemophilia B	Giannelli et al., 1984[71]
Huntington's disease	Gusella et al., 1983[53]
Lung cancers	Kok et al., 1987[72]
Miller–Dieker syndrome	van Tuinen et al., 1988[73]
Ornithine transcarbamylase deficiency	Rosen et al., 1985[74]
Phenylketonuria	Lidsky et al., 1985[75]
Prader–Willi syndrome	Donlon et al., 1986[76]
Wilson disease	Bowcock et al., 1987[77]
Tay–Sachs disease	Korneluk et al., 1986[78]
Tuberous sclerosis	Smith et al., 1990[79]
X-chromosome-linked ichthyosis	Ballabio et al., 1987[80]

specific, base-paired duplexes between the probe and the corresponding genomic sequence. The existence of that duplex is then determined by assaying for the reporter group. Some practical considerations on the rate of association and duplex stability are summarized below. (For more complete discussions, see references 5 and 6.)

7.2.1.1. The Rate of Duplex Formation and Duplex Stability

A useful measure of the stability of a duplex is the melting temperature, or T_m, at which nucleic acid strands are half denatured or disassociated. Important variables that alter the stability and/or the rate of duplex formation are base pair mismatches, fragment length, salt concentration, base composition, and formamide concentration. The temperature of the reaction also affects the rate of association. Each of these is discussed briefly.

Temperature. For a typical DNA reassociation reaction, the rate shows a broad maximum at about 25°C below the T_m. Optimal rates of reassociation are achieved at 25°C below the T_m.

Base Pair Mismatches. Base pair mismatches slow the rate of reassociation. Thus, for each 10% of mismatch, the renaturation rate is reduced by a factor of about 2 at the optimal temperature for the reassociation (i.e., about 25°C below the T_m).

During reassociation, sequences that are not perfectly complementary can form a duplex structure. Mismatched sequences are less stable than exactly matched ones. The effect of mismatching on T_m is approximately a 1° to 1.5°C decrease for each 1% mismatch.

Many diagnostic applications require the detection of single nucleotide differences. If an oligonucleotide of about 20 nucleotides in length is used as a probe, the decrease in stability of a hybrid mismatched at one site would be 5° to 7.5°C (1/20 nucleotides or 5% mismatch $\times$ [1°–1.5°C]). Careful selection of hybridization and washing temperatures thus permits discrimination between perfectly matched and internally mismatched hybrids. If two oligonucleotides are synthesized differing at only a single position, the presence or absence of hybridization using each of these probes can be used to identify which of the two sequences is present in the target DNA. These specific oligonucleotides are called allele-specific oligonucleotides (ASOs). (For an example of this application, see Section 7.3.1.1.)

Fragment Length. The rate of reassociation is greater in long fragments because of the greater possibility for nucleation, or formation of an initial short duplex structure, in the fragment. Once nucleation has occurred, the formation of a complete duplex is very rapid compared with the rate of nucleation.

The length of a base-paired duplex will affect the T_m. Longer duplexes are more stable in the size range below about 500 base pairs.

Salt Concentration. The concentration of salt affects the rate of reassociation. Below 0.1 M NaCl, a 2-fold increase in concentration increases the rate by 5- to 10-fold.

The concentration of salt also has a considerable effect on T_m, particularly at concentrations below 0.1 M. A higher salt concentration stabilizes duplex DNA by electronically shielding the negatively charged phosphodiester backbone.

Base Composition. In normal salt solutions G-C base pairs are more stable than A-T base pairs. Thus, the T_m for DNA with a high GC content will be higher.

Formamide. Formamide decreases the T_m of nucleic acid duplexes. By using 30 to 50% formamide in the hybridization solution, the hybridization temperature can be reduced to 30° to 42°C.

7.2.1.2. Applications Using Hybridization

For most of the examples discussed in this chapter, hybridizations are carried out with target DNA denatured and bound to a membrane. If the DNA is bound after having been separated electrophoretically, the process is known as a Southern blot[5] (Figure 7.1). Denatured DNA can also be applied to membranes as circular dots (dot blot) or as elongated dots (slot blot).

Nucleic acid probes can also be hybridized to target sequences in cellular material containing morphologically intact chromosomes. This is known as in situ hybridization, and can be used to map probes to particular chromosomes or to detect amplification of sequences in chromosomal regions.

7.2.2. Detecting Sequence Variation

7.2.2.1. Restriction Fragment Length Polymorphisms (RFLPs)

Restriction endonucleases are enzymes that cut both strands of DNA at a specific nucleotide sequence. The sequence recognized by the restriction enzyme is typically from four to eight base pairs in size. A base pair change anywhere in the recognition sequence will prevent enzyme cuts at that site. When total genomic DNA is cut with such a restriction enzyme, the fragments that are generated can be separated by size electrophoretically in an agarose or polyacrylamide gel. For complex genomes, there are too many individual DNA fragments to be resolved, but a fragment containing a particular sequence can be identified by Southern blotting using a nucleic acid probe for that sequence[2] (Figure 7.1).

There is a significant amount of sequence variation in humans that does not alter genetic expression. Estimates vary, but these DNA polymorphisms may occur on the average of about one in every hundred nucleotides.[7] In 1980 Botstein et al.[7] showed how these heritable polymorphisms could be exploited to generate DNA markers for use in linkage studies to map the human genome. If, for example, samples of genomic DNA from different individuals are digested with a particular restriction enzyme, separated electrophoretically, transferred to a membrane, and hybridized to a nucleic acid probe (for which the sequence is unique in the genome), the sizes of the fragments that the probe detects may vary from individual to individual (Figure 7.2). These sizes of the restriction fragments that are detected depend on the probe-enzyme combination and may result because of a polymorphism affecting the recognition sequence for that particular enzyme (Figure 7.2A), a small insertion or deletion (Figure 7.2B), a variable number of tandem repeat elements[8] (VNTRs) (Figure 7.2C), or a chromosomal re-arrangement (Figure 7.2D). Because such polymorphisms are detected by the different lengths of DNA fragments generated by enzyme cuts at or adjacent to the polymorphic site, they have been termed restriction fragment length polymorphisms (RFLPs). In some cases these RFLPs are disease-related, as in sickle cell anemia, where the coding change from GAG to GTG in codon 6 of β-globin changes the recognition sequence for the restriction enzyme *Cvn*I. (This will be discussed in more detail in Section 7.3.1.1.)

SOUTHERN BLOTTING

(a method of detecting DNA fragments that
share homology with a DNA probe)

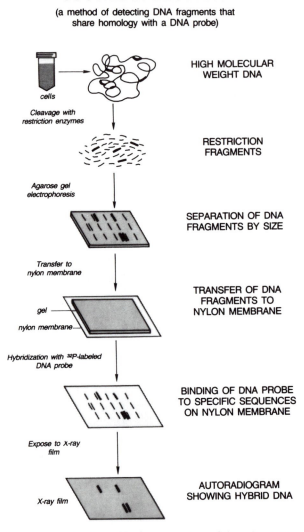

Figure 7.1. Southern Blotting: a method of detecting DNA fragments that share homology with a DNA probe.

7.2.2.2. Other Sequence Variants

Many inherited sequence variations are not detectable by restriction fragment analysis. There are, for example, polymorphisms that are generated by varying numbers of tandem repeats of $(dC\text{-}dA)_n \times (cG\text{-}dT)_n$ or $(CA)_n$ dinucleotides[9] or by variation in lengths of poly(A) tracts at the ends of *AIu*I repetitive elements.[10]

Because the length variation is at the nucleotide level, it must be detected electrophoretically on gels capable of resolving single nucleotide differences. As described below, if PCR is used to amplify the particular sequence containing the length variation, the length of the repeat in the sample can be determined by electrophoresis. Since that length is a heritable DNA marker, it can be used in linkage studies.

7.2.3. Determining the Chromosome Location of DNA Markers

7.2.3.1. Hybridization

Panels of Hybrid Cell Lines. For DNA polymorphisms to be useful in genetic studies, they must be mapped to particular chromosomes. One method of assigning polymorphisms to chromosomes is to use hybridization to identify the chromosomal origin of the probe detecting that polymorphism. Sets of characterized interspecific hybrid cell lines have been developed for this purpose. When human cells are fused with rodent cells, the complete complement of human chromosomes is not retained. By characterizing the human chromosome or chromosomes that are maintained, it is possible to generate a panel or group of hybrid cell lines in which each line contains a defined subset of the human genome and for which the set collectively retains a complete genome. The probe can be hybridized to DNA from each of these hybrid cell lines, and the pattern of hybridization across the panel can be used to determine chromosome assignment. Assignments to particular chromosome regions can also be made if a cell line in the panel retains only a fragment of a particular chromosome. More recently, hybrid lines retaining only one human chromosome, or a fragment of only one human chromosome, have been developed and exploited for mapping purposes.[11]

Hybridization in Situ. Probe DNA can be hybridized to metaphase or prometaphase chromosomes, and visualized by autoradiography or fluorescence. Combining the use of prometaphase chromosomes, fluorescently labeled probes, and chromosomal suppression hybridization[12] has also permitted the relative positioning of individual probes along chromosomes.[13]

7.2.3.2. Linkage Analysis

Another method for mapping polymorphisms is that of linkage analysis. If DNA samples from individuals of an extended family are analyzed, the inheritance of a set of polymorphic alleles within that pedigree can be compared with the inheritance of markers that already have been mapped to particular chromosomes. For example, probe X is unmapped but detects an RFLP. The gene for enzyme E and probe P have already been mapped to chromosome 1. If the inheritance of one isozyme allele of E, one fragment length allele revealed using P, and one fragment length allele revealed using X are always coinherited in the family, the

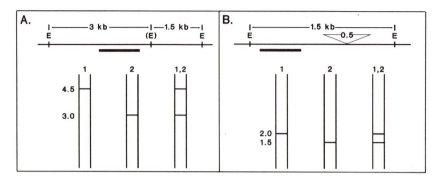

Figure 7.2AB. Types of RFLPs with patterns produced on Southern blots. The thin horizontal straight line in each panel represents genomic DNA that can be cut by a restriction enzyme at the sites indicated by *E* separated by the distances given in kilobase pairs (kb). The darker bar underneath represents the sequence of the genomic DNA above which is included in a probe. The set of vertical tracks in each panel represents different hybridization patterns resulting when the genomic DNA is cut with enzyme *E* and the resultant fragments are separated by electrophoresis, transferred to a membrane, and detected by hybridization with the labeled probe (see Figure 7.1). (**A.**) Site polymorphism. The sequence variation is at site (*E*). Thus, if DNA from an individual whose DNA on both homologous chromosomes is not cut at (*E*) but whose DNA is cut at *E* is digested and analyzed as described, the fragment detected by the probe will be 4.5 kb (allele *1*). If the site (*E*) is present on both chromosomes, the smaller 3.0-kb fragment will be detected (allele *2*). In an individual with one chromosome containing the site and the other not, both fragments will be detected (*1, 2*). (**B.**) Insertion/deletion polymorphism. In this example, a 0.5-kb sequence indicated by the expansion bar is present in some chromosomes but absent from others. When genomic DNA is cut at the sites indicated and analyzed by Southern blotting, a 2.0-kb fragment is detected if the sequence is present (allele *1*) and a 1.5-kb fragment is detected if the sequence is absent (allele *2*). Both fragments are detected in the heterozygote (*1, 2*).

most likely explanation is that the polymorphism detected by *X* maps to chromosome 1. Such analysis requires DNA samples from large, well-characterized families. Fortunately, centralized resources to facilitate such analyses have been developed. The Centre d'Étude du Polymorphisme Humain (Paris, France) and the Mutant Cell Repository (Camden, N.J.) maintain cell lines for large multigenerational pedigrees. Also, the determination of linkage relationships has been facilitated by the development of computer software such as LIPED or LINKAGE[14, 15]).

7.2.3.3. Identifying Useful Probes

The set of mapped RFLPs, and the probes that can be used to detect them, has been increasing rapidly over the past few years. According to the most recent

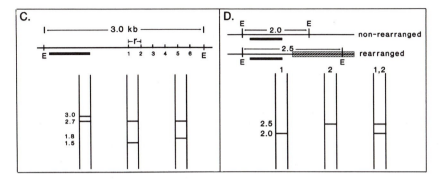

Figure 7.2CD. (**C.**) VNTR polymorphism. In this example there is a tandem repeat unit *r* of 0.3 kb, which may exist up to five times between enzymes sites *E*. If all five repeat units are present, the size fragment detected by the probe is 3.0 kb. If one or more are missing, however, the fragment sizes are correspondingly reduced (1 missing, 2.7 kb; 2 missing, 2.4 kb; 3 missing, 2.1 kb, etc.). Because there is significant variation among individuals for this type of polymorphism, the probability that both homologous chromosomes in an individual have the same number of tandem repeats at a site is rather small. Thus, the diagrams indicate different possible heterozygous conditions. (**D.**) Rearrangement. The expanded bar indicates the DNA originating from a previous distal site but now juxtaposed to the sequence detected by the probe. In this example, the rearrangement can be detected because the size of the restriction fragment generated from the non-rearranged chromosome is different from the rearranged one.

report of the DNA committee of the Human Gene Mapping Workshops (HGM), the number of DNA probes that detect polymorphism has increased from 130 in 1983 (HGM7) to 1886 in 1989 (HGM10).[16] Information about the availability and uses of these probes is maintained in several public data bases, including the Genome Data Base at Johns Hopkins University (Baltimore, Md.) and the NIH/ATCC Repository of Human and Mouse DNA Probes and Libraries at the American Type Culture Collection (Rockville, Md.). Reports of the HGM Workshops are published in *Cytogenetics and Cell Genetics* (Karger).

7.2.4. The Polymerase Chain Reaction (PCR) and Sequence Variation

The ability to amplify specific segments of DNA by the polymerase chain reaction (PCR) has changed the way biologists approach basic and applied problems in molecular biology. This is dramatically true in diagnostic applications. Because PCR produces large amounts of specific DNA fragments from small amounts of a complex template, it can be used to detect point mutations, deletions, and sequence polymorphisms for linkage analysis. These applications are discussed in the next section.

The polymerase chain reaction is used to amplify DNA that lies between two regions of known sequence, as diagrammed in Figure 7.3. DNA polymerases

Polymerase
Chain Reaction

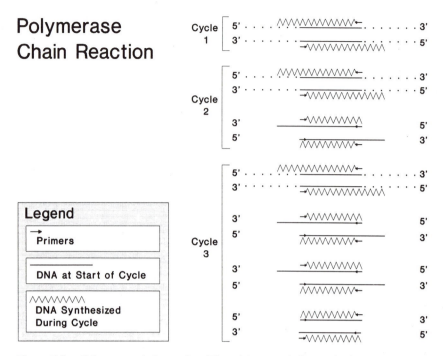

Figure 7.3. Polymerase chain reaction. The polymerase chain reaction is used to amplify DNA that lies between two regions of known sequence. Two oligonucleotides to prime DNA synthesis are selected to be complementary to sequences on opposite strands of the template and localized at boundaries of the region to be amplified. The 3′ ends of the primers must also be oriented toward the center. By the use of temperature shifts, cycles involving template denaturation, primer annealing to template, and synthesis by DNA polymerase extending from the annealed primers result in the accumulation of a double-stranded DNA fragment. The termini of the fragment are defined by the 5′ ends of the primers.

cannot synthesize DNA starting anywhere on a template; they must extend the free 3′ end at a duplex region. Thus, two oligonucleotides (amplimers) are used as primers for a DNA synthesis reaction performed in vitro. The sequences are chosen so that the oligonucleotides are complementary to sequences on opposite strands of the DNA and oriented so that when the amplimers bind to the template, their 3′ ends point toward the segment of DNA to be amplified. The primer pair is selected to be unique to these sites in the genome. By the use of temperature shifts, cycles involving (1) template denaturation, (2) primer annealing to the template, and (3) synthesis by DNA polymerase extending from the annealed primers result in the accumulation of a specific double-stranded DNA fragment. The termini of the fragment are defined by the 5′ ends of the primers. Because the products of one round of amplification serve as templates for the next, each successive cycle doubles the amount of the amplified fragment.

Early protocols for PCR used the Klenow fragment of *Escherichia coli* DNA polymerase I to catalyze DNA synthesis.[1, 17] The reaction did not find widespread application until the thermostable DNA polymerase from *Thermus aquaticus* was used. This enzyme, which can survive extended incubation at 95°C, is not inactivated by the heat denaturation step and does not need to be replaced at every round of the amplification cycle.

Although a relatively new technique, PCR has found extensive application in the diagnosis of genetic disorders (reviewed in reference 18). In general, PCR is used to amplify a specific sequence from patients' genomic DNA that may contain a known mutation. This amplified sequence is then analyzed for its size by electrophoresis or its sequence content by hybridization or direct sequencing. Because only a small amount of template DNA is required for the amplification, an adequate sample can be obtained by amniocentesis, chorionic villus sampling, or a simple mouth rinse.[19] Another major advantage of PCR is the speed of the analysis. Amplification and electrophoretic determination of the size of the amplified product can be completed in half a day, as opposed to a week for electrophoresis, Southern transfer, and pattern detection by autoradiography. Point mutations, deletions, length variations, and rearrangements can be readily detected by analysis of amplified DNA as follows.

7.2.4.1. Detection of Single Nucleotide Changes

Analysis of single base pair changes by PCR requires one of two approaches, depending on the location of that change. Sequence variation within a site recognized by a restriction enzyme can be detected by amplifying a DNA fragment that contains the site, digesting with the appropriate enzyme, and separating the fragments electrophoretically. In the example diagrammed in Figure 7.4, primers are selected that amplify a fragment of 723 bp containing several *Cvn*I sites, including one that is known to be absent in individuals carrying the sickle mutation, β^S. After digestion with *Cvn*I, amplified DNA with the restriction site (β^A allele) gives two fragments of 180 and 201 bp, along with the invariant bands of 256 and 86 bp. Amplified DNA without the site (β^S allele) gives the two invariant bands plus a fragment of 381 bp instead of the two of 180 and 201 bp. Amplified genomic DNA from an individual heterozygous for this polymorphism ($\beta^A\beta^S$) will show all five bands.

If the base pair difference is not within a restriction site, the alleles can be distinguished by hybridization. For this analysis a dot blot hybridization using allele-specific oligonucleotides (ASOs) may be used (Figure 7.5). Again, PCR primers are selected to amplify DNA containing the polymorphic site. The amplified DNA is denatured and transferred to a hybridization membrane. Labeled

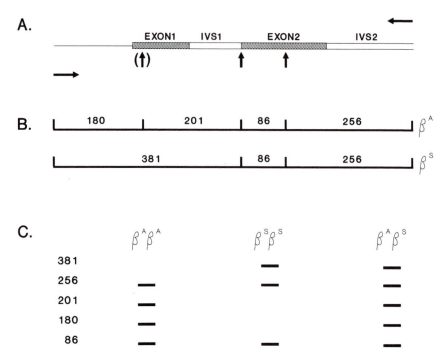

Figure 7.4. Detection of point mutations by PCR in sickle cell disease: restriction analysis. (**A.**) Exon organization of the β-globin locus around the sickle cell mutation. The vertical arrows indicate the positions of the *Cvn*I sites, including the polymorphic one in exon 1 (in parentheses) at the sickle mutation. The horizontal arrows show the positions of the oligonucleotide PCR primers. (**B.**) DNA distances in base pairs between *Cvn*I sites around the sickle mutation site. (**C.**) Representation of DNA fragments on gel after PCR amplification and digestion with *Cvn*I. $\beta^A\beta^A$, homozygous normal individual; $\beta^S\beta^S$, homozygous sickle individual; $\beta^A\beta^S$, heterozygous individual.

oligonucleotides with sequences identical to each of the alleles are hybridized separately to the DNA on the membrane. Conditions are selected so that hybridization occurs only when there is complete sequence identity (Section 7.2.1.1). As is shown in Figure 7.5, the heterozygotes show hybridization with both allele-specific probes (β-sickle, β-normal), the homozygotes only with one. Allele-specific oligonucleotides can also be used after electrophoretic separation of the amplified product and Southern transfer to ensure that the hybridization signal is resulting from a fragment of the expected size. (The use of ASOs is also discussed in Section 7.3.1.1.)

7.2.4.2. Detection of Length Variation

PCR is especially useful for detecting length variation generated by insertions, deletions, VNTRs, $(CA)_n$ dinucleotide repeats, or variable poly(A) tracts at the

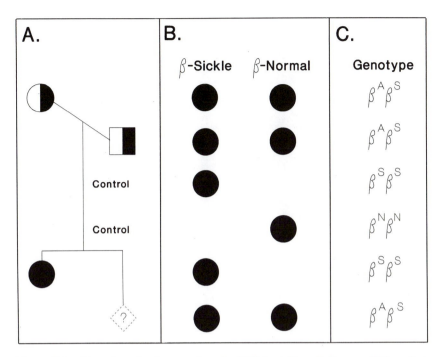

Figure 7.5. Detection of point mutations by PCR in sickle cell disease: ASO analysis and dot blots. Genomic β-globin DNA from the individuals in the pedigree (**A**) are amplified as in Figure 7.4, and the denatured, amplified DNA is dotted to a hybridization membrane and probed with ³²P-labeled ASO specific for the sickle DNA sequence (β-Sickle) or the normal sequence at the sickle site (β-Normal). (**B.**) Hybridization signal revealed by autoradiography. (**C.**) Tabulation of genotypes.

ends of *Alu*I repetitive elements. Primer sequences are selected from single copy DNA flanking the region containing the length polymorphism. By amplifying across the repeats and determining the sizes of the amplified fragments, allele assignments can be made (Figure 7.6). If the variation is at the nucleotide level, high-resolution gels must be used.

7.2.4.3. Detection of Deletions

Deletions can also be detected by PCR (Figure 7.7). Primers are selected that either flank or lie within the region of the potential deletion. The amplified product will be full length in the absence of a deletion. If primers flank a deletion, a smaller product will be amplified. In a heterozygote, a full length and a shorter amplified product will be observed. If the primer is chosen within the potential deletion region, no product will be generated from the deletion-containing chromosome.

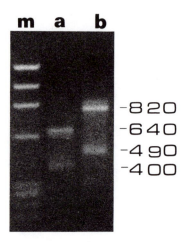

Figure 7.6. Detection of DNA length variation by PCR. Genomic DNA from two individuals was amplified by PCR using primers that specifically amplify the apo B VNTR region.[81, 82] The amplified DNA was analyzed by agarose gel electrophoresis. Lane m. DNA size markers. Lane a. Amplified DNA from individual showing alleles of 640 and 400 bp. Lane b. Amplified DNA from individual showing alleles of 820 and 490 bp.

7.2.4.4. Detection of Translocations

The synthesis of a PCR product depends on the presence of the primer sequences on the same molecule of template DNA. Thus, if two amplimers are generated for sequences that normally exist on different chromosomes, or on different mRNA molecules, the only time a product will be synthesized by PCR is when those sequences have become part of the same chromosome or DNA copy of a mRNA, as may result from translocation. Southern blotting can be used to detect translocations that occur at relatively invariant translocation breakpoints (Figure 7.2D), but when breakpoints occur over a 100 to 200 kb range of DNA, a different strategy must be used. If the translocation breakpoints occur in the introns of two different expressed sequences, for example, a complementary DNA copy of a fusion mRNA is a good template for amplification (Figure 7.8). Because processing of messenger removes intronic sequences, the fusion mRNA shows less length variation from one translocation patient to another than the DNA from

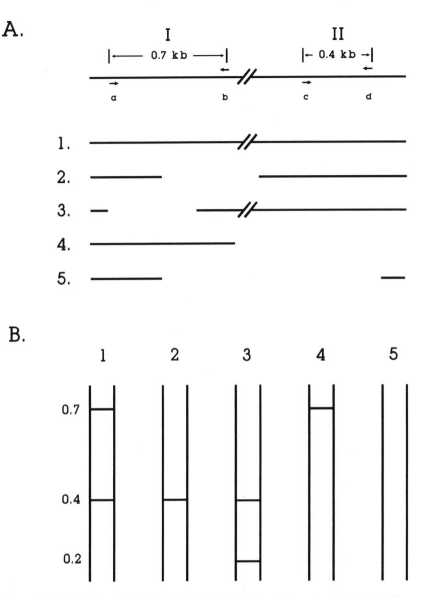

Figure 7.7. Detection of deletion by PCR. (**A.**) Representation of genomic DNA for a potential deletion region to be screened at two areas bounded by PCR primer pairs (small arrows) a and b (I) and c and d (II). 1 to 5. Representation of DNA from different deletion patients: 1, no deletion; 2, deletion removing genomic DNA containing sequence b; 3, 0.5-kb deletion between sequences a and b; 4, deletion removing sequences c and d; and 5, deletion removing sequences b, c, and d. (**B.**) Electrophoretic analyses of the amplification products from DNA in configurations 1 to 5. Deletion of sequences complementary to one or both members of a primer pair results in no amplification of that product. The 0.7-kb band is amplified from I whenever there is no deletion between or involving the region bounded by primers a and b, and the 0.4-kb band is amplified from region II whenever there is no deletion between or involving the region bounded by primers c and d. In 3, a shorter fragment is amplified because there is a small deletion between the sequences recognized by primers a and b.

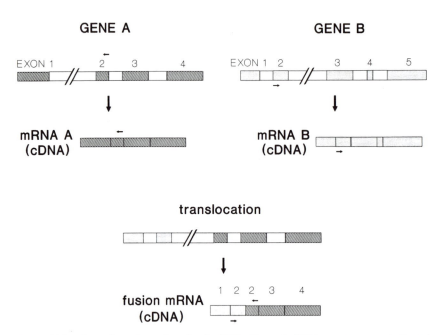

Figure 7.8. Detection of translocation by PCR. Genomic DNA structures are indicated above the mRNAs transcribed and processed from them. Exons (numbered) are indicated by dark shading. Slashes indicate potential translocation breakpoints. The small arrows above and below the diagrams indicate the hybridization positions of PCR primers. The sequence content of the mRNAs (and their cDNA copies) derived from their template DNAs are as indicated. Note that the only cDNA structure that can serve as a template for PCR amplification (because it contains sequences complementary to both PCR primers) is that resulting from the chimeric gene resulting after translocation.

which it was transcribed. Thus, one application of PCR to the diagnosis of translocation is first to make complementary DNA (cDNA) copies of mRNA and then to add PCR primers for amplification. The presence or absence of a PCR product indicates whether a translocation occurred within the boundaries of the PCR primers. The size and sequence of the amplified product can also be used to characterize the site of the translocation.[20, 21] If a particular type of translocation has more defined breakpoints, PCR can be used directly on genomic DNA.[22]

7.3. Disease-Related Diagnosis

Kan et al. demonstrated that disease-associated mutations could be characterized at the molecular level, and that once characterized, the disease could be diagnosed prenatally.[23] In 1982 a diagnostic test for sickle cell disease was developed based on recombinant DNA technology.[24, 25]

With the rapid advances in human gene mapping and the cloning of genes responsible for human genetic diseases have come diagnostic tests for an increasing number of diseases. As described below, those diseases for which the gene is mapped and cloned and the disease-causing mutation has been characterized are detected directly by determining the presence or absence of the mutant allele. Those diseases for which the gene has been only mapped are detected indirectly by identifying and tracing the passage of the disease-carrying chromosome region in a family.

7.3.1. Direct Detection

7.3.1.1. Detection of Base Pair Differences

When a particular nucleotide sequence variant is known to lead to disease, the presence or absence of that sequence can be detected directly. This includes changes at a single base pair or small number of adjacent base pairs, deletions of genetic regions (Sections 7.2.4.3. and 7.3.1.2.), localized amplifications, and translocations (Section 7.2.4.4). Direct detection is the simplest form of genetic analysis; it gives the least ambiguous results, and it may not require characterizing DNA of other family members.

Direct detection can be accomplished by any of several different protocols. If, for example, the disease is caused by a single base pair change at the recognition site for a particular restriction endonuclease, one approach is to use that enzyme in the diagnosis. In the classic case of sickle cell anemia, the aberrant hemoglobin results when there is an A to T substitution in the sixth codon of the normal β-globin gene (GAG to GTG).[23] This leads to the replacement of a glutamic acid for a valine in the protein and to the loss of a recognition site in the DNA for the restriction enzyme *Cvn*I.

PCR. The presence of the sickle cell mutation can be detected directly without a nucleic acid probe by using PCR to amplify in vitro that portion of the β-globin gene containing the sickle cell mutation site (Section 7.2.4.1 and Figure 7.4). This technique permits diagnosis from a very small sample of DNA. The amplified DNA is digested with *Cvn*I before being analyzed electrophoretically, and the presence of the sickle cell mutation is indicated by the undigested larger (381 bp) band. Two smaller 180 and 201 bp bands result from *Cvn*I digestion of DNA amplified from a normal β-globin gene (Figure 7.4). The carrier state (a heterozygote with one mutant β-globin gene on one chromosome and the normal gene on the other) is recognized by all three bands (381, 201, and 180 bp) on the gel, the largest from the amplified uncut DNA from the sickle cell gene and the two smaller ones from the amplified normal sequence containing the *Cvn*I site.

Southern Blotting. The same mutation can also be detected directly using Southern blotting. For this procedure total genomic DNA is digested with *Cvn*I and analyzed as described previously (Figures 7.1 and 7.2A), using a probe close to the *Cvn*I site. If the normal β-globin gene is present (i.e., the restriction site is present), the probe will hybridize to a smaller *Cvn*I fragment than when the sickle cell gene is present and the *Cvn*I site is absent. The carrier is revealed by the probe hybridizing to *Cvn*I fragments of both sizes in DNA from a single individual, indicating one allele with the restriction site and one lacking the site.

Oligonucleotide Probes. A third direct method to detect single base pair changes is to probe genomic DNA with oligonucleotides complementary to all possible sequences at a mutation site (ASOs). This is a more general technique than restriction fragment analysis because it does not require that the sequence change involve a restriction site. A dot blot is one application of ASOs. A small amount of total denatured genomic DNA is placed as a dot on a nitrocellulose or nylon membrane without digesting it with a restriction enzyme or analyzing it electrophoretically. Alternatively, the region of interest can be amplified by PCR and blotted to a membrane. The membrane is then hybridized to the single-stranded labeled oligonucleotide probe under conditions in which complete complementarity is required for the probe to form a double-stranded structure and thus bind to the genomic DNA. The presence or absence of the hybridized probe on the membrane is detected by the label on the probe and indicates the presence or absence of a particular sequence in the genomic DNA. By using oligonucleotide probes complementary to the normal and sickle cell sequences of the β-globin gene at the mutation site, the sickle cell genotype can be determined (Figure 7.5). Genomic DNA from the homozygous normal or homozygous sickle cell patient will hybridize with only the normal or sickle cell probe, respectively. DNA from heterozygous individuals will hybridize with both probes. Hybridization with ASOs can also be done after Southern transfer of genomic or amplified DNA to determine that the hybridization signal has been derived from the DNA fragment of the correct size.

Hybridization with allele-specific oligonucleotides can also be used diagnostically when the sequence alteration involves more than one base pair, as long as that alteration lies within the range encompassed by the oligonucleotide probe. An example of this is the use of ASOs to detect the 3-base-pair deletion in exon 10 of the cystic fibrosis (CF) transmembrane conductance regulator[26] that is present in the majority of cystic fibrosis families.[27]

Sickle cell and the predominant form of CF are examples of diseases that result from single characterized mutations. Numerous other disease phenotypes involving known genes, however, may result from one of many possible site mutations within the gene. The first occurrence of such a disease in a family may thus have been caused by one of several mutations and cannot be diagnosed by DNA methods until the specific mutation in that family has been identified. Once the mutation has been determined, the procedures described previously for detecting point mutations or in the following section for detecting deletions can be used to determine the presence of the mutation.

7.3.1.2. Detection of Deletions

Diseases that result from either characterized or new deletions can be detected by PCR or by Southern blotting using a probe that hybridizes to the potential deleted region. Duchenne muscular dystrophy (DMD), with its milder phenotype Becker muscular dystrophy (BMD), is a well-known example of a disease that is frequently caused by deletion in the gene for the cytoskeletal protein dystrophin. DMD is X-linked with an incidence of 1 in 3,300 male births.[28] One-third of all cases result from new mutation, with another third resulting from the inheritance of the DMD gene from a carrier mother who is herself a new mutant. The remainder result from the inheritance of a DMD gene present in earlier genera-tions. About 70% of mutations in the DMD gene are deletions.[29] Another much less common type of mutation is intragenic duplication.[30] (For a review on the molecular genetics of DMD see reference 31.)

Dystrophin has been mapped to the Xp21 band of the X-chromosome, and the cDNA for the entire 14 kb transcript has been cloned and sequenced.[32, 33] The availability of this cDNA has permitted direct diagnosis by Southern blots of DMD in many families in which a deletion is present. Using as probes a set of clones corresponding to the complete cDNA, genomic fragments covering approximately 400 kb can be examined for deletions.[34, 35] Because the dystrophin gene is so large, however, with current estimates of 2.3 Mb,[36] not all deletions can be detected with the cDNA probes.

The restriction fragment patterns revealed by cDNA probes on Southern blots are quite complex. Each probe may detect 10 or more genomic bands. When patterns of hybridization to Southern blots from normal and potential DMD individuals are compared, a deletion in the dystrophin gene may be detected by the absence of one or more bands present in the intact gene or by the presence of new bands corresponding to DNA from deletion boundaries. Recently standard patterns of restriction fragments that are detected when normal human DNA cleaved with either *Hind*II or *Bgl*II is hybridized with probes comprising the complete cDNA have been reported.[34, 36]

A deletion may be detected as illustrated by the autoradiogram in Figure 7.9. In this example the genomic DNA from a DMD patient (lane 1) and the normal control genomic DNA (lane 2) are digested with *Bgl*II, electrophoresed on an agarose gel, blotted to a hybridization membrane, and probed with ^{32}P-labeled probe containing the cDNA for two exons from the dystrophin gene. Note the absence of one band and the presence of a novel band in lane 1 (DMD) as compared with lane 2 (normal), indicative of a deletion.

Deletions causing DMD frequently span several hundred kilobases of the

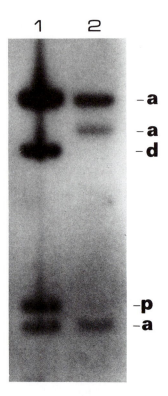

Figure 7.9. Detection of deletions and polymorphisms with cDNA probes. Genomic DNA from a DMD patient (lane 1) and a normal control (lane 2) was digested with *Bgl*II and analyzed by Southern blotting. Normal bands (a), a smaller band from the deletion site (d), and a polymorphic band (p) are readily detected. Note that the second normal band is missing from the DMD patient.

dystrophin gene and are not randomly distributed, typically overlapping two specific regions of the gene.[29] Because of these characteristics, a significant fraction of DMD deletions can be detected by using the polymerase chain reaction as an alternative to Southern analysis (Figure 7.7). In this method, sequences from the deletion-prone exons are simultaneously amplified via PCR. Any of these regions deleted in a patient's DNA will fail to amplify and are readily identified by the absence of an amplified fragment when the PCR reaction products are analyzed by agarose gel electrophoresis.[37] This multiplex PCR approach is able to detect approximately 70% of all DMD deletions. Furthermore, this technique is inexpensive, fast, requires only nanogram quantities of genomic DNA, and does not involve the use of radioisotopes.

7.3.1.3. Somatic Genetic Changes

Another application of molecular methods to aid diagnosis is the characterization of neoplasms for somatic mutations, or naturally occurring gene rearrangements. There are an increasing number of tumors in which amplification, deletion, site mutation, or rearrangement of known gene sequences has been demonstrated and may be of prognostic significance (reviewed in references 38 to 40).

Amplification. For example, amplification of ERBB2 (HER2, c-*erb*B-2) occurs in breast cancer, and this amplification shows an association with the course of the disease.[41, 42] In addition, amplification of N-*myc* sequences has been correlated to prognosis in neuroblastoma (reviewed in reference 43). Amplification of a genomic sequence can be detected by hybridization with a probe specific to that sequence in either dot blots, slot blots, or Southern blots. Increased expression, which can also be of diagnostic significance, can be determined by RNA blots. To detect genomic amplification, the hybridization signal obtained with DNA from tumor tissue is compared with that from normal tissue. Because this is a quantitative rather than a qualitative test, the normal tissue must be separated from the tumor sample before the DNA is extracted, and controls must be incorporated to evaluate the efficiency of DNA transfer to the membrane after electrophoresis if Southern blots are used. For example, Figure 7.10 schematizes an autoradiogram of a Southern blot analysis of equal amounts of genomic DNA from tumor and normal tissue of three individuals. After transfer, radioactively labeled probes for the potentially amplified sequence (which would hybridize to a 6 kb restriction fragment) and a control single-copy sequence (which would hybridize to a 3 kb restriction fragment) were mixed and hybridized to the samples on the membrane. The consistent level of signal across the control 3 kb band indicates that no sample showed significant degradation and that transfer was uniform. The signals for the 6 kb band are proportional to the genomic copy number of the sequence detected by the probe. Tumor tissue from case 2 showed no amplification, whereas the sequence from tumors 1 and 3 was amplified to different levels.

Amplification can also be detected cytogenetically as double minute chromosomes (DM), homogeneously staining regions (HSRs), and abnormal banding regions (ABRs). (For a review, see reference 44.) The sequence content of these regions can be determined by hybridization in situ with specific probes.

Deletion. There is increasing evidence that some tumors arise because of complete inactivation of a gene product that normally suppresses tumorigenesis (reviewed

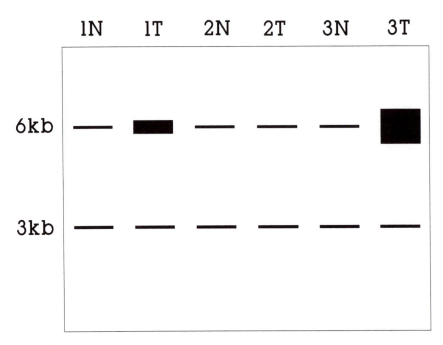

Figure 7.10. Detection of gene amplification by Southern blots. DNA from normal (N) and tumor (T) tissue from individuals 1, 2, and 3 is analyzed as described in the text. The greater signal at 6 kb in lanes 1T and 3T as compared with the normal tissue controls (1N and 3N, respectively) suggest specific amplification of that sequence in the tumors from individuals 1 and 3, but not 2. The 3-kb signal is from a different single-copy sequence probed separately as a control for integrity of genomic DNA and transfer efficiency.

in reference 45). Perhaps the best characterized of these tumor suppressor genes is the retinoblastoma gene on 13q14. Other putative tumor suppressors deleted in cancers include p53, located at 17p13,[46, 47] and DCC, located at 18q.[48] Diagnosis of these specific deletion events can be made as described in the previous sections.

Point Mutations. Specific point mutations in human tumors have been particularly well characterized for oncogenes such as the ras gene family (reviewed in reference 49), although they have been demonstrated in tumor suppressor genes as well.[47, 48] The presence of these point mutations can be determined by oligonucleotide-specific hybridization as described previously.

Translocations and Rearrangements. Several tumors are known to be associated with specific translocation events. Perhaps the best characterized of these is the reciprocal translocation involving the long arms of chromosome 9 and 22, t(9;22)(q34;q11), which generates the Philadelphia chromosome. This translocation occurs in more than 90% of chronic myelogenous leukemias (CML)[50] and in 10 to 25% of acute lymphoblastic leukemias (ALL).[51] Although it is known that the translocation generates a novel chimeric protein product from the 5' sequences of the BCR gene on chromosome 22 and the 3' sequences of the ABL gene on chromosome 9, the actual position of the breakage events on both chromosomes can occur over extensive genomic regions. In the ABL gene, for example, the break occurs over a 200 kb region in the intron between exons IB and II. The breakpoint in BCR is more heterogeneous, occurring either 5' to the gene, in the first intron, or within what is known as the breakpoint cluster region, which includes five small exons of BCR. As was discussed previously (Section 7.2.4.4, Figure 7.8), PCR with a cDNA of the fusion mRNA as template can be used to detect and characterize individual translocation events.[20, 21] Once that translocation has been characterized in a patient, the existence of the specific chimeric mRNA can also be used to diagnose relapse after treatment.

The diagnosis of some cancers is aided by determining whether they are clonal or whether they arose from several different mutational events. For disorders involving B and T cells, this diagnosis involves identification of whether all cells have the same immunoglobulin or T cell receptor rearrangement, respectively. Characterizing the rearrangement can also yield valuable information about the differentiation status of the tumor cells.

7.3.2. Indirect Detection

7.3.2.1. Indirect Diagnosis When the Gene Is Known but the Precise Defect Is Unknown

There are several diseases for which multiple mutations have arisen in the disease-associated gene over time. We have already discussed the example of DMD. Another example is the cystic fibrosis gene. Although it appears that the majority of CF patients have a 3-base-pair deletion encoding phenylalanine 508, haplotype analysis suggests that there have been at least seven other mutational events.[27] The deletion can be detected by direct methods such as PCR amplification and specific oligonucleotide hybridization, but diagnosis of cystic fibrosis in families transmitting any of the uncharacterized mutations cannot be done directly because the actual mutation site is not known. Instead, what must be determined is the inheritance from each parent of the chromosome region that is assumed to transmit the CF phenotype.

The inheritance of disease-causing alleles can be evaluated by following the coinheritance of DNA markers and the disease in a family. This analysis is most reliable when markers are closely linked to the disease (low frequencies of recombination) and have been mapped on both sides of the disease locus. Undetected recombination events between the markers and the disease locus are thus much less probable.

Consider the case of an autosomal recessive disease with a low rate of new mutations such as CF. Both parents of an affected child must be heterozygous for the disease-causing gene. Diagnosis depends on determining which chromosome from each parent has transmitted the disease allele to an affected child. For example, Figure 7.11 diagrams the simplified case of a family in which a second child (II.2) is known to have the disease, the first child (II.1) is unaffected, and the possibility that a fetus (II.3) is affected needs to be determined. Analyzing the inheritance of alleles of the polymorphic DNA markers P1, P2, and P3 that flank the disease locus can be used to estimate the likelihood that the disease gene is also inherited. This analysis can be done in several ways. One method involves Southern blotting (Figures 7.1 and 7.2). For simplicity, assume that the DNA markers are all RFLPs, each detected when genomic DNA is digested with the same enzyme (E), and that the fragments generated are sufficiently distinct to be resolved on the same gel (Table 7.2). Then DNA samples can be obtained from the family members and the fetus, digested with restriction enzyme E, and the fragments separated electrophoretically. The fragments could then be transferred to a membrane and hybridized to a mixture of labeled probes p1, p2, and p3 detecting polymorphisms P1, P2, and P3, respectively. The results could be as schematized in Figure 7.11C.

Each band on the Southern blot can be attributed to an allele on the polymorphisms detected by probes p1, p2, and p3. For example, the affected child with bands at 9.4, 4.4, 1.3, and 1.0 kb is homozygous for P1 (9.4 and 9.4 kb), homozygous for P2 (1.0 and 1.0 kb), and heterozygous for P3 (4.4 and 1.3 kb) (Figure 7.11C).

Analysis of these polymorphisms could also have been done by PCR (Figure 7.12). Amplimers could be designed to flank each of the polymorphic sites P1, P2, and P3, as was discussed previously (Section 7.2.4, Figure 7.4). DNA samples could then be obtained from each of the family members and the fetus for PCR amplification. To simplify the presentation, assume that the amplimers and the amplification products have been designed so that different alleles can be resolved in one amplification-digestion reaction (Table 7.2). Then a mix of amplimers for the polymorphisms P1, P2, and P3 could be added to the DNA sample. Each specific portion of the genomic DNA would then be amplified in the same reaction. An aliquot of the amplified DNA could then be digested with enzyme E, and the resultant fragments separated by agarose gel electrophoresis and visualized by ethidium bromide staining. Again, analysis of the fragment sizes for the affected child indicate that she is homozygous for P1 (400 bp), P2 (190 and 110 bp) and heterozygous for P3 (220, 140, and 80 bp). The genotypes of the rest of the family members for these markers can be determined similarly.

Independent of the method used to determine the genotypes, they are next used to identify the inheritance of the chromosomes carrying a disease-causing allele. In part, this depends on what is called the informativeness of the markers in a particular family. For example, the potential polymorphic site P2 is not informa-

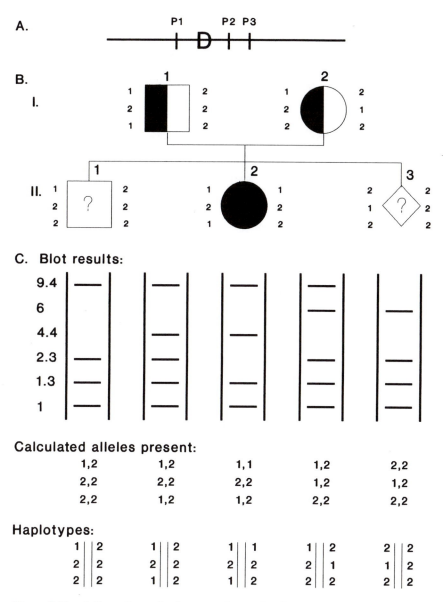

Figure 7.11. Indirect diagnosis of autosomal recessive diseases by Southern blots. (**A.**) A representation of the positions of the polymorphic marker sites P1, P2, and P3 relative to the disease locus *D*. (**B.**) Pedigree diagram showing the parental (I) and second (II) generations. The shading indicates the chromosome carrying the disease allele, with the numbers representing the alleles for P1, P2, and P3 in order on the corresponding sides. The numbers over each circle, square, or diamond indicate the individuals of the generation. (**C.**) Southern blot analyses giving the patterns expected after digesting genomic DNA from each of the individuals diagrammed over the lane with enzyme E and hybridizing with a mixture of labeled probes for the P1, P2, and P3 polymorphisms. The numbers at the left give the sizes in kb. Under each lane diagram is a summary of the alleles detected for the P1, P2, and P3 polymorphisms and the calculated haplotypes corresponding to those in *B*.

Table 7.2. Sizes Characteristic of Alleles for Polymorphisms P1, P2, and P3[a]

Polymorphic Site	Probe	Sizes Detected on Southern Blots (kb) Alleles		Amplimers	Sizes Detected by PCR (bp) Alleles	
		1	2		1	2
P1	p1	9.4	2.3	A1P1,A2P1	400	240, 160
P2	p2	6.5	1.0	A1P2,A2P2	300	190, 110
P3	p3	4.4	1.3	A1P3,A2P3	220	140, 80

[a]This table lists the allelic fragment sizes expected for polymorphisms P1, P2, and P3 from both Southern blot and PCR analyses. The genomic DNA and the amplification products are digested with enzyme E before being separated electrophoretically.

tive in this family. The affected child (II.2), her father (I.1), and her unaffected brother (II.1) are all homozygous for the P2-2 allele, so which paternal chromosome has been inherited by which child cannot be determined by analyzing the P2 polymorphism alone. In contrast, the affected child is homozygous for the P1-1 allele, and both her parents are heterozygous for this allele. Thus, assuming no recombination between P1 and the disease allele, the chromosomes carrying the P1-1 allele also carry the disease allele.

When this analysis is completed for all polymorphisms, probable haplotypes (the alleles carried on one of the two homologous chromosomes) for P1, P2, the disease locus, and P3 can be assigned. In this example, the chromosome bearing the P1-1, P2-2, and P3-1 alleles in the father and the P1-1, P2-2, and P3-2 alleles in the mother were inherited by the affected child. Her older brother, inheriting the P1-2, P2-2, and P3-2 paternal chromosome and the P1-1, P2-2, and P3-2 maternal chromosome (also carrying the mutant-bearing CF allele) is probably a carrier, and the fetus (paternal P1-2, P2-2, P3-2 and maternal P1-2, P2-1, P3-2) is probably disease-free and not a carrier.

Another example in which indirect diagnosis may be required for a disease associated with a known gene product is muscular dystrophy. As discussed earlier, 30 to 40% of DMD patients have no detectable deletions and the families must be studied by linkage analysis. The cDNA clones that are so useful for deletion can also serve as probes for intragenic RFLPs, using the same Southern blots (Figure 7.9). Intragenic and extragenic genomic probes have also been used.

Indirect diagnosis of DMD also involves determining the linkage of DNA markers and the disease. The difference in this case is that the disease is X-linked, so that males are hemizygous and the haplotype of the X chromosome is easily determined. Without repeating the details of how to determine haplotypes, consider the case shown in Figure 7.13. In this family, the woman indicated by II-1 has a brother (II-4) with muscular dystrophy. She is thus at risk for being a carrier. In screening DNA samples from family members, no deletion is detected. Haplotypes are established by Southern blots or PCR methods. The affected male

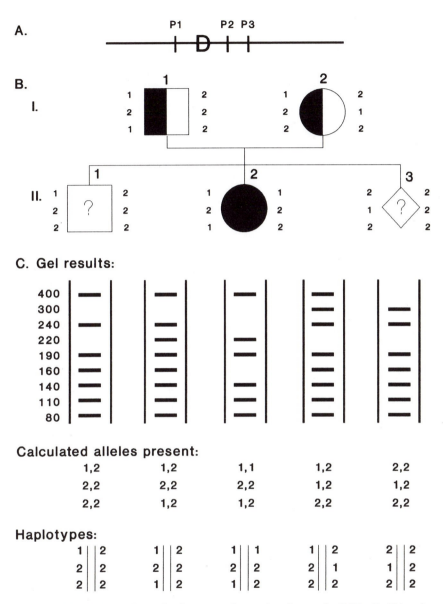

Figure 7.12. Indirect diagnosis of autosomal recessive diseases by PCR. (**A, B.**) as in Figure 7.11. (**C.**) Electrophoretic analysis of amplification products from DNA of the diagrammed individuals (Table 7.2) with the alleles and haplotypes summarized as in Figure 7.11.

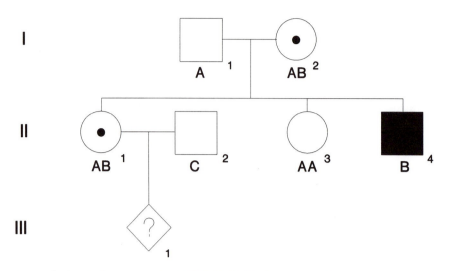

Figure 7.13. Indirect diagnosis of DMD/BMD. Haplotypes for polymorphic markers on the X chromosome are determined as described in the text. Males, with only one X chromosome, have only one haplotype. II.1 (AB), by sharing the B haplotype with her affected brother (II.4), and her mother (I.2, AB) is a probable carrier. A fetus (III.1) with the B haplotype would likely either be affected (male) or a carrier (female).

(II-4) and the first sister (II-1) have inherited the B haplotype from the mother. Again, assuming no recombination or new mutations, II-1 is likely to be a carrier. If the fetus (III-1) is male and has inherited the B haplotype, the risk for inheriting DMD is quite high. A female fetus could be a carrier (haplotype BC) or unaffected (haplotype AC).

7.3.2.2. Indirect Diagnosis When the Disease Is Mapped but the Gene Is Unknown

There are many genetic diseases that have been mapped to particular chromosomes but for which the gene containing the disease-causing mutation has not been characterized. Prenatal diagnosis or the determination of carrier status for these diseases must also be done indirectly. When the disease is autosomal and recessive, the analysis is as described in the first case in Section 7.3.2.1 (Figure 7.11). DNA markers are chosen that flank the disease locus if possible, and the inheritance of these markers is determined to infer the inheritance of the disease.

When the disease is autosomal dominant, inheritance of markers is still used to infer inheritance of the disease, but the analysis of the pedigree and the choice of samples to analyze may be different. An example is the diagnosis of Huntington's disease (HD). This is a progressive disease of the central nervous system inherited as an autosomal dominant trait with complete penetrance (reviewed in reference 52). The onset of symptoms is usually delayed until the fourth or fifth decade. Any child of an affected parent has a 50% chance of developing the disease but will not manifest that disease until after decisions may be made about childbearing.

Huntington's disease is known to map to the distal region of chromosome 4p. Several DNA markers have been characterized in that region that can be used diagnostically to determine whether the at-risk individual has the haplotype that in his family is associated with the HD gene.[53–56] Since the exact position of the HD gene is uncertain at this time, these probes probably do not flank the gene but map proximally to it. In this disease the diagnostic service offered may be of two categories: the determination of risk for someone with an affected parent (presymptomatic diagnosis), and the determination of risk for someone with an affected grandparent without determination of the genotype of the parental generation (exclusion testing). The latter becomes important in those cases in which one prospective parent has an affected parent, that prospective parent does not want to know whether he or she is at risk, but does want to know if a fetus may be excluded from carrying the disease gene. By careful selection of family members to be tested, exclusion of inheritance by the fetus can be determined without determining the status of the parent.[57, 58] The ethical questions presented by presymptomatic testing for late-onset disease are complex.[59, 60]

The pedigree analyses are similar to those previously discussed for indirect diagnoses. In the example shown in Figure 7.14A, II-2 has requested a presymptomatic diagnosis. DNA samples are analyzed and haplotypes are determined. For ease in diagramming, these haplotypes are assigned letter codes of A, B, C, and D. Because he inherited completely different chromosomes from those of his affected brother, barring recombination, he can be diagnosed as disease free. Note that samples from his affected mother were not required for this analysis. Figure 7.14B diagrams the results for an exclusion test. Again, barring recombination, the fetus can be excluded from having inherited HD because it did not inherit either grandpaternal chromosome. The disease status of his father cannot be determined from this analysis because there are insufficient data.

7.3.2.3. Sources of Uncertainty

Uncertainties in indirect diagnosis arise from several sources. Whenever it is not possible to determine the presence of the actual mutation, there is always a chance of recombination between the disease locus and the DNA markers. If the markers are closely linked and are chosen to flank the disease locus so that undetected recombination would require a double crossover, this uncertainty can be reduced. Indeed, for some diseases in which closely flanking markers have been determined (and there is thus little recombination), and in which new mutations are rare, the haplotype for certain DNA markers can be closely associated with the disease.

Analysis of a limited number of family members with a limited number of

A.   B.

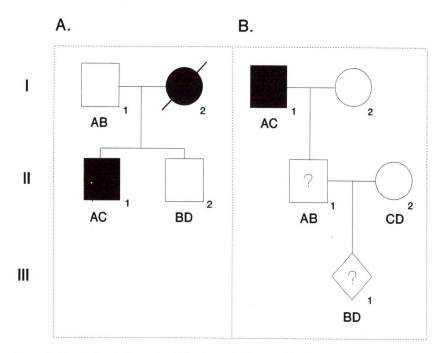

Figure 7.14. Indirect diagnosis of Huntington's disease. (**A.**) Presymptomatic diagnosis of Huntington's disease. (**A.**) Presymptomatic diagnosis for a child of an affected parent. DNA samples are obtained from I.1, II.1, and II.2, and haplotypes are determined for a set of polymorphic markers. The haplotypes A and B are detected in I.1, A and C in II.1, and B and D in II.2. Since II.2 (BD) does not share a haplotype with his affected brother, II.1 (AC), it is not probable that he has inherited the chromosome from his mother (I.1) that carries Huntington's disease. (**B.**) Exclusion testing. II.1 has a parent with HD (I.1) and wants to know whether the fetus (III.1) may inherit the HD-carrying chromosome. Determination of haplotypes by molecular mechanisms demonstrates that the fetus (BD) does not share a haplotype with the grandfather (AC) and most likely has not inherited HD. Since DNA from the grandmother (I.2) was not analyzed, the likelihood that II.1 has inherited HD cannot be determined.

probes also increases the possibility of misdiagnosis owing to undetected paternity questions or germ line mosaicism. The latter is suspected, for example, whenever more than one child has the same haplotype which is inconsistent with parental tissue samples.

Another problem that may arise is that caused by the difficulty of diagnosis of the disease so that its occurrence in the family can be followed. For CF this is usually not a problem, but for some diseases with genetic components, the determination of whether a family member has the disease may be difficult because of delayed onset (as in Huntington's disease) or because of a range of phenotypes resulting from incomplete penetrance or the multiple genes associated with the disease.

7.3.2.4. Indirect Diagnosis of Somatic Changes

An increasing number of neoplasms have been associated with chromosomal deletions. These deletions are assumed to be one of the causes of the neoplastic transformation, by leading to the loss of function of a putative tumor suppressor gene. If that gene has not been identified, only the deletion event can be diagnosed. Polymorphic markers at potential deletion sites can be used to detect such deletions. If DNA from normal tissue in the patient shows heterozygosity for a particular marker but the tumor tissue demonstrates only one allele, the loss of heterozygosity indicates a deletion at the marker region.

Another example of indirect diagnosis is in bone marrow transplantation. It is frequently important to determine whether host or donor or both host and donor cells are being maintained in the bone marrow. Probes detecting VNTR polymorphisms have been used to determine the alleles characteristic of the host and the donor and to characterize those retained in the marrow.[61]

7.4. Future Prospects

The previous sections have demonstrated how molecular diagnosis depends on a certain level of knowledge about the genetic component of a disease. That knowledge usually develops by the following progression: (1) recognition that a disease has a genetic component by its incidence in families; (2) mapping of the disease to a particular region on a specific chromosome; (3) cloning the gene itself; and (4) characterization of the mutation or mutations responsible for the altered activity of the gene. The tools available to aid diagnosis depend on the level of knowledge concerning the gene responsible for the disease. Even if the gene is not known, as long as it is mapped it can be detected indirectly by analysis of the coinheritance of the disease with a mapped DNA marker. Once the disease-causing gene has been cloned and the mutation characterized, the presence of the mutation in an individual can be determined directly.

The number of genetic diseases that can be analyzed using DNA methods presently represent only a small fraction of the known genetic phenotypes in humans. The number of genetic disorders in the current printed version of *Mendelian Inheritance in Man*[62] is 4,344. As of the 1989 Human Gene Mapping Workshop, only about 1,800 human genes and fragile sites had been mapped.[63] Of these, only about 950 have been cloned,[16] and only a few of these are known to be associated with a disease. Thus, the application of this technology for diagnosis is in its infancy.

As genetic maps with higher resolution become available, discrete genetic factors responsible for diseases that have multiple genetic components can be identified.[64] The future may therefore see molecular characterization of the genetic components of more diseases,[65] such as hypertension, alcoholism, atherosclerosis, diabetes, and mental illnesses. Advance indications of these conditions may permit early therapeutic or preventative intervention.

Today, the characterization of genetic components of disease using molecular methods is usually accomplished in a research laboratory associated with a clinical staff in a university teaching hospital. Southern blots or dot blots and radioisotopes are frequently required. As the technology develops, however, the protocols will become more suitable for routine clinical use. Such developments will probably incorporate nonisotopic DNA probes, PCR, and greater automation.

References

1. Saiki, R.K., Scharf, S., Falbona, F., Mullis, K.B., Horn, G.T., Erlich, H.A., and Arnheim, N. 1985. Enzymatic amplification of β-globin genomic sequences and restriction site analysis for diagnosis of sickle cell anemia. *Science* 230:1350–1354.

2. Southern, E.M. 1975. Detection of specific sequences among DNA fragments separated by gel electrophoresis. *J. Mol. Biol.* 98:503–517.

3. Trent, J.M., Kaneko, Y., and Mitelman, F. 1989. Report of the committee on structural chromosome changes in neoplasia. *Cytogenet. Cell Genet.* 51:533–562.

4. Harper, P.S., Frézal, J., Ferguson-Smith, M.A., and Schinzel, A. 1989. Report of the committee on clinical disorders and chromosomal deletion syndromes. *Cytogenet. Cell. Genet.* 51:563–611.

5. Hames, B.D., and Hiddins, S.J., eds. 1985. *Nucleic acid hybridisation: A practical approach.* Oxford: IRL.

6. Keller, G.H., and Manak, M.M. 1989. *DNA probes.* New York: Stockton.

7. Botstein, D.R., White, R.L., Skolnick, M.., and Davis, R.W. 1980. Construction of a genetic linkage map in man using restriction fragment length polymorphisms. *Am. J. Hum. Genet.* 32:314–331.

8. Nakamura, Y., Leppert, M., O'Connell, P., Wolff, R., Holm, T., Culver, M., Martin, C., Fujimoto, E., Hoff, M., Kumlin, E., and White, R. 1987. Variable number of tandem repeat (VNTR) markers for human gene mapping. *Science* 235:1616–1622.

9. Weber, J.L., and May, P.E. 1989. Abundant class of human DNA polymorphisms which can be typed using the polymerase chain reaction. *Am. J. Hum. Genet.* 44:388–396.

10. Economou, E.P., Bergen, A.W., Warren, A.C., and Antonarakis, S.E. 1990. The polydeoxyadenylate tract of *Alu* repetitive elements is polymorphic in the human genome. *Proc. Natl. Acad. Sci. USA* 87:2951–2954.

11. Pritchard, C.A., Casher, D., Uglum, E., Cox, D.R., and Myers, R.M. 1989. Isolation and field-inversion gel electrophoresis analysis of DNA markers located close to the Huntington disease gene. *Genomics* 4:408–418.

12. Lichter, P., Cremer, T., Borden, J., Manuelidis, L., and Ward, D.C. 1988. Delineation of individual human chromosomes in metaphase and interphase cells by in situ suppression hybridization using recombinant DNA libraries. *Hum. Genet.* 80:224–234.

13. Lichter, P., Tang, C.C., Call, K., Hermanson, G., Evans, G.A., Housman, D., and Ward, D.C. 1990. High-resolution mapping of human chromosome 11 by in situ hybridization with cosmid clones. *Science* 247:64–69.

14. Ott, J. 1984. Estimation of the recombination fraction in human pedigrees: Efficient computation of the likelihood for human linkage studies. *Genet. Epidemiol.* (suppl. 1):241–246.

15. Lathrop, G.M., Lalouel, J.M., Julier, C., and Ott, J. 1985. Multilocus linkage analysis in humans: Detection of linkage and estimation of recombination. *Am. J. Hum. Genet.* 37:482–498.

16. Kidd, K.K., Bowcock, A.M., Schmidtke, J., Track, R.K., Ricciuti, F., Hutchings, G., Bale, A., Pearson, P., and Willard, H. 1989. Report of the DNA committee and catalogs of cloned and mapped genes and DNA polymorphisms. *Cytogenet. Cell Genet.* 51:622–947.

17. Mullis, K., Faloona, F., Scharf, S., Saiki, R., Horn, G., and Erlich, H. 1986. Specific enzymatic amplification of DNA in vitro: The polymerase chain reaction. *Cold Spring Harbor Symp. Quant. Biol.* 51:263.

18. Erlich, H.A., ed. 1989. *PCR technology: Principles and applications for DNA amplification.* New York: Stockton.

19. Lench, N., Stanier, P., and Williamson, R. 1988. Simple non-invasive method to obtain DNA for gene analysis. *Lancet* 1:1356–1358.

20. Hooberman, A.L., Carino, J.J., Leibowitz, D., Rowley, J.D., Le Beau, M.M., Arlin, Z.A., and Westbrook, C.A. 1989. Unexpected heterogeneity of BCR-ABL fusion mRNA detected by polymerase chain reaction in Philadelphia chromosome-positive acute lymphoblastic leukemia. *Proc. Natl. Acad. Sci. U.S.A.* 86:4259–4263.

21. Roth, M.S., Antin, J.H., Bingham, E.L., and Ginsburg, D. 1989. Detection of Philadelphia chromosome-positive cells by the polymerase chain reaction following bone marrow transplant for chronic myelogenous leukemia. *Blood* 74:882–885.

22. Crescenzi, M., Seto, M., Herzig, G.P., Weiss, P.D., Griffith, R.C., and Korsmeyer, S.J. 1988. Thermostable DNA polymerase chain amplification of t(14,18) chromosome breakpoints and detection of minimal residual disease. *Proc. Natl. Acad. Sci. U.S.A.* 85:4869–4873.

23. Kan, Y.W., Golbus, M.S., and Dozy, A.M. 1976. Prenatal diagnosis of alpha-thalassemia: Clinical application of molecular hybridization. *N. Engl. J. Med.* 295:1165–1167.

24. Chang, J.C., and Kan, Y.W. 1982. A sensitive new prenatal test for sickle-cell anemia. *N. Engl. J. Med.* 307:30–32.

25. Orkin, S.H., Little, P.F.R., Kazazian, H.H., and Boehm, C.D. 1982. Improved detection of the sickle mutation by DNA analysis: Application to prenatal diagnosis. *N. Engl. J. Med.* 307:32–36.

26. Ballabio, A., Gibbs, R.A., and Caskey, C.T. 1990. PCR test for cystic fibrosis detection. *Nature (London)* 343:220.

27. Kerem, B., Rommens, J.M., Buchanan, J.A., Markiewicz, D., Cox, T.K., Chakravarti, A., Buchwald, M., and Tsui, L.-C. 1989. Identification of the cystic fibrosis gene: Genetic analysis. *Science* 245:1073–1080.

28. Mostacciuolo, M.L., Lombardi, A., Cambissa, V., Danieli, G.A., and Angelini, C. 1987. Population data on benign and severe forms of X-linked muscular dystrophy. *Hum. Genet.* 75:217–220.

29. Forrest, S.M., Cross, G.S., Speer, A., Gardner-Medwin, D., Burn, J., and Davies, K. 1987. Preferential deletion of exons in Duchenne and Becker muscular dystrophies. *Nature (London)* 329:638–640.

30. Hu, X., Burghes, A.H.M., Bulman, D.E., Ray, P.N., and Worton, R.G. 1989. Evidence for mutation by unequal sister chromatid exchange in the Duchenne muscular dystrophy gene. *Am. J. Hum. Genet.* 44:855–863.

31. Worton, R.G., and Thompson, M.W. 1988. Genetics of Duchenne muscular dystrophy. *Annu. Rev. Genet.* 22:601–629.

32. Koenig, M., Hoffman, E.P., Bertelson, C.J., Monaco, A.P., Feener, C., and Kunkel, L.M. 1987. Complete cloning of the Duchenne muscular dystrophy (DMD) cDNA and preliminary genomic organization of the DMD gene in normal and affected individuals. *Cell* 50:509–517.

33. Koenig, M., Monaco, A.P., and Kunkel, L.M. 1988. The complete sequence of dystrophin predicts a rod-shaped cytoskeletal protein. *Cell* 53:219–226.

34. Darras, B.T., and Francke, U. 1988. Normal human genomic restriction-fragment patterns and polymorphisms revealed by hybridization with the entire dystrophin cDNA. *Am. J. Hum. Genet.* 43:612–619.

35. Darras, B.T., Blattner, P., Harper, J.F., Spiro, A.J., Alter, S., and Francke, U. 1988. Intragenic deletions in 21 Duchenne muscular dystrophy (DMD)/Becker muscular dystrophy (BMD) families studied with the dystrophin cDNA: Location of breakpoints on *Hin*dIII and *Bgl*II exon-containing fragments maps, meiotic and mitotic origin of the mutations. *Am. J. Hum. Genet.* 43:620–629.

36. den Dunnen, J.T., Grootscholten, P.M., Bakker, E., Blonden, L.A.J., Ginjaar, H.B., Wapenaar, M.C., van Paassen, H.M.B., van Broeckhoven, C., Pearson, P.L., and van Ommen, G.J.B. 1989. Topography of the Duchenne muscular dystrophy (DMD) gene: FIGE and cDNA analysis of 194 cases reveals 115 deletions and 13 duplications. *Am. J. Hum. Genet.* 45:835–847.

37. Chamberlain, J.S., Gibbs, R.A., Ranier, J.E., Nguyen, P.N., and Caskey, C.T. 1988. Deletion screening of the Duchenne muscular dystrophy locus via multiplex DNA amplification. *Nucl. Acids Res.* 16:11141–11156.

38. Bishop, J.M. 1987. The molecular genetics of cancer. *Science* 235:305–311.

39. Riou, G.F. 1988. Proto-oncogenes and prognosis in early carcinoma of the uterine cervix. *Cancer Surv.* 7:441–456.

40. Saez, R.A., McGuire, W.L., and Clark, G.M. 1989. Prognostic factors in breast cancer. *Semin. Surg. Oncol.* 5:102–110.

41. Slamon, D.J., Godolphin, W., Jones, L.A., Holt, J.A., Wong, S.G., Keith, D.E., Levin, W.J., Stuart, S.G., Udove, J., Ullrich, A., and Press, M.F. 1989. Studies of the Her-2/neu protooncogene in human breast and ovarian cancer. *Science* 244:707–712.

42. Tsuda, H., Hirohashi, S., Shimosato, Y., Hirota, T., Tsugane, S., Yamamoto, H., Miyajima, N., Toyoshima, K., Yamamoto, T., Yokota, J., Yoshida, T., Sakamoto, H., Terada, M., and Sugimura, T. 1989. Correlation between long-term survival in breast cancer patients and amplification of two putative oncogene-coamplification units: *hst-1/int-2* and c-*erb*B-2/*ear*-1. *Cancer Res.* 49:3104–3108.

43. Brodeur, G.M., and Fong, C. 1989. Molecular biology and genetics of human neuroblastoma. *Cancer Genet. Cytogenet.* 41:153–174.

44. Wohlman, S.R., and Henderson, A.S. 1989. Chromosomal aberrations as markers of oncogene amplification. *Hum. Pathol.* 20:308–315.

45. Skuse, G.R., and Rowley, P.T. 1989. Tumor suppressor genes and inherited predisposition to malignancy. *Semin. Oncol.* 16:128–137.

46. Baker, S.J., Fearon, E.R., Nigro, J.M., Hamilton, S.R., Preisinger, A.C., Jessup, J.M., vanTuinen, P., Ledbetter, D.H., Barker, D.F., Nakamura, Y., White, R., and Vogelstein, B. 1989. Chromosome 17 deletions and p53 gene mutations in colorectal carcinomas. *Science* 244:217–221.

47. Nigro, J.M., Baker, S.J., Preisinger, A.C., Jessup, J.M., Hostetter, R., Cleary, K., Bigner, S.H., Davidson, N., Baylin, S., Devilee, P., Glover, T., Collins, F.S., Weston, A., Modali, R., Harris, C.C., and Vogelstein, B. 1989. Mutations in the p53 gene occur in diverse human tumor types. *Nature (London)* 342: 705–707.

48. Fearon, E.R., Cho, K.R., Nigro, J.M., Kern, S.E., Simons, J.W., Ruppert, J.M., Hamilton, S.R., Preisinger, A.C., Thomas, G., Kinzler, K.W., and Vogelstein, B. 1990. Identification of a chromosome 18q gene that is altered in colorectal cancers. *Science* 247:49–56.

49. Varmus, H.E. 1984. The molecular genetics of cellular oncogenes. *Annu. Rev. Genet.* 18:553–612.

50. Rowley, J.D., and Testa, J.R. 1983. Chromosome abnormalities in malignant hematologic diseases. *Adv. Cancer Res.* 36:103–148.

51. Third International Workshop on Chromosomes in Leukemia. 1983. Chromosomal abnormalities and their clinical significance in acute lymphoblastic leukemia. *Cancer Res.* 43:868–873.

52. Jenkins, J.B., and Conneally, P.M. 1989. The paradigm of Huntington disease. *Am. J. Hum. Genet.* 45:169–175.

53. Gusella, J.F., Wexler, N.S., Conneally, P.M., Naylor, S.L., Anderson, M.A., Tanzi, R.E., Watkins, P.C., Ottina, K., Wallace, M.R., Sakaguchi, A.Y., Young, A.B., Shoulson, I., Bonilla, E., and Martin, J.B. 1983. A polymorphic DNA marker linked to Huntington's disease. *Nature (London)* 306:234–238.

54. Gilliam, T.C., Bucan, M., MacDonald, M.E., Zimmer, M., Haines, J.L., Cheng, S.V., Pohl, T.M., Meyers, R.H., Whaley, W.L., Alletto, B.A., Faryniarz, A., Wasmuth, J.J., Frischauf, A.M., Conneally, P.M., Lehrach, H., and Gusella, J.F.

1987. A DNA segment encoding two genes very tightly linked to Huntington's disease. *Science* 238:950–952.

55. Hayden, M.R., Hewitt, J., Wasmuth, J.J., Kastelein, J.J., Langlois, S., Conneally, M., Haines, J., Smith, B., Hilbert, C., and Allard, D. 1988. A polymorphic DNA marker that represents a conserved expressed sequence in the region of the Huntington disease gene. *Am. J. Hum. Genet.* 42:125–131.

56. Wasmuth, J.J., Hewitt, J., Smith, B., Allard, D., Haines, J.L., Skarechy, D., Partlow, E., and Hayden, M.R. 1988. A highly polymorphic locus very tightly linked to the Huntington's disease gene. *Nature (London)* 332:734–736.

57. Harper, P. 1986. The prevention of Huntington's chorea. *J. R. Coll. Physicians London* 20:7–14.

58. Millan, F.A., Curtis, A., Mennie, M., Holloway, S., Boxer, M., Faed, M.J. W., Crawford, J.W., Liston, W.A., and Brock, D.J.H. 1989. Prenatal exclusion testing for Huntington's disease: A problem of too much information. *J. Med. Genet.* 26:83–85.

59. Fahy, M., Robbins, C., Bloch, M., Turnell, R.W., and Hayden, M.R. 1989. Different options for prenatal testing for Huntington disease using DNA probes. *J. Med. Genet.* 26:353–357.

60. Bloch, M., and Hayden, M.R. 1990. Opinion: Predictive testing for Huntington disease in childhood—Challenges and implications. *Am. J. Hum. Genet.* 46:1–4.

61. Gatti, R.A., Nakamura, Y., Nussmeier, M., Susi, E., Shan, W., and Grody, W.W. 1989. Informativeness of VNTR genetic markers for detecting chimerism after bone marrow transplantation. *Disease Markers* 7:105–112.

62. McKusick, V.A., ed. 1988. *Mendelian inheritance in man: Catalogs of autosomal dominant, autosomal recessive, and X-linked phenotypes,* 8th ed. Baltimore: Johns Hopkins University Press.

63. McAlpine, P.J., Shows, T.B., Boucheix, C., Stranc, L.C., Berent, T.G., Pakstis, A.J., and Douté, R. 1989. Report of the nomenclature committee and the 1989 catalog of mapped genes. *Cytogenet. Cell Genet.* 51:13–66.

64. Lander, E.S., and Botstein, D. 1989. Mapping Mendelian factors underlying quantitative traits using RFLP linkage maps. *Genetics* 121:185–199.

65. Friedman, T. 1990. Opinion: The human genome project—Some implications of extensive "reverse genetic" medicine. *Am. J. Hum. Genet.* 46:407–414.

66. Kidd, V.J., Wallace, R.B., Itakura, K., et al. 1983. Alpha1-antitrypsin deficiency detection by direct analysis of the mutation in the gene. *Nature (London)* 304:230–234.

67. Suthers, G.K., Callen, D.F., Hyland, V.J., Kozman, H.M., Baker, E., Eyre, H., Harper, P.S., Roberts, S.H., Hors-Cayla, M.C., Davies, K.E., Bell, M.V., and Sutherland, G.R. 1989. A new DNA marker tightly linked to the fragile X locus (FRAXA). *Science* 246:1298–1300.

68. Geever, R.F., Wilson, L.B., Nallaseth, F.S., Milner, P.F., Bittner, M., and Wilson, J.T. 1981. Direct identification of sickle cell anemia by blot hybridization. *Proc. Natl. Acad. Sci. U.S.A.* 78:5081–5985.

69. Ottolenghi, S., Lanyon, W.G., and Paul, J. 1974. The severe form of alpha-thalassemia is caused by a haemoglobin gene deletion. *Nature (London)* 251:389–391.

70. Antonarakis, A.S., Waber, P.G., Kittur, S.D., et al. 1985. Hemophilia A: Detection of molecular defects and of carriers by DNA analysis, *N. Engl. J. Med.* 313:842–848.

71. Giannelli, G., Anson, D.S., Choo, K.H., Rees, D.J., Winship, P.R., Ferrari, N., Rizza, C.R., and Brownlee, G.G. 1984. Characterization and use of an intragenic polymorphic marker for detection of carriers of haemophilia B (factor IX deficiency). *Lancet* 1:239–241.

72. Kok, K., Osinga, J., Carritt, B., Davis, M.B., van der Hout, A.H., van der Veen, A.Y., Landsvater, R.M., De Liej, M.L. M. H., Berendsen, H.H., Postmus, P.E., Poppema, S., and Buys, C.H.C. M. 1987. Deletion of a DNA sequence at the chromosomal region 3p21 in all major types of lung cancer. *Nature (London)* 330:578–581.

73. vanTuinen, P., Dobyns, W.B., Rich, D.C., Summers, K.M., Robinson, T.J., Nakamura, Y., and Ledbetter, D.H. 1988. Molecular detection of microscopic and submicroscopic deletions associated with Miller-Dieker syndrome. *Am. J. Hum. Genet.* 43:587–596.

74. Rosen, R., Fox, J., Fenton, W.A., Horwich, A.L., and Rosenberg, L.E. 1985. Gene deletion and restriction fragment length polymorphisms at the human ornithine transcarbamylase locus. *Nature (London)* 313:815–817.

75. Lidsky, A.S., Guttler, F., and Woo, S.L.C. 1985. Prenatal diagnosis of classical phenylketonuria by DNA analysis. *Lancet* 1:549–551.

76. Donlon, T.A., Lalande, M., Wyman, A., Bruns, G., and Latt, S.A. 1986. Isolation of molecular probes associated with the chromosome 15 instability in the Prader-Willi syndrome. *Proc. Natl. Acad. Sci. U.S.A.* 83:4408–4412.

77. Bowcock, A.M., Farrer, L.A., Cavalli-Sforza, L.L., Hebert, J.M., Kidd, K.K., Frydman, M., and Bonne-Tamir, B. 1987. Mapping the Wilson disease locus to a cluster of linked polymorphic markers on chromosome 13. *Am. J. Hum. Genet.* 41:27–35.

78. Korneluk, R.G., Mahuran, D.J., Neote, K., Klavins, M.H., O'Dowd, B.F., Tropak, M., Willard, H.F., Anderson, M.J., Lowden, J.A., and Gravel, R.A. 1986. Isolation of cDNA clones coding for the alpha-subunit of human beta-hexosaminidase: Extensive homology between the alpha-and beta-subunits and studies on Tay-Sachs disease. *J. Biol. Chem.* 261:8407–8413.

79. Smith, M., Smalley, S., Cantor, R., Pandolfo, M., Gomez, M.I., Baumann, R., Flodman, P., Yoshiyama, K., Nakamura, Y., Julier, C., Dumars, K., Haines, J., Trofatter, J., Spence, M.A., Weeks, D., and Conneally, M. 1990. Mapping of a gene determining tuberous sclerosis to human chromosome 11q14-11q23. *Genomics* 6:105–114.

80. Ballabio, A., Parenti, G., Carrozzo, R., Sebastio, G., Andria, G., Buckle, V., Fraser, N., Craig, I., Rocchi, M., Romeo, G., Jobsis, A.C., and Persico, M.G. 1987. Isolation and characterization of a steroid sulfatase cDNA clone: Genomic

deletions in patients with X-chromosome-linked ichthyosis. *Proc. Natl. Acad. Sci. U.S.A.* 84:4519–4523.

81. Boerwinkle, E., Xiong, W., Fourest, E., and Chan, L. 1989. Rapid typing of tandemly repeated hypervariable loci by the polymerase chain reaction: Application to the apolipoprotein B 3′ hypervariable region. *Proc. Natl. Acad. Sci. U.S.A.* 86:212–216.

82. Ludwig, E.H., Friedl, W., and McCarthy, B.J. 1989. High resolution analysis of a hypervariable region in the human apolipoprotein B gene. *Am. J. Hum. Genet.* 45:458–464.

8

Human Genome Mapping and Sequencing: Applications in Pharmaceutical Science

C. E. Hildebrand, R. L. Stallings, D. C. Torney,
J. W. Fickett, N. A. Doggett, D. A. Nelson,
A. A. Ford, and R. K. Moyzis

8.1. Introduction

Basic and applied sciences have led to enormous progress in both the prevention and treatment of a spectrum of human diseases. Many of these accomplishments are reviewed and discussed elsewhere in this volume. Some of the most notable successes have come in preventing or treating diseases of bacterial or viral etiology. In this context the relatively recent increased incidence of AIDS (acquired immunodeficiency syndrome) and viral hepatitis (caused by the hepatitis B virus) has presented new and urgent challenges to the pharmaceutical sciences and the biotechnology industry (see Chapters 10 and 11, this volume).

While studies of the etiology, ontogeny, prophylaxis, and therapeutics for these diseases must remain a high priority, the recognition of human diseases or predisposition to diseases having a genetic or inherited component will present new opportunities for the design and development of therapeutic or prophylactic agents. It is in the latter areas that human genome mapping and sequencing can provide the fundamental information and reagents for genetic disease diagnosis, prevention, or treatment, especially in common diseases with complex causes (involving multiple genes) such as cancer, heart disease, hypertension, rheumatoid arthritis, schizophrenia, and manic depression.[1-6] The focus of this chapter will be on the value of, and fundamental need for, high-resolution physical maps of the human genome.[7] Physical mapping will be presented as a step toward obtaining rapidly and efficiently the DNA sequence of any region of interest or importance from a clinical or a basic science viewpoint. We will present in detail how physical maps can be assembled and how they are used. Finally, we will briefly review approaches to rapidly acquiring the DNA sequence of a target region.

8.2. Background

During the past two decades, dramatic advances in recombinant DNA technology and cell biology have created the framework for furthering our understanding of the molecular basis of human genetic diseases.[8, 9] Molecular-level explanations have been provided for the hemoglobinopathies (sickle cell anemia and α- and β-thalassemias) as well as for several of the "classical" inherited disorders, for example, phenylketonuria (a defect in the phenylalanine hydroxylase gene), hemophilia (a defect in clotting factor VIII), color blindness (associated with alterations in the genes encoding rhodopsin-like pigments), familial hypercholesterolemia and heart disease (associated with the gene encoding the low-density lipoprotein receptor), Duchenne muscular dystrophy (traced to deletions in the dystrophin gene), and, more recently, cystic fibrosis,[10–12] neurofibromatosis 1,[13–15] and familial adenomatous polyposis.[16–18] These are a few of the genes recently mapped and cloned using a variety of genetic and molecular techniques and requiring the active collaboration of many laboratories. The molecular basis and mechanistic understanding of several of these genetic disorders have provided insights that contribute to the design of potential therapeutic measures.[8, 19] If high-resolution maps of the human genome had been available at the time the genetic linkage of these diseases was discovered, it is likely that several years of research effort and thousands of investigator-years could have been saved in the search for and isolation of the disease genes.[20]

To provide the background and rationale for constructing a map and, ultimately, the complete sequence of the human genome, we first will review briefly the more classical or conventional approaches to determining the genetic basis of human disease, designated "forward genetics"[21] or, more recently, "functional cloning" (see "Note added in proof"). In general, in these cases a defective or mutant gene or gene product protein has been identified and isolated, possibly through an understanding of an aberrant metabolic pathway, and a variety of procedures have been used to identify and isolate the relevant gene(s) encoding the aberrant protein product (e.g., Hb and phenylalanine hydroxylase; also see reference 8).

An alternative strategy—designated "reverse genetics"—in which the gene product has not been identified but in which a defined fragment of DNA is consistently coinherited with the disease, is effective for pursuing the molecular basis of genetic diseases.[8, 21]

8.3. Forward Genetics/Functional Cloning

Forward-genetics approaches to determining the molecular basis of phenotype (in some cases, a specific disease phenotype) have been used successfully in identifying and isolating genes linked to enzyme deficiencies [e.g., hypoxanthine-guanine phosphoribosyl-transferase (HPRT) in Lesch-Nyhan syndrome and phenylalanine hydroxylase in phenylketonuria] or to familial coronary artery disease [associated with a defective low-density lipoprotein (LDL) receptor], for example. In general, this approach starts with the search for, and isolation of, a protein or an enzyme associated with the phenotype. A variety of techniques have been used to identify candidate gene products associated with disease phenotypes. One strategy looks for protein mutations linked to the disease. Once a protein has been identified and purified, at least two routes can be taken to isolating the corresponding gene. In one approach the purified protein is used to produce an antibody. The antibody is used to trap nascent protein chains on the translation complex containing mRNA and ribosomes (or polysomes) obtained from appropriate cell cultures or from primary tissue explants. The mRNA can be extracted and used to synthesize, by reverse transcription, a complementary DNA (cDNA) strand, which can then be made double stranded and cloned into a plasmid vector.[22] The cloned cDNA is used as a probe for identifying the gene in recombinant genomic libraries (i.e., recombinant libraries constructed from partially digested genomic DNA). Antibodies to a purified protein can be used directly to identify the cognate cDNA cloned into an expression vector.[22]

Alternatively, the candidate protein product can be partially sequenced and the amino acid sequence used to "reverse" translate into the possible cognate nucleotide coding sequences. A set of synthetic oligonucleotides is then generated and can be used as probes for screening genomic libraries for the desired gene.[22]

8.4. Reverse Genetics/Positional Cloning: The Need for High-Resolution Genetic and Physical Maps

In this section we explore in detail the methods of reverse genetics (positional cloning) and illustrate the power of these methods in identifying and isolating disease genes.

The approaches of forward genetics (functional cloning) are based on a candidate gene product implicated in the molecular basis of the disease. However, for many genetic diseases (e.g., cystic fibrosis, Duchenne muscular dystrophy, Huntington's chorea, and polycystic kidney disease) a candidate gene product could not or cannot readily be identified. In addition to these single-gene disorders, the inability to associate specific gene product(s) in any mechanistic way with complex genetic diseases (e.g., cancer, diabetes, hypertension, rheumatoid arthritis, and schizophrenia) has required entirely new approaches. To pursue the suspect genes in such conditions, a strategy that has become known as reverse genetics (positional cloning) has been introduced and used effectively in several disorders[8, 11, 21] (see Chapter 7, this volume).

A major conceptual leap in this new area of human genetics was made by Botstein, White, Skolnick, and Davis over a decade ago.[23] They proposed that the methods of molecular biology and recombinant DNA technology could be combined to address the problem of tracking the inheritance of genotypes (genetic loci, or anonymous DNA fragments) with inherited phenotypic traits, specifically human genetic diseases. Initially, genotypic differences, which could be identified with individuals within a pedigree, were detected as restriction fragment length polymorphisms (RFLPs) identified by a DNA marker (see Chapter 7, this volume). DNA markers that are located close to a disease locus are frequently inherited with the disease. Genetic distance between markers is related to the meiotic recombination frequency between the markers. The location of a disease gene can be approached initially by identifying polymorphic DNA markers, which recombine more frequently with one another than each one individually recombines with the disease locus. This characteristic would locate the disease gene between the two DNA markers. In general, the polymorphic DNA markers can be localized to a specific chromosome and, in many cases, to a specific subregion of the chromosome using a variety of cytogenetic or somatic cell genetic methods.[25] This measurement of genetic linkage is a first step in locating and eventually isolating a disease gene. However, the resolution of current genetic linkage maps is relatively coarse. The average spacing between markers is approximately 10 centimorgans (cM), or approximately 10% recombination, although there are many markers separated by less than 1% recombination.[2, 3, 24, 26]

Although there is no fixed relationship to convert recombination frequency to physical distance along the genome (or chromosome), a common metric for the conversion of genetic linkage to physical distance is 1 cM $\sim 10^6$ base pairs.[3] Therefore, if a disease locus is identified to lie between two flanking markers, say ~ 1 cM from each marker, a disease gene would be located somewhere within an approximately 2×10^6 bp region. If the gene were to turn out to be an average-sized gene, say $\sim 30,000$ bp (30 kb), 60 to 70 genes could be located in the 2-Mb region. At the extremes there could be as many as 500 to 1,000 small genes or as few as 5 to 10 large genes in a 2-Mb region.

In Figure 8.1 different types of maps of the human genome are illustrated using a model chromosome. This figure shows the increased resolution of various kinds of maps, from the very coarse cytogenetic map revealing cytochemical staining differences represented as bands along the chromosome (an average light or dark band contains ~ 10 Mb of DNA), to the DNA sequence, the ultimate level of physical mapping resolution. The cytogenetic map can include information relating to the position of a specific marker along the chromosome identified by a technique called in situ hybridization.[27–28] The genetic linkage map[24] provides information relating to the *order* of markers along the chromosome and is derived by following the inheritance of DNA markers, or specific phenotypic traits (including genetic disease phenotypes), in multigeneration families. A radiation hybrid map[29] is a type of physical map generated by following the frequency with which human DNA markers or genes remain linked on the same DNA fragment produced by radiation fragmentation of "normal" genomic DNA. In the radiation hybrid experiment, an irradiated human cell is fused with a rodent cell line, and random segregation of the radiation-fragmented human DNA fragments is allowed to proceed. The resolution of this type of map is dependent on the radiation dose to the donor human cell (higher resolution with larger doses), but the approximate resolution is of the order of several hundred kilobases between markers. This type of map does not yield physical distance but gives information

DIFFERENT KINDS OF MAPS

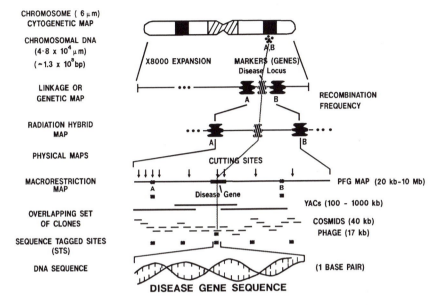

Figure 8.1. Different kinds of physical and genetic maps. The lowest level of resolution of the human genome is the 22 pairs of autosomal chromosomes and the X and Y sex chromosomes visible at mitosis. Techniques have been devised for resolving chromosomal subregions by differential staining with DNA-specific dyes.[28] Additional levels of expanded resolution of the order of and distances between markers (genes or anonymous DNA sequences; e.g., A and B in the figure) are discussed in the text. A proposed system for interrelating the different types of maps using short regions of DNA sequences and polymerase chain reaction technology is indicated by the stippled boxes, which represent sequence tagged sites (STS).

on the precise order of markers along the chromosome. Macrorestriction maps are generated by long-range restriction mapping using rare-cutting restriction endonucleases to fragment genomic DNA and pulsed-field gel electrophoresis[30,31] to resolve the resulting large DNA fragments (50–5,000 kb). The gels resulting from these analyses are handled similarly to conventional Southern blots for DNA transfer and hybridization (see Chapter 7, this volume). Macrorestriction maps also do not yield cloned DNA but define physical distances between markers in terms of kilobases.

Physical maps consisting of overlapping cloned DNA fragments can be constructed from multiple types of recombinant DNA libraries. Cosmid and phage recombinants can be used directly for DNA sequencing, whereas large fragment clones require subcloning and subsequent overlap mapping to provide the appropriate input materials for DNA sequencing. The annotated DNA sequence is the ultimate physical map (i.e., the DNA sequence with locations and boundaries of coding and noncoding regions of genes as well as the locations of other sequence features important in the regulation of gene expression or in the replication and packaging of chromatin and chromosomes).

The rationale for constructing high-resolution physical maps that are anchored to high-resolution genetic and cytogenetic maps is to provide a genomic roadmap so that any region or combination of regions (e.g., multiple loci involved in polygenic diseases) can be examined rapidly at the DNA-sequence level without having to employ the tedious procedures of conventional reverse genetics to identify and isolate a disease gene or other target region.

In the absence of high-resolution physical and genetic linkage maps, the objective of the reverse genetic approach is to proceed rapidly from genetic linkage of a phenotype (e.g., a disease) to identification, isolation, and sequencing of the DNA in the region containing the target gene. In this process it is essential to narrow the search for the target gene as rapidly and efficiently as possible while generating materials (i.e., cloned DNA) that can be used for eventual sequencing of the target region.

A variety of methods have been invented for identifying and isolating a gene (or specific DNA region) once genetic linkage to flanking markers has been established.[3, 10, 24, 32] These methods include chromosome walking and chromosome jumping techniques[10, 32] used interactively with genetic linkage analysis to determine progress in getting closer to the target gene. In general, these approaches are sequential in nature and therefore require multiple steps to move from flanking marker loci to the target gene. The rate of progress in identifying and isolating a target gene is dependent on several factors. A key factor at the outset of the process is the proximity of genetic markers to the target gene. Following from the previous example, if a disease gene is located between two polymorphic markers separated by 2 cM and (for ease of illustration) the disease gene is 1 cM from each marker, the target gene (of average size, ~ 30 kb) would be located approximately equidistant from the flanking markers. As was noted previously, the actual physical distance can only be estimated from linkage data. In this case we assume that the target gene lies approximately 1 Mb from the flanking markers. If standard chromosome walking is to be employed, a well-represented total genomic recombinant DNA library is used to identify cloned DNA fragments identified by the flanking linkage markers, which serve as the starting points for the walk. If a total genomic library constructed in a cosmid vector is used, the cloned fragments will be 35 to 45 kb long, and approximately 75,000 recombinants are required for 1-fold coverage of the human genome if the cosmids were laid end to end. We discuss the use of cosmid libraries initially because cosmids provide a substrate for rapidly obtaining DNA sequence. The use of larger fragment cloning vectors will be discussed later.

In a 5-fold redundant genomic cosmid library (i.e., ~ 375,000 cosmid clones), the probability of the library containing at least one clone representing a specific region is 0.99 (from Poisson statistics, $1 - e^{-5} = 0.99$).[33] Alternatively, if the chromosomal assignment of the target gene is known, a chromosome-specific library would considerably reduce the effort required in the initial screening and subsequent walking steps. Since the human genome is naturally divided into 22 autosomal pairs plus the X and Y sex chromosomes, there are 24 chromosome types (an average chromosome contains approximately 130 Mb of DNA), so that a well-represented chromosome-specific library would reduce by 24-fold (for an average chromosome) the effort involved in screening. Thus, in the first step of screening a chromosome-specific library for clones identified by the linkage markers, only 16,000 cosmid clones would have to be screened. The screening effort can be streamlined further by using two-dimensional, high-density, gridded arrays of primary recombinants from a chromosome-specific library. This approach is being used for the construction of a global physical map for human chromosome 16 and will be discussed below. Returning to the problem of chromosome walking to move from a linkage marker to the target gene, each step in the walk would gain ~ 20 kb on average. Therefore, if it were possible to walk unidirectionally from both flanking linkage markers toward the target gene, ~50 walking steps would be required. This estimate takes into account the fact that the initial walking steps do not provide unambiguous directionality for the walk. Thus, several additional initial walking steps are necessary to determine if the walk is proceeding from the genetic linkage marker toward the target gene.

The sequential and tedious nature of chromosome walking is complicated by the fact that certain regions of the genome, or an individual chromosome, are unclonable for a variety of reasons, including the presence of repetitive sequences that have a significant underrepresentation of restriction enzyme sites for the restriction endonuclease used for constructing the library. The stability of certain cloned regions will vary depending on the host/vector system used. Thus, multiple libraries constructed using various restriction endonucleases and different cloning vectors and hosts are required to maximize coverage of and ensure continuity in chromosome walking experiments.[34]

A novel technique called chromosome jumping has been used to accelerate the progress of chromosome walking methods and to circumvent the problems presented by regions of "nonclonable" DNA. This technique uses a specialized "jumping" library constructed from very high molecular weight total genomic DNA by cutting the DNA with an infrequently cutting restriction enzyme (e.g., *Not*I) which cuts every 0.5 to 1.0 Mb. The resulting fragments are circularized in the presence of a linearized plasmid having a selectable marker.[32, 35] The product of this step is a large double-stranded circular DNA molecule containing 0.5 to 1.0 Mb of genomic DNA with the short (5 kb) plasmid molecule inserted between the *Not*I sites at the ends of the genomic fragment. The resulting circles are cut with a second restriction enzyme—*Bam*HI—and cloned into the *Bam*HI sites of a phage vector. Thus, the recombinants obtained from this cloning step contain the ends of the genomic *Not*I fragment, separated by the plasmid insert.

These jumping clones can then be used as probes in a total genomic or chromosome-specific library to identify new walking clones. Thus, jumping libraries are an important resource in bridging "nonclonable" regions and in spanning large regions to complement and accelerate the chromosome walking process.

Jumping libraries can also be used with a complementary "linking" library that contains *Not*I restriction sites and adjacent DNA as plasmid clones.[32] These libraries can be constructed from total genomic DNA or other source DNA containing a single human chromosome or chromosome subregion. Since linking clones contain a rare-cutter site (e.g., *Not*I) and adjacent DNA, they can be used as probes to identify adjacent large DNA fragments generated by cutting genomic DNA with the same rare cutter (*Not*I). In these experiments genomic DNA is digested with a rare-cutter restriction enzyme (e.g., *Not*I) and separated by pulsed field gel electrophoresis to resolve the large DNA fragments. The DNA is transferred to a hybridization membrane, as in standard Southern blotting (see Chapter 7, this volume), and the blot is analyzed with a linking clone as the probe. In the ideal case the linking clone identifies the two fragments adjacent to the *Not*I site contained in the linking clone.[32]

Alternating application of jumping and linking libraries can provide a large-scale restriction (or macrorestriction) map of the order and distances between rare-cutter sites. These approaches provide the advantage that large regions of the genome can be characterized rapidly, so that progress in identifying and isolating a target gene is substantially accelerated. However, these methods yield only analytical maps and do not provide cloned DNA for the regions characterized. Furthermore, these procedures are sequential and require specialized resources (jumping and linking libraries) that are subject to some of the constraints mentioned previously for other genomic recombinant libraries.

The development of new techniques for cloning large DNA fragments (~ 100–1,000 kb inserts) into yeast artificial chromosomes (YACs) represents a major breakthrough in chromosome mapping, in general, and provides an essential new tool in the reverse genetics approach to identifying and isolating disease genes.[36] Although YAC cloning provides a rapid method for isolating a large genomic region that is almost certain to contain the target gene, the effort involved in finding the gene in the several-hundred-kilobase human insert in the YAC clone presents a significant challenge. Recall that the DNA sequence of the target region is the ultimate goal of the mapping phase of the reverse genetics approach. In most cases a YAC represents only a few percent of the total yeast host DNA, and the human insert is too large to be accommodated directly by current DNA sequencing techniques, so that subcloning of the YACs into cosmids or into a sequencing vector is required. To circumvent the need for random (or shotgun) sequencing of these subclones, it is advantageous to construct an ordered map of overlapping subclones spanning the human insert in the YAC. It is then possible to proceed with a directed sequencing strategy of a selected subregion of the YAC insert that, based on independent evidence, is likely to contain the target gene.

New techniques have recently been developed for rapid screening of YAC libraries to detect regions likely to contain a target (disease) gene.[37] These approaches have been used effectively in demonstrating that a specific target gene can be rapidly identified in a YAC library using a small amount of sequence information from a marker flanking a target gene. The specific experimental techniques include the synthesis of short oligonucleotides (based on the sequence data) that can be used as primers in a polymerase chain reaction (PCR; see Chapter 7, this volume) for a "pooled" screening of a YAC library.[37]

All of the preceding techniques and methods have made, or are beginning to make, significant contributions to the reverse genetics approach for identifying and isolating human disease genes or other targets genes. However, all of these methods are sequential and require multiple steps to achieve isolation of the target gene, starting with genetic linkage markers that flank the region of interest. The availability of high-resolution genetic and physical maps of the human genome will circumvent the need for sequential walking and jumping. Multilevel high-resolution maps will accelerate the identification and isolation of genes for the single-locus disorders and will be essential for unraveling the complex molecular nature of polygenic disorders.

8.5. The Scale of the Human Genome Physical Mapping Problem

Estimates of the size of the human genome range from 3 to 3.5×10^9 base pairs.[20, 38] Figure 8.2 provides a comparison of the sizes of other genomes for which physical maps are currently being constructed and summarizes the level of resolution of many of the tools of cellular and molecular biology that are being used to approach the mapping of the human and other complex genomes. Overlapping phage clone maps have been constructed for an *Escherichia coli* genome (~ 5 Mb)[39–40] and for a large fraction of the yeast *Saccharomyces cerevisiae* genome (~ 15 Mb).[41] A macrorestriction map of the *E. coli* genome has also been completed.[42] A physical map of the nematode *Caenorhabditis elegans* genome (~ 80 Mb) consisting of sets of overlapping cosmids (or contigs) linked by YAC clones is nearing completion.[43–44] In addition to the important information and resources generated by these mapping projects, they have provided valuable insights and technological innovations for approaching large-scale physical mapping problems. One significant insight is that the genetic map of the target genome is essential for making rapid progress on the physical map. It is important even at early stages of physical mapping to establish the interrelationship between the order of phenotypic markers on the genetic map and their positions on the physical map.

The 3×10^9 bp human genome (and the genomes of other mammalian species) represents a major challenge in scale. However, the complexity of the human genome can be reduced by approaching the mapping problem chromosome by

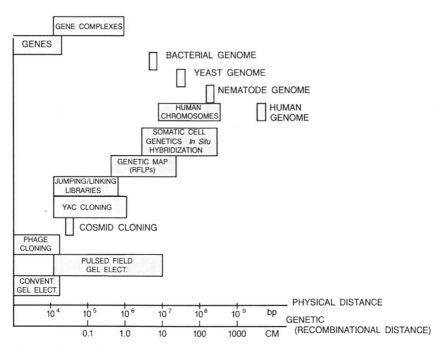

Figure 8.2. Scale of genomic physical mapping problem (genome sites) compared with sites of functional units (genes, gene complexes) and various technologies for physical mapping.

chromosome. The smallest human chromosome (#21) contains approximately 50 Mb of DNA. This is of the order of the size of the nematode genome and is within the capabilities of existing mapping techniques (Figure 8.2). Thus, if the human genome physical mapping problem is approached chromosome by chromosome, the mapping effort is tractable with an arsenal of cellular and molecular genetics technologies that can be used to establish physical order and distance between loci at various levels of resolution. Some of these have been discussed previously in the context of reverse-genetics approaches to identifying and isolating specific genes.

In the mapping of human or other mammalian genomes, somatic cell hybrids have unique roles and have been used extensively in assigning the chromosomal location of genes or anonymous DNA markers.[25] The power of this approach is that single human chromosomes or a subregion of an individual chromosome can be propagated in a heterologous cell, in almost all cases a rodent (mouse or hamster) cell line. The approaches for producing these hybrid cell lines are described in detail elsewhere.[45] Cell lines containing a single human chromosome (monochromosome hybrid) have been used as a source of specific human chromosomes for fluorescence-activated chromosome sorting.[46] Flow-sorted human chromosomes provide a high-purity source of chromosome-specific DNA for the construction of chromosome-specific recombinant libraries, which have multiple applications in genome mapping. The levels of resolution of the other techniques listed in Figure 8.2 have been described previously.

The construction of any physical map containing more than a few tens of elements requires modern computing resources. A data base can provide secure storage and convenient, selective retrieval for the large amounts of information gathered in high-resolution physical mapping. The "Laboratory Notebook" database used in the physical mapping of human chromosome 16, for example, contains sizes and hybridization signals for roughly 100,000 restriction fragments of about 4,000 clones, and overlap likelihoods for approximately 7 million clone pairs. For efficiency, the collection and interpretation should be computer assisted as well.[47, 48]

8.6. Paradigms for the Construction of Large-Scale Physical Maps

Several strategies have been used in constructing large-scale physical maps for the bacterial, yeast, and nematode genomes.[39, 41, 43] In general, these strategies have been directed toward constructing physical maps consisting of overlapping sets of genomic clones in phage ($\sim$ 17 kb) or cosmid ($\sim$ 40 kb) vectors. The requirements for these approaches include unbiased genomic libraries (i.e., libraries showing minimal bias with respect to sequences that are cloned and having the expected representation of single-copy sequences) and a robust procedure for identifying overlapping clones using a variety of "fingerprinting" methods. Most of these fingerprinting methods have been based on restriction enzyme digestions with four- or six-base-pair recognition enzymes (which cut genomic DNA approximately every 256 or 4,000 base pairs, respectively) followed by electrophoretic separation of the restriction fragments and recording of the image of the fragment profile. Fragment sizes are determined and the data are entered into an electronic data base. The fragment-size data from many randomly selected clones are compared computationally, and overlaps are determined using statistical techniques.[39, 41, 43, 49–51] A simple analogy is the assignment and comparison of bar codes for each of many hundreds or thousands of clones.

An illustration of these "bottom-up" approaches is shown in Figure 8.3. In the case of the *E. coli* physical map, complete restriction maps were constructed for each randomly chosen clone. This *restriction mapping* approach is advantageous in that it produces information on the order of fragments for each clone, which permits identification of overlapping clones with a smaller region of overlap than is required by a *restriction fingerprinting* approach, in which only restriction fragment sizes are compared with tests for clone overlap.[39, 40]

Other methods for identifying overlapping clones (including phage, cosmids, YACs and other large-fragment cloning vectors) have been developed and are

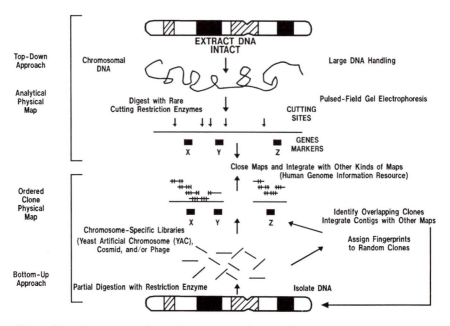

Figure 8.3. Comparison of complementary top-down and bottom-up strategies for genomic physical mapping. Multiple complementary strategies are required to construct complete physical maps, as discussed in the text.

being used primarily in completing maps of yeast and nematode genomes and for initiating the construction of physical maps of human chromosomes and other complex genomes.[32, 37, 44, 52–53] The method proposed by Poustka et al. uses oligonucleotide hybridization-based approaches on two-dimensional arrays of phage or cosmid clones on hybridization membranes. In this procedure a "binary fingerprint" is generated based on the hybridization, or lack thereof, of each of many (~ 60–100) probes applied independently to the array of recombinant clones. The binary fingerprints are analyzed using statistical methods, and overlapping clones are identified. Although this method has not been implemented in large-scale physical mapping of mammalian genomes, it has been used in a pilot study to map the Herpes simplex virus genome.[52]

In other hybridization-based mapping techniques, multiplexing procedures are employed to reduce the number of operations required to identify overlapping clones in two-dimensional filter arrays of individual clones[53] or to select a single YAC clone from a large pool of clones representing an entire genomic library.[37] This latter procedure utilizes synthetic short oligonucleotides designed to recognize a specific genomic region as primers for the polymerase chain reaction, which can identify the targeted genomic region with high sensitivity and specificity (see Chapter 7, this volume). This procedure will be discussed in more detail below.

In approaching the construction of a global physical map(s) of an entire genome, it is advantageous to subdivide the mapping problem to exploit the natural organization of the target genome. In the case of the human genome and the genomes of other higher eukaryotic species, the natural subdivision of the genome into 24 discrete types of chromosomes (22 autonomes and the X and Y sex chromosomes), coupled with instrumentation capabilities (e.g., fluorescence-activated chromosome sorting) and biological resources (e.g., monochromosome somatic cell hybrids), provides the opportunity to approach the physical (and genetic) mapping problems chromosome by chromosome. In addition to minimizing the size of the target, combinations of complementary strategies for producing physical maps of an entire genome or chromosome will be necessary to compensate for a number of interfering factors: (1) bias in sequence representation in recombinant libraries in various vector-host systems; (2) the presence of unclonable regions or regions that are prone to rearrangement or deletion during propagation; and (3) the presence of large regions (1–10 Mb) of simple sequence DNA (e.g., heterochromatic DNA containing various classes of tandemly repeated satellites), which would be difficult, if not impossible, to map by any of the fingerprinting or multiplex hybridization procedures. In Figure 8.3 one of the fundamental applications of macrorestriction mapping is in characterizing physical distances across regions that are unclonable or contain extended repetitive sequence arrays.

8.7. General Features of Human Genome Organization

The human genome and the complex genomes of other higher eukaryotic species (both animal and plant) differ in several respects from the smaller genomes of lower eukaryotic species. It is estimated that the $\sim 3 \times 10^9$ bp of the human genome encodes 50,000 to 100,000 genes distributed among 22 autosomal chromosomes and the X and Y sex chromosomes. In contrast to prokaryotic and lower eukaryotic genomes, a hallmark of higher eukaryotic genomes is the presence of multiple classes of repetitive DNA interspersed within and between these single-copy genes and multigene families.[55] It has been shown from studies of the abundance and distribution of DNA sequences in the human genome that there are at least four different major classes of interspersed repetitive sequences designated *Alu,* L1, THE, and $(GT:AC)_n$.[56–58] In addition to these interspersed "repeats," large tandem arrays of repetitive sequence have been identified at the centric constriction (centromere) of metaphase chromosomes and in the regions directly adjacent to the centromere.[59–61] These repetitive sequence classes are all designated satellite DNA but are subdivided into five distinct classes: alpha,[62] beta,[63] and satellites I, II, and III.[61]

The alpha satellite is found at the centromeres of all human chromosomes and occurs as a hierarchy of tandem arrays of a simple pentameric sequence.[63] The sizes of the alpha satellite arrays on different chromosomes in a single individual and among many individuals are highly variable, but in general, the alpha satellite region encompasses at least 1 Mb of DNA.[65] The variant centromeric satellite DNAs are found in varying abundances on specific chromosomes.[59] These satellite classes of repetitive sequence represent regions of the genome that may be intractable for the construction of overlapping clone maps but are accessible to physical mapping by pulsed field gel techniques.

Another discrete class of repetitive DNA sequence has been identified at the telomeres of all vertebrate chromosomes.[60, 66] The properties of this telomeric repetitive DNA have permitted the cloning of human telomeres in specially designed yeast artificial chromosome vectors (containing a yeast centromere, an autonomously replicating sequence, a yeast-selectable marker, a cloning site, and a trypanosome telomere, and also a bacterial origin of replication and bacterial-selectable marker), designated half-YACs because they have only one telomere and therefore require ligation with a human fragment containing a telomeric repeat in order to be propagated stably in the yeast host.[67] The identification and cloning of human telomeres represent a major milestone in mapping the human genome. This achievement provides landmarks for the physical ends of the 24 types of human chromosomes as well as cloned fragments from which polymorphic telomeric DNA sequences can be isolated to cap the genetic map.[68]

For the construction of physical maps of *interstitial* regions of chromosomes, alternative random mapping strategies that yield sets of overlapping cloned fragments can be implemented. Additional directed steps are required to link together and "anchor" or localize the cloned contigs to the genetic and cytogenetic maps. In addition to these contigs and information about their locations, it is desirable to use a strategy, or combination of strategies, that will produce information about the global organization of various classes of repetitive sequences in the human chromosome—information that can be used to exploit the properties of a chromosome for modifying or streamlining future mapping strategies and technologies.

8.8. Construction of a Physical Map of Human Chromosome 16

In our laboratory new approaches to genomic physical mapping have been developed and implemented to construct a physical map of an entire human chromosome. Chromosome 16 was selected as a target chromosome for physical mapping for multiple reasons: (1) the presence of several genetic disease loci, including the locus for polycystic kidney disease and a centric inversion of the chromosome involving specific breakpoints on short and long arms frequently observed in acute myelomonocytic leukemia (maps of human chromosome 16 are summarized by Reeders and Hildebrand[69]); (2) the availability of a large number (> 250) of regionally mapped anonymous DNA fragments, many of which are polymorphic genetic linkage markers; and (3) the availability of panels of somatic cell hybrids for regionally assigning the locations of contigs.[25]

Our approach to constructing a physical map of human chromosome 16 encompasses several phases that have been implemented in parallel. The first phase is focused on construction of both (1) long-range restriction maps of the centromeric satellite repetitive sequence regions and (2) overlapping cosmid contig maps of the interstitial regions of the chromosome. The second phase will extend and link up the cosmid contigs with additional cosmid or YAC (or other large-fragment insert) clones using combinations of random and directed methods with the goal of achieving 2-Mb contiguous cloned DNA for the entire interstitial, euchromatic regions of chromosome 16 (i.e., the regions between telomeres and the centromeric heterochromatin). The final phase of mapping will require (1) closure of the remaining gaps by directed approaches such as macrorestriction mapping, targeted YAC cloning of pulsed-field gel fragments, or directed screening of YAC or other large-fragment libraries and (2) integration of the overlapping clone maps with the cytogenetic and genetic maps. The first and second phases are heavily dependent on the construction of high-quality cosmid and YAC recombinant libraries and associated technologies for efficient handling of large numbers of samples for DNA preparation, restriction enzyme digestion, gel electrophoresis, and data acquisition, analysis, and management. The requirements for data analysis and management place demands on the development and implementation of both algorithms for computational analysis and data base management tools for handling the large amounts of data from other levels of mapping (Figure 8.1). The computational and data management tools developed for early phases of physical mapping are extensible to the final phase involving the integration of various types of mapping data.

In the first phase of the construction of a physical map of chromosome 16, the assembly of a "bottom-up" cosmid contig map is based on insights into genomic organization.[56] Our strategy has been developed to use specific classes of repetitive sequences in a fingerprinting scheme to increase the efficiency of identifying overlapping cosmid clones, thereby accelerating the progress in assembly of sets of contigs spanning 100-to-200-kb regions.[51, 56] Another advantage of this strategy is that it can be implemented in most standard molecular biology laboratories and does not require expensive instrumentation. The conceptual and experimental details of this strategy are illustrated in the next four figures.

An essential resource for initiating the cosmid contig mapping was a high-quality chromosome-specific cosmid library. A chromosome 16–specific cosmid library was constructed as part of the National Laboratory Gene Library Project.[46] Another fundamental component was the source of chromosomes for fluorescence-activated chromosome sorting, a mouse-human hybrid cell line designated CY 18 containing a single copy of human chromosome 16.[25] DNA isolated from the flow-sorted chromosomes was used to construct a cosmid library in the vector cCOS-1.[53] A 67-fold represented library was produced and a 10-fold coverage (26,000 recombinants) of the primary recombinants was arrayed in 96-well microtiter dishes, as illustrated in Figure 8.4. Similar arrays have been produced for several libraries representing other human chromosomes. Initial formatting of primary recombinants in arrays, or grids, is an essential step in maintaining an archival reference resource, since each clone has an identity given by the microtiter plate number and the position of the clone in the array. Information relating

CHROMOSOME ORGANIZATION AT THE MOLECULAR LEVEL

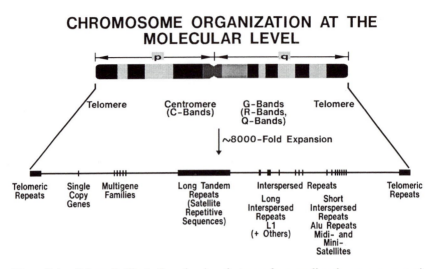

Figure 8.4. Schematic illustration of various features of mammalian chromosome organization.

to each clone is maintained in entries into a relational data base management system designated the Laboratory Notebook data base. This data resource is noted as the "reference materials data base for probes and contigs" in Figure 8.5.

The first step in the mapping effort involves the assessment of the quality of the cosmid library in terms of its purity and the representation of single-copy sequences in the library. With respect to the purity of the library, the major concern is with the presence of recombinants containing mouse genomic DNA, since the chromosomes 16 were sorted from a monochromosome #16 mouse-human hybrid. This feature underscores the value of sorting specific human chromosomes from monochromosome hybrids in that the only human DNA containing recombinants are from the target chromosome; all other recombinants contain nonhuman (rodent) DNA. It is straightforward to screen for mouse recombinants using total genomic mouse as a probe on replicas of the microtiter arrays of cosmid colonies that have been transferred to membrane filters. In the library described here the frequency of nonhuman recombinants was <8%. Multiple single-copy probes for loci known to occur on human chromosome 16 were used to screen subsets of the arrayed library, and all loci were found to be present at their expected frequencies.[51]

The advantages provided by the initial archival formatting of primary recombinant libraries in microtiter arrays are generally applicable whether the libraries are in cosmid, phage, YAC, or some other vector (Figure 8.5). This format permits rapid copying of arrays to additional microtiter dishes, as well as the replication of multiple primary arrays to a single membrane filter using laboratory automation to produce high-density arrays.[46, 52] The production and maintenance of microtiter and filter arrays provide excellent methods for tracking individual

HUMAN GENOME REFERENCE MATERIALS

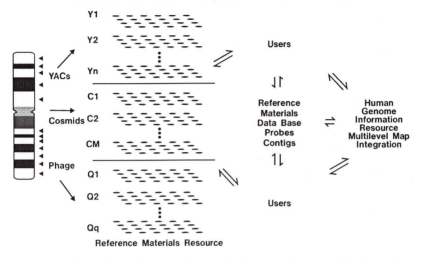

Figure 8.5. Illustration of a general approach for managing materials (recombinant clones in various vectors) and information for physical mapping project. The two-dimensional arrays of dots represent individual clones stored cryogenically on microtiter dishes or as DNA on filter membranes.

clones used in multiple mapping applications at several collaborating laboratories so that the various levels of mapping data (Figure 8.1) can be integrated rapidly. The utility of the cosmid arrays is demonstrated in our chromosome 16 project.

As was noted previously, the quality assurance step is essential in evaluating a recombinant library for any large-scale mapping application. In our physical mapping strategy the next step is to identify centromeric satellite DNA-containing clones in arrays of 7,500 of the original 26,000 recombinants. The identification of these clones allows them to be excluded from further analysis, since it would be difficult, if not impossible, to construct unambiguous cosmid contig maps for regions of tandemly repeated DNA. Independent studies have identified large blocks of both alpha satellite DNA at the centromeric constriction and a variant of satellite II DNA adjacent to the centromere on the long arm.[59] Pulsed-field gel electrophoresis measurements have shown that these blocks of satellite DNA encompass at least 5 Mb of DNA in the centromeric and pericentric regions of chromosome 16.[70] As was discussed previously, the construction of long-range restriction maps is a useful approach for collecting physical mapping information for regions that are unclonable as well as for regions in which cosmid contig maps may not be needed (or desired) for proceeding to the DNA sequence.

After satellite DNA-containing clones are identified in the arrays, the next steps use the interspersed repetitive sequences, that is, the $(GT)_n$ repeats are used to select clones for fingerprinting from the arrays. These repeats appear to be uniformly distributed along the euchromatic arms of human chromosomes. The

REPEAT FINGERPRINTING DATA ACQUISITION

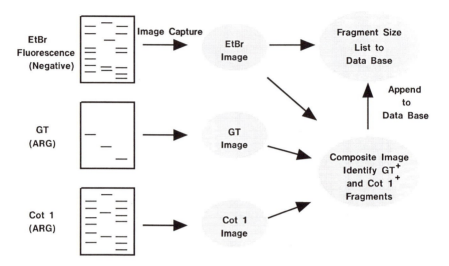

Figure 8.6. Schematic representation of cosmid clone fingerprint data acquisition and analysis using the repetitive DNA sequence fingerprinting strategy.[51]

$(GT)_n$ repeats are also used as probes to generate restriction fragment fingerprints that produce both restriction fragment size profiles and information on the presence or absence of these $(GT)_n$ repetitive sequences (as well as *Alu* and L1 repeats) on individual fragments.[51] The essential features of this fingerprinting strategy are illustrated in Figures 8.6, 8.7, and 8.8.

In the general case, the restriction fragment fingerprinting scheme developed in our laboratory uses both restriction fragment length and the presence or absence of specific repetitive sequence classes (e.g., GT, *Alu,* and L1 repeats) identified by hybridization with labeled repetitive sequence probes. The additional information provided by the combined fragment length and hybridization status significantly reduces the number of shared fragments between two clones required to declare that the clones overlap with high probability.[51]

In the specific case as implemented in our laboratory, the initial set of clones selected for fingerprinting are identified as having $(GT)_n$ repeats by screening the 7,500-clone subset of the initial 26,000-clone array. The rationale for preselecting the $(GT)_n$ positive clones for fingerprinting is 2-fold: (1) preselection increases progress in identifying overlapping pairs of cosmids that share at least one $(GT)_n$-containing fragment; and (2) more important, the information provided by superimposing the hybridization status onto each restriction fragment allows assignment of overlapping cosmids with high probability even if two cosmids share only one $(GT)_n$ positive fragment and overlap by <20% [i.e., the hybridization status is defined as the presence of a $(GT)_n$ repeat or some combination of the more abundant *Alu* and L1 repeats identified by a mixed probe designated

REPEAT FINGERPRINTING DATA FLOW

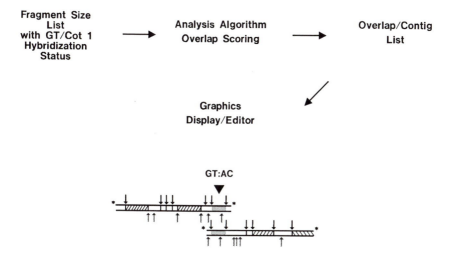

Fragment Size List with GT/Cot 1 Hybridization Status ⟶ Analysis Algorithm Overlap Scoring ⟶ Overlap/Contig List

Graphics Display/Editor

GT:AC

Figure 8.7. Analysis of repetitive sequence fingerprinting data to identify clones containing a shared (GT) repeat detected by the overlap detection algorithm.

Cot 1 containing the most abundant total genomic DNA sequences isolated by kinetic fractionation to a Cot value of 1].[51, 56] The preselection of $(GT)_n$ repeats and subsequent application of the repetitive sequence fingerprinting approach achieve significantly accelerated progress in contig assembly in the early phase of the mapping project. In fact the progress in contig growth is substantially greater (4-fold) than expected by any purely random bottom-up fingerprinting strategy.[49, 51]

The basic steps in the fingerprinting process are outlined in Figure 8.6. DNA is isolated from individual cosmid clones and three separate aliquots of each DNA isolate are digested with the restriction enzymes *Eco*RI, *Hind*III, and a combination of *Eco* + *Hin*d. The restriction digests are analyzed by standard agarose gel electrophoresis. Following electrophoresis the gel is stained with the fluorescent dye, ethidium bromide, and an image of the DNA fragments is recorded on positive or negative film. DNA is transferred to membrane filters by conventional Southern blotting techniques (see Chapter 7, this volume). Filters are hybridized with radioisotropically labeled probes to detect (1) $(GT)_n$ repeats and (2) Cot 1 repeats (i.e., primarily *Alu* and L1 repeats). Autoradiographic (ARG) images of the hybridized Southern blots are acquired on X-ray film. Filters are treated to remove the radiolabeled repeat probes and are reprobed with probes for the cosmid vector so that the end fragments containing both genomic DNA and vector DNA can be identified.

The negative from the fluorescence photograph (Figure 8.6) is captured using an image-processing workstation, and the image is processed to generate files of

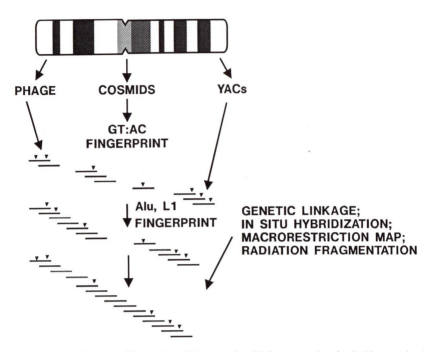

Figure 8.8. Schematic illustration of the use of multiple approaches for linking up short cosmid contigs into contigs that span 1 to 2 Mb regions and are ordered and oriented along the chromosome (more detail is provided in Figure 8.9).

fragment sizes for each lane of the gel based on the sizes of known markers in reference lanes. These image files also contain information on the optical densities of the fragments, which can be used in editing the image to identify doublets or triplets, that is, fragments of similar size that comigrate in the gel.

Most of the information management for the physical mapping project takes place on a network of Sun workstations. The server on this network, a Sun 4/280, supports a Laboratory Notebook data base, implemented as a relational data base using the Sybase Relational Database Management System (RDBMS). Because Sybase supports a client–server architecture, both the data in the Laboratory Notebook and the Sybase user interface are available on every workstation on the network (in fact, subject to security constraints, on every workstation worldwide connected to the Internet).

The report from the image-processing workstation, containing restriction fragment sizes and other information from the fluorescence photograph, is transferred to a Sun workstation, and parsed, and the information is placed in the data base as soon as it is generated. Raster images of the autoradiographs are made on a desktop scanner and stored on the Sun network as well.

The next operational step requires the comparison of the ethidium bromide (EtBr) image with ARG images to acquire and record the hybridization status of each restriction fragment. For a few clones and images this operation can be handled manually by overlaying photographic prints of the EtBr image with the ARG films. For the large number of clones involved in high-resolution physical mapping, handling the films on a light box and recording the data by hand is too time-consuming and error-prone. Thus, we developed a program called SCORE to compare the images and record the hybridization scores.[47] SCORE first consults the data base and retrieves the fragment sizes for all lanes of a particular gel. Using these, it constructs a cartoon of the EtBr image in green, on a color monitor. Next it retrieves a named ARG image and shows it, in grayscale, roughly aligned with the EtBr image. The user stretches the ARG image to align it with the EtBr image using the workstation mouse. Next, the user points to each fragment (again using the mouse), is given a menu of possible scores, and records the hybridization score. When the user finishes, SCORE records all the hybridization scores in the Laboratory Notebook.

Analysis of the data for clone-fragment size and hybridization status to determine pairwise likelihoods for overlap is performed by Bayesian statistical methods developed at Los Alamos.[71] The basic elements of this approach and the algorithm have been presented.[51] An illustration of the data flow path for the repeat fingerprinting strategy is shown in Figure 8.7. The output of this program is a list of pairs of overlapping clones. This list is subsequently sorted to compile a list of contigs.[72] A schematic representation of an overlapping pair with a shared $(GT)_n$ fragment indicated is shown in Figure 8.7.

The likelihood of overlap derived from clone fingerprint data is, for every possible pair of clones, stored in the Laboratory Notebook data base. It is important to note that low likelihoods provide information just as important as that from high ones. In addition to the likelihoods from fingerprint data, a much smaller body of overlap data derived from other sources (e.g., grid hybridization) is also stored in the same table of the data base. A genetic algorithm has been developed[73] to find contig maps that fit all of the pairwise overlap data as well as possible.

The rate of progress of the cosmid contig assembly using the repeat fingerprinting strategy exceeds, by at least a factor of four, any of the other bottom-up fingerprinting schemes.[51] Currently, after fingerprinting 3,145 $(GT)_n$ positive cosmids, the statistical analysis assigned these to 460 contigs with an average contig size of 106 kb. Thus, approximately 47.8 Mb (or 54%) of the euchromatic arms of human chromosome 16 is assigned to contigs. The longest contig spans 240 kb. Pulsed-field gel electrophoresis has been used to confirm that the cosmid contig maps are consistent and colinear with the genomic DNA from the mono-chromosome 16 hybrid, CY 18, from which chromosomes were sorted for construction of the cosmid library. In all cases examined, including the 240-kb contig, the macrorestriction data confirm the colinearity of the cosmid contig maps with the genomic DNA.[51]

Integration of known genes and anonymous DNA markers, including polymorphic genetic linkage markers, into the cosmid contig maps is accomplished by hybridizing probes for these loci to the primary cosmid arrays or reformatted subsets of the contigs that have been assembled. In analyzing the results of these measurements, the Laboratory Notebook again plays a pivotal role in integrating the cosmid contig physical maps with the known loci, for example, α-globin (HBA), metallothionein (MT), and polycystic kidney disease (PKD1).[69]

To assign individual contigs to the cytogenetic map, several approaches are being used including (1) the regional mapping panels of somatic cell hybrids containing specific subregions of chromosome 16,[25] and (2) in situ hybridization of single cosmids from contigs to human metaphase chromosomes.[74]

In interrelating different levels of maps (Figure 8.1), it is essential to link the cosmid contig maps to the genetic map. This step is fundamental to placing the physical maps in order along the chromosome and locating the cloned physical maps relative to disease loci, or loci of other phenotypic importance (e.g., chromosome translocation breakpoints). In this application, the repeat fingerprinting strategy provides the resources and information for integrating the physical cosmid contig maps with the genetic linkage map using the highly polymorphic $(GT)_n$ repeats.[75] Detailed contig mapping information can be used to select cosmids having GT repeats at the ends of the largest contigs for use in genetic linkage analysis. The value of this approach to integrating genetic and physical maps is that new tools [i.e., sequence tagged sites (STSs), discussed in some detail presently, for contig linkup and map closure can be produced as part of the GT polymorphism analysis.[76]

The approach to detection of GT polymorphisms requires DNA sequence information flanking both sides of the GT repeat. This information is needed to produce unique, short, synthetic DNA sequence primers for the polymerase chain reaction (PCR) to amplify the region containing the GT repeat. (The PCR technology as applied to linkage analysis is described in chapter 7 of this volume.) In detecting GT polymorphisms, the PCR product from DNA samples from random individuals, multigenerational families, or disease families is analyzed on a high-resolution polyacrylamide gel to determine length variation associated with a single locus in specific GT repeats. This information can then be used to determine linkage of the polymorphism relative to other closely linked markers.[75]

A schematic representation of a low-resolution physical map is shown in Figure 8.8 to illustrate the results of cosmid contig mapping, ordering of contigs by genetic linkage, in situ hybridization, or radiation hybrid mapping coupled with directed approaches to link up contigs. A more detailed illustration of the second phase of physical mapping, the extension of short contigs and linking up of contigs separated by both small and large gaps, is shown in Figure 8.9. Figure 8.9 also illustrates the power of STS markers as a "common language" for interrelating maps produced by different technical approaches.

Figure 8.9 summarizes the essential aspects of the second phase in constructing a complete physical map of cloned, contiguous DNA for the euchromatic arms of chromosome 16. For extension of existing contigs and closure of small gaps between cosmid cortigs, additional fingerprinting of GT positive clones, as well as GT negative clones, is expected to yield significant progress. For larger

CONTIG EXTENSION AND MAP CLOSURE

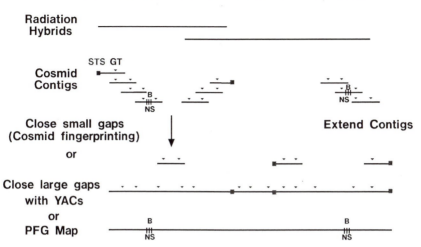

Figure 8.9. Multiple approaches for contig extension and map closure. The value of sequence tagged sites (STS) is again represented here as stippled boxes.

gaps, perhaps including regions not represented in the cosmid library, it will be necessary to incorporate YAC clones into our mapping approach. YAC clones that span large gaps could be identified by pooled screening techniques[37] using the PCR method with primers produced from sequence information obtained from the ends of the longest cosmid contigs. Depending on the location of a single-copy sequence (that could be used as an STS) relative to the GT repeat(s), it may be possible to acquire in one sequencing analysis sufficient DNA sequence information for both an STS and a polymorphic marker for use in merging physical and genetic linkage maps. These STS markers can also serve in PCR approaches to closing large gaps using radiation hybrids,[77] although this approach does not provide cloned DNA that can be used for further analysis.

The third phase of the physical mapping effort involves anchoring the cloned physical map with respect to the cytogenetic and genetic linkage maps. While this phase is being incorporated in parallel with the earlier phases of cosmid contig mapping, the combinatorics involved in integrating and representing these multiple levels of maps as the number of data elements increases will require additional advances in the areas of data management and analysis.

8.9. The Rapid Evolution of Physical Mapping Technology and Considerations for the Construction of Physical Maps of Other Human Chromosomes and Other Genomes

The invention of large-fragment cloning using YACs and the recent developments in both polymerase chain reaction technology and applications have provided new tools that have revolutionized conceptual approaches to physical mapping and the integration of maps at different levels of resolution. The construction of well-represented total human genomic YAC libraries[36] and the application of pooled PCR screening using PCR primer pairs[37] for specific genomic regions have led to rapid methods for isolating large regions of the human genome adjacent to or containing a disease gene or other genes or features of interest.[76] Further developments in the area of large-fragment cloning and the multiplexing of PCR screening procedures promise to yield new strategies for constructing long-range physical maps of cloned DNA representing megabase stretches of the genome. Although large-fragment cloning vehicles have the advantage of providing rapid access to large regions of the genome in one step, there remains the problem of going from a large-fragment clone to the DNA sequence of specific subregions within the large-fragment clone. Subcloning YAC clones into vectors from which DNA sequencing can proceed is a significant task, since only a small percent of a YAC clone is human DNA insert. Efficient methods for obtaining DNA sequence from YAC clones will be needed. As the area of large-scale genomic physical mapping matures, several important features are beginning to emerge as characteristic of successful physical mapping strategies. These features include modularity, parallelism, portability, connectivity, and distributivity.

Modularity is represented by the availability of materials that subdivide the genome into entities or modules of various sizes ranging from a single chromosome to a chromosome subregion to a 500-kb YAC clone. Modules may be available as somatic cell hybrids, chromosome-specific libraries, or total genomic YAC libraries. *Parallelism* can be achieved by multiple laboratories working on distinct but interrelated aspects of a mapping problem simultaneously—for example, the cytogenetic map, the genetic linkage map, the macrorestriction map, the cloned contig map, or the DNA sequence. *Portability* can be achieved through the availability of DNA sequence information as STS markers, which can be communicated electronically between laboratories for use in the selection of specific YAC clones. The desired YAC clones can be transferred to the requesting laboratory. *Connectivity* is attained rapidly over long ranges by identifying overlapping YAC clones by some type of multiplexing strategy or by directed walking from markers (again, a high density of genetic linkage markers or STS markers plays an important role). The *distributivity* of both materials and information is essential to merging the various types of data and the multiple levels of maps.

A number of factors should be considered before an investigator (or laboratory) embarks on a global physical mapping effort of an entire genome or even a single chromosome of a complex organism. These include the size of the target genome or chromosome and the desired rate of progress required to obtain the DNA sequence of the region of interest. Quantitative models and simulations of various mapping strategies applied to specific genomes will make it possible to compare predicted mapping progress with actual experience from laboratories that have undertaken large-scale mapping and sequencing projects. Such comparisons will be useful in deciding how to modify mapping (or sequencing) efforts to make them more efficient or when to shift from one mapping strategy to another (e.g., from a random, bottom-up approach to directed schemes to achieve closure).

8.10. Integration of Physical Maps with Other Kinds of Maps

Multiple paths for constructing physical maps of complex genomes will continue to employ a variety of ingenious molecular genetics tools and technological advances, many of which remain to be invented. In addition, communications and computing will be key components in bringing the mapping information together, analyzing and managing the information, and disseminating the information to users. Paradigms for information management in areas of DNA sequence,[79] in human gene mapping,[80] and in electronic data publishing,[81] have been developed and their value demonstrated. As the magnitude of physical mapping data and DNA sequence information increases, there will be an even greater need for high-speed communications and large-scale, flexible data base management systems.[82]

8.11. Obtaining the DNA Sequence: The Ultimate Physical Map

With the initiation of multiple large-scale mapping projects on several human chromosomes and the accumulation of cloned DNA representing large regions ($\sim$ 1–2 Mb) of the human genome, rapid methods for acquiring DNA sequence from these regions will be needed. Numerous approaches to this problem have been presented in several previous reviews.[83–84]

In the initial phases of discussion of the approaches to sequencing the human genome, there was a tendency to view the project primarily as an engineering challenge to improve DNA sequencing speeds and decrease costs several orders of magnitude so that sequencing the 3 billion base pairs within the next 15 years could be accomplished by random, or shotgun, sequencing of short cloned segments, which would be stitched together to form the final sequence.[20, 38] As was discussed in the beginning of this chapter, cloned physical maps for individual human chromosomes coupled with high-resolution genetic linkage maps provide essential tools for rapidly accessing a region containing a target disease gene or, perhaps, a gene involved in disease predisposition. With high-resolution physical maps available, it is possible to implement directed DNA sequencing strategies in the target region. With current automated sequencing technology, approximately 10,000 bases of "raw" (i.e., unedited and in contiguous stretches of not more than 800 bases) can be produced by a single sequenator in one day. Much more raw sequence is required to produce finished (i.e., contiguous, overlapping sequence) for large regions. Multiple subcloning or direct sequencing approaches can be taken to obtain the DNA sequence of a 40-kb cosmid insert. In the most optimal situation a cosmid insert could be sequenced in a few weeks. Therefore, with high-resolution physical maps consisting of overlapping cosmid clones, acquisition of the DNA sequence for regions encompassing several hundred

MULTIDIMENSIONAL MAPS

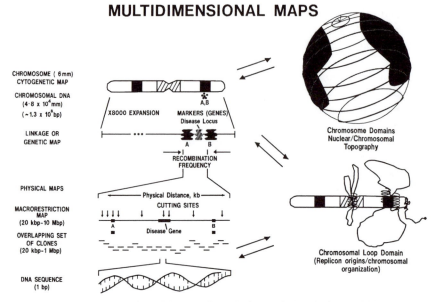

Figure 8.10. Representation of the use of physical maps in developing multidimensional maps of genome organization. Various techniques can be applied to utilize physical mapping material and information to map the organization of large genomic regions in interphase nuclei and as mitotic chromosomes. These applications can be extended to reveal changes in the genome organization associated with cellular differentiation states or with aberrant disease states (e.g., neoplasia).

kilobases in searching for a gene should be feasible in a period of months, especially if several sequenators were utilized in parallel.

In addition to the value of physical maps in providing rapid access to a target gene, these maps provide the resources to explore features of chromosome structure and dynamics in novel ways. Instead of viewing a chromosome as a monolithic structure, detailed physical maps can be coupled with new technologies for visualizing the organization of the genome in nuclei and in chromosomes. This will allow the definition of previously inaccessible features of chromosome organization, such as the topographical organization of a chromosome in an interphase nucleus and the hierarchies of organization and packaging of metaphase chromosomes (Figure 8.10). It will be possible to understand specific relationships between the distribution of the various classes of repetitive sequences and chromosome architecture and dynamics.

8.12. Interpreting the DNA Sequence and Potential Applications in the Diagnosis, Prevention, and Treatment of Human Genetic Disease

As mapping projects proceed, high-resolution physical maps of the human genome will be correlated with the genetic linkage map. Targeted sequencing of specific regions of the genome will lead to the identification of the gene(s) whose altered structure or function are the cause of a genetic disorder or predisposition to disease.[1, 8, 19] Numerous specific genes or multigene families identified by genomic mapping and sequencing will be important in disease diagnosis and the design of novel therapeutic agents. Among these will be genes encoding peptide hormones, cell surface receptors, and genes that function in modulating cell growth and division. Defects in these classes of genes are associated with diseases such as cystic fibrosis, asthma, schizophrenia, myasthenia gravis, and various cancers.[85]

High-resolution genetic and physical maps will rapidly narrow the search for the disease gene to a region of perhaps 1 Mb. Using the high-resolution physical map coupled with directed and automated DNA sequencing approaches, it is conceivable that the DNA sequence for a region of this size could be completed in less than a year. The task remains to identify the gene(s) and regulatory regions within this sequenced region. Once the target gene associated with the disease phenotype is identified, rapid sequencing technologies or other DNA diagnostic techniques[86] can be used to compare this gene in DNA from normal and affected individuals. Identification of consistent differences in the genes between normal and affected individuals can be used to define the specific mutation or mutational spectra characteristic of the disease. Definition of the molecular nature of the mutation might be used directly with computational techniques to determine whether the mutation is likely to affect a specific structural or functional domain of the gene product (enzyme or structural protein). Knowledge of the defect at the molecular structure and function levels can provide the framework for understanding the molecular pathogenesis and indicate paths toward treatment or early diagnosis coupled with preventative therapy. In some cases of inherited enzyme deficiencies and hemoglobinopathies, the possibilities for using somatic gene transfer therapies are under intense investigation. Although we have not addressed in this chapter the importance of constructing physical maps of other complex species, notably the mouse, it will be essential to identify mouse genes having homology with their human "disease" gene counterparts so that the mouse can be exploited as an experimental system to study the molecular pathogenesis of, and potential therapies for, a specific disease or deficiency.

It will remain a challenge for future pharmaceutical scientists, molecular biologists, geneticists, computational scientists, and genetic engineers working independently and in multidisciplinary teams to invent new approaches to designing, developing, testing, and applying therapeutic agents and strategies based on a molecular understanding of defective structural proteins (e.g., cell surface receptors) or enzymes involved in human pathogenesis.

Acknowledgments

The authors acknowledge the dedicated and expert technical assistance from the physical mapping team, including T. G. Tesmer, L. M. Clark, N. C. Brown,

L. M. Meincke, L. Saunders, A. C. Munk, S. Anzick, and J. Graeber. We are grateful for valuable discussions with National Laboratory Gene Library Project investigators Dr. L. L. Deaven (Director), and J. L. Longmire. Computing systems administration support by Dr. J. H. Jett and M. Wilder and information management by the Human Genome Information Resource team at Los Alamos National Laboratory are gratefully acknowledged. We are indebted to Monica Fink for preparation of the manuscript through many revisions.

Note Added in Proof: During the preparation of this volume, significant advances have been made in the technologies and resources for assembling physical maps of complex genomes.[87–89] Genomic libraries, notably for the human genome, have been constructed in yeast artificial chromosome (YAC) vectors carrying genomic inserts ranging in size from a few hundred kb to greater than 1 megabase. The increasing availability of these libraries coupled with the generation of an increasing number of STS markers, many of which are highly polymorphic and closely spaced genetic markers,[90] has made it possible to construct contig maps of overlapping YAC clones spanning the entire euchromatic region of the human Y chromosome[91] and genomic regions up to 8 Mb for the human X chromosome.[88] While these low-resolution cloned physical maps are valuable tools in the subsequent search for disease genes, substantial effort is still required to identify genes in even one member of a YAC contig since subcloning into cosmids followed by direct sequencing, or further subcloning into specially designed sequencing vectors and then sequencing, is necessary to identify disease-causing genes and the specific mutations characteristic of the disease. Hence, the high-resolution cosmid maps being constructed for extended genomic regions (such as described in this chapter) will be very useful in the disease gene searches and in understanding chromosome structure at high resolution, as well.[51] Furthermore, consistent problems with chimeric clones reported for the currently available total genomic YAC libraries[88] suggest that problems may be encountered in identifying a disease gene in an inaccurate map. The higher-resolution cosmid contig maps derived from cosmid libraries constructed directly from either total genomic DNA or flow-sorted chromosomes would avoid errors associated with potential problems in the use of YAC contig maps.[51]

New approaches to cloning a disease gene, once its approximate location is known, have also been developed. These approaches are now collectively designated "positional cloning."[92] Previously, these approaches were called reverse genetics. Another change in nomenclature has been applied to the cloning of a disease gene when the gene product (protein or enzyme) is known. This process, previously designated forward genetics, is now called "functional cloning."[92]

Dramatic developments in DNA sequencing technology have shown the feasibility of hybridization based approaches for obtaining DNA sequence information and for comparing DNA sequences.[93,94]

Finally, the recently founded Genome Data Base provides informational resources for acquiring, managing, and disseminating the enormous amounts of new genomic mapping data for the scientific community.[95]

References

1. Bodmer, W.F. 1986. Human genetics: The molecular challenge. *Cold Spring Harbor Symp. Quant. Biol.* 51:1–13.

2. White, R., Leppert, M., O. Connell, P., Nakamura, Y., Julier, C., Woodward, S., Silva, A., Wolff, R., Lathrop, M., and Lalouel, J.-M. 1986. Construction of human genetic linkage maps. I. Progress and perspectives. *Cold Spring Harbor Symp. Quant. Biol.* 51:29–38.

3. Watkins, P.C. 1988. Restriction fragment length polymorphism (RFLP): Applications in human chromosome mapping and genetic disease research. *BioTechniques* 6:310–320.

4. Lander, E.S., and Botstein, D. 1989. Mapping Mendelian factors underlying quantitative traits using RFLP linkage maps. *Genetics* 121:185–199.

5. Knudson, A.G. 1986. Genetics of human cancer. *Annu. Rev. Genet.* 20:231–251.

6. Russell, D.W., Lehrman, M.A., Sudhof, T.C., Yamamoto, T., Davis, C.G., Hobbs, H.H., Brown, M.S., and Goldstein, J.L. 1986. The LDL receptor in familial hypercholesterolemia: Use of human mutations to dissect a membrane protein. *Cold Spring Harbor Symp. Quant. Biol.* 2:811–819.

7. McKusick, V.A., and Ruddle, F.H. 1987. Editorial: Toward a complete map of the human genome. *Genomics* 1:103–106.

8. Friedmann, T. 1990. Opinion: The human genome project—Some implications of extensive reverse genetic medicine. *Am. J. Hum. Genet.* 46:407–414.

9. McKusick, V. 1986. *Mendelian inheritance in man,* 7th ed. Baltimore: Johns Hopkins University Press.

10. Rommens, J.M., Iannuzzi, M.C., Kerem, B., Drumm, M.L., Melmer, G., Dean, M., Rozmahel, R., Cole, J.L., Kennedy, D., Hidaka, N., Zsiga, M., Buchwald, M., Riordan, J.R., Tsui, L., and Collins, F.S. 1989. Identification of the cystic fibrosis gene: Chromosome walking and jumping. *Science* 245:1059–1065.

11. Riordan, J.R., Rommens, J.M., Kerem, B., Alon, N., Rozmahel, R., Grzelczak, Z., Zielenski, J., Lok, S., Plavsic, N., Chou, J., Drumm, M.L., Iannuzzi, M.C., Collins, F.S., and Tsui, L. 1989. Identification of the cystic fibrosis gene: Cloning and characterization of complementary DNA. *Science* 245:1066–1073.

12. Kerem, B., Rommens, J.M., Buchanan, J.A., Markiewicz, D., Cox, T.K., Chakravarti, A., Buchwald, M., and Tsui, L. 1989. Identification of the cystic fibrosis gene: Genetic analysis. *Science* 245:1073–1080.

13. Cawthon, R.M., Weiss, R., Xu, G., Viskochil, D., Culver, M., Stevens, J., Robertson, M., Dunn, D., Gesteland, R., O'Connell, P., and White, R. 1990. A major segment of the neurofibromatosis type 1 gene: cDNA sequence, genomic structure, and point mutations. *Cell* 62:193–201.

14. Viskochil, D., Buchberg, A.M., Xu, G., Cawthon, R.M., Stevens, J., Wilff, R.K., Culver, M., Carey, J.C., Copeland, N.G., Jenkins, N.A., White, R., and O'Connell, P. 1990. Deletions and a translocation interrupt a cloned gene at the neurofibromatosis type 1 locus. *Cell* 62:187–192.

15. Wallace, M.R., Marchuk, D.A., Andersen, L.B., Letcher, R., Odeh, H.M., Saulino, A.M., Fountain, J.W., Brereton, A., Nicholson, J., Mitchell, A.L., Brownstein,

B.H., and Collins, F.S. 1990. Type n1 neurofibromatosis gene: Identification of a large transcript disrupted in three NF1 patients. *Science* 249:181–186.

16. Groden, J., Thliveris, A., Samowitz, W., Carlson, M., Gelbert, L., Albertsen, H., Joslyn, G., Stevens, J., Spirio, L., Robertson, M., Sargeant, L., Krapcho, K., Wolff, E., Burt, R., Hughes, J.P., Warrington, J., McPherson, J., Wasmuth, J., LePaslier, D., Abderrahim, H., Cohen, D., Leppert, M., and White, R. 1991. Identification and characterization of the familial adenomatous polyposis coli gene. *Cell* 66:589–600.

17. Joslyn, G., Carlson, M., Thliveris, A., Albertsen, H., Gelbert, L., Samowitz, W., Groden, J., Stevens, J., Spirio, L., Robertson, M., Sargeant, L., Krapcho, K., Wolff, E., Burt, R., Hughes, J.P., Warrington, J., McPherson, J., Wasmuth, J., LePaslier, D., Abderrahim, H., Cohen, D., Leppert, M., and White, R. 1991. Identification of deletion mutations and three new genes at the familial polyposis locus. *Cell* 66:601–613.

18. Kinzler, K.W., Nilbert, M.C., Su, L.K., Vogelstein, B., Bryan, T.M., Levy, D.B., Smith, K.J., Preisinger, A.C., Hedge, P., McKechnie, D., Finniear, R., Markham, A., Groffen, J., Boguski, M.S., Altschul, S.F., Horii, A., Ando, H., Miyoshi, Y., Miki, Y., Nishisho, I., and Nakamura, Y. 1991. Identification of FAP locus genes from chromosome 5q21. *Science* 253:661–665.

19. Caskey, C.T. 1986. Summary: A milestone in human genetics. *Cold Spring Harbor Symp. Quant. Biol.* 2:1115–1119.

20. Office of Technology Assessment 1988. *Mapping our genes—the genome projects: How big, how fast?* Washington, D.C.: U.S. Congress, Office of Technology Assessment.

21. Orkin, S.H. 1986. Reverse genetics and human disease. *Cell* 47:845–850.

22. Sambrook, J., Fritsch, E.T., and Maniatis, T. 1989. *Molecular cloning: A laboratory manual*, 2nd ed. Cold Spring Harbor, N.Y.: Cold Spring Harbor Laboratory.

23. Botstein, D., White, R.L., Skolnick, M., and Davis, R.W. 1980. Construction of a genetic linkage map in man using restriction fragment length polymorphisms. *Am. J. Hum. Genet.* 32:314–331.

24. White, R., and Lalouel, J.-M. 1988. Chromosome mapping with DNA markers. *Sci. Am.* 258(2):40–48.

25. Callen, D., Hyland, V.J., Baker, E.G., Fratini, A., Simmers, R.N., Mulley, J.C., and Sutherland, G.R. 1986. Fine mapping of gene probes and anonymous DNA fragments to the long arm of chromosome 16. *Genomics* 2:144–153.

26. Donis-Keller, H., Green, P., et al. 1987. A genetic linkage map of the human genome. *Cell* 51:319–337.

27. Gall, J.G., and Pardue, M.L. 1969. Formation and detection of RNA-DNA hybrid molecules in cytological preparations. *Proc. Natl. Acad. Sci. U.S.A.* 78:3755–3759.

28. Verma, R.S., and Babu, A. 1989. *Human chromosomes: A manual of basic techniques*. New York: Pergamon.

29. Cox, D.R., Pritchard, C.A., Uglum, E., Casher, D., Kobori, J., and Myers, R.M. 1989. Segregation of the Huntington disease region of human chromosome 4 in a somatic cell hybrid. *Genomics* 4:397–407.

30. Schwartz, D.C., and Cantor, C.R. 1984. Separation of yeast chromosome-sized DNAs by pulsed-field gel electrophoresis. *Cell* 37:67–75.

31. Cantor, C.R., Smith, C.L., and Mathew, M.D. 1988. Pulsed-field gel electrophoresis of very large DNA molecules. *Annu. Rev. Biophys. Biophys. Chem.* 17:287–304.

32. Poustka, A., Pohl, T., Barlow, D.P., Zehetner, G., Craig, A., Michiels, F., Ehrich, E., Frischauf, A.M., and Lehrach, H. 1986. Molecular approaches to mammalian genetics. *Cold Spring Harbor Symp. Quant. Biol.* 51:159–167.

33. Clark, L., and Carbon, J. 1976. A colony bank containing synthetic ColE1 hybrid plasmids representative of the entire *E. coli* genome. *Cell* 9:91–101.

34. Seed, B., Parker, R.C., and Davidson, N. 1982. Representation of DNA sequences in recombinant DNA libraries prepared by restriction enzyme partial digestion. *Gene* 19:201–209.

35. Collins, F.S., and Weissman, S.M. 1984. Directional cloning of DNA fragments at a large distance from an initial probe: A circularization method. *Proc. Natl. Acad. Sci. U.S.A.* 81:6812–6816.

36. Burke, D.T., Carle, G.F., and Olson, M.V. 1987. Cloning of large segments of exogenous DNA into yeast by means of artificial chromosome vectors. *Science* 236:806–812.

37. Green, E.D., and Olson, M.V. 1990. Systematic screening of yeast artificial-chromosome libraries by use of the polymerase chain reaction. *Proc. Natl. Acad. Sci. U.S.A.* 87:1213–1217.

38. National Research Council. 1988. *Mapping and sequencing the human genome.* Report of Committee on Mapping and Sequencing the Human Genome. National Research Council, National Academy of Sciences, Washington, D.C.

39. Kohara, Y., Akiyama, A., and Isono, K. 1987. The physical map of the whole *E. coli* chromosome: Application of a new strategy for rapid analysis and sorting of a large genomic library. *Cell* 50:495–508.

40. Daniels, D.L., and Blattner, F.R. 1987. Mapping using gene encyclopedias. *Nature* (London) 325:831–832.

41. Olson, M.V., Dutchik, J.E., Graham, M.Y., Brodeur, G.M., Helms, C., Frank, M., MacCollin, M., Scheinman, R., and Frank, T. 1986. Random-clone strategy for genomic restriction mapping in yeast. *Proc. Natl. Acad. Sci. U.S.A.* 83:7826–7830.

42. Smith, C.L., Econome, J.G., Schutt, A., Klco, S., and Cantor, C.R. 1987. A physical map of the *Escherichia coli* K12 genome. *Science* 236:1448–1453.

43. Coulson, A., Sulston, J., Brenner, S., and Karn, J. 1986. Toward a physical map of the genome of the nematode *Caenorhabditis elegans. Proc. Natl. Acad. Sci. U.S.A.* 83:7821–7825.

44. Coulson, A., Waterston, R., Kiff, J., Salston, J., and Kohara, Y. 1988. Genome linking with yeast artificial chromosomes. *Nature (London)* 335:184–186.

45. Vogel, F., and Motulsky, A.G. 1986. *Human genetics: Problems and approaches.* New York: Springer-Verlag.

46. Deaven, L.L., VanDilla, M.A., Bartholdi, M.F., Carrano, A.V., Cram, L.S., Fuscoe, J.C., Gray, J.W., Hildebrand, C.E., Moyzis, R.K., and Perlman, J. 1986. Construction of human chromosome-specific DNA libraries from flow-sorted chromosomes. *Cold Spring Harbor Symp. Quant. Biol.* 51:159–167.

47. Cannon, T.M., Koskela, R.J., Burks, C., Stallings, R.L., Ford, A.A., Hempfner, P.E., Brown, H.T., and Fickett, J.W. 1991. A program for computer-assisted scoring of Southern blots. *BioTechniques* 10:764–767.

48. White, S.W., Torney, D.C., and Whittaker, C.C. 1990. A parallel computational approach using a cluster of IBM ES/3090 600 Js for physical mapping of chromosomes. *Proc. Supercomputing 90: IEEE Computing Society and ACM SIGARCH.* New York, N.Y., Nov. 12–16.

49. Lander, E.S., and Waterman, M.S. 1988. Genomic mapping by fingerprinting random clones: A mathematical analysis. *Genomics* 2:231–239.

50. Carrano, A.V., Lamerdin, J., Ashworth, L.K., Watskins, B., Branscomb, E., Slezak, T., deJong, P.J., Keith, D., Raff, M., McBride, L., Meister, S., and Kronick, M. 1989. A high resolution, fluorescence-based, semi-automated method for DNA fingerprinting. *Genomics* 4:129–136.

51. Stallings, R.L., Torney, D.C., Hildebrand, C.E., Longmire, J.L., Deaven, L.L., Jett, J.H., Doggett, N.A., and Moyzis, R.K. 1990. Physical mapping of human chromosomes by repetitive sequence fingerprinting. *Proc. Natl. Acad. Sci. U.S.A.* 87:6218–6222.

52. Craig, A., Nizetic, D., Hoheisel, J.D., Zehetner, G., and Lehrach, H. 1990. Ordering of cosmid clones covering the Herpes simplex type I (HSV-I) genome: A test case for fingerprinting by hybridization. *Nucl. Acids Res.* 18:2653–2660.

53. Evans, G.A., and Lewis, K.A. 1989. Physical mapping of complex genomes by cosmid multiplex analysis. *Proc. Natl. Acad. Sci. U.S.A.* 86:5030–5034.

54. Nelson, D.L., Ledbetter, S.A., Corbo, L., Victoria, M.F., Ramirez-Solis, R., Webster, T.D., Ledbetter, D.H., and Caskey, C.T. 1989. *Alu* polymerase chain reaction: A method for rapid isolation of human-specific sequences from complex DNA sources. *Proc. Natl. Acad. Sci. U.S.A.* 86:6686–6690.

55. Singer, M.F. 1982. Highly repeated sequences in mammalian genomes. *Int. Rev. Cytol.* 76:67–112.

56. Moyzis, R.K., Torney, D.C., Meyne, J., Buckingham, J.M., Wu, J.-R., Burks, C., Sirotkin, K.M., and Goad, W.B. 1989. The distribution of interspersed repetitive DNA sequences in the human genome. *Genomics* 4:273–289.

57. Weiner, A.M., Deininger, P.L., and Efstratiadis, A. 1986. Nonviral retroposons: Genes, pseudogenes, and transposable elements generated by the reverse flow of genetic information. *Annu. Rev. Biochem.* 55:631–661.

58. Scott, A.F., Schmeckpeper, B.J., Abdelrazik, M., Comey, C.T., O'Hara, B., Rossiter, J.P., Cooley, T., Heath, P., Smith, K.D., and Margolet, L. 1987. Origin of the human L1 elements: Proposed progenitor genes deduced from a consensus DNA sequence. *Genomics* 1:113–125.

59. Moyzis, R.K., Albright, K.L., Bartholdi, M.F., Cram, L.S., Deaven, L.L., Hildebrand, C.E., Joste, N.E., Longmire, J.L., Meyne, J., and Schwarzacher-Robinson, T. 1987. Human chromosome-specific repetitive DNA sequences: Novel markers for genetic analysis. *Chromosoma* 95:375–386.

60. Meyne, J., Littlefield, L.G., and Moyzis, R.K. 1989. Labeling of human centromeres using an alphoid DNA consensus sequence: Application to the scoring of chromosome aberrations. *Mut. Res.* 226:75–79.

61. Prosser, J., Frommer, M., Paul, C., and Vincent, P.C. 1986. Sequence relationships of three human satellite DNAs. *J. Mol. Biol.* 187:145–155.

62. Waye, J., and Willard, H.F. 1989. Human beta satellite DNA: Genomic organization and sequence definition of a class of highly repetitive tandem DNA. *Proc. Natl. Acad. Sci. U.S.A.* 86:6250–6254.

63. Willard, H.F., and Waye, J.S. 1987. Hierarchical order in chromosome-specific human alpha satellite DNA. *Trends Genet.* 3:192–198.

64. Moyzis, R.K., Buckingham, J.M., Cram, L.S., Dani, M., Deaven, L.L., Jones, M.D., Meyne, J., Ratliff, R.L., and Wu, J.-R. 1988. A highly conserved repetitive DNA sequence (TTAGGG)n, present at the telomeres of human chromosomes. *Proc. Natl. Acad. Sci. U.S.A.* 85:6622–6626.

65. Greig, G.M., England, S.B., Bedford, H.M., and Willard, H.F. 1989. Chromosome-specific alpha satellite DNA from the centromere of human chromosome 16. *Am. J. Hum. Genet.* 45:862–872.

66. Meyne, J., Ratliff, R.L., and Moyzis, R.K. 1989. Conservation of the human telomere sequence (TTAGGG)$_n$ among vertebrates. *Proc. Natl. Acad. Sci. U.S.A.* 86:7049–7053.

67. Riethman, H.D., Moyzis, R.K., Meyne, J., Burke, D.T., and Olson, M.V. 1989. Cloning human telomeric DNA fragments into *Saccharomyces cerevisiae* using a yeast-artificial-chromosome vector. *Proc. Natl. Acad. Sci. U.S.A.* 86:6240–6244.

68. Dietz-Band, J., Riethman, H.C., Hildebrand, C.E., and Moyzis, R.K. 1990. Characterization of polymorphic loci on a telomeric fragment of DNA from the long arm of human chromosome 7. *Genomics* 8:168–170.

69. Reeders, S., and Hildebrand, C.E. 1989. Report of the committee on the genetic constitution of chromosome 16. *Cytogenet. Cell Genet.* 51:299–318.

70. Doggett, N.A., Clark, L.M., Hildebrand, G.E. and Moyzis, R.K. Long range organization of human alpha satellite and satellite II sequences on chromosome 16. Cold Spring Harbor Meeting on Genome Mapping and Sequencing. p. 50 (1990).

71. Balding, D.J., and Torney, D.C. 1991. Statistical analysis of DNA fingerprint data for ordered clone physical mapping of human chromosomes. *Bull. Math. Biol.* 53:853–879.

72. Jett, J.H., and Torney, D.C. Unpublished results.

73. Sorensen, D., Cinkosky, M.J., and Fickett, J.W. In preparation.

74. Licher, P., Tang, C.-J., Call, K., Hermanson, G., Evans, G.A., Housman, D., and Ward, D.C. 1990. High-resolution mapping of human chromosome 11 by *in situ* hybridization with cosmid clones. *Science* 247:64–69.

75. Weber, J.L., and May, P.E. 1989. Abundant class of human DNA polymorphisms which can be typed using the polymerase chain reaction. *Am. J. Hum. Genet.* 44 388–396.

76. Olson, M.V., Hood, L., Cantor, C., and Botstein, D. 1989. A common language for physical mapping of the human genome. *Science* 245:1434–1435.

77. Cox, D.R. Personal communication.

78. Wells, R.D., Collier, D.A., Hanvey, J.C., Shimizu, M., and Wohlrab, F. 1988.

The chemistry and biology of unusual DNA structures adopted by oligopurine—oligopyrimidine sequences. *FASEB J.* 2:2939–2949.

79. Burks, C., Fickett, J.W., Goad, W.B., Kanehisa, M., Lewitter, F.I., Rindone, W.P., Swindell, C.D., Tung, C.-S., and Bilofsky, H.A. 1985. The GenBank nucleic acid sequence database. *Cabios Rev.* 1:225–233.

80. Ruddle, F., and Kidd, K.K. 1989. Human gene mapping workshops in transition. *Cytogenet. Cell Genet.* 51:1–2.

81. Cinkosky, M.J., Fickett, J.W., Gilna, P., and Burks, C. 1991. Electronic data publishing and GenBank. *Science* 252:1273–1277.

82. DeLisi, C. 1988. Computers in molecular biology: Current applications and emerging trends. *Science* 240:47–52.

83. Church, G.M., and Kieffer-Higgins, S. 1988. Multiplex DNA sequencing. *Science* 240:185–188.

84. Smith, L., and Hood, L. 1987. Mapping and sequencing the human genome: How to proceed. *Biotechnology* 5:933–939.

85. Snyder, S.H., and Narahashi, T. 1990. Receptor-channel alterations in disease: Many clues, few causes. *FASEB J.* 4:2707–2708.

86. Landegren, U., Kaiser, R., Caskey, C.T., and Hood, L. 1988. DNA diagnostics—Molecular techniques and automation. *Science* 242:229–237.

87. Green, E.D., and Green, Phillip. 1991. Sequence-tagged site (STS) content mapping of human chromosomes: Theoretical considerations and early experiences. *PCR Methods and Applications* 1:77–90.

88. Schlessinger, D., Little, R.D., Freije, D., Abidi, F., Zucchi, I., Porta, G., Pilia, G., Nagaraja, R., Johnson, S.K., Yoon, J-Y., Srivastava, A., Kere, J., Palmeri, G., Ciccodicola, A., Montanaro, V., Romano, G., Casamassimi, A., and D'Urso, M. 1991. Yeast artificial chromosome-based genome mapping: Some lessons from Xq24-q28. *Genomics* 11:783–793.

89. Anderson, C. 1992. New French genome centre aims to prove that bigger really is better. *Nature* 357:526–527.

90. Weissenbach, J., Gyapay, G., Dib, C., Vignal, A., Morissette, J., Vaysseix, G., and Lathrop, M. 1992. A second-generation linkage map of the human genome. *Nature* 359:794–801.

91. Foote, S., Vollrath, D., Hilton, A., and Page, D.C. 1992. The human Y chromosome: Overlapping DNA clones spanning the euchromatic region. *Science* 258:60–66.

92. Collins, F.S. 1992. Positional cloning: Let's not call it reverse anymore. *Nature Genetics* 1:3–6.

93. Stretzoska, Z., Paunesku, T., Radosavlijevic, D., Labat, I., Dramanac, R., and Crkvenjakov, R. 1991. DNA sequencing by hybridization: 100 bases read by a non-gel based method. *Proc. Natl. Acad. Sci. U.S.A.* 88:10,089–10,093.

94. Cantor, C.R. Mirzabekov, A., and Southern, E.M. 1992. Report on the sequencing by hybridization workshop. *Genomics* 13:1378–1383.

95. Pearson, P.L. 1991. The genome data base (GDB)—a human gene mapping repository. *Nucl. Acids Res.* 19:2237–2239.

9

Clinical Use of Monoclonal Antibodies

John M. Brown

9.1. Introduction

The in vivo use of monoclonal antibodies (MAbs) has been under evaluation now for over 10 years. The approaches have fallen into two major categories: diagnostic imaging and therapy. Although the early studies focused primarily on cancer, the field has branched out to include most major forms of human disease. The early optimism that the clinical arena would be quickly conquered has been tempered by reality. Although some MAbs have demonstrated clear clinical utility and are now commercial products or in advanced stages of development, many others are still undergoing preclinical and early clinical development and evaluation. The problems have been numerous, including difficulties with antibody characteristics, the need for new technologies to conjugate other molecules to MAbs, and the biochemical and physiological complexities of the diseases being tackled. Major advances have now been made in many of these areas, and it is likely that the next decade will witness the fulfillment of much of the early promise concerning the clinical use of MAbs.

   The following is a review of the current status of the parenteral use of MAbs to diagnose and treat human disease. An attempt has been made to draw some general principles from the large body of research data that has been generated to date. First is presented a discussion of the commercial production of pharmaceutical-grade MAbs and the associated issues of quality control and regulatory concerns. Next is presented a review of clinical uses, starting with diagnostic imaging and then proceeding to therapeutic applications. Several MAbs that are in advanced stages of development at Centocor are discussed in greater detail to illustrate some of the issues involved. Last is a discussion of the problem of the host antibody response to administered MAbs, which has significant implications for the use of repeated doses.

9.2. Commercial Manufacture

9.2.1. MAb Production

9.2.1.1. Ascites

Various aspects of MAb manufacture have been recently reviewed.[1, 2] MAbs are generally produced by one of two methods: ascites production in mice or in vitro fermentation. Ascites production involves injecting hybridoma cells into the peritoneal cavity of histocompatible mice. The mice are pretreated by i.p. injection of Pristane to irritate the peritoneal cavity and establish a conditioned environment that can support the growth of ascitic tumors. The fluid produced can contain a high concentration of secreted MAb, 3 to 15 mg/ml, and 3 to 5 ml or more can be harvested per mouse. Many pharmaceutical-grade MAbs have been produced using this technique, especially those MAbs developed in small amounts for initial clinical trials. The first MAb approved by the FDA for therapeutic use, OKT3, is produced by ascites. Ascites production has several drawbacks, however. It is costly and not entirely reliable, product is contaminated with low levels of normal mouse immunoglobulin as well as other mouse proteins, and adventitious agents from the mice, such as viruses, may be introduced. Human MAbs are very difficult and expensive to produce as ascites in mice. The human hybridomas may require special immunodeficient mice, and the antibody yield in ascites is often lower than with murine hybridomas.

9.2.1.2. Fermentation

Because of the problems with ascites-produced MAbs, mammalian cell culture fermentation has become the accepted method for large-scale production of clinical MAbs. An outline of a fermentation-based manufacturing process is presented in Table 9.1. Fermentation of hybridoma cells has the advantages that there is no contamination with normal mouse immunoglobulin; the process can

Table 9.1. Outline of Potential Steps in a Fermentation-Based Manufacturing Process

1. Preparation of cell banks	3. Downstream processing
a. Master cell bank (MCB)	a. Initial purification
b. Manufacturer's working cell bank (MWCB)	b. Digestion to fragments (pepsin, papain)
	c. Further purification
2. Fermentation	d. Conjugation (radioisotope chelator, toxin, drug)
a. Preculture of MWCB	e. Final purification
b. Fermentation	
c. Harvest of culture medium	4. Pharmaceutical manufacturing
d. Clarification and concentration	a. Formulation
	b. Vialing
	c. Lyophilization

be cost-effective, especially as serum requirements in the culture medium are reduced; it is reliable; it can be directly scaled up from small pilot bioreactors to very large production-scale; and the problem of contamination by adventitious agents is greatly reduced. Production levels, although significantly lower than in ascites, can be relatively high, 0.1 to 0.5 mg/ml. Human MAbs generally produce at lower levels. Because the fermentation medium is supplemented with serum or protein-based growth factors as well as other constituents such as antibiotics, crude product will be contaminated with these agents and methods must be devised for their removal.

Bioreactors (fermentors) fall into three major categories: stirred-tank and airlift systems, hollow fiber systems, and other matrix-based systems such as collagen beads, alginate beads, and ceramic cartridges. The stirred-tank and airlift systems are most suitable for commercial production because they can be readily scaled up to very large sizes ($>$ 1000 l). Regardless of the system employed, the harvested culture broth containing the MAb product will require the removal of contaminating cells and cell debris. This clarification step can be accomplished by centrifugation or filtration. The product is then usually concentrated (typically 10- to 20-fold) by ultrafiltration to reduce volume prior to subsequent downstream processing.

9.2.2. Processing

The subsequent purification of MAbs from ascites or fermentation production, as well as any further processing, is usually referred to as downstream processing. The goal of the purification is to remove contaminants from the production process, such as other proteins, nucleic acids, endotoxin, and adventitious agents, and any other process additives that are introduced during subsequent downstream processing, such as enzymes for digestion and reagents for producing conjugates. The process devised must reproducibly remove these contaminants to very low levels as demanded for pharmaceutical products, while maintaining the immuno-reactivity of the MAb.

9.2.2.1. Chromatography

Various chromatographic techniques are the main components of a purification process. The initial purification of a MAb is often performed by affinity chromato-graphy, by which a high degree of purification can be obtained in a single step. For most IgG MAbs, this step utilizes protein A-agarose. Subsequent downstream processing procedures for further polishing of the product may involve ion ex-change, gel filtration, and hydrophobic interaction chromatography. Both a ca-tion-exchange column and an anion exchange column may be used in the same process. Anion exchange is often used as a method to remove endotoxin and DNA contamination. Gel filtration chromatography is often used as a final polishing step to remove both high- and low-molecular-weight contaminants, including unwanted forms of the MAb such as aggregates and small fragments.

9.2.2.2. Fragmentation

As will be discussed in subsequent sections of this chapter, fragments of IgG MAbs such as $F(ab')_2$, Fab, and Fab', produced by enzymatic digestion, are often used for the final clinical product.

9.2.2.3. Conjugation

The final form of the MAb, whether intact or as a fragment, may be conjugated with a variety of agents, as is discussed in the following sections. These agents may be chelates, which allow the product to form a stable bond with radioisotopes, and cytotoxic agents such as toxins and chemotherapy drugs. The conjugation chemistry must be relatively easy to perform under large-scale manufacturing conditions and must not damage the binding activity of the MAb, and by-products of the reaction must be removed by subsequent purification steps.

9.2.2.4. Other Processing Techniques

Controlled precipitation, such as with certain salts and other agents, may be used both to purify and to concentrate the product. Ultrafiltration can also be utilized to concentrate MAbs, and dialysis via ultrafiltration (diafiltration) is used to remove low-molecular-weight contaminants and to exchange product into different buffers. Special treatments may be added to the process to remove or inactivate adventitious agents such as viruses. These treatments may involve extremes of temperature or pH, inactivating agents, or ultrafiltration. Sterilization by filtration, generally with 0.2 μm filters, is normally performed at the end of all processing steps.

9.2.3. Formulation

The terminal stage of processing is termed pharmaceutical manufacturing. It involves formulation and the subsequent steps of vial filling, stoppering, and lyophilization (if performed). The formulation of proteins to ensure stability and activity has been recently reviewed.[3] The formulation process itself may simply consist of concentrating the product by ultrafiltration, buffer exchange into the final formulation matrix by diafiltration or gel filtration, and then dilution to the required protein concentration. More complex formulations, such as those for radiolabeled products, may require additional steps for the blending of specialized ingredients, and may need special precautions such as anaerobic conditions to preserve the sulfhydryls of Fab' products.

Prior to filling, the product is passed through a final sterile filter. The filling itself can use standard automated pharmaceutical filling equipment but may require specialized conditions such as filling under an inert gas to preserve Fab' sulfhydryls. The effects of the container and stopper on the product also need to be considered. Lyophilization, although used infrequently for MAb products at the present time, is a viable option, especially when shelf-life stability of liquid formulations is a problem.

9.2.4. Quality Control

Quality control of the product is designed to assess its purity, potency, safety, stability, and consistency of manufacture. Because MAbs are produced from living cells in animals or by in vitro fermentation, there is special concern for residual impurities from these sources, as well as contamination with adventitious agents. Extensive testing must be performed at every stage in the manufacturing process to determine that known impurities are removed and that potential adventitious agents will be inactivated or removed. Testing begins at the cell bank stage and focuses on cell line stability, identity, and the presence of adventitious agents such as viruses and myocoplasma. Testing during downstream processing focuses on the removal of impurities derived from the cell line and fermentation medium, with particular attention on the ability to remove or inactivate adventitious viruses and reduce levels of DNA released from hybridoma cells during the production process. Testing of the final product focuses on the concerns of sterility, safety, apyrogenicity, and biochemical or immunological characterization of the MAb for identity, purity, and potency.

The U.S. Food and Drug Administration has issued a number of guidelines with respect to the manufacturer of MAb products. Of particular interest are the guidelines for the characterization of cell lines[4] and MAb products.[5] As was previously mentioned, there is special concern for virus and DNA contamination in the final product. Viruses and viral DNA are potential vectors for human disease. Cellular DNA, because it is derived from a hybridoma cell with a malignant phenotype, may have potential for integration into normal host DNA with subsequent cell transformation. The FDA has suggested a limit of 10 pg of DNA per dose,[5] and the World Health Organization has concluded that 100 pg of DNA or less per dose poses a negligible risk.[6]

9.3. Clinical Use

9.3.1. General Concepts

A number of recent compendia have covered the major aspects of both diagnostic imaging and therapeutic applications.[7-11] Before dealing with specific clinical uses, it would be helpful to discuss some concepts that apply in general to MAb localization (Table 9.2). Many of these immunologic and pharmacologic concepts

Table 9.2. Critical Factors for MAbs Used in Vivo

1. Antibody-related	2. Target-related
a. Affinity	a. Vascularity
b. Specificity	b. Distribution of antigens
c. Class and subclass	c. Antigenic heterogeneity
d. Murine vs. human	d. Antigenic modulation
e. Intact vs. fragments	e. Circulating antigen
f. Dose	
g. Route of administration	
h. Immunogenicity	

have been recently reviewed.[12-14] For effective localization, the MAb should have high affinity for the target antigen. There should be minimal cross-reactivity with nontarget tissues, although the specificity need not be absolute. The class and subclass of the immunoglobulin are important factors, in that they will affect both the biodistribution and blood clearance of the MAb, as well as the interaction with immune mechanisms of the host. Blood clearance and catabolism are generally faster with heterologous murine immunoglobulins than with homologous human MAbs.

The lower the molecular weight of the antibody, the faster the blood clearance. Intact murine IgG has a blood half-life of approximately 30 hours and is catabolized in liver and spleen, as well as other sites in the body. $F(ab')_2$ and Fab/Fab' have half-lives of approximately 20 hours and 2 hours, respectively, and accumulate in the kidneys. Because of their smaller molecular size, fragments are better able to diffuse out of capillaries and into target tissues. However, for interaction with immune effector functions of the host, intact immunoglobulin with the Fc portion present is required. Antibody dose also has an effect on blood clearance and biodistribution. Higher doses, generally at 10 mg or above, appear to saturate nonspecific sites in the liver and other organs. As a consequence, more antibody remains in the blood and clearance is slower. Routes of injection other than intravenous have also been shown to alter biodistribution. Intralymphatic and intraperitoneal administration tend to concentrate and prolong the dose at these sites and lead to lower systemic levels. The immunogenicity of MAb products will be discussed later in this chapter.

Characteristics of the target tissue are also important in MAb localization. The tissue should be well vascularized or directly exposed to the blood system. This is often a problem with solid tumors, which may be poorly vascularized. The antigens in the target tissue should be densely and uniformly expressed, and accessible to antibody binding. Viable cells must express target antigen on the cell surface, whereas MAbs that localize dead cells may depend on reactivity with intracellular antigens. Tumor cells often demonstrate antigenic heterogeneity. That is, antigen expression will vary from cell to cell within a tumor as well as between different tumors. In some instances, antigenic modulation has also

Table 9.3. Diagnostic Imaging Applications

1. Cardiovascular disease	3. Cancer
a. Myocardial infarction	a. Solid tumors (anti-tumor-associated antigens)
b. Deep-vein thrombosis	b. Locating and sizing
c. Atherosclerosis	c. Detection of occult tumors
2. Sites of bacterial infection	d. Determining suitability for MAb therapy
	e. Monitoring response to therapy

been demonstrated, whereby the process of antibody binding to cell-surface antigen causes the complex to be shed or internalized, rendering the cell devoid of expressed antigen. Cross-reactive target antigen may also be found in the circulation, often as a consequence of tumor cell shedding. This circulating antigen is able to complex the injected MAb and block it from binding to the target site.

9.3.2. Diagnostic Imaging

9.3.2.1. General Concepts

A number of recent compendia have addressed the topic of diagnostic imaging with radiolabeled MAbs.[9, 10, 15] The technique, also referred to as immunoscintigraphy, employs the use of a planar gamma camera to detect the two-dimensional distribution in the body of gamma-emitting radioisotopes conjugated to MAbs. Recently, cameras utilizing single photon emission computed tomography (SPECT) have been used to give more sensitive, three-dimensional evaluations. Diagnostic imaging has been applied to cardiovascular disease, infectious disease, and cancer (Table 9.3). In addition to the general critical factors listed in Table 9.2, some other issues are also important in imaging applications[16] (Table 9.4). The choice of radioisotope is a central issue.[17] The isotope must have a gamma emission energy suitable for patient safety and current gamma camera technology. The half-life must be long enough to allow time for MAb localization, but short enough not to pose a safety issue. The labeling chemistry must be practical and yield a stable linkage. The isotope should be relatively inexpensive and readily available.

Table 9.5 lists the isotopes most commonly used in imaging applications. The

Table 9.4. Additional Critical Factors Specific to Imaging MAbs

1. Radioisotope	2. Pharmacokinetics	3. Target
a. Energy	a. Blood clearance	a. Size
b. Half-life	b. Uptake in normal	b. Attenuation
c. Labeling chemistry	tissues	c. Target-to-nontarget ratio
d. Stability of linkage	c. Dosimetry	
e. Availability and cost		

Table 9.5. *Radioisotopes Used in Imaging*

Isotope	Half-life	Gamma Emmission (keV)
^{123}I	13 hours	159
^{125}I	60 days	35
^{131}I	8 days	364
^{111}In	2.8 days	173
		247
^{99m}Tc	6 hours	140

iodine isotopes suffer as a group from linkage instability owing to dehalogenation in vivo. ^{131}I has a gamma energy that is too high for optimal gamma camera use, as well as an associated beta particle emission that can lead to unfavorable radiation dosimetry. The low energy of ^{125}I has made it suitable only for a specialized technique involving detection with a hand-held probe during surgical operations.[18] ^{123}I has near-ideal half-life and energy properties, but is very expensive and not readily available. ^{111}In, which uses chelation chemistry for MAb labeling, has been successfully used in many imaging applications. It suffers from being relatively expensive and having a tendency to localize in the liver, although the latter problem has been decreased with some recent improvements in chelation chemistry. ^{99m}Tc is ideal in most aspects. Along with good half-life and energy properties, ^{99m}Tc is inexpensive and readily available from a generator source common in clinical nuclear medicine facilities. Its major drawback, difficult labeling chemistry, has recently been overcome at Centocor with the development of several rapid and convenient methods for stable ^{99m}Tc attachment.[19, 20]

Pharmacokinetic aspects are also critical to imaging applications. The labeled antibody must clear from the blood sufficiently that target uptake is not obscured. The choice of isotope half-life is an important factor. If the imaging MAb has a relatively slow blood clearance, because it is an intact IgG or a large dose is used, an isotope with a long half-life should be employed. ^{99m}Tc and ^{123}I are essentially restricted to use with rapidly clearing MAbs, such as fragments. Uptake in normal tissues is also a concern, because it might directly obscure target uptake and can produce false-positive images that reduce the clinical efficacy of the technique. Localization in normal tissues also dictates which isotopes and how much activity can be used, so that the radiation dose to critical organs remains within safe levels. ^{99m}Tc conjugation chemistry has been developed at Centocor, utilizing a cleavable ester-linker, which may enhance renal clearance and thus reduce the radiation dose to kidneys associated with Fab' fragments.[21]

Characteristics of the target also play an important role in imaging. The target must be of sufficient size to be imaged by current technology, generally 0.5 to 1.0 cm in diameter or greater. Overlaying tissue and distance from the gamma camera detector will cause attenuation of the signal. Overlaying bone is especially a problem. The uptake of antibody at the target site must be significantly greater than the surrounding nontarget tissue to provide a usable image. It is not necessary that the MAb be absolutely specific for the target, as long as the cross-reactive antigens on nontarget tissues are expressed at a sufficiently low level.

9.3.2.2. Cardiovascular Disease

Myocardial Infarction. Centocor has been involved in developing a MAb for imaging myocardial necrosis associated with acute myocardial infarction.[22, 23] Myoscint™ is the first MAb-based imaging agent to receive approval from the Committee for Proprietary Medicinal Products of the European Economic Community. It is now on the market in several European countries. The product consists of a kit containing 0.5 mg of antimyosin Fab fragment conjugated with the chelator DTPA (diethylenetriaminepentaacetic acid). The liquid formulated Fab-DTPA is labeled by mixture with approximately 2 mCi of ^{111}In chloride. After incubation for about 10 minutes, the ^{111}In-labeled MAb is ready without further processing for iv injection. Imaging is performed after 24 to 48 hours using either a planar gamma camera or SPECT.

The antimyosin MAb is specific for human cardiac myosin and binds to intracellular myosin exposed as a consequence of myocardial necrosis. Figure 9.1 demonstrates the localization of ^{111}In-labeled antimyosin Fab in a typical patient with an anterolateral myocardial infarction. To date, over 600 patients have been studied. No adverse reactions have occurred. The product has shown a high degree of sensitivity for detecting infarction and specificity for excluding a recent ischemic event in patients admitted with chest pain syndrome. Indium antimyosin is able to detect the location and extent of necrotic heart tissue and should be useful in the diagnosis of heart attack and in the early risk stratification of patients demonstrated to have myocardial damage.

Deep-Vein Thrombosis. Deep-vein thrombosis (DVT) is a disease that would appear to have several aspects that make it ideal for imaging applications. Foremost is the fact that the target clot lies within the blood system, having direct access to iv administered MAb. Secondly, DVT is located primarily in the lower extremities and thus well away from nonspecific uptake in the blood pool and organs of the central body. The use of MAbs in DVT imaging has been recently reviewed.[24, 25] The approaches have been divided into the use of labeled MAbs directed against platelets, which are components of actively developing thrombi, and against fibrin. Although the clinical experience with antiplatelet MAbs has been limited, the technique appears useful during the acute phase of thrombogenesis but may not be effective with clots older than 24 to 48 hours. Heparin therapy also appears to decrease clot uptake of platelet-specific MAbs.

Centocor has been involved in developing DVT-imaging MAbs directed against the fibrin component of thrombi.[24, 25] Several MAbs are being evaluated that are specific for fibrin and that do not cross-react with the precursor molecule,

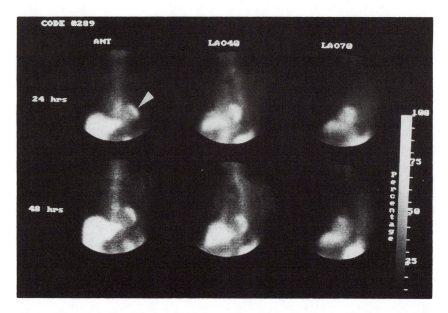

Figure 9.1. Anterior, 40°, and 70° left anterior oblique images from a patient with an acute myocardial infarction. The images were obtained 24 and 48 hours after injection of ^{111}In antimyosin. The infarct in the 24 hour anterior view is indicated by an arrow. Uptake in liver, kidneys, and bone marrow can also be seen.

fibrinogen. Studies have evaluated an ^{111}In-labeled Fab-DTPA that showed a sensitivity of 86 to 100% in detecting thrombi in the extremities of patients. Specificity for identifying clot-negative regions was also excellent. The scans could be evaluated as early as four hours after injection, and concurrent heparin therapy had little effect on diagnostic accuracy.

More recently, a MAb with similar fibrin specificity has been evaluated as a ^{99m}Tc-Fab' fragment. The ^{99m}Tc conjugation chemistry involves site-specific labeling of sulfhydryl groups in the hinge region of the Fab', thus not affecting binding activity at the opposite end of the molecule.[19] The product has been developed as a single-vial lyophilized kit that is labeled by reconstitution with pertechnetate from a ^{99m}Tc generator.[2] After 30 minutes of incubation at room temperature, labeling is complete and the product is ready for injection without further purification. Initial trials in DVT patients administered 0.5 mg of Fab' labeled with 15 to 20 mCi of ^{99m}Tc have demonstrated a high sensitivity and specificity for detecting clots in the regions of pelvis, thigh, knee, and calf.[25] Scanning could be performed at 4 to 6 hours and may also be diagnostic as early as 30 to 60 minutes after injection. Imaging was not affected by concurrent heparin therapy. Recent in vitro studies have indicated that the concurrent use of other anticoagulant and thrombolytic agents (warfarin, tissue plasminogen activator, streptokinase, and urokinase) also will not directly interfere with clot imaging. Thus, the ^{99m}Tc-labeled antifibrin product has many characteristics, including ease of labeling and use in the clinic, that make it ideal as an imaging agent.

Atherosclerosis. Atherosclerotic plaque, a cause of occlusive disease in coronary and peripheral arteries, is being studied as a target for imaging MAbs. Several antibodies, among which is a MAb directed against activated platelets found in ruptured plaque,[26] are currently being evaluated at Centocor. Such plaque-imaging MAbs may be useful in screening for patients at risk of developing myocardial infarction, and in identifying patients who may go on to reocclude after an angioplasty procedure.

9.3.2.3. Infectious Disease

Recently, research efforts have been directed at MAb imaging of sites of infection. Antibodies directed against specific bacterial antigens have proven successful in an experimental model to localize abscesses.[27] MAbs directed against granulocytes, inflammatory leukocytes that accumulate at sites of bacterial infection, have shown high sensitivity and specificity for detecting localized infections in clinical trials.[28] Nonspecific IgG, by virtue of its Fc portion, has been shown in clinical trials to successfully localize sites of infection owing to reactivity with Fc receptor-bearing inflammatory cells.[27]

9.3.2.4. Cancer

A number of studies over the last decade have evaluated the clinical use of MAbs for scintigraphic detection of cancer.[8-11, 15] The investigations have covered all major types of solid tumors, including melanoma,[29] colorectal carcinoma,[30] ovarian carcinoma,[31, 32] breast carcinoma,[32] sarcoma,[33] and lung carcinoma.[34] The antibodies have been directed against a wide array of tumor markers (Table 9.6). Centocor has been involved in clinical trials of a number of antibodies for imaging cancers such as ovarian, breast, and colorectal carcinoma.

Tumor-imaging studies have shown that MAbs are able to successfully detect tumors as small as 1 cm in diameter, and previously unknown lesions are often discovered. Although some clinical trials have demonstrated diagnostic sensitivity and specificity in excess of 80%, few products have yet been approved for

Table 9.6. Tumor Markers Used in MAb Imaging

Carcinoembryonic antigen	Transferrin receptor
α-Fetoprotein	Epidermal growth factor receptor
Human chorionic gonadotropin	Oncogene products
Human milk fat globule	Tumor-associated cell surface antigens
Prostatic acid phosphatase	Gangliosides
Ferritin	Necrosis-associated intracellular antigens

commercial use. Most of the critical factors outlined in Tables 9.2 and 9.4 have been of special importance in the development of tumor-imaging applications. Selecting MAbs with proper affinity and specificity has been difficult. Virtually none of the antigens used as targets for imaging are truly tumor specific; they are expressed at certain levels in at least some normal tissues. This lack of specificity has caused problems with uptake in normal tissues and organs. Additionally, most tumor types demonstrate antigenic heterogeneity such that cells within a tumor as well as different tumors of the same histologic type do not all express the same antigens. Most tumors are not well vascularized and thus are not readily accessible to antibody binding. Much of a given tumor mass may be necrotic and not expressing antigen. The presence of circulating antigen, often associated with tumors, has been found to inhibit localization in some cases but not in others.[30]

Many of the earlier studies were not optimal in terms of MAb size and radiolabel. It now appears that antibody fragments, particularly Fab and Fab' will give the best tumor-to-background ratios. Smaller fragments will penetrate the target site better, and at the same time will clear from blood and normal tissues more readily. The application of fast-clearing fragments allows the use of short-lived ^{99m}Tc, a radiolabel with nearly ideal scintigraphic characteristics whose conjugation chemistry has recently been optimized. The size of the MAb dose is still being evaluated. While many studies have demonstrated successful imaging with doses of 0.1 to 1.0 mg, other studies have found improved localization with doses in excess of 10 mg.[14]

As many of the issues pertaining to tumor imaging are being resolved, it is clear that commercial applications will come to fruition in the near future. Immunoscintigraphy will be able to locate tumors and define their size and, by virtue of the MAb specificity, yield additional information not available from conventional diagnostic modalities. Many studies have already shown that occult tumors not detectable by other techniques have been discovered by MAb imaging. Imaging after tumor therapy can be used to monitor therapeutic responses, detect recurrences, and guide subsequent clinical decisions. As will be described below, MAbs may be increasingly used as therapeutic agents. An initial scintigraphic evaluation with the same MAb may help to determine its suitability for therapy. More specialized applications, such as intraperitoneal administration to image ovarian tumors[32] and intralymphatic injection to detect tumorous lymph nodes,[35] may become increasingly useful. The use of mixtures of MAbs (termed "cocktails") directed against different antigens has been applied with some success in colorectal carcinoma[30] and may address the central problem of antigenic heterogeneity.

9.3.3. Therapy

9.3.3.1. General Concepts

The therapeutic applications of MAbs have been addressed in a number of recent compendia.[7-11] The disease states currently being evaluated are listed in Table

Table 9.7.   Therapeutic Applications

1. Transplantation	3. Cardiovascular disease (thromboembolic)
a. Organ	4. Autoimmune disease
b. Bone marrow	5. Cancer
2. Infectious disease (bacterial, viral)	

9.7. While the majority of efforts have been directed against cancer, many other conditions are under study. Before addressing the specific diseases, a brief discussion of the general concepts involved in MAb therapy is in order. Although the critical factors listed in Table 9.2 still apply, Table 9.8 lists the general categories of therapeutic mechanisms and the additional critical factors that apply in these specific cases.

MAbs Alone. Under certain conditions, MAbs have therapeutic activity when used without the attachment of cytotoxic agents.[36, 37] Binding to an antigen may neutralize its biologic effects or may block the activity of a growth factor receptor. Intact MAbs bearing the Fc portion are capable of recruiting the immune effector functions of the host. Circulating antigens and antigen-bearing cells can be opsonized and phagocytosed by cells of the reticuloendothelial system. Complement-dependent cytotoxicity and antibody-dependent cytotoxicity involving host cytotoxic lymphocytes and monocytes may also be operative in the destruction of target cells. The species source of the MAb and its immunoglobulin class and subclass are critical for successful interaction with host immune functions. The use of intact MAbs also has the advantage of slower clearance from the blood, thus allowing more time for penetration and binding to target tissues. A major benefit of using MAbs alone is that there should be no toxicity for nontarget tissues, even those in which the MAbs might accumulate nonspecifically.

Radioisotope Immunoconjugates. These MAb conjugates are designed to deliver cytotoxic doses of radioactivity to target cells.[13, 36, 37] Most of the critical factors

Table 9.8.   Additional Critical Factors Specific to Therapeutic MAbs

1. Use of MAbs alone	2. Radioisotope conjugates	3. Toxin and drug conjugates
a. Mechanisms	a. Choice of isotope	a. Choice of agent
i. Neutralization of antigen	b. Type of radioactive	b. Conjugation chemistry
ii. Opsonization and	emission	c. Antibody
phagocytosis	c. Antibody	internalization
iii. Complement-dependent	internalization	d. Toxicity
cytotoxicity	d. Dosimetry in target	e. Immunogenicity of
iv. Antibody-dependent	vs. nontarget tissues	toxin
cytotoxicity		
b. Generally requires intact		
immunoglobulin		

Table 9.9. Some Potential Radioisotopes for Therapy

Isotope	Half-life	Emission
^{125}I	60 days	Auger election
^{131}I	8 days	β (0.6 MeV)
^{186}Re	3.5 days	β (1.1 MeV)
^{188}Re	17 hours	β (2.1 MeV)
^{90}Y	2.5 days	β (2.3 MeV)
^{211}At	7 hours	α (5.9, 7.5 MeV)

for imaging MAbs (Table 9.4) also apply to radiotherapy. The choice of isotope and energy of emission is critical. Table 9.9 lists some of the isotopes currently being evaluated. Most studies have used β emitters. These medium-energy particles are able to penetrate multiple cell diameters in thickness. MAbs carrying these isotopes are able to kill the cells they are bound to as well as nearby bystander cells. This is useful when antigenic heterogeneity of tumor cells is an issue but may be a problem for damage to surrounding nontarget cells.

High-energy α emitters, although more dangerous and difficult to work with, are also being evaluated. α particles have a path length of about one cell diameter, but each emission has a high likelihood of killing the cell in its path. When bound to a target cell surface, the particles should kill that and immediately adjacent cells but spare more distant nontarget cells. Low-energy Auger electrons emitted by ^{125}I, with an extremely short path length, are also being studied.[38] This approach requires that the MAb conjugate be internalized and the ^{125}I enter the nucleus. Once there, the Auger electrons emitted have a high probability of damaging DNA and killing the cell. Adjacent cells not binding the MAb will be unharmed, and nonspecific localization of the isotope in nontarget tissues should not lead to internalization, thus sparing them from damage.

The chemistry of conjugation is a critical factor, and therapeutic isotopes are usually selected based on similarity in attachment chemistry to imaging isotopes. Astatine chemistry is similar to iodine, yttrium is similar to indium, and rhenium is similar to technetium. Another critical factor is the relative radiation dose delivered to target and nontarget tissues. With the medium- and high-energy isotopes, damage to normal tissues due to nonspecific localization of MAb conjugates is a limiting factor for the maximum dose that can be administered to a patient. Imaging studies have indicated that, at best, only about 0.01% of the injected dose accumulates in a gram of target tissue. The remainder of the dose, therefore, must be kept from harming normal tissues. For this reason, MAb fragments, which clear more rapidly from blood and normal tissues, may be the form of choice for radioimmunotherapy applications.

Toxin and Drug Immunoconjugates. MAbs are also being evaluated for the delivery of potent toxins and drugs to target cells while reducing toxicity to nontarget tissues.[36, 37, 39, 40] The toxins have included ricin, abrin, pokeweed antiviral protein, gelonin, diphtherin toxin, and *Pseudomonas* endotoxin. Chemotherapy drugs have included doxorubicin, daunorubicin, methotrexate, melphalan, chlorambucil, vinca alkaloids, and mitomycin-C. A critical factor is the conjugation chemistry. After binding of the MAb to the cell surface, the toxin or drug must be internalized and made available for damaging the cell. Some, but not all, antigens will internalize after antibody is bound, carrying the conjugate inside the cell. Often, the conjugation chemistry involves a labile bond that is cleaved intracellularly, releasing free toxin or drug. The bond, however, must remain relatively stable in blood. MAb conjugates spare toxicity to normal tissues because less of the dose reaches these sites than if free drug were administered, thus improving the therapeutic index. Additionally, nonspecific localization in normal tissues may not involve antigen-antibody interactions. Therefore the conjugates will not be as readily internalized and normal-tissue toxicity will be lessened. A potential problem with toxins is that they are relatively large, foreign proteins and can induce a strong immune response in the host.

9.3.3.2. Transplantation

Organ Transplantation. Several MAbs against T cell antigens have been evaluated for the treatment of solid organ transplant rejection. The most successful of these is OKT3, which was the first MAb to be licensed for human use.[41] The murine intact IgG antibody, directed against the CD3 antigen on T lymphocytes, is used unconjugated to effectively reverse acute rejection episodes with renal, hepatic, cardiac, and combined kidney-pancreas transplants. MAb binding blocks T cell function, in part because of opsonization and the removal of T cells by the reticuloendothelial system. OKT3 may also be useful prophylactically in the immediate posttransplantation period. Acute side effects such as fever and chills are common and probably a consequence of the T cell damage.

Bone Marrow Transplantation. MAbs are being evaluated for the treatment of graft-versus-host disease in bone marrow transplantation. OKT3 has shown promise in this regard.[41] A ricin toxin immunoconjugate directed against the CD5 antigen of T lymphocytes has demonstrated in vivo effectiveness in both treating and preventing graft-versus-host disease with histoincompatible marrow grafts.[36, 42] Side effects associated with the toxin were well tolerated or absent.

9.3.3.3. Infectious Disease

Several MAbs against endotoxin have been evaluated as aids in antibiotic therapy of bacteremia and sepsis.[43, 44] Centocor has developed a human IgM MAb, HA-1A, directed against the lipid A component of endotoxin.[44] The MAb has been shown to reduce the lethal effects of gram-negative bacteria in animal models. HA-1A was the first human MAb to be tested in the clinic. Safety and initial efficacy trials have now been completed. In a double-blind, placebo-controlled study of 543 patients, a single 100 mg iv dose produced a statistically significant decrease in mortality in patients with gram-negative bacteremia and sepsis, and in patients whose gram-negative bacteremia had progressed to septic shock. The MAb was well tolerated in the patients. Although the exact mechanism is unknown, HA-1A is hypothesized to block the toxic effects of circulating endotoxin, which includes induction and release of the mediators of shock and tissue damage. Because of the high mortality associated with gram-negative sepsis, HA-1A should be an important new agent in the treatment of patients with this disease.

Viral infections are also being targeted for MAb therapy. Antibodies against cytomegalovirus are in early clinical trials.[45]

9.3.3.4. Cardiovascular Disease

Centocor has been evaluating a murine MAb, 7E3, specific for the glycoprotein IIb/IIIa fibrinogen receptor of platelets.[46] The $F(ab')_2$ and Fab fragments alone, by binding to the receptor, are potent inhibitors of in vitro platelet aggregation and in vivo platelet thrombus formation in animal studies. Clinical studies with single doses of $F(ab')_2$ up to 0.2 mg/kg have been effective in profoundly inhibiting platelet aggregation measured in vitro and significantly prolonging bleeding time. Hemorrhagic toxicity has not been noted. Clinical trials are currently underway to determine the potential role of 7E3 in thrombolytic therapy of coronary artery disease.

9.3.3.5. Autoimmune Disease

MAbs directed against B- and T-lymphocyte antigens are currently being evaluated for therapy of autoimmune diseases such as rheumatoid arthritis and multiple sclerosis.[47, 48] Initial clinical trials have demonstrated significant therapeutic effects. In some patients with rheumatoid arthritis, improvement in clinical symptoms has persisted for at least five months after a single 7-day course with an intact, unconjugated MAb.[48]

9.3.3.6. Cancer

The use of MAbs in cancer therapy has been recently reviewed.[37] Although a multitude of studies have been performed, involving nearly every type of cancer, lasting therapeutic effects have been rare. The problems have centered on antigenic heterogeneity, the lack of cytotoxicity of MAbs used alone and as conjugates, the chemistry of conjugation, the proper dosage forms and regimens, and the immune response to mouse antibodies. Some of the more promising studies will be discussed below. The problem of immunogenicity will be addressed in the following section.

MAbs Used Alone. Antitumor effects have been demonstrated in B-cell lymphoma with antiidiotype MAbs and in T cell lymphoma. Complete responses have been rare. Antiganglioside MAbs have shown some effectiveness in melanoma, osteosarcoma, and neuroblastoma. Centocor has participated in the development and evaluation of MAb 17-1A against a gastrointestinal carcinoma-associated antigen.[49] Antitumor effects have been noted in patients with colorectal and pancreatic carcinoma.

Radioisotope Conjugates. ^{131}I-labeled MAbs have demonstrated therapeutic activity with hepatomas, Hodgkin's disease, and B and T cell lymphomas. Intraperitoneal administration of ^{131}I-MAbs in ovarian carcinoma has also been effective. Because of the retention of radioactivity in the blood pool and normal tissues, bone marrow toxicity has been dose limiting with most intravenously administered MAbs. Other isotopes such as ^{90}Y are currently being evaluated.

Toxin Conjugates. Modest antitumor effects have been demonstrated with ricin immunoconjugates in melanoma and chronic lymphocytic leukemia. The clinical responses have been hampered by immune responses to both the murine MAb and the toxin, and by ricin-mediated normal tissue toxicity (capillary leak syndrome).

Drug Conjugates. The few clinical studies reported to date, involving methotrexate and doxorubicin immunoconjugates, have not clearly demonstrated efficacy. Further research is needed to identify optimal drugs and conjugation chemistries such that the immunoreactivity of the MAb is not damaged and drug is delivered in active form to the tumor cell interior, while sparing normal tissues.

9.3.4. Immunogenicity

The side effects associated with MAb injections have not been a major problem.[37, 50] The most often reported side effects are fever and chills, some of which may be a consequence of MAb reactivity with target cells. Immune complexes with circulating antigen may lead to elevations in liver transaminase levels. Mild allergic reactions have occurred in about 10% of patients receiving mouse MAbs. However, less than 1% of these patients developed anaphylactic shock mediated by IgE antibodies. Steroids and epinephrine treatment have allowed repeated MAb administration in anaphylactic patients.

Despite the relatively low levels of allergic reactions and anaphylaxis, many studies have documented the development of human antimouse antibodies (HAMA) of the IgG and IgM classes. This topic has been recently reviewed.[50, 51] HAMA may arise in 50% or more of patients treated. It can be detected as early as two weeks after first injection and may gradually decline or remain elevated for six months to a year. Repeated MAb injections cause a more rapid response and will induce higher HAMA levels. The major problem of HAMA is that it will complex with administered MAb and prevent it from reaching the target site. Thus, the effectiveness of continuous MAb administration over prolonged periods of time, or the use of repeated doses, may be compromised. As was already discussed, biological toxins have also been shown to provoke a potent immune response that can neutralize subsequent treatments.

Early in the course of HAMA development, the majority of the response is directed against the immunoglobulin constant regions, primarily the Fc portion. Some of the response is also antiidiotypic, directed against unique determinants of the variable region. Repeated administration may increase the proportion of the antiidiotypic response. Such antibodies can functionally mimic the original target antigen and serve as a means of immunization in therapeutic applications. Studies have also shown that antiidiotypic antibodies against the original antiidiotypic antibodies can occur.[52] Such a mechanism has been postulated to explain the delayed clinical responses often observed in cancer patients treated with murine MAbs alone.[52]

A number of possibilities exist for minimizing the impact of HAMA.[50] Antibody fragments, lacking an Fc portion and clearing rapidly from the body, may be less immunogenic. Lower doses and nonrepeated doses may also have a similar effect on minimizing HAMA. An analysis of over 600 patients imaged with a single dose of 0.5 mg of murine antimyosin Fab-DTPA could detect no HAMA responses.[53] It may be possible to overcome preexisting HAMA by injecting a larger dose. Preexisting HAMA has also been removed by plasmapheresis prior to reinjection.[54] The use of an immunosuppressive drug such as cyclosporin-A has been effective in some cases in blunting the HAMA response. Mouse–human chimeric antibodies, containing murine variable regions and human constant regions, have been less immunogenic than their murine counterparts.[55] However, responses to the murine portion still occur. Lastly, fully human MAbs may be nonimmunogenic, as has been shown for HA-1A,[44] although there is still a potential for antiidiotypic responses to occur.

9.4. Summary

Large-scale preparation of monoclonal antibodies (MAbs) for clinical use begins with their production from animal ascites or from in vitro fermentation of the hybridoma cells. The MAbs are then processed and purified by various biochemical techniques. The resultant product must meet accepted standards for pharmaceutical products, with particular emphasis on the removal of potential contamination with viruses and DNA. MAbs have been used in diagnostic imaging applications after conjugation with gamma-emitting radioisotopes. Successful imaging has been performed in cardiovascular diseases such as myocardial infarction and deep-vein thrombosis, and in cancer. Therapeutic applications have involved the use of MAbs alone and conjugated with particulate-emitting radioisotopes, biological toxins, and chemotherapy drugs. Therapeutic uses have involved organ and bone marrow transplantation, septic shock, thromboembolic disease, autoimmune disease, and cancer. A major obstacle to the clinical use of MAbs is the generation of human antimouse antibodies. A number of approaches to minimize this problem are being evaluated, including the use of humanized chimeric antibodies and fully human MAbs. The technical gains and knowledge acquired over the last decade of MAb use should lead to many improvements in clinical applications in the near future.

References

1. Seaver, S.S., ed. 1987. *Commercial production of monoclonal antibodies: A guide for scale-up*. New York: Dekker.

2. Bogard, W.C., Dean, R.T., Deo, Y., Fuchs, R., Mattis, J.A., McLean, A.A., and Berger, H.J. 1989. Practical considerations in the production, purification, and formulation of monoclonal antibodies for immunoscintigraphy and immunotherapy. *Semin. Nucl. Med.* 19:202–220.

3. Wang, W.J., and Hanson, M.A. 1988. Parenteral formulations of proteins and peptides: stability and stabilizers. *J. Parent. Sci. Technol.* 42 suppl.:S2–S26.

4. Esber, E.C. 1987. *Points to consider in the characterization of cell lines used to produce biologicals. Washington, D.C.: Center for Drugs and Biologics, FDA.*

5. Hoffman, T. 1987. *Points to consider in the manufacture and testing of monoclonal antibody products for human use*. Washington, D.C.: Center for Drugs and Biologics, FDA.

6. WHO Study Group. 1987. *Acceptability of cell substrates for production of biologicals*. World Health Organization Technical Report Series 747. Geneva: WHO.

7. Roth, J.A., ed. 1986. *Monoclonal antibodies in cancer: Advances in diagnosis and treatment*. Mount Kisco, N.Y.: Futura.

8. Strelkauskas, A.J., ed. 1987. *Human hybridomas: Diagnostic and therapeutic applications*. New York: Dekker.

9. Vogel, C. ed. 1987. *Immunoconjugates: Antibody conjugates in radioimaging and therapy of cancer*. New York: Oxford University Press.

10. Srivastava, S.C., ed. 1988. *Radiolabeled monoclonal antibodies for imaging and therapy*. New York: Plenum.

11. Borrebaeck, C.A.K., and Larrick, J.W., eds. 1990. *Therapeutic monoclonal antibodies*. New York: Stockton.

12. Halpern, S.E., and Dillman, R.O. 1987. Problems associated with radioimmunodetection and possibilities for future solutions. *J. Biol. Response Modif.* 6:235–262.

13. Sands, H. 1988. Radioimmunoconjugates: An overview of problems and promises. *Antibody Immunoconj. Radiopharm.* 1:213–226.

14. Zuckier, L.S., Rodriguez, L.D., and Scharff, M.D. 1989. Immunologic and pharmacologic concepts of monoclonal antibodies. *Semin. Nucl. Med.* 19:166–186.

15. Chatal, J., ed. 1989. *Monoclonal antibodies in immunoscintigraphy*. Boca Raton, Fla.: CRC.

16. Serafini, A.N., Garty, I., Jabir, A.M., Friden, A., Dewanjee, M., Ganz, W., and Sfakianakis, G.N. 1989. Monoclonal antibody imaging: Clinical and technical perspectives. *Antibody Immunoconj. Radiopharm.* 2:225–234.

17. Bhargava, K.K., and Acharya, S.A. 1989. Labeling of monoclonal antibodies with radionuclides. *Semin. Nucl. Med.* 19:187–201.

18. Martin, E.W., Mojzisik, C.M., Hinkle, G.H., Sampsel, J., Siddiqi, M.A., Tuttle, S.E., Sickle-Santanello, B., Colcher, D., Thurston, M.O., Bell, J.G., Farrar, W.B., and Schlom, J. 1988. Radioimmunoguided surgery using monoclonal antibody. *Am. J. Surg.* 156:386–392.

19. Pak, K.Y., Nedelman, M.A., Kanke, M., Khaw, B.A., Mattis, J.A., Strauss, H.W., Dean, R.T., and Berger, H.J. 1992. An instant kit method for labeling antimyosin Fab' with technetium-99m: Evaluation in an experimental myocardial infarct model. *J. Nucl. Med.,* 33:144–149.

20. Dean, R.T., Weber, R., Pak, K., Boutin, R., Buttram, S., Nedelman, M., and Lister-James, J. 1990. New facile methods for stably labeling antibodies with technetium-99m. In *Technetium and rhenium in chemistry and nuclear medicine*, Vol 3. M. Nicolini, G. Bandoli, and U. Mazzi, eds. New York: Raven. Pp. 605–607.

21. Weber, R.W., Boutin, R.H., Nedelman, M.A., Lister-James, J., and Dean, R.T. 1990. Enhanced kidney clearance with an ester-linked ^{99m}Tc-radiolabeled antibody Fab'-chelator conjugate. *Bioconjugate Chem.* 1:431–437.

22. Khaw, B.A., Strauss, H.W., and Haber, E. 1989. Production and characterization of monoclonal antimyosin antibody: Immunoscintigraphic visualization of necrotic myocardium. In *Monoclonal antibodies in immunoscintigraphy*. (J. Chatal, ed. Boca Raton, Fl.: CRC. Pp. 339–355.

23. Johnson, L.L., and Seldin, D.W. 1989. The role of antimyosin antibodies in acute myocardial infarction. *Semin. Nucl. Med.* 19:238–246.

24. Schaible, T.F., Alavi, A., and Berger, H.J. 1989. Preliminary studies with radiolabeled monoclonal antibodies of clot imaging in patients with deep-vein thrombosis. In *Monoclonal antibodies in immunoscintigraphy*. J. Chatal, ed. Boca Raton, FL.: CRC. Pp. 399–410.

25. Koblik, P.D., DeNardo, G.L., and Berger, H.J. 1989. Current status of immunoscintigraphy in the detection of thrombosis and thromboembolism. *Semin. Nucl. Med.* 19:221–237.

26. Miller, D.D., Boulet, A., Garcia, D., Heyl, B., Straw, J., Chaudhuri, T., Palmaz, J., McEver, R., Daddona, P., Neblock, D., Pak, K., and Berger, H.J. 1989. Technetium-99m monoclonal S-12 antibody imaging of in vivo platelet activation after balloon arterial injury in an experimental atherosclerotic model. *J. Nucl. Med.* 30:787.

27. Strauss, H.W., Fischman, A.J., Khaw, B.A., Callahan, R.J., Nedelman, M., Wilkinson, R., Keech, F., Kramer, P., Hanson, W.P., and Rubin, R. 1989. Detection of acute inflammation with immune imaging. In *Monoclonal antibodies in immunoscintigraphy*. J. Chatal, ed. Boca Raton, Fl.: CRC. Pp. 325–338.

28. Lind, P., Langsteger, W., Koltringer, P., Dimai, H.P., Passel, R., and Eber, O. 1990. Immunoscintigraphy of inflammatory processes with a technetium-99m-labeled monoclonal antigranulocyte antibody (MAb BW 250/183). *J. Nucl. Med.* 31:417–423.

29. Dvigi, C.R., and Larson, S.M. 1989. Radiolabeled monoclonal antibodies in the diagnosis and treatment of malignant melanoma. *Semin. Nucl. Med.* 19:252–261.

30. Goldenberg, D.M., Goldenberg, H., Sharkey, R.M., Lee, R.E., Higginbotham-Ford, E., Horowitz, J.A., Hall, T.C., Pinsky, C.M., and Hansen, H.J. 1989. Imaging of colorectal carcinoma with radiolabeled antibodies. *Semin. Nucl. Med.* 19:262–281.

31. Bast, R.C., and Knapp, R.C. 1986. Monoclonal reagents in immunodiagnosis and immunotherapy of human epithelial ovarian carcinoma. In *Monoclonal antibodies in cancer: Advances in diagnosis and treatment.* J.A. Roth, ed. Mount Kisco, N.Y.: Futura. Pp. 61–86.

32. Thor, A.D., and Edgerton, S.M. 1989. Monoclonal antibodies reactive with human breast or ovarian carcinoma: In vivo applications. *Semin. Nucl. Med.* 19:295–308.

33. Greager, J.A., Brown, J.M., Pavel, D.G., Garcia, J.L., Blend, M.J., and Das Gupta, T.K. 1986. Localization of human sarcoma with radiolabeled monoclonal antibody: A preliminary study, part 1. *Cancer Immunol. Immunother.* 23:148–154.

34. Bourguet, P., Dazord, L., and Herry, J.Y. 1989. Immunoscintigraphy of primary lung cancer. In *Monoclonal antibodies in immunoscintigraphy.* J. Chatal, ed. Boca Raton, Fla.: CRC. Pp. 311–321.

35. Keenan, A.M. 1989. Immunolymphoscintigraphy. *Semin. Nucl. Med.* 19:322–331.

36. Byers, V.S., and Baldwin, R.W. 1988. Therapeutic strategies with monoclonal antibodies and immunoconjugates. *Immunology* 65:329–335.

37. Dillman, R.O. 1989. Monoclonal antibodies for treating cancer. *Ann. Internal Med.* 111:592–603.

38. Woo, D.V., Li, D., Mattis, J.A., and Steplewski, Z. 1989. Selective chromosomal damage and cytotoxicity of ^{125}I-labeled monoclonal antibody 17-1a in human cancer cells. *Cancer Res.* 49:2952–2958.

39. Blakey, D.C., Wawrzynczak, E.J., Wallace, P.M., and Thorpe, P.E. 1988. Antibody toxin conjugates: A perspective. *Prog. Allergy* 45:50–90.

40. Pietersz, G.A., Smyth, M.J., Kanellos, J., Cunningham, Z., Sacks, N.P.M., and McKenzie, I.F.C. 1988. Preclinical and clinical studies with a variety of immunoconjugates. *Antibody Immunoconj. Radiopharm.* 1:79–103.

41. Todd, P.A., and Brogden, R.N. 1989. Muromonab CD3: A review of its pharmacology and therapeutic potential. *Drugs* 37:871–899.

42. Henslee, P.J., Byers, V.S., Jennings, C.D., Marciniak, E., Thompson, J.S., Macdonald, J.S., Romond, E.H., Messino, M.J., and Scannon, P.S. 1989. A new approach to the prevention of graft-versus-host disease using XomaZyme-H65 following histo-incompatible partially T-depleted marrow grafts. *Transplant. Proc.* 21:3004–3007.

43. Gorelick, K., Scannon, P.J., Hannigan, J., Wedel, N., and Ackerman, S.K. 1990. Randomized placebo-controlled study of E5 monoclonal antiendotoxin antibody. In *Therapeutic monoclonal antibodies*. C.A.K. Borrebaeck, and J.W. Larrick, eds. New York: Stockton. Pp. 253–261.

44. Ziegler, E.J., Fisher, C.J., Sprung, C.L., Straube, R.C., Sadoff, J.C., Foulke, G.E., Wortel, C.H., Fink, M.P., Dellinger, R.P., Teng, N.N.H., Allen, I.E., Berger, H.J., Knatterud, G.L., LoBuglio, A.F., Smith, C.R., and the HA-1A Sepsis Study Group. 1991. Treatment of gram-negative bacteremia and septic shock with HA-1A human monoclonal antibody against endotoxin. *N. Engl. J. Med.* 324:429–436.

45. Foung, S.K.H., Bradshaw, P.A., and Emanuel, D. 1990. Uses of human monoclonal antibodies to human cytomegalovirus and varicella-zoster virus. In *Therapeutic monoclonal antibodies*. C.A.K. Borrebaeck and J.W. Larrick, eds. New York: Stockton. Pp. 173–185.

46. Gold, H.K., Gimple, L.W., Yasuda, T., Leinbach, R.C., Werner, W., Holt, R., Jordan, R., Berger,H., Collen, D. and Coller, B.S. 1990. Pharmacodynamic study of $F(ab')_2$ fragments of murine monoclonal antibody 7E3 directed against human platelet glycoprotein IIb/IIIa in patients with unstable angina pectoris. *J. Clin. Invest.* 86:651–659.

47. Hafler, D.A., Ritz, J., Schlossman, S.F., and Weiner, H.L. 1988. Anti-CD4 and anti-CD2 monoclonal antibody infusions in subjects with multiple sclerosis. *J. Immunol.* 141:131–138.

48. Herzog, C., Walker, C., Muller, W., Rieber, P., Reiter, C., Riethmuller, G., Wassmer, P., Stockinger, H., Madic, O., and Pichler, W.J. 1989. Anti-CD4 antibody treatment of patients with rheumatoid arthritis. Effect on clinical course and circulating T cells. *J. Autoimmun.* 2:627–642.

49. LoBuglio, A.F., Saleh, M.N., Lee, J., Khazaeli, M.B., Carrano, R., Holden, H., and Wheeler, R.H. 1988. Phase I trial of multiple large doses of murine monoclonal antibody CO17-1A.I. Clinical aspects. *J. Natl. Cancer Inst.* 80:932–936.

50. Dillman, R.O. 1990. Human antimouse and antiglobulin responses to monoclonal antibodies. *Antibody Immunoconj. Radiopharm.* 3:1–15.

51. Van Kroonenburgh, M.J.P.G., and Pauwels, E.K.J. 1988. Human immunological response to mouse monoclonal antibodies in the treatment or diagnosis of malignant diseases. *Nucl. Med. Commun.* 9:919–930.

52. Wettendorff, M., Iliopoulos, D., Tempero, M., Kay, D., DeFreitas, E., Koprowski, H., and Herlyn, D. 1989. Idiotypic cascades in cancer patients treated with monoclonal antibody CO17-1A. *Proc. Natl. Acad. Sci. U.S.A.* 86:3787–3791.

53. Brown, J.M., Dean, R.T., Kaplan, P., Eden, P.J., Fuccello, A.J., Cox, P.H., and Berger, H. 1988. Absence of human antimouse antibody (HAMA) response in patients given antimyosin Fab-DTPA monoclonal antibody. *J. Nucl. Med.* 29:851.

54. Zimmer, A.M., Rosen, S.T., Spies, S.M., Goldman-Leikin, R., Kazikiewicz, J.M., Silverstein, E.A., and Kaplan, E.H. 1988. Radioimmunotherapy of patients with

cutaneous T-cell lymphoma using an iodine-131-labeled monoclonal antibody: Analysis of retreatment following plasmapheresis. *J. Nucl. Med.* 29:174–180.

55. LoBuglio, A.F., Wheeler, R.H., Trang, J., Haynes, A., Rogers, K., Harvey, E.B., Sun, L., Ghrayeb, J., and Khazaeli, M.B. 1989. Mouse/human chimeric monoclonal antibody in man: Kinetics and immune response. *Proc. Natl. Acad. Sci. U.S.A.* 86:4220–4224.

10

Anti-AIDS Drug Development

Prem Mohan

10.1. List of Abbreviations

The following is a list of abbreviations that appear in this chapter.

AIDS: acquired immunodeficiency syndrome

Ara-A: arabinosyladenine

ARC: AIDS-related complex

AZT: 3'-azido-2',3'-dideoxythymidine

C-Ara-A: carbocyclic arabinosyladenine

CPF: *N*-carbomethoxycarbonyl-propyl-phenylalanyl benzyl ester

DDA: 2',3'-dideoxyadenosine

DDC: 2',3'-dideoxycytidine

DDI: 2',3'-dideoxyinosine

D4T: 2',3'-dideoxy-2',3'-didehydrothymidine

DNA: deoxyribonucleic acid

DNJ: deoxynojirimycin

env: envelope

gag: group-specific antigen

gp: glycoprotein

HEPT: 1-[(2-hydroxyethoxy)methyl]-6-(phenylthio)thymine

HIV: human immunodeficiency virus

HTLV: human T-lymphotropic virus

MAb: monoclonal antibody

nef: negative regulatory factor (uncertain)

PAVAS: sulfated copolymer of acrylic acid and vinyl alcohol

PMEA: 9-(2-phosphonylmethoxyethyl)adenine

PMEDAP: 9-(2-phosphonylmethoxyethyl)-2,6-diaminopurine

pol: polymerase

PSS: poly(4-styrenesulfonc acid)

PVAS: sulfated polymer of vinyl alcohol

PVS: poly(vinylsulfonic acid)

rev: differential regulator

RNA: ribonucleic acid

RT: reverse transcriptase

TIBO: tetrahydroimidazo[4,5,1-*jk*][1,4]-benzodiazepin-2(1*H*)-one and -thione

tat: transactivator

vif: virion infectivity factor

vpr: weak transcriptional activator

vpu: required for efficient virion budding

10.2. Introduction

Acquired immunodeficiency syndrome (AIDS) is a fatal retroviral disease that is caused by the human immunodeficiency virus (HIV).[1-3] AIDS still continues to be resistant to all forms of curative therapy. After it was established that AIDS was caused by an RNA virus, many research groups rushed to find agents that would inhibit in vitro multiplication of the virus. One approach that was followed was the random screening of a wide variety of nucleoside candidates, since this class of derivatives had shown antiviral activity against other human viruses. A second approach was targeted at an enzyme that was known to be present in RNA viruses, namely reverse transcriptase (RT). Since many animal retroviral diseases were known, and agents that inhibited these animal RTs were available, these agents were tested against the RT of HIV-1. A third approach was the random screening of already available chemical substances. Indeed, a combination of these approaches did provide the initial drug leads in the early years of the epidemic.

10.3. Genes and Life Cycle of the AIDS Virus

The genomes of many animal retroviruses have been shown to contain the *gag*, *pol*, and *env* genes.[4] The fact that HIV-1 contains at least six additional genes provides unprecedented complexity to the HIV-1 genome. These genes are *vif, nef, rev, vpu, vpr,* and *tat.*[5] In addition, while the function of these new genes begins to become completely understood, their presence probably explains why the virus evades immune surveillance and curative therapy.

The life cycle of HIV-1 begins with its attachment to the target cell. The receptor for this attachment on HIV-1 is the viral envelope protein gp 120, and the corresponding receptor on the target cell is the CD4 receptor. Therefore, any cell that has the CD4 receptor is a target for the virus, and this is true for cells of the macrophage lineage. The macrophage serves as a reservoir for multiplication and harboring of the AIDS virus and is less susceptible to the cytopathic effects of the virus. On the other hand, the T helper-inducer cell is destroyed, leading to a gradual immunodeficiency and the onset of AIDS.[6]

The affinity of the viral gp 120 and the host cell CD4 receptor has been considered an important interaction for entry into the T4 helper-inducer cell. However, it is also known that non-CD4-containing cells can also be infected,[7] which suggests that other receptors may also be involved. After viral attachment and fusion, the virion uncoats and empties the genomic RNA and the RT into the cytoplasm of the target cell. At this stage, RT serves three functions. Using the RNA as a template, it catalyzes an RNA-dependent DNA synthesis to produce a single strand of DNA. Using the ribonuclease H section of RT, the enzyme systematically degrades the genomic RNA. Finally, using the newly synthesized DNA as a template, RT catalyzes a DNA-dependent DNA synthesis of a complementary DNA strand. The DNA double helix, also called proviral DNA, is now ready to enter the nucleus of the host cell. After translocation into the nucleus, the proviral DNA is integrated into the host genome by the viral enzyme integrase.

At this stage the blueprint for production of more virions is available. However, most commonly the virus may remain latent for a period of time until it is activated by host factors or gene products of other viruses, some of which are cytomegalovirus, Epstein–Barr virus, HTLV-I, and herpes simplex virus.[8] Expression of the viral genes leads to the production of viral genomic RNA and messenger RNA, followed by protein synthesis, assembly, and viral budding.

10.4. Targets for Drug Design

Anti-AIDS drug design may be pursued on two specific lines: (1) the design of new specific inhibitors and (2) the development of known inhibitors. The design of new specific inhibitors can be pursued after a detailed study of the virus life cycle, which reveals that many sites may be envisioned as potential targets for drug action. Generally, the most popular targets are enzymes, and this approach is potentially promising only if the enzyme is specific for the virus. This approach is strengthened if the relevant enzyme has been functionally characterized and its structural parameters confirmed by X-ray analysis. If all these data are available, the medicinal chemist is well positioned to design inhibitors. However, it must be realized that until these data become available, it is still possible to conduct a study involving the design of inhibitors. This is only possible after the discovery of a lead compound. A pertinent example is the discovery of many potent inhibitors of HIV-1 RT that were designed without the X-ray crystallographic data of the enzyme. Up until now, in terms of HIV-1, three enzymes have been targeted. They are RT, protease, and glucosidase. On the other hand, new specific inhibitors may be designed to inhibit viral processes like binding, giant cell (syncytia) formation, gene expression, or any characteristic function of the virus.

The development of known inhibitors is an area that is commonly pursued extensively because the initial lead compound is seldom structurally optimized for biological activity. These modifications are undertaken to prolong half-life, decrease toxicity, increase specificity, and optimize distribution or a parameter that would enhance activity. Another more challenging approach is to perform structural manipulations on a derivative that is inactive. Obviously, this is not a popular approach but can and should be pursued only after the scientist has thoroughly understood the structural parameters that govern the biological activity at the directed target.

10.5. Potential Anti-AIDS Agents

10.5.1. Nucleosides

After the retroviral cause of AIDS was discovered, many nucleoside derivatives were tested for their activity against HIV. The first nucleoside that was used against patients with AIDS was 3'-azido-2',3'-dideoxythymidine (AZT) (1). AZT was first synthesized as an antitumor agent in 1964 with failing biological results.[9] Ten years later it was shown that AZT had an antiretroviral effect against Friend leukemia virus.[10] Another decade later AZT demonstrated in vitro inhibition of the cytopathogenic effect of the AIDS virus.[11] This was later followed by the administration of AZT to patients with AIDS.[12] Other nucleosides, such as 2',3'-dideoxycytidine (DDC) (2), 2',3'-dideoxyadenosine (DDA) (3), 2',3'-dideoxy-inosine (DDI) (4), and 2',3'-dideoxy-2',3'-didehydrothymidine (D4T) (5), have been in clinical trials, and the FDA approval of one or more of these agents seems imminent. So far, only AZT, DDC, and DDI are federally licensed for wide-scale use in patients with AIDS and ARC.

The mechanism of action of these nucleoside agents involves the sequential in vivo phosphorylation of the 5'-hydroxyl group to the active species, the triphosphate derivative. The triphosphate inhibits the retroviral RT and also acts as a chain terminator.[13] The performance of these nucleoside agents in the clinic has revealed various forms of toxicity. These are bone marrow suppression and a severe anemia (AZT), peripheral neuropathy (DDC, DDI, and D4T), and pancreatitis (DDI).[14] These side effects may be dose limiting and may preclude the long-

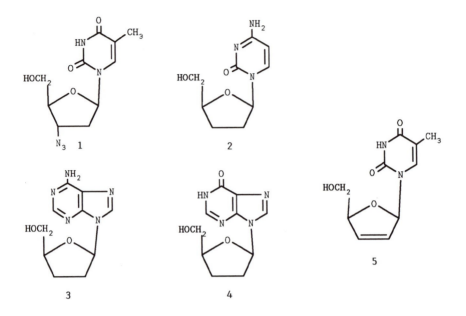

term use of these nucleosides in many patients. In addition, AZT-resistant HIV-1 mutants have been reported.[15]

The shortcomings associated with the foregoing nucleosides have led to many structure activity relationship studies in this area.[14] One problem with purine nucleosides (e.g., DDA and DDI) was their short half-life owing to acid lability. This property was circumvented by preparing 2'-fluoro analogues, which were as active as the parent analogues but exhibited increased toxicity.[16]

10.5.2. Acyclic and Carbocyclic Nucleoside Analogues

The concept of preparing acyclic analogues as antiviral agents was strengthened after the success of a guanosine analogue, acyclovir, as a treatment for herpes.[17] Reports have appeared on allene derivatives of adenine and cytosine,[18] and phosphomethoxyalkyl derivatives of adenine, namely 9-(2-phosphonylmethoxyethyl)adenine (PMEA) (6) and 9-(2-phosphonylmethoxyethyl)-2,6-diaminopurine (PMEDAP) (7). PMEA and PMEDAP are more potent than AZT in certain animal models[19, 20] and have the advantage of also having antiviral activity against other viruses such as Epstein–Barr, cytomegalovirus, and herpes simplex. Interestingly, after a single dose of these agents, in vivo antiviral activity can be demonstrated for several days.[21]

Carbocyclic derivatives have the sugar heterocyclic oxygen [-O-] replaced by a methylene group [-CH2-]. This modification for arabinosyladenine (Ara-A) led to the discovery of carbocyclic arabinosyladenine (C-Ara-A, cyclaridine) with activity against genital herpes in a guinea pig model.[22] One problem with Ara-A

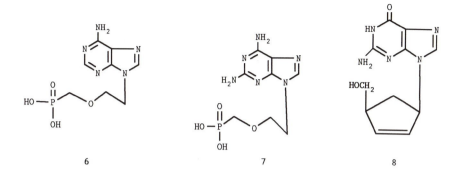

6   7   8

was its rapid deamination to the hypoxanthine derivative, which is markedly less active than the parent compound.[23] Cyclaridine is resistant to the deamination reaction by adenosine deaminase.[24] A variety of carbocyclic analogues having a double bond between the 2′- and 3′-position have also been examined but did not show promising activity.[25] However, the carbocyclic 2′,3′-dideoxy-2′,3′-didehydronucleoside derivative of guanosine (carbovir [**8**]) was shown to have in vitro anti-HIV-1 activity.[26]

10.5.3. Protease Inhibitors

The determination of the X-ray crystal structure of the HIV aspartyl protease[27–29] stimulated the design and synthesis of a wide variety of protease inhibitors. The viral protease processes the *gag* and *pol* proteins into components that are needed for maturation of the virus particles. Inhibition of the viral protease results in noninfectious viral particles. Further, since the function of the protease is a postintegration event, the quest for protease inhibitors has remained an attractive approach.

A vast majority of inhibitors have been designed by synthesizing compounds that are modeled around a selected scissile bond. In the inhibitors, this bond is replaced by a nonhydrolyzable isostere having tetrahedral geometry in the transition state. This concept has produced several inhibitors.[30–32] One challenge that faced the design of protease inhibitors was the need for selective activity against the viral protease over the human aspartic proteases, renin, pepsin, gastricsin, cathepsin D, and cathepsin E. Indeed, this has been achieved with inhibitor **9**.[33]

Another approach to protease inhibitors has capitalized on the C_2 symmetry of the protease active site. Consequently, symmetrical inhibitors have been designed, synthesized, and evaluated. One of these agents, derivative **10,** has revealed potent activity.[34] The 2.8 Å crystal structure of the inhibitor–enzyme complex demonstrated that the inhibitor was bound to the enzyme in a highly symmetric fashion.[35] Since many of the potent inhibitors are largely peptide in character, the problems associated with peptide drugs (e.g., metabolism,

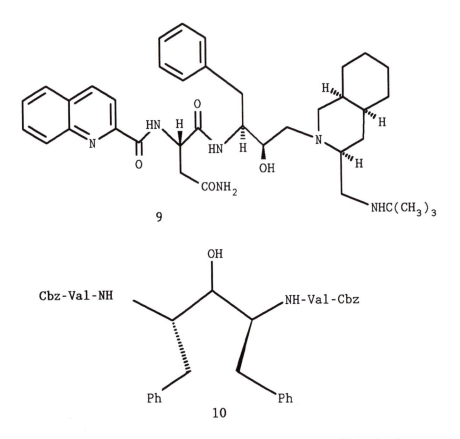

9

10

distribution, and bioavailability) will have to be overcome to aid the development of these agents.

10.5.4. Antisense Oligonucleotides

The antisense concept[36] has been popular because, theoretically, it has the potential to block gene expression. This may be achieved by designing a complementary (antisense) oligonucleotide strand that is targeted to a specific portion of the genome or messenger RNA. After hybridization, these agents block expression of the genome or the messenger RNA.

An initial concern in the use of these derivatives was the rapid in vivo degradation of these agents by nucleases. This problem was approached by substituting the negatively charged oxygen on the phosphodiester linkages with methyl, sulfur, or amine groups, to produce methylphosphonates (**11**), phosphorothioates (**12**), or phosphoamidate (**13**) derivatives. Using these approaches, several agents have been synthesized and shown to possess anti-HIV-1 activity.[37–39] Since cellular uptake can be envisioned as a problem for these agents, linkage with poly-L-

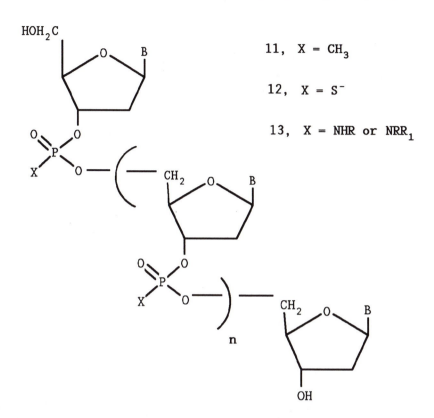

11, X = CH₃

12, X = S⁻

13, X = NHR or NRR₁

lysine[40] or cholesteryl[41] has been suggested to aid the uptake process. Oligonucleo-tides have also been targeted to the rev gene.[42] The added advantage of antisense oligonucleotides is that they inhibit syncytia formation, RT, and HIV-1 replication in chronically infected cells.[43]

10.5.5. Glucosidase Inhibitors

α-Glucosidases I and II are responsible for trimming the glycosylated envelope protein gp 120. The action of these enzymes, which involves the sequential removal of one and two glucose units, respectively, is necessary to facilitate maturation and maintain the infectivity of the virion particle. The fact that these events occur after the proviral integration stage made the glucosidase enzymes interesting targets for inhibition.

A natural product and a derivative, castanospermine (**14**) and *N*-butyldeoxyno-jirimycin (*N*-butyl-DNJ) (**15**), are inhibitors of these enzymes.[44] Structurally, these compounds are aminosugar derivatives that have a chiral disposition related to glucose.[45] Many other similar derivatives have also been evaluated for anti-HIV-1 activity.[46, 47]

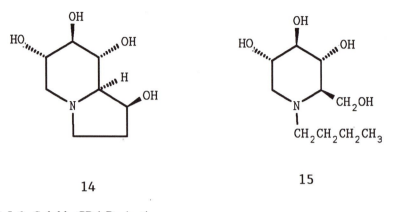

14

15

10.5.6. Soluble CD4 Derivatives

The realization that the CD4–gp 120 interaction was important for the pathogenic process led to the realization that administering the CD4 receptor in a soluble form may be a useful form of intervention. Many research groups have published the results of studies in this area.[48–52]

One concern with using soluble CD4 was its in vivo half-life of approximately 45 minutes.[53] This was remedied by linking the soluble CD4 derivative to an immunoglobulin to make immunoadhesins.[54] An immunoadhesin has been shown to be able to cross the placenta of a primate.[55] It is advantageous that HIV-1 infection of macrophages is also prevented by soluble CD4 derivatives.[56] It has also been demonstrated that HIV-2 is less susceptible to soluble CD4 than HIV-1.[57, 58] It is important that the in vitro activity of soluble CD4 be able to be duplicated with clinical strains of the AIDS virus. One study has shown that neutralization of HIV-1 clinical isolates requires 200 to 2,700 times more soluble CD4 than is needed for inhibiting laboratory strains of HIV-1.[59] Unfortunately, the performance of soluble CD4 in the clinic has not been encouraging.[60] Finally, the wide-scale utility of this mode of therapy remains questionable, since the AIDS virus can also infect CD4-negative cells.[7]

10.5.7. Anionic Compounds

The first drug used in AIDS patients, suramin (**16**),[61] belonged to the anionic class of potential anti-HIV-1 agents. However, suramin did not show any clinical or immunological improvement in patients and was not further pursued.[62] However, interest in suramin as a biologically active compound has not waned. It has been shown to have a half-life of 44 to 54 days with little or no degradation in vivo.[63] Activity against protein kinase C has been reported,[64] and there is interest in investigating suramin as an anticancer agent.[65]

The discovery of the in vitro anti-HIV-1 activity of suramin[66] focused serious attention on the sulfonic acid group. Anti-HIV-1 activity was also demonstrated

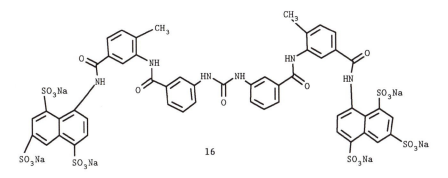

16

in sulfonic acid dyes containing the naphthalene,[67] anthraquinone,[68] and triphenyl-methane[69] nuclei. Two representatives of the naphthalene dyes are Direct yellow 50 (**17**) and Evans blue (**18**). Other anti-HIV-1 sulfonic acid derivatives belong to the stilbene[70] and natural product[71, 72] classes. Suramin analogues more potent and less toxic than suramin have also been synthesized and evaluated for anti-HIV-1 activity.[73]

Being cognizant of the carcinogenic potential of certain azo sulfonic acid dyes, we have prepared and evaluated a series of naphthalenesulfonic acid spacer derivatives utilizing a variety of HIV-1 and HIV-2 assays.[74–76] Compound **19** emerged as the most potent compound in these studies.[76] Activity was not limited to the bis series of compounds. Potent activity was demonstrated in a small naphthalenedisulfonic acid derivative (**20**) against a clinical isolate of HIV-1.[77] Using molecular modeling parameters for a flexible bis naphthalenesulfonic acid derivative (**21**) and suramin (**16**), we have shown that both these agents can mimic the helical turn of B-DNA.[78]

Various sulfated polysaccharides, such as dextran sulfate (**22**), are potent

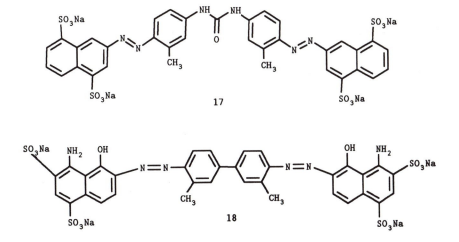

17

18

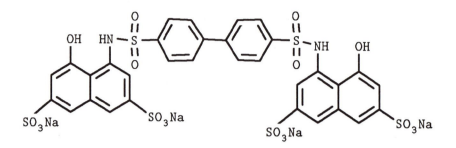

19

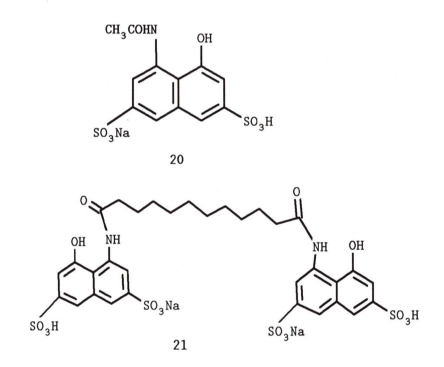

20

21

inhibitors of HIV-1.[79] Oral clinical evaluation of dextran sulfate revealed no anti-HIV-1 activity.[80] This may have been predicted owing to the polar nature of this derivative and its inability to cross cellular membranes. While ongoing studies will evaluate the potential of parenterally administering dextran sulfate, another problem remains to be addressed. Owing to the polysaccharide structure of dextran sulfate, this derivative will be susceptible to in vivo glycosidic bond

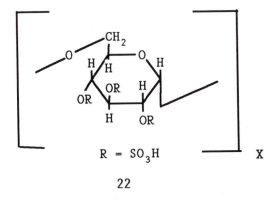

R = SO₃H

22

cleavage. Indeed, this has been demonstrated in an animal model. Further, it was shown that dextran sulfate is degraded into smaller molecular weight fragments that are known to be inactive.[81]

In one study it has been shown that the polysaccharide structure is not required for activity in these polysulfated compounds. Two compounds having a hydrocarbon backbone are the sulfated copolymer of acrylic acid and vinyl alcohol (PAVAS [23]) and the sulfated polymer of vinyl alcohol (PVAS [24]). Both PAVAS and PVAS are potent inhibitors of syncytia formation.[82] Although these compounds have the potential to be more stable than dextran sulfate in vivo, they may still be prone to desulfation reactions catalyzed by sulfatase enzymes.

In order to address this issue we chose to evaluate several novel polymers containing the sulfonic acid group. The sulfonic acid group is unreactive to metabolizing enzymes and is excreted largely unchanged with negligible desulfonation.[83, 84] Indeed, we have shown that these sulfonic acid polymers demonstrated potent and selective in vitro anti-HIV-1 activity. The most active agent in this

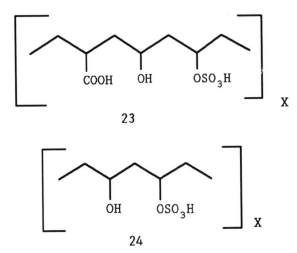

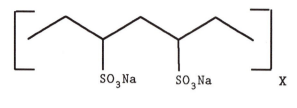

25

family is an aliphatic polymer that demonstrates an in vitro therapeutic index of
>1,750. However, in the HIV-1 RT assay, poly(4-styrenesulfonic acid) (PSS)
exhibited a potency that was 230 times greater than suramin, a known potent
inhibitor of RT. Another aromatic polymer, poly(vinylsulfonic acid) (PVS) (25)
was 161 times more active than suramin and is an attractive candidate, since it has
a molecular weight of 2,000. These polymers are inhibitors of viral adsorption.[85]

Recently we have conducted a series of experiments to determine the nature
of viral-binding inhibition exhibited by the sulfonic acid polymers. These studies
revealed that while the polymers inhibited binding of anti-gp 120 MAb (directed
to the V3 fusion region), only PSS and PVS (25) interfered with the binding
of OKT4A/Leu3a MAb to the CD4 receptor.[86] Based on all of the preceding
observations, these sulfonic acid derivatives are a promising class of new anti-
HIV-1 agents. The activity of these agents may be questioned on the basis of
their polar nature and their unlikely ability to enter cells. It first needs to be
determined whether the anti-RT activity or the inhibition of viral adsorption is
the required activity. Depending on the activity to be targeted, suitable prodrugs
or analogues may be prepared. It should also be noted that inhibitors of viral
adsorption have the potential to be used as chemoprophylactic agents against
AIDS.

10.5.8. Nonnucleoside Inhibitors of Reverse Transcriptase

The retroviral RT has remained an attractive target for therapeutic intervention.
With the known drawbacks of the nucleoside RT inhibitors, many laboratories
have searched for nonnucleoside RT inhibitors. These studies have unveiled
potent anti-HIV-1 activity in many derivatives exhibited at nanomolar concentrations. One class of novel derivatives are the tetrahydroimidazo[4,5,1-jk][1,4]-
benzodiazepin-2(1H)-one and -thione (TIBO) derivatives. These compounds
were discovered after a search of 600 different classes of compounds. It is
interesting that the TIBO derivatives inhibit the replication of HIV-1 but not of
HIV-2, or any other DNA or RNA viruses. One compound in this series, R82150
(26), is almost as potent as AZT (1) and has a therapeutic index that is five times
higher than AZT (1).[87] Several 1-[(2-hydroxyethoxy)methy]-6-(phenylthio)thym-
ine (HEPT) derivatives have been synthesized and evaluated against HIV.[88, 89]
The HEPT derivatives have selective activity against HIV-1 replication, but no

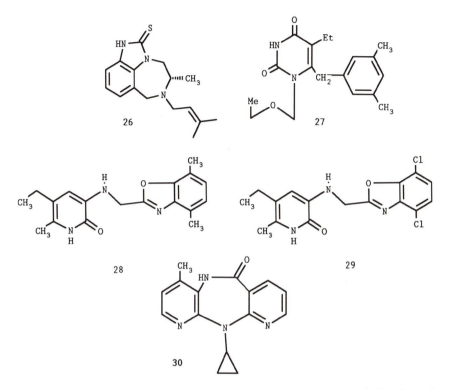

activity against HIV-2 or any other retroviruses. One derivative of this class of compounds, **27,** has demonstrated an in vitro therapeutic index of over 100,000.[90] These derivatives act by inhibiting RT.[91] Recently, two pyridinone derivatives, L-697,639 (**28**) and L-697,661 (**29**), have also shown RT inhibition and are being actively pursued for further development.[92] A dipyridodiazepinone derivative, BI-RG-587 (**30**),[93] is also undergoing evaluation; it has been shown that it shares the same RT binding site with a TIBO derivative.[94] It has also been demonstrated that derivatives from two structurally diverse classes, TIBO and HEPT, bind to the same site on HIV-1 RT.[95]

10.5.9. Miscellaneous Compounds

There are many derivatives that have exhibited anti-HIV-1 activity. For many of these agents, the activity needs to be confirmed. This section will highlight the activity of a few derivatives that have been shown, or are suggested, to act by mechanisms other than those discussed in the previous sections. Oxathiin carboxanilide (**31**), an agent originally synthesized as a fungicide, has been found to act at an early stage of HIV-1 reproduction.[96] *N*-Carbomethoxycarbonyl-propyl-phenylalanyl benzyl esters (CPFs) such as **32** bind with the envelope protein gp 120 and inhibit binding of the virus to the target cell.[97] Since the *gag*

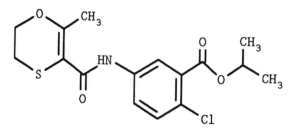

31

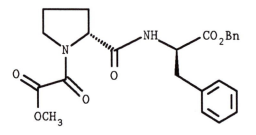

32

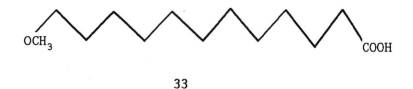

33

precursor protein and its p17 product are myristoylated before the virus particles can be assembled are released from the cell, it has been suggested that inhibition of the myristoylation process can be a good target for obtaining anti-HIV activity.[98] 12-Methoxydodecanoic acid (33), a heteroatom analogue of myristic acid, has shown activity against HIV-1.[99] Since protein kinase C induces phosphorylation of the CD4 receptor during the pathogenic process,[100] inhibitors of protein kinase C may be valuable as potential anti-HIV-1 agents. The antiretroviral activity of hypericin and pseudohypericin[101] has been linked to their ability to inhibit protein kinase C. In this context, since suramin (16) is also known to inhibit protein kinase C,[64] the design of analogues targeted toward this end may prove worthy of study.

10.6. Future Prospects

The AIDS virus is the most studied, yet most complex, pathogenic virus that has challenged drug research. Since its discovery, a wide array of chemical derivatives have been shown to inhibit the AIDS virus in culture. Further, these agents belong to different chemical classes and have exhibited activity at different sites in the viral life cycle. Since only a few of these agents have reached the stage of clinical trials, a comprehensive evaluation of the beneficial effects of these inhibition sites is not possible.

Only the nucleoside RT inhibitors have been studied in a large-scale clinical setting and it seems that they will not be useful as singular anti-AIDS agents. However, the potential of anti-HIV nucleosides may be enhanced if used in combination with other agents that have a different mechanism of action. On the other hand, since RT inhibition involves a preintegration event, the nucleoside derivatives may be useful in asymptomatic patients. Indeed, these investigations are being actively pursued. However, RT may still be a viable target, and what is needed is a class of agents that demonstrates a different mode of binding to the enzyme. In this regard, the nonnucleoside RT inhibitors are promising agents, and their clinical evaluation is anxiously awaited.

Glucosidase inhibitors are challenged by the fact that the glucosidase enzymes are not specific for the virus life cycle. The design of peptide or peptide-like drugs (for example, protease inhibitors) would be faced with the need to convert them to small-molecule nonpeptide mimics that maintain the peptide property for biological activity but are different enough to resist degradation. For the soluble CD4 molecule mimics to be most practical, it would require the design of small molecules that interact with the attachment or fusion site(s), or both. In both the aforementioned cases, the discovery of agents that belong to a totally different structural class but that have the functions of the peptide prototype would revolutionize this area of research. Anionic compounds will have to be designed to be more selective, as well as to be able to reach the required target site. These goals can be achieved by undertaking rigorous and rational structural modification studies.

In the ongoing search for new anti-HIV agents, the researcher should not forget to continue to investigate new sources for potential derivatives. Many laboratories are conducting the random or semirandom screening of a vast number of synthetic derivatives. This strategy led to the discovery of many anti-HIV-1 agents. Worthy of note are the highly potent and selective TIBO derivatives.[87] While this approach is still necessary for any disease, it is also necessary to investigate sources that can furnish new and novel chemical structures. In this regard, natural products are a large reservoir of as yet still untapped chemicals with a well-established track record of diverse and potent biological activities.[102]

In the study of drug design strategies to halt viral replication, the scientist has to constantly evaluate the potential of new targets so that suitable inhibitors may be prepared. This need becomes more relevant when agents with known mechanisms of action fail to cure a disease in the clinic. The evolving molecular biology of the AIDS virus has presented us with new and novel putative targets. The regulatory genes, *tat* and *rev,* are still attractive targets for the medicinal chemist. The recent determination of the X-ray crystal structure of the ribonuclease H domain of RT[103] will definitely provide the impetus to design and synthesize inhibitors. Other viable targets are the function of the integrase enzyme. It remains to be determined whether these targets, other new targets, or new modes to inhibit previously known sites in the HIV-1 life cycle, will have the unprecedented ability to finally cure AIDS.

Acknowledgment

The author gratefully acknowledges a Scholar Award (700065-5-RF) from the American Foundation for AIDS Research.

References

1. Barre-Sinoussi, F., Chermann, J.C., Rey, F., Nugeyre, M.T., Chamaret, S., Gruest, T., Dauguet, C., Axler-Blin, C., Bezinet-Brun, F., Rouzioux, C., Rozenbaum, W., and Montagnier, L. 1983. Isolation of a T cell lymphotropic virus from a patient at risk for the acquired immunodeficiency syndrome (AIDS). *Science* 220:868–871.

2. Gallo, R.C., Salahuddin, S.Z., Popovic, M., Shearer, G.M., Kaplan, M., Haynes, B.F., Palker, T.J., Redfield, R., Oleska, J., Safai, B., White, G., Foster, P., and Markham, P.D. 1984. Frequent detection and isolation of pathogenic retroviruses (HTLV-III) from patients with AIDS and at risk from AIDS. *Science* 224:500–503.

3. Levy, J.A., Hoffman, A.D., Kramer, S.M., Landis, J.A., Shimabukuro, J.M., and Oshiro, L.S. 1984. Isolation of lymphocytopathic retroviruses from San Francisco patients with AIDS. *Science* 225:840–842.

4. Varmus, H. 1988. Retroviruses. *Science* 240:1427–1435.

5. Greene, W.C. 1991. The molecular biology of human immunodeficiency virus type 1 infection. *N. Engl. J. Med.* 324:308–317.

6. Fauci, A.S. 1988. The human immunodeficiency virus: Infectivity and mechanisms of pathogenesis. *Science* 239:617–622.

7. Tateno, M., Gonzalez-Scarano, F., and Levy, J.A. 1989. Human immunodeficiency virus can infect CD4-negative human fibroblastoid cells. *Proc. Natl. Acad. Sci. U.S.A.* 86:4287–4290.

8. Rosenberg, Z.F., and Fauci, A.S. 1990. Activation of latent HIV infection. *J. NIH Res.* 2:41–45.

9. Horowitz, J.P., Chua, J., and Noel, M. 1964. Nucleosides. 5. The monomesylates of 1-(2-deoxy-β-D-lyxofuranosyl)thymine. *J. Org. Chem.* 29:2076–2078.

10. Ostertag, W., Roesler, G., Krieg, C.J., Kind, J., Cole, T., Crozier, T., Gaedicke,

G., Steinheider, G., Kluge, N., and Dube, S. 1974. Induction of endogenous virus and of thymidine kinase by bromodeoxyuridine in cell cultures transformed by Friend virus. *Proc. Natl. Acad Sci. U.S.A.* 71:4980–4985.

11. Mitsuya, H., Weinhold, K.J., Furman, P.A., St. Clair, M.H., Lehrman, S.N., Gallo, R.C., Bolognesi, D., Barry, D.W., and Broder, S. 1985. 3'-Azido-3'-deoxythymidine (BW A509U): An antiviral agent that inhibits the infectivity and cytopathic effect of human T-lymphotropic virus type III/lymphadenopathy-associated virus in vitro. *Proc. Natl. Acad. Sci. U.S.A.* 82:7096–7100.

12. Yarchoan, R., Klecker, R.W., Weinhold, K.J., Markham, P.D., Lyerly, H.K., Durack, D.T., Gelmann, E., Lehrman, S.N., Blum, R.M., Barry, D.W., Shearer, G.M., Fischl, M.A., Mitsuya, H., Gallo, R.C., Collins, J.M., Bolognesi, D.P., Myers, C.E., and Broder, S. 1986. Administration of 3'-azido-3'-deoxythymidine, an inhibitor of HTLV-III/LAV replication to patients with AIDS or AIDS-related complex. *Lancet* 1:575–580.

13. Yarchoan, R., and Broder, S. 1989. Anti-retroviral therapy of AIDS and related disorders: General principles and specific development of dideoxynucleosides. *Pharmacol. Ther.* 40:329–348.

14. Mitsuya, H., Yarchoan, R., and Broder, S. 1990. Molecular targets for AIDS therapy. *Science* 249:1533–1544.

15. Larder, B.A., Darby, G., and Richman, D.D. 1989. HIV with reduced sensitivity to Zidovudine (AZT) isolated during prolonged therapy. *Science* 243:1731–1734.

16. Marquez, V.E., Tseng, C.K. -H., Hitsuya, H., Aoki, S., Kelley, J., Ford, H., Jr., Roth, J.S., Broder, S., Johns, D.G., and Driscoll, J.S. 1990. Acid-Stable 2'fluoro purine dideoxynucleosides as active agents against HIV. *J. Med. Chem.* 33:978–985.

17. Reardon, J.E., and Spector, T. 1989. Herpes simplex virus type 1 DNA polymerase. *J. Biol. Chem.* 264:7405–7411.

18. Hayashi, S., Phadtare, S., Zemlicka, J., Matsukura, M., Mitsuya, H., and Broder, S. 1988. Adenallene and cytallene: Acyclic nucleoside analogues that inhibit replication and cytopathic effect of human immunodeficiency virus in vitro. *Proc. Natl. Acad. Sci. U.S.A.* 85:6127–6131.

19. Balzarini, J., Naesens, L., Herdewijn, P., Rosenberg, I., Holy, A., Pauwels, R., Baba, M., Johns, D.G., and de Clercq, E. 1989. Marked *in vivo* antiretrovirus activity of 9-(2-phosphonylmethoxyethyl)adenine, a selective anti-human immunodeficiency virus agent. *Proc. Natl. Acad. Sci. U.S.A.* 86:332–336.

20. Naesens, L., Balzarini, J., Rosenberg, I., Holy, A., and De Clercq, E. 1989. 9-(2-Phosphonylmethoxyethyl)-2,6-diaminopurine (PMEDAP): A novel agent with anti-human immunodeficiency virus activity in vitro and potent anti-Moloney murine sarcoma virus activity in vivo. *Eur. J. Clin. Microbiol. Infect. Dis.* 8:1043–1047.

21. De Clercq, E. 1991. Basic approaches to anti-retroviral treatment. *J. Acq. Immune Defic. Syndr.* 4:207–218.

22. Vince, R., Daluge, S., Lee, H., Shannon, W.M., Arnett, G., Schafer, T.W., Nagabhushan, T.L., Reichert, P., and Tsai, H. 1983. Carbocyclic arabinofuranosy-

ladenine (cyclaridine): Efficacy against genital herpes in guinea pigs. *Science* 221:1405–1406.

23. De Clercq, E. 1988. Recent advances in the search for selective antiviral agents. *Adv. Drug Res.* 17:1–59.

24. Vince, R., and Daluge, S. 1977. Carbocyclic arabinosyladenine, an adenosine deaminase resistant antiviral agent. *J. Med. Chem.* 20:612–613.

25. De Clercq, E., Van Aerschot, A., Herdwijn, P., Baba, M., Pauwels, R., and Balzarini, J. 1989. Anti-HIV-1 activity of 2′,3′-dideoxynucleoside analogues: Structural activity relationship. *Nucleosides Nucleotides* 8:659–671.

26. Vince, R., Hua, M., Brownell, J., Daluge, S., Lee, F., Shannon, W.M., Lavelle, G.C., Qualls, J., Weislow, O.S., Kiser, R., Canonico, P.G., Schultz, R.H., Narayanan, V.L., Mayo, J.G., Shoemaker, R.H., and Boyd, M.R. 1988. Potent and selective activity of a new carbocyclic nucleoside analog (carbovir: NSC 614846) against human immunodeficiency virus in vitro. *Biochem. Biophys. Res. Commun.* 156:1046–1053.

27. Navia, M.A., Fitzgerald, P.M.D., McKeever, B.M., Leu, C. -T., Heimbach, J.C., Herber, W.K., Sigal, I.S., Darke, P.L., and Springer, J.P. 1989. Three-dimensional structure of aspartyl protease from human immunodeficiency virus HIV-1. *Nature (London)* 337:615–620.

28. Wlodawer, A., Miller, M., Jaskolski, M., Sathyanarayana, B.K., Baldwin, E., Weber, I.T., Selk, I.M., Clawson, L., Schneider, J., and Kent, S.B.H. 1989. Conserved folding in retroviral proteases: crystal structure of a synthetic HIV-1 protease. *Science* 245:616–621.

29. Lapatto, R., Blundell, T., Hemmings, A., Overington, J., Wilderspin, A., Wood, S., Merson, J.R., Whittle, P.J., Danley, D.E., Geoghegan, K.F., Hawrylik, S.J., Lee, S.E., Scheld, K.G., and Hobart, P.M. 1989. X-ray analysis of HIV-1 proteinase at 2.7 Å resolution confirms structural homology among retroviral enzymes. *Nature (London)* 342:299–302.

30. Dreyer, G.B., Metcalf, B.W., Tomaszek, T.A., Jr., Carr, T.J., Chandler, A.C.C., III, Hyland, L., Fakhoury, S.A., Magaard, V.W., Moore, M.L., Strickler, J.E., Debouck, C., and Meek, T.D. 1989. Inhibition of human immunodeficiency virus 1 protease *in vitro:* Rational design of substrate analogue inhibitors. *Proc. Natl. Acad Sci. U.S.A.* 86:9752–9756.

31. Overton, H.A., McMillan, D.J., Gridley, S.J., Brenner, J., Redshaw, S., and Mills, J.S. 1990. Effect of two novel inhibitors of the human immunodeficiency virus protease inn the maturation of the HIV gag and gag-pol polyproteins. *Virology* 179:508–511.

32. McQuade, T.J., Tomasselli, A.G., Liu, L., Karacostas, V., Moss, B., Sawyer, T.K., Heinrikson, R.L., and Tarpley, W.G. 1990. A synthetic HIV-1 protease inhibitor with antiviral activity arrests HIV-like particle maturation. *Science* 247:454–456.

33. Roberts, N.A., Martin, J.A., Kinchington, D., Broadhurst, A.V., Craig, J.C., Duncan, I.B., Galpin, S.A., Handa, B.K., Kay, J., Krohn, A., Lambert, R.W., Merrett, J.H., Mills, J.S., Parkes, K.E.B., Redshaw, S., Ritchie, A.J., Taylor,

D.L., Thomas, G.J., and Machin, P.J. 1990. Rational design of peptide-based HIV proteinase inhibitors. *Science* 248:358–361.

34. Kempf, D.J., Norbeck, D.W., Codacovi, L.M., Wang, X.C., Kohlbrenner, W.E., Wideburg, N.E., Paul, D.A., Knigge, M.F., Vasavanonda, S., Craig-Kennard, A., Saldivar, A., Rosenbrook, W., Jr., Clement, J.J., Plattner, J.J., and Erickson, J. 1990. Structure-based C_2 symmetry inhibitors of HIV protease. *J. Med. Chem.* 33:2687–2689.

35. Erickson, J., Neidhart, D.J., VanDrie, J., Kempf, D.J., Wang, X.C., Norbeck, D.W., Plattner, J.J., Rittenhouse, J.W., Turon, M., Wideburg, N., Kohlbrenner, W.E., Simmer, R., Helfrich, R., Paul, D.A., and Knigge, M. 1990. Design, activity and 2.8 Å crystal structure of a C_2 symmetric inhibitor complexed to HIV-1 protease. *Science* 249:527–533.

36. Zamecnik, P.C., and Stephenson, M.L. 1978. Inhibition of rous sarcoma virus replication and cell transformation by a specific oligodeoxynucleotide. *Proc. Natl. Acad. Science. U.S.A.* 75:280–284.

37. Stein, C.A., Matsukura, M., Subasinghe, C., Broder, S., and Cohen, J.S. 1989. Phosphorothioate oligodeoxynucleotides are potent sequence nonspecific inhibitors of *de novo* infection by HIV. *AIDS Res. Human Retrovir.* 5:639–646.

38. Montefiori, D.C., Sobol, R.W., Jr., Li, S.W., Reichenbach, N.L., Suhadolnik, R.J., Charubala, R., Pfleiderer, W., Modliszewski, A., Robinson, W.E., Jr., and Mitchell, W.M. 1989. Phosphorothioate and cordycepin analogues of 2′,5′-oligoadenylate: Inhibition of human immunodeficiency virus type 1 reverse transcriptase and infection *in vitro*. *Proc. Natl. Acad. Sci. U.S.A.* 86:7191–7194.

39. Sarin, P.S., Agrawal, S., Civeira, M.P., Goodchild, J., Ikeuchi, T., and Zamecnik, P.C. 1988. Inhibition of acquired immunodeficiency syndrome virus by oligodeoxynucleoside methylphosphonates. *Proc. Natl. Acad. Sci. U.S.A.* 85:7448–7451.

40. LeMaitre, M., Bayard, B., and Lebleu, B. 1987. Specific antiviral activity of a poly(L-lysine)-conjugated oligodeoxyribonucleotide sequence complimentary to vesicular stomatitis virus N protein mRNA initiation site. *Proc. Natl. Acad. Sci. U.S.A.* 84:648–652.

41. Letsinger, R.L., Zhang, G., Sun, D.K., Ikeuchi, T., and Sarin, P.S. 1989. Cholesteryl-conjugated oligonucleotides: Synthesis, properties, and activity as inhibitors of replication of human immunodeficiency virus in cell culture. *Proc. Natl. Acad. Sci. U.S.A.* 86:6553–6556.

42. Matsukura, M., Zon, G., Shinozuka, K., Robert-Guroff, M., Shimada, T., Stein, C.A., Mitsuya, H., Wong-Staal, F., Cohen, J.S., and Broder, S. 1989. Regulation of viral expression of human immunodeficiency virus *in vitro* by an antisense phosphorothioate oligodeoxynucleotide against *rev* (*art/trs*) in chronically infected cells. *Proc. Natl. Acad. Sci. U.S.A.* 86:4244–4248.

43. Agrawal, S., Ikeuchi, T., Sun, D., Sarin, P.S., Konopka, A., Maizel, J., and Zamecnik, P.C. 1989. Inhibition of human immunodeficiency virus in early infected and chronically infected cells by antisense oligodeoxynucleotides and their phosphorothioate analogues. *Proc. Natl. Acad. Sci. U.S.A.* 86:7790–7794.

44. Fleet, G.W.J., Karpas, A., Dwek, R.A., Fellows, L.E., Tyms, A.S., Petursson,

S., Namgoong, S.K., Ramsden, N.G., Smith, P.W., Son, J.C., Wilson, F., Witty, D.R., Jacob, G.S., and Rademacher, T.W. 1988. Inhibition of HIV replication by amino-sugar derivatives. *FEBS-Lett.* 237:128–132.

45. Saul, R., Molyneux, R.J., and Elbein, A.D. 1984. Studies on the mechanism of castanospermine inhibition of α- and β-glucosidases. *Arch. Biochem. Biophys.* 230:668–675.

46. Karpas, A., Fleet, G.W.J., Dwek, R.A., Petursson, S., Namgoong, S.K., Ramsden, N.G., Jacob, G.S., and Rademacher, T.W. 1988. Aminosugar derivatives as potential anti-human immunodeficiency virus agents. *Proc. Natl. Acad. Sci. U.S.A.* 85:9229–9233.

47. Gruters, R.A., Neefjes, J.J., Tersmette, M., De Goede, R.E.Y., Tulp, A., Huisman, H.G., Miedema, F., and Ploegh, H.L. 1987. Interference with HIV-induced syncytium formation and viral infectivity by inhibitors of trimming glucosidase. *Nature (London)* 330:74–77.

48. Hussey, R.E., Richardson, N.E., Kowalski, M., Brown, N.R., Chang, H.-C., Siliciano, R.F., Dorfman, T., Walker, B., Sodroski, J., and Reinherz, E.L. 1988. A soluble CD4 protein selectively inhibits HIV replication and syncytium formation. *Nature (London)* 331:78–81.

49. Smith, D.H., Byrn, R.A., Marsters, S.A., Gregory, T., Groopman, J.E., and Capon, D.J. 1987. Blocking of HIV-1 infectivity by a soluble, secreted form of the CD4 antigen. *Science* 238:1704–1707.

50. Fisher, R.A., Bertonis, J.M., Meier, W., Johnson, V.A., Costopoulos, D.S., Liu, T., Tizard, R., Walker, B.D., Hirsch, M.S., Schooley, R.T., and Flavell, R.A. 1988. HIV infection is blocked in vitro by recombinant soluble CD4. *Nature (London)* 331:76–78.

51. Deen, K.C., McDougal, J.S., Inacker, R., Folena-Wasserman, G., Arthos, J., Rosenberg, J., Maddon, P.J., Axel, R., and Sweet, R.W. 1988. A soluble form of CD4 (T4) protein inhibits AIDS virus infection. *Nature (London)* 331:82–84.

52. Traunecker, A., Luke, W., Karjalainen, K. 1988. Soluble CD4 molecules neutralize human immunodeficiency virus type 1. *Nature (London)* 331:84–86.

53. Broder, S., Mitsuya, H., Yarchoan, R., Pavlakis, G.N. 1990. Antiretroviral therapy in AIDS. *Ann. Intern. Med.* 113:604–618.

54. Capon, D.J., Chamow, S.M., Mordenti, J., Marsters, S.A., Gregory, T., Mitsuya, H., Byrn, R.A., Lucas, C., Wurm, F.M., Groopman, J.E., Broder, S., and Smith, D.H. 1989. Designing CD4 immunoadhesins for AIDS therapy. *Nature (London)* 337:525–531.

55. Byrn, R.A., Mordenti, J., Lucas, C., Smith, D., Marsters, S.A., Johnson, J.S., Cossum, P., Chamow, S.M., Wurm, F.M., Gregory, T., Groopman, J.E., and Capon, D.J. 1990. Biological properties of a CD4 immunoadhesin. *Nature (London)* 344:667–670.

56. Perno, C.F., Baseler, M.W., Broder, S., Yarchoan, R. 1990. Infection of monocytes by human immunodeficiency virus I blocked by inhibitors of CD4-gp 120 binding, even in the presence of enhancing antibodies. *J. Exp. Med.* 171:1043–1056.

57. Clapham, P.R., Weber, J.N., Whitby, D., McIntosh, K., Dalgleish, A.G., Maddon, P.J., Deen, K.C., Sweet, R.W., and Weiss, R. 1989. Soluble CD4 blocks the infectivity of diverse strains of HIV and SIV for T cells and monocytes but not for brain and muscle cells. *Nature (London)* 337:368–370.

58. Looney, D.J., Hayashi, S., Nicklas, M., Redfield, R.R., Broder, S., Wong-Staal, F., and Mitsuya, H. 1990. Differences in the interaction of HIV-1 and HIV-2 with soluble CD4. *J. Acq. Immune Defic. Syndr.* 3:649–657.

59. Daar, E.S., Li, X.L., Moudgil, T., and Ho, D.D. 1990. High concentrations of recombinant soluble CD4 are required to neutralize primary human immunodeficiency virus type 1 isolates. *Proc. Natl. Acad. Sci. U.S.A.* 87:6574–6578.

60. Kahn, J.O., Allan, J.D., Hodges, T.L., Kaplan, L.D., Arri, C.J., Fitch, H.F., Izu, A.E., Mordenti, J., Sherwin, S.A., Groopman, J.E., and Volberding, P.A. 1990. The safety and pharmackinetics of recombinant soluble CD4 (rCD4) in subjects with the acquired immunodeficiency syndrome (AIDS) and AIDS-related complex. A phase I study. *Ann Intern. Med.* 112:254–261.

61. Broder, S., Yarchoan, R., Collins, J.M., Lane, H.C., Markham, P.D., Klecker, R.W., Redfield, R.R., Mitsuya, H., Hoth, D.F., Gelman, E., Groopman, J.E., Resnick, L., Gallo, R.C., Myers, C.E., and Fauci, A.S. 1985. Effects of suramin on HTLV-III/LAV infection presenting as Kaposi's sarcoma or AIDS-related complex: Clinical pharmacology and suppression of viral replication *in vivo*. *Lancet* 2:627–630.

62. Kaplan, L.D., Wolfe, P.R., Volberding, P.A., Feorino, P., Levy, J.A., Abrams, D.I., Kiprov, D., Wong, R., and Kaufman, L. 1987. Lack of response to suramin in patients with AIDS and AIDS-related complex. *Am. J. Med.* 82:615–620.

63. Collins, J.M., Klecker, R.W., Jr., Yarchoan, R., Lane, H.C., Fauci, A.S., Redfield, R.R., Broder, S., and Myers, C.E. 1986. Clinical pharmacokinetics of suramin in patients with HTLV-III/LAV infection. *J. Clin. Pharmacol.* 26:22–26.

64. Mahoney, C.W., Azzi, A., and Huang, K.-P. 1990. Effects of suramin, an anti-human immunodeficiency virus reverse transcriptase agent, on protein kinase C. *J. Biol. Chem.* 265:5424–5428.

65. Stein, C.A., La Rocca, R.V., Thomas, R., McAtee, N., and Myers, C.E. 1989. Suramin: An anticancer drug with a unique mechanism of action. *J. Clin. Oncol.* 7:499–508.

66. Mitsuya, H., Popovic, M., Yarchoan, R., Matsushita, S., Gallo, R.C., and Broder, S. 1984. Suramin protection of T cells in vitro against infectivity and cytopathic effect of HTLV-III. *Science* 226:172–174.

67. Balzarini, J., Mitsuya, H., De Clercq, E., and Broder, S. 1986. Comparative inhibitory effects of suramin and other selected compounds on the infectivity and replication of human T-cell lymphotropic virus (HTLV-III)/lymphadenopathy-associated virus (LAV). *Int. J. Cancer* 37:451–457.

68. Schinazi, R.F., Chu, C.K., Babu, R., Oswald, B.J., Saalmann, V., Cannon, D.L., Eriksson, B.F.H., and Nasr, M. 1990. Anthraquinones as a new class of antiviral agents against human immunodeficiency virus. *Antiviral Res.* 13:265–272.

69. Baba, M., Schols, D., Pauwels, R., Balzarini, J., and De Clercq, E. 1988. Fuchsin acid selectively inhibits human immunodeficiency virus (HIV) *in vitro*. *Biochem. Biophys. Res. Commun.* 155:1404–1411.

70. Cardin, A.D., Smith, P.L., Hyde, L., Blankenship, D.T., Bowlin, T.L., Schroder, K., Stauderman, K.A., Taylor, D.L., and Tyms, A.S. 1991. Stilbene disulfonic acids. CD4 antagonists that block human immunodeficiency virus type-1 growth at multiple stages of the virus life cycle. *J. Biol. Chem.* 266:13355–13363.

71. Suzuki, H., Tochikura, T.S., Iiyama, K., Yamazaki, S., Yamamoto, N., and Toda, S. 1989. Lignosulfonate, a water soluble lignin from the waste liquor of the pulping process, inhibits the infectivity and cytopathic effects of human immunodeficiency virus in vitro. *Agr. Biol. Chem.* 53:3369–3380.

72. Gustafson, K.R., Cardellina, J.H., II, Fuller, R.W., Weislow, O.S., Kiser, R.F., Snader, K.M., Patterson, G.M.L., and Boyd, M.R. 1989. AIDS-antiviral sulfolipids from cyanobacteria (blue-green algae). *J. Natl. Cancer. Inst.* 81:1254–1258.

73. Jentsch, K.D., Hunsmann, G., Hartmann, H., and Nickel, P. 1987. Inhibition of human immunodeficiency virus type 1 reverse transcriptase by suramin-related compounds. *J. Gen. Virol.* 68:2183–2192.

74. Mohan, P., Singh, R., Wepsiec, J., Gonzalez, I., Sun, D.K., and Sarin, P.S. 1990. Inhibition of HIV replication by naphthalenemonosulfonic acid derivatives and a bis naphthalenedisulfonic acid compound. *Life Sci.* 47:993–999.

75. Mohan, P., Singh, R., and Baba, M. 1991. Potential anti-AIDS agents. Synthesis and antiviral activity of naphthalenesulfonic acid derivatives against HIV-1 and HIV-2. *J. Med. Chem.* 34:212–217.

76. Mohan, P., Singh, R., and Baba, M. 1991. Anti-HIV-1 and HIV-2 activity of naphthalenedisulfonic acid derivatives. Inhibition of cytopathogenesis, giant cell formation and reverse transcriptase activity. *Biochem. Pharmacol.* 41:642–646.

77. Mohan, P., Singh, R., and Baba, M. 1991. Novel naphthalenedisulfonic acid anti-HIV-1 agents. Synthesis and activity against reverse transcriptase, virus replication and syncytia formation. *Drug Design Disc.* 8:69–82.

78. Mohan, P., Hopfinger, A.J., and Baba, M. 1991. Naphthalenesulfonic acid derivatives as potential anti-HIV-1 agents. Chemistry, biology and molecular modeling of their inhibition of reverse transcriptase. *Antiviral Chem. Chemother.* 2:215–222.

79. Baba, M., Pauwels, R., Balzarini, J., Arnout, J., Desmyter, J., and De Clercq, E. 1988. Mechanism of inhibitory effect of dextran sulfate and heparin on replication of human immunodeficiency virus *in vitro*. *Proc. Natl. Acad. Sci. U.S.A.* 85:6132–6136.

80. Lorentsen, K.J., Hendrix, C.W., Collins, J.M., Kornhauser, D.M., Petty, B.G., Klecker, R.W., Flexner, C., Eckel, R.H., and Lietman, P.S. 1989. Dextran sulfate is poorly absorbed after oral administration. *Ann. Intern. Med.* 111:561–566.

81. Hartman, N.R., Johns, D.G., and Mitsuya, H. 1990. Pharmacokinetic analysis of dextran sulfate in rats as pertains to its clinical usefulness for therapy of HIV infection. *AIDS Res. Human Retrovir.* 6:805–812.

82. Baba, M., Schols, D., De Clercq, E., Pauwels, R., Nagy, M., Gyorgyi-Edelenyi, J., Low, M., and Gorog, S. 1990. Novel sulfated polymers as highly potent and selective inhibitors of human immunodeficiency virus replication and giant cell formation. *Antimicrob. Agents Chemother.* 34:134–138.

83. Williams, R.T. 1959. *Detoxication mechanisms. The metabolism and detoxication of drugs, toxic substance and other inorganic compounds.* London: Chapman and Hall. Pp. 497–500.

84. Batten, P.L. 1979. Metabolism of 2-naphthylamine sulfonic acids. *Toxicol. Appl. Pharmacol.* 48:A171.

85. Mohan, P., and Baba, M. 1991. Novel sulfonic acid polymers as a new class of potent and highly selective anti-HIV-1 agents. Abstracts volume, 7th International Conference on AIDS, Florence, Italy, June 16–21, 1991. Abstr. TU.A.64.

86. Mohan, P., Schols, D., Baba, M., and De Clercq, E. 1991. Sulfonic acid polymers as a new class of human immunodeficiency virus inhibitors. *Antiviral Res.* 18:139–150.

87. Pauwels, R., Andries, K., Desmyter, J., Schols, D., Kukla, M.J., Breslin, H.J., Raemaeckers, A., Van Gelder, J., Woestenborghs, R., Heykants, J., Schellekens, K., Janssen, M.A.C., De Clercq, E., and Janssen, P.A.J. 1990. Potent and selective inhibition of HIV-1 replication *in vitro* by a novel series of TIBO derivatives. *Nature (London)* 343:470–474.

88. Tanaka, H., Baba, M., Hayakawa, H., Sakamaki, T., Miyasaka, T., Ubasawa, M., Takashima, H., Sekiya, K., Nitta, I., Shigeta, S., Walker, R.T., Balzarini, J., and De Clercq, E. 1991. A new class of HIV-1-specific 6-substituted acyclouridine derivatives: Synthesis and anti-HIV-1 activity of 5- or 6-substituted analogues of 1-[(2-hydroxyethoxy)methyl]-6-(phenylthio)thymine (HEPT). *J. Med. Chem.* 34:349–357.

89. Baba, M., De Clercq, E., Iida, S., Tanaka, H., Nitta, I., Ubasawa, M., Takashima, H., Sekiya, K., Umezu, K., Nakashima, H., Shigeta, S., Walker, R.T., and Miyasaka, T. 1990. Anti-human immunodeficiency virus type 1 activities and pharmacokinetics of novel 6-substituted acyclouridine derivatives. *Antimicrob. Agents Chemother.* 34:2358–2363.

90. Baba, M., Shigeta, S., De Clercq, E., Tanaka, H., Miyasaka, T., Ubasawa, M., Umezu, K., and Walker, R.T. 1991. Highly potent and selective inhibition of HIV-1 replication by a novel series of 6-substituted acyclouridine derivatives. Abstracts volume, 7th International Conference on AIDS, Florence, Italy, June 16–21. Abstr. TU.A.63.

91. Baba, M., De Clercq, E., Tanaka, H., Ubasawa, M., Takashima, H., Sekiya, K., Nitta, I., Umezu, K., Nakashima, H., Mori, S., Shigeta, S., Walker, R.T., and Miyasaka, T. 1991. Potent and selective inhibition of human immunodeficiency virus type 1 (HIV-1) by 5-ethyl-6-phenylthiouracil derivatives through their interaction with the HIV-1 reverse transcriptase. *Proc. Natl. Acad. Sci. U.S.A.* 88:2356–2360.

92. Goldman, M.E., O'Brien, J.A., Ruffing, T.L., Stern, A.M., Gaul, S.L., Saari, W.S., Wai, J.S., Hoffman, J., Rooney, C.S., Quintero, J.C., Schleif, W.A.,

Emini, E.A., and Nunberg, J.H. 1991. HIV-1 specific pyridinone RT inhibitors. 1. Preclinical biological characterization of two investigational new drugs. Abstracts volume, 7th International Conference on AIDS, Florence, Italy, June 16–21. Abstr. TU.A.67.

93. Merluzzi, V.J., Hargrave, K.D., Labadia, M., Grozinger, K., Skoog, M., Wu, J.C., Shih, C.-K., Eckner, K., Hattox, S., Adams, J., Rosehthal, A.S., Faanes, R., Eckner, R.J., Koup, R.A., and Sullivan, J.L. 1990. Inhibition of HIV-1 replication by a nonnucleoside reverse transcriptase inhibitor. *Science* 250:1411–1413.

94. Wu, J.C., Warren, T.C., Adams, J., Proudfoot, J., Skiles, J., Raghavan, P., Perry, C., Potocki, I., Farina, P.R., and Grob, P.M. 1991. A novel dihydrodiazepinone inhibitor of HIV-1 reverse transcriptase acts through a nonsubstrate binding site. *Biochemistry* 30:2022–2026.

95. Pauwels, R., Debyser, Z., Andries, K., Yamamoto, N., Baba, M., Schols, D., Kukla, M., Desmyter, J., Janssen, P.A.J., and De Clercq, E. 1991. Interaction of TIBO and TIBO-like compounds with HIV-1 RT. Abstracts volume, 7th International Conference on AIDS, Florence, Italy, June 16–21. Abstr. W.A. 1007.

96. Bader, J.P., McMahon, Schultz, R.J., Narayanan, V.L., Pierce, J.B., Harrison, W.A., Weislow, O.S., Midelfort, C.F., Stinson, S.F., and Boyd, M.R. 1991. Oxathiin carboxanilide, a potent inhibitor of human immunodeficiency virus reproduction. *Proc. Natl. Acad. Sci. U.S.A.* 88:6740–6744.

97. Finberg, R.W., Diamond, D.C., Mitchell, D.B., Rosenstein, Y., Soman, G., Norman, T.C., Schreiber, S.L., and Burakoff, S.J. 1990. Prevention of HIV-1 infection and preservation of CD4 function by the binding of CPFs to gp 120. *Science* 249:287–291.

98. Saermark, T., and Bex, F. 1989. Acylation of HIV proteins. *Biochem. Soc. Trans.* I 17:869–871.

99. Bryant, M.L., Heuckeroth, R.O., Kimata, J.T., Ratner, L., and Gordon, J.I. 1989. Replication of human immunodeficiency virus 1 and moloney murine leukemia virus is inhibited by different heteroatom containing analogs of myristic acid. *Proc. Natl. Acad. Sci. U.S.A.* 86:8655–8659.

100. Fields, A.P., Bednarik, D.P., Hess, A., and May, W.S. 1988. Human immunodeficiency virus induces phosphorylation of its cell surface receptor. *Nature (London)* 333:278–280.

101. Takahashi, I., Nakanishi, S., Kobayashi, E., Nakano, H., Suzuki, K., and Tamaoki, T. 1989. Hypericin and pseudohypericin specifically inhibit protein kinase C: possible relation to their antiretroviral activity. *Biochem. Biophys. Res. Commun.* 165:1207–1212.

102. Farnsworth, N.R., Akerele, O., Bingel, A.S., Soejarto, D.D., and Guo, Z.-G. (1985). Medicinal plants in therapy. *Bull. WHO* 63:965–981.

103. Davies, J.F., II, Hostomska, Z., Hostomsky, Z., Jordan, S.R., and Matthews, D.A. 1991. Crystal structure of the ribonuclease H domain of HIV-1 reverse transcriptase. *Science* 252:88–95.

11

Oral Adenoviruses as the Carriers for Human Immunodeficiency Virus or Hepatitis B Virus Surface Antigen Genes

Michael D. Lubeck, Satoshi Mizutani, Alan R. Davis, and Paul P. Hung

11.1. Introduction

The development of virus vaccines based on live recombinant adenoviruses has received increasing attention in the recent past because live virus-vectored vaccines possess significant advantages for immunization against viral diseases relative to inactivated or subunit vaccines. A major strength of the adenovirus vector approach is immunization by oral administration of vaccine. An adenovirus recombinant vaccine is administered as an enteric-coated tablet, which ensures safe passage of virus to the intestines where asymptomatic replication occurs. Foreign viral proteins are thus expressed in the context of an infection of the gut. We believe that presentation of viral antigens in this manner is likely to evoke substantial humoral immunity, including secretory responses, as well as significant cell-mediated immunity. In addition, adenoviruses represent an attractive viral vector system for vaccine development because they exhibit several characteristics that allow for high expression of foreign viral genes. In this chapter several features of the adenovirus vector system as applied to its use for vaccination purposes will be reviewed, with an emphasis on the use of adenoviruses as vaccine vectors for diseases caused by hepatitis B virus (HBV) and human immunodeficiency virus (HIV).

11.2. Adenoviruses

Adenoviruses are nonenveloped double-stranded DNA viruses (for a general review of adenoviruses, see reference 1). The adenovirus genome consists of approximately 36,000 base pairs, which encodes for approximately 50 viral genes. The capsid, which contains 252 capsomeres, shows icosahedral symmetry and is 60 to 90 nm in diameter. Antigenic diversity among human adenoviruses is extensive. Forty-two distinct serotypes have been recognized[2] and five new candidate types have been recently described.[3] Classification in subgenera is based on oncogenicity in newborn rodents. Members of subgenus A exhibit high oncogenicity, subgenus B members generally show slight oncogenicity, and viruses of other subgenera (C, D, E, and F) are nononcogenic. Viruses under development as vectors for vaccines include Ad4 (group E), Ad5 (group C), and a nononcogenic strain of Ad7 (group B).

Adenoviruses are endemic throughout the world. Most children have been exposed to several adenovirus serotypes by the age of 6. Adenoviruses have been associated most frequently with upper and lower respiratory tract infections (pharyngitis, bronchitis, croup, pneumonia) but also cause gastrointestinal, ocular, and genitourinary tract infections. Adenovirus infections caused by serotype 1, 2, 5, and 6 generally occur in an endemic form, whereas adenovirus serotypes 3, 4, 7, 8, 14, and 21 more frequently occur in epidemics. Serotypes 40 and 41 cause acute enteric infections, and Ad8 is the primary cause of epidemic keratoconjunctivitis.

11.3. Adenovirus Vaccines

Epidemics of acute respiratory disease (ARD) in military populations have been frequently caused by adenovirus types 4, 7, and 21. A live, orally administered vaccine for Ad4 was developed in the 1960s to control the substantial morbidity in the military caused by this illness.[4, 5] Vaccination is accomplished by the delivery of wild-type (unattenuated) Ad4 in the form of an enteric-coated tablet that establishes a selective infection of the intestinal tract. The intestinal infection is asymptomatic and induces a type-specific immunological response that protects against subsequent ARD caused by Ad4. A live oral Ad7 vaccine was later developed for use in the military to protect against ARD caused by this adenovirus serotype.[4, 5] These vaccines have been administered to tens of millions of military personnel over the past three decades and have established an enviable record of safety and efficacy. No adverse reactions have been reported following the use of the vaccine, and vaccination affords protection against ARD in greater than 95% of vaccine recipients. Following the vaccination of military personnel, transmissibility of Ad4 and Ad7 viruses among recruits was not observed, although transmission to spouses was documented.[6] In general, following vaccination, virus is excreted in stools of vaccinees for from 4 to 13 days. A majority of vaccinees experience a 4-fold increase in type-specific neutralizing serum antibodies. Although vaccination of high-risk nonmilitary personnel has been suggested (e.g., institutionalized adults or children), general use of these vaccines in civilian populations is not warranted owing to the low incidence of acute disease in this population.

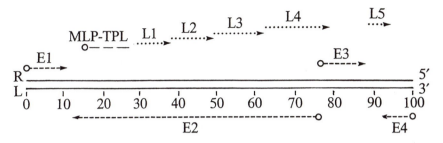

Figure 11.1. Simplified diagram of Ad2 transcript map. Each map unit is approximately 360 base pairs. *MLP*, major late promoter, *TPL*, tripartite leader. In general, early regions E1 through E4 are transcribed from initial infectious DNA and late regions L1 through L5 are expressed from amplified viral DNA.

11.4. Adenovirus Genomic Organization

The genomic organizations of the three human adenoviruses being used for vector development (Ad4, Ad5, and Ad7) are basically similar to that of Ad2, which is presented in Figure 11.1. Two phases of gene expression, early and late, have been identified, which are separated temporally at about eight hours post-infection. Early (E) regions 1 through 4 are driven by at least seven promoters. Also active in the early phase is the late (L) region L1, which is regulated by the major late promoter (MLP). The E1 region has been intensively studied (for a review see reference 7) and has been found to serve as a major transcriptional regulator (E1A) that transactivates transcription of both cellular and viral genes and represses transcription of other regions. The E2 region encodes proteins involved in virus replication (e.g., DNA polymerase and DNA binding proteins), whereas the E4 region exerts control over viral assembly, host protein synthesis shutoff, viral DNA replication, and late mRNA synthesis.

The E3 region encodes proteins that influence the interaction of the immune system with virus-infected cells. This region is nonessential for virus replication in vitro, as viruses containing large deletions in this region have been shown to be replication competent in cell culture. Moreover, this region has been shown to be nonessential for the replication of virus in vivo. Deletion mutants as well as viruses containing substituted E3 regions replicate in the lungs of certain rodent species[8] or in the gut of chimpanzees[9] (discussed in detail below). Deletion of this region allows the introduction of larger foreign gene inserts and thus may be an important aspect of strategies for recombinant vaccine construction.

Nevertheless, the functions of the E3 genes are not completely understood and deletion of this region may result in undesirable consequences. One E3-encoded protein, gp19K, associates with Class I MHC antigens in the endoplasmic reticulum, where it blocks glycosylation and cell surface expression of these molecules.[10] Because cytotoxic T lymphocytes (CTL) require Class I MHC expression as a restriction element to mediate recognition of virally infected cells, the block in Class I MHC expression results in functional (cytolytic) impairment of CTL. It has been speculated that this protein, which is conserved among the B, C, D, E, and F subgenera, serves a critical function in protecting virus-infected cells from immune destruction by Class I MHC-restricted effector cells.[11] A 14,700 MW protein, also encoded by E3, has also been demonstrated to exert protective effects against host antiviral mechanisms. Gooding et al. demonstrated that the 14.7K protein blocks cytolysis of Ad-infected cells by tumor necrosis factor (TNF).[12] A role for the E3 region in the pathogenesis of adenovirus disease has been described by Ginsberg et al.[13] In this study, Ad2/Ad5 recombinant adenoviruses containing E3 region deletions were observed to exhibit enhanced virulence and pathogenecity in cotton rats relative to wild-type virus. A complete correlation was reported between mutants with increased pathogenic effects and mutants with ablated gp 19K function. On the other hand, oral administration to chimpanzees of an adenohepatitis B vaccine that contained a large E3 deletion, including deletion of the gene encoding gp19K, did not result in observable clinical disease.[9]

Late adenovirus genes, which encode proteins required for virion assembly, are driven by the powerful Ad MLP. A single message extending from 17 to greater than 98 map units is processed to form five late mRNA families. All late messages from these families carry a tripartite leader at their 5' terminus, which serves as an initiation site for ribosome binding. Virus assembly occurs in the cell nucleus. In highly permissive infections, burst sizes of up to 5,000 plaque-forming units (PFU) per cell are produced by 48 hours.

11.5. Strategies for the Construction of Adenovirus-Vectored Vaccines

Adenovirus-vectored vaccines have some advantages (mentioned previously) but also have several constraints to be overcome. The adenovirus genome codes for viral proteins in both DNA strands. As was mentioned previously, the genome is arranged into four early and one late transcriptional units. Each transcription unit codes for several proteins that are expressed via complex splicing and polyadenylation events. Transcription and posttranscriptional processing are tightly controlled in adenovirus replication. Therefore, there are not many sites in the genome for insertion of either foreign gene(s) or an independent transcription unit (expression cassette) without interfering with other virus functions.

The packaging limitations for the viral genome during virus assembly restrict the size of foreign DNA that can be inserted. In cell culture, different cell types can accommodate varying amounts of excessive DNA (i.e., overpackaged recombinants exhibit greater genetic stability in A549 cells relative to WI-38 cells). As was mentioned earlier, E3 proteins are not essential for virus replication. However, at least four of these proteins are thought to help the growth of adenoviruses in humans. The overall effect of these E3-coded proteins on the virus growth in humans must be dependent on an individual's genetic background and immunological history. The influence of the E3 region is likely to be most apparent in adult populations, who have had extensive previous experience with adenoviruses. Another constraint involves the selection of the appropriate adenovirus serotypes for use as vaccine vectors. The ideal vectors will be highly attenuated (causing no disease) with a low prevalence among the general population. In any event, recombinant adenovirus-vectored vaccines must likely grow well and express large quantities of viral antigens to effectively immunize vaccine recipients.

11.5.1. E3 deletion and Insertion of Foreign Genes

Deletion of the nonessential E3 region allows for the accommodation of foreign genes. The maximum size of the Ad deletion is variable, depending on the Ad serotype. Roughly 3 kbp can be removed from the E3 region. Foreign genes can then be expressed by either inserting a gene between E3 promoter elements and its polyadenylation signal or by inserting a foreign gene elsewhere in the genome as an independent transcription unit (expression cassette) using adenovirus major late promoter (MLP) elements.

The former approach was successfully applied to the construction of adenovirus recombinant with hepatitis B virus surface antigen gene,[8, 9] HIV envelope gene,[14] and respiratory syncytial virus (RSV) F gene.[15] In all cases, the E3 promoter at the insertion site was used for gene expression. Therefore, the expression of inserted gene(s) is limited to only the early time of the virus replication cycle unless either the gene is inserted within the late native leader sequence (i.e., y leader of Ad5) or a splice acceptor of one of the late genes is placed in front of the inserted gene. The DNA sequence downstream of the inserted gene should be carefully engineered to avoid an unexpected splicing event, which could result in the loss of the inserted sequence.

The latter approach was also successfully tried by several investigators.[16–20] A new expression cassette consists of a major late promoter (MLP), a cDNA copy of tripartite leader (TPL), a gene, and a poly(A) signal. Late expression has a definite advantage for high-level expression of gene products because the inserted gene is amplified by 10^4- to 10^5-fold. Mansour et al. found that the intervening sequence (IVS) between the first and the second leader enhanced transcription from the MLP.[21] The insertion of a part of the IVS significantly increased the expression level of hepatitis B surface antigen.[20]

The insertion site and orientation of the expression cassette in a recombinant virus genome have a significant impact on virus growth. Therefore, the insertion site should be carefully selected by analyzing the DNA sequence in the area surrounding a possible site. The site between E4 and the right-end inverted terminal repeat (ITR) was chosen for insertion of the expression cassette.[16, 20] However, it is interesting that small changes of the microenvironment for E4

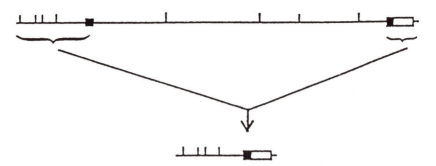

Figure 11.2. Structure of the minichromosomes formed during replication of Ad7 recombinant with the expression cassette. *Bam*HI restriction site maps of Ad7 recombinant and its minichromosomes, indicating which regions of the parental genome are retained in the minichromosomes. The black boxes represent the duplicate copies of the MLP; the open box represents the remainder of the genetic expression cassette containing the HBsAg gene.

promoter upstream elements significantly affect the virus growth (B. Mason, B. Bhat, unpublished observation). Insertion of an expression cassette at the terminal ends may be the best approach unless the expression cassette has a nearly complete transcription termination structure. This is because transcription beyond the end of the cassette may create antisense RNA, which may then interfere with normal viral messages of the opposite strand. This may explain the failure of a recent attempt to construct a recombinant adenovirus with an expression cassette inserted in the E3 region with a leftward orientation (J. Morin, unpublished observation). HBsAg was expressed well from an expression cassette inserted with the rightward orientation at the right-hand terminal end between the E4 promoter upstream elements and the ITR.[20]

As was mentioned, the expression cassette contains the same MLP and TPL sequences as the viral MLP and TPL. This recombinant virus structure creates the possibility of intermolecular homologous recombination of viral DNAs during virus replication. The resulting short segment of DNA, minichromosome ITR-MLP-TPL-gene-poly(A)-ITR (Figure 11.2), has been detected in cells infected with a recombinant of the foregoing structure; moreover, detectable amounts of the minichromosome were assembled into the virion structure (B. Mason, J. Morin, unpublished observation). As the E1a and E1b regions of adenovirus are included in the minichromosome DNA, which is encapsidated and which can thus be delivered into cells, it is important to eliminate this event. Insertion of an expression cassette with a reverse orientation eliminates minichromosome formation.

However, when an expression cassette was inserted in one recombinant in the left-ward orientation, virus growth was significantly impaired, possibly because transcription termination was not complete. Therefore, it may be that a better insertion site for an expression cassette will be at the left-terminal end between the left ITR and E1a promoter elements, with insertion engineered in the leftward orientation. This type of recombinant has not yet been tested.

11.5.2. Consideration of E3 Proteins Included in Adenovirus-Vectored Vaccines

As was discussed before, the importance of E3-coded proteins in the context of adenovirus-vectored vaccines is not known. In the event that recombinant virus growth is impaired in the host owing to lost E3 functions, it would clearly be better to include the relevant genes. This situation would then require the deletion of some other part of the genome to make additional space for the insertion of foreign genes. Recently, it was shown that ORF6 in the E4 region is the only ORF in that region that is required for Ad5 replication.[22] If ORF1 through ORF4 can be deleted without an effect on virus growth, it is possible to make space for an additional 1.4 kbp. Restoration of a part of the E3 region and deletion of a portion of the E4 region could be another approach to the construction of a helper independent adenovirus-vectored vaccine. This approach is currently being tested. Other approaches may be possible when the functions of additional viral-coded proteins are elucidated.

With these considerations and experimental results, several adenovirus type 7 and type 4 recombinants with the hepatitis B surface antigen gene (HBsAg), HIV envelope gene, respiratory syncytial virus F and G genes, and parainfluenza 3 virus HN and F genes have been constructed. In vitro expression of the foreign viral genes has been demonstrated for the recombinants. Animal experiments with some of these recombinant viruses are discussed in the next section.

11.6. Animal Models for Adenovirus-Vectored Vaccines

Human adenoviruses exhibit a highly restricted host range and, in general, replicate poorly in nonhuman species. Over the past few decades, models for adenovirus disease have been extensively sought. Although models for acute clinical disease have not been reported, experimental animals that show asymptomatic infections have been identified. The cotton rat model was developed by Pacini et al.,[23] who demonstrated that following intranasal instillation of virus in this rodent species, replication occurred in nasal mucosa and lungs. However, this model is permissive for only certain adenovirus serotypes. The highest levels of replication were observed with adenovirus types 1, 2, 5, and 6. However, cotton rats were not permissive for adenovirus types 3, 4, or 7, the latter two of which represent our primary candidate vectors. The hamster was later shown to support replication of Ad5 in the respiratory tract.[8] Initial reports of asymptomatic infection of experimental animals by primarily group C adenoviruses have been described,[24, 25] but these models have not been further studied. With regard to animal models

Table 11.1. *Seroresponses of Dogs Following Immunization with Adeno-Hepatitis B Vaccines*

| Animal Number | Primary Immununization: Ad7-HBsAg | | | | Booster-Immunization: Ad4-HBsAg | | | |
| | Anti-Ad7[a] | | Anti-HBs[b] | | Anti-Ad4 | | Anti-HBs | |
	0 wk	4 wk	0 wk	4 wk	0 wk	4 wk	0 wk	4 wk
536	<10	160	0	90	0	>1024	190	23.000
539	<10	>640	0	20	2	>1024	52	18,000
545	<10	20	0	34	8	>1024	18	505

[a]Reciprocal neutralization titers.

[b]Anti-HBs responses expressed in mIU.

for the testing of Ad4- and Ad7-vectored vaccines, Sinha et al. reported that dogs experienced silent infections by these adenovirus serotypes as well as by Ad2 and Ad3.[26] We have confirmed that Ad4 and Ad7 establish respiratory infections in dogs and induce seroconversion to their respective adenovirus serotype. As this is the only animal in which we can reproducibly induce an immune response to these Ad serotypes, we have made extensive use of this model for the evaluation of Ad4- and Ad7-vectored vaccines (see Section 11.7).

The chimpanzee has recently been shown to sustain enteric infection by Ad4 and Ad7 viruses[9] and thus represents a useful model for oral-enteric immunization by Ad4- and Ad7-vectored vaccines. The feasibility of oral-enteric immunization against foreign virus proteins expressed by adenovirus vectors has been recently demonstrated using recombinant adeno-hepatitis B vaccines.

11.7. Evaluation of Adeno-Hepatitis B Vaccines in Dogs

Adeno-hepatitis B vaccines containing the HBsAg gene inserted in the E3 region were evaluated in dogs. In one experiment, 6- to 12-month-old beagles were vaccinated by intratracheal inoculation with an Ad7 hepatitis B recombinant vaccine at doses of greater than 10^6 PFU. Table 11.1 shows that following this primary immunization, animals seroconverted to both Ad7 and HBsAg. At four weeks postvaccination, anti-HBs serum antibody titers ranged from 20 to 90 mIU. (A response of 10 mIU in humans following vaccination with conventional subunit vaccines is considered to be protective.) Eighteen weeks after the primary immunization the dogs received a booster immunization with an Ad4-HBsAg vaccine. This virus was of similar construction to the Ad7-HBsAg recombinant and contained the HBsAg gene in the E3 region. One month following the booster immunizations, the dogs had seroconverted to Ad4 and showed very significant booster responses to HBsAg. These data illustrate the potential of the adenovirus

Table 11.2. Immunization of Chimpanzees with Adeno-hepatitis B Recombinant Viruses: Seroresponses to Adenovirus and Vaccine Take

Chimpanzee Number	Primary Immunization: Ad7-HBsAg				Booster Immunization: Ad4-HBsAd[c]			
	Dose (PFU)	Gut Replication[a]	Anti-Ad7[b]		Dose (PFU)	Gut Replication[a]	Anti-Ad4	
			0 wk	4 wk			0 wk	4 wk
1373	3×10^7	6 wks	<4	256	8×10^9	2 wks	<4	128
1374	1.5×10^7	6 wks	<4	64	8×10^9	5 wks	<4	128
1375	2×10^9	6 wks	<4	256	8×10^9	2 wks	<4	128

[a]Recombinant virus detected in stools.
[b]Reciprocal anti-Ad neutralization titers.
[c]Eleven weeks postprimary immunization.

system to induce substantial humoral immune responses to foreign viral gene products.

11.8. Evaluation of Adeno-Hepatitis B Vaccine in Chimpanzees

Chimpanzees were recently identified as a model for oral-enteric immunization with Ad4- and Ad7-vectored vaccines.[9] Because this model is key in establishing the feasibility of the adenovirus vector approach and in testing future recombinant Ad vaccines, this study will be presented detail. Two colonies of chimpanzees were initially screened for preexisting antibody titers to adenovirus types 4 and 7. A sampling of 10 chimpanzees housed at Buckshire (Perkasie, Pa.) indicated that all of the animals had significant preexisting neutralizing antibody titers to Ad4 and that 9 of the 10 animals had significant antibody titers to Ad7. On the other hand, 1 of 31 chimpanzees housed at Sema (Rockville, Md.) had preexisting antibody titers to Ad7, and 23% (7 of 31) were seronegative for Ad4. It was thus apparent that chimpanzees are either infected by agents that cross-react with these human adenoviruses or are infected in captivity with human adenovirus. Differences in seroprevalence rates between the two chimpanzee colonies were attributed to differences in conditions under which the chimpanzees were maintained. Chimpanzees at Buckshire were housed together and intermingled freely, whereas chimpanzees at Sema were housed in individual isolators. Thus, the testing of recombinant vaccines requires careful prescreening of chimpanzees, a considerable problem considering the limited availability of chimpanzees and high seroprevalence rates to Ad4.

In this study, one chimpanzee (1375) received the vaccine strain of Ad7 followed by an immunization with an Ad4-hepatitis B vaccine, and two chimpanzees (1373 and 1374) received sequential oral immunizations with recombinant Ad7- and Ad4-hepatitis B vaccines (Table 11.2). Ad7 vaccine was administered at a dose of 10^9 pfu in an enteric-coated gelatin capsule (the vaccine dosage used

Table 11.3. Immunization of Chimpanzees with Adeno-Hepatitis B Recombinant
Viruses: Anti-HBs Serology and HBV Challenge

Chimpanzee Number	Anti-HBs		HBV Challenge[a]	Elevated Liver ALT in Serum	HBSAg Antigenemia (onset/duration)	Anti-HBc Response
	Primary Immun.	Booster Immun.				
1373	−	+	+	−	+/2 wks	+
1374	+	+	+	−	−/−	+
1375	−	+	+	+	+/11 wks	+++
1376 (control)	n/a	n/a	+	+	+/10 wks	+++

[a] $10^{3.5}$ 50% chimpanzee infectivity doses administered intravenously 2 months after booster immunization.

for immunization of military recruits is approximately 10^6 PFU). This chimpanzee subsequently shed infectious Ad7 in feces over a period of six weeks. An anti-Ad antibody seroresponse was first evident at about 2 weeks postinoculation and remained at high titer over the next several months. At 11 weeks postimmunization, this animal was inoculated with an Ad4-hepatitis B virus recombinant. In the face of a strong anti-Ad7 antibody response, the Ad4 recombinant replicated in the gut for at least 2 to 3 weeks (as determined by recovery of infectious virus in feces) and induced a strong anti-Ad4 neutralizing antibody response (Table 11.2). This finding demonstrated that Ad4 recombinant virus can be used as a booster vehicle following primary immunization with an Ad7-vectored vaccine. This animal did not develop a detectable seroresponse to HBsAg (anti-HBs).

Two chimpanzees received a primary dose of an Ad7-HBsAg recombinant virus (Wy-Ad7HZ6-1) followed by an Ad4-HBsAg recombinant virus booster. The Wy-Ad7HZ6-1 vaccine was administered orally at approximately 10^7 PFU. Infectious recombinant virus was subsequently shed in feces for a 6- to 7-week period and induced a strong anti-Ad7 type-specific neutralizing serum antibody response that was first detectable at 2 weeks postinoculation (Table 11.2). One of the two chimpanzees also developed a modest but significant anti-HBs antibody response that peaked at 5 weeks postinoculation at 9 mIU (Table 11.3). This response rapidly waned to undetectable levels by 10 weeks postimmunization. Following the Ad4 recombinant booster immunization at 11 weeks postprimary immunization, both animals rapidly developed anti-HBs seroresponses first detected at 2 weeks postbooster immunization. The peak secondary responses ranged from about 10 to 20 mIU and were both largely of the IgG isotype. These immunizations were followed by HBV challenge with $10^{3.5}$ 50% chimpanzee infectious units 2 months later. Following HBV challenge, one of the two chimpanzees experienced a brief period of HBsAg antigenemia (1–2 weeks) and both animals had detectable antibody responses to HBcAg, indicating that both animals experienced HBV infections. Neither animal developed acute liver disease as measured by elevated serum levels of liver transaminases. Thus, immunization with Ad-HBsAg recombinants did not protect the animals from HBV infection but did protect one animal from the development of acute hepatitis while the second chimpanzee experienced modified disease. This study provided information in three critical areas. First, it established the chimpanzee as a permissive model for oral-enteric immunization with Ad4- and Ad7-vectored vaccines. Second, it demonstrated that recombinant adenoviruses can induce significant seroresponses to foreign viral genes following enteric replication of viruses. And third, it established that heterotypic recombinant adenoviruses can be utilized through oral administrations to induce booster immune responses.

11.9. Evaluation of Adeno-HIV Vaccines

Dewar et al. constructed an Ad5 vector containing the HIV-1 envelope glycoprotein (gp 160) gene inserted in the Ad5 E3 region, under control of the E3 promoter.[14] Infection of human cells with this virus resulted in the expression of the HIV-1 envelope precursor, gp 160, which was subsequently processed to envelope glycoproteins gp 120 and gp 41. Syncytia formation of infected MOLT-4 cells demonstrated retention of biological function of the envelope molecule. Cotton rats were inoculated intranasally with 10^8 PFU of this recombinant. Seroresponses were evaluated by Western blot analysis and enzyme-linked immunosorbent assay. Antibodies reactive with envelope proteins were first detected at the third week postinoculation and the response peaked at 50 days postimmunization.

11.10. In Vivo Evaluation of Other Recombinant Adenovirus Vaccines

An Ad5–herpes simplex virus (HSV) recombinant virus was constructed by insertion into the E3 region of the coding sequence for glycoprotein B (gB), a major component of the HSV envelope that plays a role in the entry of virus into cells.[27] Infection of murine and human cells by this virus, termed AdgB2, resulted in high-level expression of gB. Mice, which are abortively infected with Ad5, were immunized with AdgB2 by the intraperitoneal route, and gB-reactive antibodies that exhibited neutralizing activity in vitro were induced. The animals were protected from later HSV lethal challenge.

In an interesting report by Prevec et al.,[28] an Ad5-based vesicular stomatis virus (VSV) recombinant was tested in several animal species to evaluate this Ad as a vector for veterinary vaccines. AdG12, which contains the VSV glycoprotein gene, was used to infect animal cell lines that were permissive, semipermissive, or nonpermissive for Ad5 replication. Bovine and canine cells were permissive and nonpermissive, respectively, for AdG12 replication. The inoculation of cows, dogs, and piglets by the subcutaneous or oral routes resulted in significant anti-

VSV neutralizing antibody responses. Although mouse cells were confirmed to be nonpermissive for Ad5 replication, mice inoculated intraperitoneally with this virus produced high titers of neutralizing antibodies and were protected from subsequent lethal VSV challenge.

11.11. Conclusions

Strategies for the construction of genetically stable adenovirus recombinants have been developed. Insertion of foreign viral genes in the adenovirus E3 region or expression cassettes in the right-hand terminus of the adenovirus genome between the E4 region and ITR has yielded recombinant viruses that express high levels of foreign viral proteins in vitro. Sequential oral administration of recombinant Ad7- and Ad4-hepatitis B vaccines to two chimpanzees resulted in detectable serum antibody responses to HBsAg. Upon HBV challenge, one chimpanzee was protected from acute HBV-induced disease and the other chimpanzee experienced modified disease. Recombinant viruses that produce several-fold more HBsAg have been constructed and will be tested for immunogenicity in the chimpanzee model in the near future. Recombinant vaccines containing genes from other viruses, such as HIV, herpes simplex virus, and vesicular stomatitis virus, have been constructed. These viruses express significant amounts of foreign viral proteins in vitro.

In spite of these early successes, several important concerns regarding the adenovirus vector approach for vaccination persist. What is the influence on vaccine take of preexisting immunity to adenoviruses? Can preexisting immunity to a particular adenovirus vector be overcome using higher doses of vaccine? Alternatively, can preexisting immunity be overcome through the engineering of relatively rare serotypically distinct adenoviruses? Also unknown at this time is the influence of E3 deletions on recombinant vaccine replication in individuals with previous exposure to adenoviruses. Are certain genes within the E3 region required to assist the virus to evade immune elimination? Also unknown is whether the gut is an optimal site for inducing strong humoral and cellular immune responses for specific antigens. If these concerns can be successfully addressed, recombinant adenovirus vaccines may prove to be a useful approach for immunization against viral disease.

11.12. Summary

Live oral vaccines based on adenovirus types 4 and 7 are currently in use in military populations for the prevention of acute respiratory disease. These vaccines have been shown to be extraordinarily safe and efficacious when employed in this setting. Viral vaccines based on live adenovirus vectors are currently under development. Strategies have been developed to insert foreign viral genes, including the human immunodeficiency virus (HIV) envelope (env) and hepatitis B virus (HBV) surface antigen (HBsAg) genes, at various locations in the adenovirus genome. Recombinant vaccines have been developed that express high levels of HIV env protein or HBsAg in cell culture. Human adenoviruses exhibit a highly restricted host range and replicate poorly in experimental animals, making preclinical testing of adenovirus-vectored vaccine difficult. Nevertheless, animal models have been identified and developed for immunogenicity testing of recombinant adenovirus vaccines. Recombinant adeno-hepatitis B vaccines have been shown to be highly immunogenic in dogs following lower respiratory tract infections. In the chimpanzee model, Ad7- and Ad4-vectored HBsAg vaccines induced immune protection from acute hepatitis following oral-enteric immunizations. This study indicates that recombinant adenoviruses undergo sufficient enteric replication in this host to induce significant immune responses to the foreign viral gene products and confirms the utility of adenovirus vectors as a general approach to immunization.

References

1. Ginsberg, H., ed. 1984. *The adenoviruses*. New York: Plenum.

2. Wigand, R., and Adrian, T. 1986. Classification and epidemiology of adenoviruses in adenovirus DNA. In *The viral genome and its expression*. W. Doerfler, ed. Boston: Martinus Nijhoff. Pp. 409–441.

3. Hierholzer, J.C., Wigand, R., Anderson, L.J., Adrian, T. and Gold, J.W.M. 1988. Adenoviruses from patients with AIDS: A plethora of subtypes and a description of five new serotypes of subgenus D (types 43-47). *J. Infect. Dis.* 158:804–813.

4. Couch, R.B., Chanock, R.M., Cate, T.R., Lang, D.L., Knight, V., and Huebner, R.J. 1963. Immunization with types 4 and 7 adenovirus by selective infection of the intestinal tract. *Am. Rev. Resp. Dis.* 88:304–403.

5. Top, F.H., Jr., Dudding, B.A., Russell, P.K., and Buscher, E.L. 1971. Control of acute respiratory disease in recruits with adenovirus type 4 and 7 vaccines. *Am. J. Epidemiol.* 94:142–146.

6. Stanley, E.D., and Jackson. G.G. 1969. Spread of enteric live adenovirus type 4 vaccine in married couples. *J. Infect. Dis.* 119:51–59.

7. Nevins, J.R. 1987. Regulation of early adenovirus gene expression. *Microbiol. Rev.* 51(4):419–430.

8. Morin, J.E., Lubeck, M.D., Barton, J.E., Conley, A.J., Davis, A.R., and Hung, P.P. 1987. Recombinant adenovirus induces antibody response to hepatitis B virus surface antigen in hamsters. *Proc. Natl. Acad. Sci. U.S.A.* 84:4626–4630.

9. Lubeck, M.D., Davis, A.R., Chengalvala, M., Natuk, R.J., Morin, J.E., Molnar-Kimber, K., Bhat, B.M., Mizutani, S., Hung, P.P., and Purcell, R.H. 1989. Immunogenicity and efficacy testing in chimpanzees of an oral hepatitis B vaccine based upon live recombinant adenovirus. *Proc. Natl. Acad. Sci. U.S.A.* 86:6763–6767.

10. Andersson, M., Pääbo, S., Nilsson, T., and Peterson, P.A. 1985. Impaired intracellular transport of class I MHC antigens as a possible means for adenoviruses to evade immune surveillance. *Cell* 43:215–222.

11. Hurgert, H., Maryanski, J.L., and Kvist, S. 1987. "E3/19K" protein of adenovirus type 2 inhibits lysis of cytolytic T lymphocytes by blocking cell-surface expression of histocompatibility class I antigens. *Proc. Natl. Acad. Sci. U.S.A.* 84:1356–1360.

12. Gooding, L.R., Elmore, L.W., Tollefson, A.E., Brady, H.A., and Wold, W.S.M. 1988. A 14,700 MW protein from the E3 region of adenovirus inhibits cytolysis by tumor necrosis factor. *Cell* 53:341–346.

13. Ginsberg, H.S., Lundholm-Beauchamp, W., Horswood, R.L., Pernis, B., Wold, W.S.M., Chanock, R.M., and Prince, G. 1989. Role of early region (E3) in pathogenesis of adenovirus disease. *Proc. Natl. Acad. Sci. U.S.A.* 86:3823–3827.

14. Dewar, R.L., Natarajan, V., Vasudevachari, M.B., and Salzman, N.P. 1989. Synthesis and processing of human immunodeficiency virus type 1 envelope proteins encoded by a recombinant human adenovirus. *J. Virol.* 63:129–136.

15. Collins, P.L., Davis, A.R., Lubeck, M.D, Mizutani, S., Hung, P.P., Prince, G.A. Camargo, E., Purcell, R.H., Chanock, R.M., and Murphy, B.R. 1990. Evaluation in chimpanzees of vaccinia virus recombinants that express the surface glycoproteins of human respiratory syncytial virus. *Vaccine* 8:164–168.

16. Saito, I., Oya, Y., Yamamoto, K., Yuasa, T., and Shimojo, H. 1985. Construction of non-defective adenovirus type 5 bearing a 2.8-kilobase hepatitis B virus DNA near the right end of its genome. *J. Virol.* 54:711–719.

17. Berkner, K.L., Schaffhausen, B.S., Roberts, T.M., and Sharp, P.A. 1987. Abundant expression of polyomavirus middle T antigen and dehydrofolate reductase in an adenovirus recombinant. *J. Virol.* 61:1213–1220.

18. Davidson, D., and Hassell, J.A. 1987. Overpopulation of polyomavirus middle T antigen in mammalian cells through the use of an adenovirus vector. *J. Virol.* 61:1226–1239.

19. Chanda, P.K., Natuk, R.J., Mason, B.B., Bhat, B.M., Greenberg, L., Dheer, S.K. Molnar-Kimber, K.L., Mizutani, S., Lubeck, M.D., Davis, A.R., and Hung, P.P. 1990. High level expression of the envelope glycoproteins of the human immunodeficiency virus type I in presence of rev gene using helper independent adenovirus type 7 recombinants. *Virology*, 175:535–547.

20. Mason, B.B., Davis, A.R., Bhat, B.M., Chengalvala, M., Lubeck, M.D., Zandle, G., Kostek, B., Cholodofsky, S., Dheer, S., Molnar-Kimber, K., Mizutani, S., and Hung, P.P. 1990. Adenovirus vaccine vectors expressing hepatitis B surface antigen: Importance of regulatory elements in the adenovirus major late intron. *Virology*, 177:452–461.

21. Mansour, S.L., Grodzicker, T., and Tjian, R. 1986. Downstream sequences affect transcription initiation from the adenovirus major late promoter. *Mol. Cell. Biol.* 6:2684–2694.

22. Bridge, E., and Ketner, G. 1989. Redundant control of adenovirus late gene expression by early region 4. *J. Virol.* 63:631–638.

23. Pacini, D.L., Dubovi, E.J., and Clyde, W.A., Jr. 1984. A new animal model for human respiratory tract disease due to adenovirus. *J. Infect. Dis.* 150:82–97.

24. Rowe, W.P., Huebner, R.J., Hartley, J.W., Ward, T.G., and Parrott, R.H. 1955. Studies of the adenoidal-pharyngeal-conjuctival (APC) group of viruses. *Am. J. Hyg.* 61:197–218.

25. Pereira, H.G., Allison, A.C., and Niven, J.S.F. 1962. Fatal infection of newborn hamsters by an adenovirus of human origin. *Nature, (London)* 196:244–245.

26. Sinha, S.K., Fleming, L.W., and Scholes, S. 1960. Current considerations in public health of the role of animals in relation to human viral diseases. *J. Am. Vet. Med. Assoc.* 136:481–485.

27. McDermott, M.R., Graham, F.L., Hanke, T., and Johnson, D.C. 1988. Protection of mice against lethal challenge with herpes simplex virus by vaccination with an adenovirus vector expressing HSV glycoprotein B. *Virology* 169:244–247.

28. Prevec, L., Schneider, M., Rosenthal, K.L., Belbeck, L.W., Derbyshire, J.B., and Graham, F.L. 1989. Use of human adenovirus-based vectors for antigen expression in animals. *J. Gen. Virol.* 70:429–434.

12

The Use of Nonclassical Techniques in the Production of Secondary Metabolites by Plant Tissue Cultures

B. O'Keefe and C. Wm. W. Beecher

12.1. Production of Endogenous Compounds

Plant tissue culture is considered to be an economically viable means for producing secondary metabolites with a commercial value of over $1,000 per kilo. The
major category of such compounds is the pharmaceuticals, where such a compound may constitute the pharmaceutically active agent or a formulation (flavor-
ing or coloring) agent. Although the isolation of these metabolites from cultivated
or collected plant material is often the method of choice, there are many cases in
which cultivation of the plant is difficult (e.g., *Chondodendron tomentosum*), or
acquisition of the desired metabolites would endanger the species (e.g., *Taxus
brevifolia*). In all such cases tissue culture represents a viable alternative for the
production of compounds that are not synthetically accessible. For these reasons,
much effort has been devoted to developing plant cell lines to produce such
compounds. Ideally plant cell culture systems would be amenable to large-scale
fermentation and produce high yields within short periods. In fact, after some 15
years worth of effort there are a number of patents that have been issued describing
plant tissue culture systems that yield various compounds[1]; in general, however,
such systems have proven less successful than had been hoped. The major reason
for this disappointment seems to be that while it is reasonably easy to generate
plant tissue cell lines, a great majority of these cell lines do not produce the
sought-after secondary metabolites of their parent plants. To circumvent these
problems a number of nonclassical plant tissue techniques have been explored
that hold great promise. These nonclassical plant tissue culture techniques include
(1) the development of specific production media, (2) the use of biotic or abiotic
elicitors, and (3) the direct manipulation of the plant genome, often termed
transformation, by the various pathogenic soil bacteria in the genus *Agrobacterium*. Since the production of plant drugs by classical tissue culture techniques

have been reviewed previously,[2, 3] we will turn our attention to the use of these nonclassical techniques as metabolite production systems in plant tissue biotechnology.

The development of specialized media is an extension of standard fermentation technology, whereby an organism is cultivated first on a medium designed to support maximum growth rate and then transferred to another medium that induces the appropriate secondary metabolism. These specialized media are often derivatives of standard media (typically, Murashige and Skooge[4] or Gamborg B-5[5] in which sugar or phosphate levels have been modified, although a number of other strategies have been employed. As an example, the use of Murashige and Skooge media in which all phytohormones have been omitted can enhance secondary metabolite production 6-fold in *Catharanthus roseus* suspension cells.[6] Other modifications that have been used are the reduction of nitrogen to increase the production of capsaicin (used in topical cream for shingles)[7] and enhanced production of rosmarinic acid in media in which both nitrogen and phosphate concentrations have been reduced.[8] Dual-media systems have traditionally been used in the production of antibiotics by large-scale fermentation of microorganisms.

The second strategy, the use of elicitors (scheme 12.1), is peculiar to plant tissue culture fermentations. The elicitation process has been frequently used and reviewed in the literature[9–11, 12, 13] yet knowledge of the processes involved remains rudimentary. While the elicitation process has been reported to occur upon contact with a number of specific chemical entities, ranging from metal ions[14] to herbicides,[15] the vast majority of reported elicitations use either a proteinaceous or polysaccharide preparation derived from a yeast or fungus. Compounds that have been used to elicit secondary metabolite production range from crude cell-wall preparations of *Penicillium* spp.[16], *Verticillium dahlia*, and *Fusarium moniliforme*[14] to specific compounds such as bacitracin, colistin, and polymyxin B.[17] In most cases it is assumed that the elicitation response represents a disease resistance mechanism that plants have developed. Recently Apostol et al. have reported intense bursts of hydrogen peroxide that are generated within minutes of the introduction of an elicitor.[18]

In the final strategy (i.e., transformation) a small piece of DNA (T-DNA), derived from an extrachromosomal circle of DNA (termed a Ti- or Ri-plasmid) within the *Agrobacterium* cell, is transferred and integrated into the plant cell genome[19] (Figure 12.1). Following integration, metabolic and morpholgical changes occur that in some cases enhance the production of medicinally important compounds. Two species of *Agrobacterium* are of particular importance in plant cell tissue culture: *A. tumefaciens*, which causes the formation of undifferentiated tumors ("crown gall"), and *A. rhizogenes*, which induces the formation of fast-growing nongeotropic, highly branched root tissue ("hairy roots"; see Figure 12.2). Plant cells transformed by *Agrobacterium* spp. are often characterized by the production of unique amino acids called opines and the ability to grow on

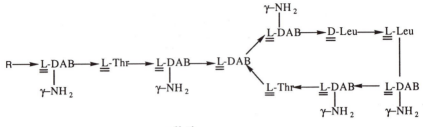

colistin

R= (+)-6-methyloctanoyl

DAB = α,γ–diaminobutyric acid

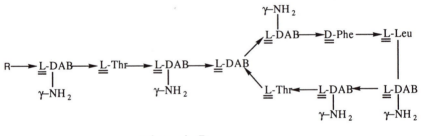

polymyxin B₁

R= (+)-6-methyloctanoyl

DAB = α,γ–diaminobutyric acid

Scheme 12.1. Antibiotic elicitation agents.

media lacking growth hormones. Both of these transformation characteristics are the result of the *Agrobacterium* plasmid's ability to incorporate itself and its genome into the host plant's genome. Specifically, the inserted plasmid carries the unregulated genes required for the production of normal growth hormones and, of course, the genes for the biosynthetic enzymes required for the opine production.

The host range of *Agrobacterium* spp. varies with each individual strain but is

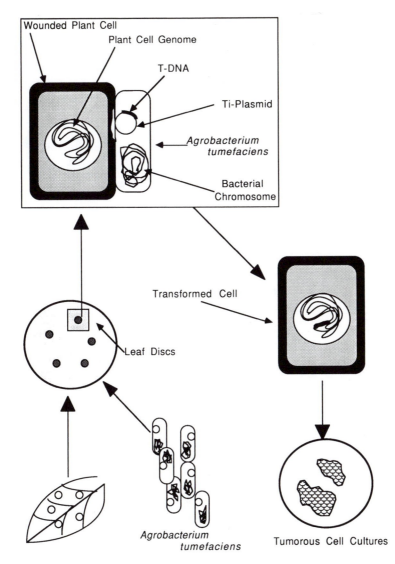

Figure 12.1. Cellular events during transformation.

generally limited to dicotyledonous plants, which limits the extent of their usage. Nonetheless, they may be more broadly useful, since some gymnosperms have been successfully transformed; monocotyledonous plants, though generally resistant to infection, may be transformed by methods such as protoplast-to-bacteria fusion or electroporation. *Agrobacterium* transformation is believed to occur by a process similar to bacterial conjugation, in which a single strand of DNA (T-

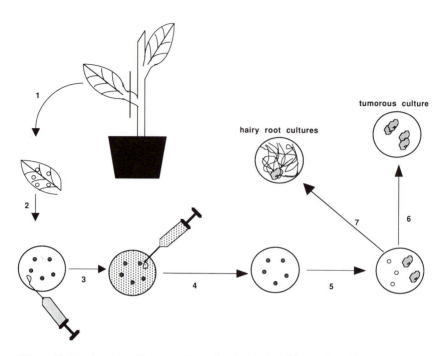

Figure 12.2. Agrobacterium transformation. 1. Leaf excision and sterilization. 2. Discs cut from sterilized leaf. 3. Innoculation with *Agrobacterium rhizogenes*. 4. Incubation on media containing antibiotics. 5. Incubation on media lacking phytohormones. 6. Culture on full strength Gambourg B5 media. 7. Culture on half strength Gambourg B5 media.

strand) is passed over to a mechanically wounded plant cell via binding to a specific membrane protein.[20] Manipulations of the Ti-plasmid have created improved systems of *Agrobacterium* transformation; vectors have been developed that allow for the regeneration of intact plants from transformed cells[21] and the efficient screening of cell lines via the insertion of genes for resistance to antibiotics.[22] Furthermore, both restriction endonuclease sites and regions homologous to the *Escherichia coli* plasmid pBR322 have been cloned into the T-DNA region of the Ti-plasmid.[23] These advances, and others, have made *Agrobacterium* a powerful tool for the insertion of foreign genes into plants.

Furthermore, since the inserted DNA is randomly placed, this technique also represents a mechanism for inducing "mutations" in the host plant genome. Instances have been reported in which the insertion of plasmid DNA altered the control that the parent plant was imposing on a specific biosynthetic pathway.[24] In short, *Agrobacterium* transformation can lead to the increased production of endogenous compounds by a number of different mechanisms: (1) by inducing noncontrolled growth by means of native hormones, (2) by insertion of foreign genes coding for nonnatural biosynthetic products, or (3) by "mutating" the host

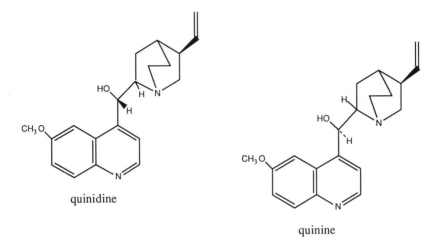

quinidine

quinine

Scheme 12.2. Quinoline alkaloids.

genome. These three strategies are frequently used on similar cell lines and may be viewed as historically related. The following discussion will focus on specific organisms and detail the successive developments needed to advance the current state of the art. In this overview, recent developments in the production of economically significant compounds by these technologies will be discussed. The discussion will be broken into two parts: in the first we will consider these techniques in the production of "native" metabolites, and the second we will describe the production of foreign gene products.

12.1.1. Quinoline Alkaloids

The production of the quinoline alkaloids, quinidine and quinine (Scheme 12.2), is of particular importance owing to their clinical use in treating malaria (quinine) and in restoring normal heart rhythms (quinidine). These alkaloids, which account for an annual trade of approximately $50 million, are found mostly in the Rutaceae, with commercial sources of quinine and quinidine coming primarily from *Cinchona* plantations in Central America, Africa, the Philippines, India, and Indonesia. Various plant tissue culture systems have been investigated with an eye toward their production.[25–27] The research group of Rhodes, Hamill, Robins, and Payne has perhaps most thoroughly studied the production of these and related quinoline alkaloids in cell cultures of *C. ledgeriana*. In early experiments they noted that although suspension and callus cultures failed to produce significant quantities of either quinine or quinidine, differentiated shoot systems demonstrated full biosynthetic capacity. This was despite the fact that other quinoline alkaloids were biosynthesized by callus culture systems at the rate of 20 micrograms per gram fresh weight.[28] Furthermore, these same authors reported on various attempts to optimize the original growth media by altering the phytohormone balance to enhance the production of quinoline alkaloids.[29] This approach met with limited success, resulting in an increase in alkaloid content to a level of 140 micrograms per gram fresh weight. The alkaloids cinchonine and cinchonidine, the biosynthetic precursors of quinine and quinidine, were the major alkaloids produced.

Trying to overcome these failures, this same group detailed the transformation of this species with A. tumefaciens to form hormone-independent, undifferentiated cell suspension cultures ("crown gall" cultures).[30] The average quinoline alkaloid content of these cells was 5 to 6 times greater than that of the previously studied untransformed cell line; however, when one particular tumorous cell line was grown in complete darkness it produced up to 50 times the quinoline alkaloid content of transformed suspension cultures grown in the light. Unfortunately, once again the major quinoline alkaloids produced were cinchonine and cinchonidine.

In a more recent report, hairy root cultures of C. ledgeriana, transformed with A. rhizogenes, were found to produce two to three times the alkaloid content of the most productive, dark-grown crown-gall cells described in their earlier work, with the methoxylated quinolines, quinine and quinidine, making up 50 to 70% of the total (as compared with 10% of the total alkaloids in crown gall cultures).[31] Although this seemingly successful report allows for some hope, it must be pointed out that the growth rate of C. ledgeriana hairy root cultures was slow, yielding only a 6- to 8-fold increase in fresh weight per month. Although this is faster than the growth rates of whole plants, it is quite slow when contrasted with typical reports of 30- to 60-fold increases for other hairy root cultures.[32] In addition to quinine and quinidine, hairy root cultures of C. ledgeriana displayed the capability to form all the major Cinchona alkaloids and also produced a spectrum of as yet unidentified indole alkaloids.

12.1.2. Indole Alkaloids

Several important theraputic drugs, such as physostigmine, ajmalicine, reserpine, vincristine, and vinblastine, are part of the class of compounds known as indole alkaloids (Scheme 12.3). These alkaloids can be found in the plant families Apocynaceae, Loganiaceae, Rubiaceae, and Fabaceae, among others, and in the hallucinogenic fungi Claviceps spp. and Psilocybe spp. Two compounds of particular interest, vincristine (VCR) and vinblastine (VLB), are very complex dimeric monoterpene indole alkaloids with a commercial value of over $6,000 per gram. They are used for their antineoplastic activity and are the drugs of choice for the treatment of childhood leukemia and Hodgkin's disease. The commercial source of these compounds is plant material from the species Catharanthus roseus, in which they are produced in minute quantities (VCR at 20 mg/

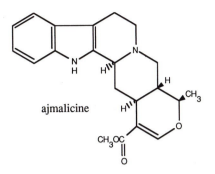

ajmalicine

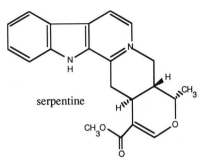

serpentine

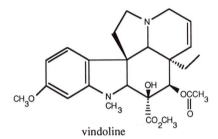

vindoline

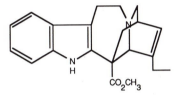

catharanthine

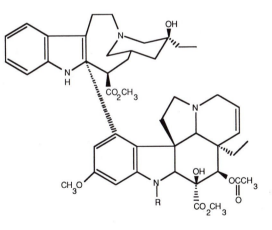

vinblastine   R= CH$_3$

vincristine   R= CHO

Scheme 12.3.   Indole alkaloids.

1,000 kg of dried plant material and VBL at 1.0 g/1,000 kg of dried plant material). *Catharanthus roseus* is also the source of the cardiovascular modulating agent ajmalicine and a spectrum of other indole alkaloids.

Although extensive work has been done to produce these compounds in plant tissue cultures, classical cell suspension cultures of *C. roseus* have thus far failed to produce VCR or VBL.[33] Indeed, even the production of vindoline, an important monomeric precursor of these drugs, has not yet been unequivocally demonstrated in undifferentiated suspension cultures. Only by the induction of differentiated shoot cultures, via the addition of the phytohormone benzyladenine into the growth media, has vindoline production been demonstrated in classical cell culture.[34]

The use of specific production media, however, has shown that increases in ajmalicine and catharanthine (the other monomeric precursor of VBL and VCR) can be achieved by transferring cells from growth media into media optimized for alkaloid production.[33] In this work, later confirmed by other researchers,[35] alkaloid accumulation (ajmalicine and serpentine) in *C. roseus* suspension cultures was increased by transferring cells from growth media[4] to media with a higher sucrose concentration (8% as compared with 3%) and deficient in nitrogen, phosphate, and phytohormones. Other successful media alterations have included the addition of bioregulators (e.g., 2-diethylaminoethyl-2,4-dichlorophenylether, which increased ajmalicine content by 270 to 570% and catharanthine by 82 to 146%)[36] or differentiation inducers like 5-azacytidine, which caused the formation of a new secondary metabolite lirioresinol B mono β-D glucoside.[37] These agents have increased the yield of specific compounds but have not resulted in the production of VLB, VCR, or vindoline.

Using a different approach, Eilert et al. transformed *C. roseus* cells with *A. tumefaciens* to produce tumorous cell lines that, though able to form ajmalicine at 0.060 mg/g fresh weight, also failed to produce VCR or VLB.[38] These crown gall cultures produced the monomeric unit, catharanthine, but did not synthesize vindoline. Yet another group transformed *C. roseus* cells with *A. rhizogenes*, producing hairy root cultures that showed a 20-fold increase in cell mass over a 28-day growth period.[39] These cultures synthesized high yields of a wide variety of monomeric indole alkaloids including catharanthine, ajmalicine, and vindolinine. Radioimmunoassay also showed them to contain very small quantities of vinblastine. The quantities of alkaloids produced in hairy root cultures were similar to those of intact roots of the plant. Alkaloid production by transformed cultures from other indole alkaloid-producing plant species have shown results similar to those mentioned previously in that they produced significant quantities of the physiologically active monomeric indole alkaloids but did not synthesize vincristine or vinblastine.[40] The dimeric indole alkaloids represent some of the most complicated chemical structures produced by plants, and their production by tissue culture methods will continue to be sought. However, the increases in production of important indole alkaloids such as ajmalicine, serpentine, and

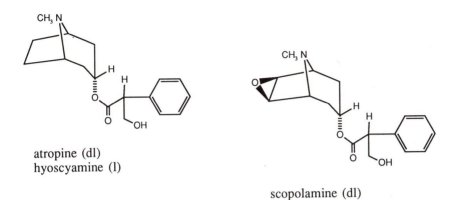

atropine (dl)
hyoscyamine (l)

scopolamine (dl)

Scheme 12.4. Tropane alkaloids.

catharanthine by transformation procedures are a definite improvement over previous results.

12.1.3. Tropane Alkaloids

Tropane alkaloids (Scheme 12.4), which are found in the Solanaceae, Erythroxylaceae and Convolvulaceae, include several important pharmaceuticals, most notably, scopolamine, hyoscyamine, atropine, and cocaine. Since species from the plant family Solanaceae are very responsive to cell culture techniques, plants producing these alkaloids have been actively investigated via classical tissue culture methods with mixed results.[41, 42]

Because these species biosynthesize the majority of their biologically active alkaloids in their roots, many attempts have been made to increase tropane alkaloid production by transformation with *A. rhizogenes*. For example, *Datura candida*, a woody solanaceous species, transformed with *A. rhizogenes* was induced to form hairy root cultures that showed a 20-fold increase in cell mass over a 1-month growth period and, upon analysis, had an alkaloid content of 0.68% of the dry weight.[43] Hyoscyamine and scopolamine were present in a ratio comparable to that of the whole plant. These results were similar to earlier reports on transformed cultures of *D. stramonium* in which 0.3% of the dry weight of hairy root cultures was hyoscyamine.[44] These *D. stramonium* cultures, however, displayed a 55-fold increase in cell mass over the same growth period, 15 times the growth rate of untransformed root cultures from the same plant. Jaziri et al., also working with *D. stramonium*, have reported hairy root cultures that synthesized scopolamine at the rate of 0.56% by dry weight (%DW).[45]

Another solanaceous species, *Atropa belladonna* (deadly nightshade), transformed with *A. rhizogenes* produced hairy root cultures that increased in cell mass 60-fold over a 28-day growth period and synthesized scopolamine at 0.024%

DW and hyoscyamine at 0.371% DW.[46] It should be noted that the yield of tropane alkaloids by these cultures remained consistent under nonselective culture conditions and was equivalent to or higher than the production by 1-year-old roots from intact plants on a percentage of dry weight basis. These results are in general agreement with other studies on *Hyoscyamus muticus*[32] and *Scopalia japonica*,[47] which have also shown increased production of tropane alkaloids in hairy root cultures as compared with normal callus or suspension cultures. By producing tropane alkaloids at levels similar to that of the intact plant while maintaining such rapid growth rates, hairy root cultures of solanaceous species demonstrate the potential of *A. rhizogenes* transformation as a means of increasing the production of pharmaceutically relevant compounds in plant tissue cultures.

12.1.4. Miscellaneous Alkaloids

In addition to the quinoline, indole, and tropane alkaloids, two other groups of alkaloids have been produced by *Agrobacterium*-transformed plant cell cultures. (Scheme 12.5). Quinolizidine alkaloids, a group of compounds including cytisine (a toxic hallucinogen) and anagyrine (an arrhythmia agent), are found primarily in the Fabaceae. In general, the quinolizidine alkaloid-producing members of the Fabaceae have proven resistant to transformation by *Agrobacterium*. Recently however, the production of these alkaloids, normally synthesized in stem and leaf tissue, was reported in hairy root cultures of *Spartium junceum* by Wink and Witte at levels comparable to or greater than those found in stem tissue from intact plants as measured in micrograms per gram fresh weight.[48]

Nicotine, a toxic compound that acts on the central nervous system, is produced in several plant species, of which *Nicotiana tabacum* tobacco is the best known. Hairy root cultures of *N. rustica*, a closely related species, produced nicotine at isolatable yields of 0.28 mg per gram fresh weight, equivalent to 80 percent of the total alkaloid content.[49] Qualitatively and quantitatively, the alkaloid spectrum found in these tissues was identical to that found in intact plants of the same species. These cultures, which increased in cell mass 35-fold over a 16-day growth period, secreted additional nicotine alkaloids, 10 mg/ml, into the medium throughout the growth period. Similar results have been obtained by Parr et al.,[50] who examined hairy root cultures from a variety of different *Nicotiana* species, and Saito et al.,[51] who were working with *A. rhizogenes*-transformed *N. tabacum* cells. As may be expected, hairy root tissues produced significant quantities of nicotine and secreted them into the media. When the researchers infected *N. tabacum* with a mutant strain of *A. tumefaciens*, they induced the formation of tumerous stem and leaf tissue ("shooty teratomas") that, in contrast, did not produce detectable quantities of nicotine alkaloids. When shooty teratomas and hairy root cultures were cocultivated in the same vessel, however, the shooty teratomas contained a high level of both nicotine and nornicotine (a breakdown

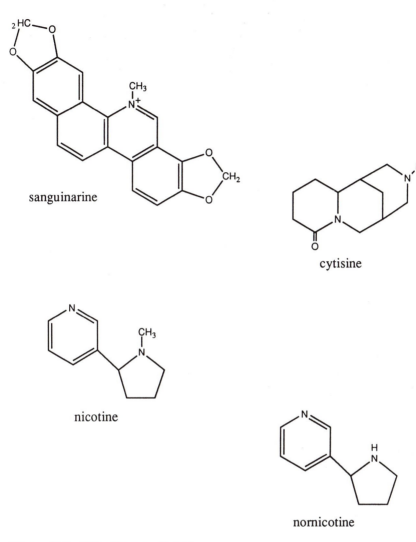

sanguinarine

cytisine

nicotine

nornicotine

Scheme 12.5. Miscellaneous alkaloids.

product of nicotine), while the concentration in both the hairy root cultures and the media decreased. The ability of shooty teratoma cultures to biotransform and store alkaloids secreted into media by hairy roots is an interesting new development made possible by the use of hormone-independent transformed cells. Furthermore, the potential for de novo synthesis of secondary metabolites, normally produced in aerial plant parts, by shooty teratomas is being examined. If these

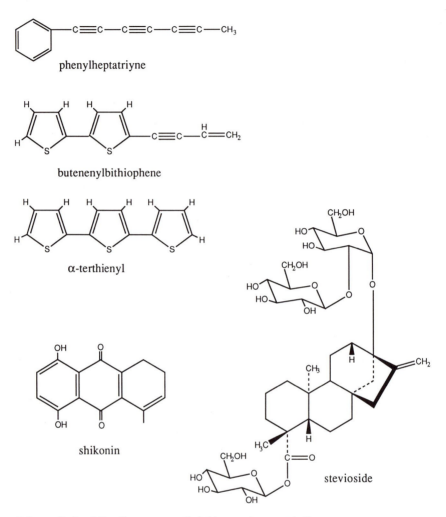

Scheme 12.6. Miscellaneous nonalkaloid secondary metabolites.

transformed tissues display biosynthetic capacities that mirror normal leaf tissue, dramatic advances in the production of important pharmaceuticals could be realized.

12.1.5. Nonalkaloidal Substances

All of the work with *Agrobacterium*-transformed cell lines so far discussed has focused on the production of biologically active alkaloids, but it is important to note that several nonalkaloidal secondary metabolites of interest have been produced in hairy root tissues (Scheme 12.6). In fact, even though the pharmaceutical importance of these compounds is often considered less than that of the alkaloids, it is within this group that tissue culture has made its greatest strides toward the commercial application of plant cell culture technology.

The production of polyacetylenes and thiophenes is widespread in the plant families Compositae and Umbelliferae. These acetate-derived compounds are antibiotic toward fungi, bacteria, and nematodes. Hairy root cultures from the genus *Tagetes* and *Bidens* were investigated by Flores et al.[52] In results similar to those reported for solanaceous species, these hairy root cultures produced the same pattern of polyacetylenes and thiophenes synthesized in the roots of intact plants. In a later report, this same group examined the effect of challenging hairy root cultures with fungal elicitors.[53] When transformed cultures from three species of Compositae—*Bidens sulphureus, B. Pilosus,* and *Carthamus tinctorius*— were challenged by fungal culture filtrates, they increased their production of polyacetylenes up to 20-fold within 24 hours of treatment and maintained their biosynthetic capacity at high levels for at least an additional 24 hours. The main polyacetylenes produced by this treatment (trideca-1,3,11-ene-5,7,9-yne for *B. sulphureus* and trideca-2,10-ene-4,6,8-yne-1-ol for *B. pilosus* and *C. tinctorius*) all showed antifungal activity. Though the increased production of polyacetylenes and thiophenes by hairy root cultures has not been seen in all cases,[54] it is clear that some cell lines have displayed significant improvements in biosynthetic capacity following transformation with *A. rhizogenes.*

Another group of biologically active compounds are the glycosides. Steviobioside, the precursor to the diterpene glycoside sweetener stevioside, has been produced in hairy root cultures of the plant species *Stevia rebaudiana.*[55] These particular cultures responded to production media containing higher concentrations of sucrose by increasing the rate of steviobioside production, a response often seen in untransformed cell lines. Although this particular compound is of little direct pharmacological interest, it has great commercial interest owing to the use of stevioside as a sweetening agent in many parts of Asia. In addition, the Ginsengosides, active constituents of the roots of *Panax ginseng*, were produced by hairy root cultures in the same ratio as roots from the intact plant but in quantities up to 2.4 times higher on a dry weight basis.[56] This production was realized after only a three-week growth period in which the hairy root cultures grew two to three times faster than untransformed root cultures. Considering that the average age of commercially marketed ginseng roots is five to seven years, it is not surprising that a large-scale fermentation process of *P. ginseng* root cultures has been reported in Japan using airlift bioreactors.[57]

The best known large-scale fermentation process of plant cell cultures is that used for the commercial production of the shikonin pigments. These red compounds, used in the manufacture of cosmetics, are found in cell cultures from the species *Lithospermum erythrorhizon* of the Boraginaceae. Enhancement of shikonin derivative production from suspension cultures was achieved by the use of a two-stage media system developed by Fujita et al.[58] This system, in which lower nitrate and phosphate levels were combined with higher levels of copper and sucrose, resulted in 6-fold increases in the yield of shikonin derived from dried cell material. Since these production media were unable to provide maximal growth rates, a system was developed whereby cells were cultured in media that supplied the highest growth rates[59] and then transferred to the production media. This two-step process resulted in the highest production of shikonin derivatives per unit time.

Hairy root cultures of *L. erythrorhizon* have also been produced.[60] These transformed cultures synthesized shikonin derivatives in quantities similar to, and often greater than, those found in root tissue from intact plants, but at levels below those produced in suspension cultures using the finely tuned process of Fujita et al.[58]

12.2. Transformations Involving Foreign Genes

Recombinant DNA technology has been successfully used to produce a number of pharmaceutically important compounds (e.g., human insulin, tissue plasminogen activator, and erythropoietin). To date the majority of this work has depended on bacterial hosts (e.g., *E. coli*) and the production of peptide products. The use of plant cells as possible hosts for the insertion of foreign genes is now being investigated. The most widely used vehicle ("vector") for the insertion of foreign genes into plants is a modified form of the Ti-plasmid from *Agrobacterium tumefaciens*.[61] This plasmid can insert segments of DNA (T-DNA) that average 9.5 kilobases but can be as large as 50 kilobases.[62] These modified plasmids take advantage of the fact that the only sequence requirement for DNA transfer are the 25 base pair segments on the right- and left-hand borders of the T-DNA and several virulence genes located outside of the T-DNA region. The rest of the T-DNA can be removed and replaced with no loss of transformation capacity, thereby allowing the transmission of DNA from bacterial cell to plant cell without the morphological and physiological changes normally associated with *Agrobacterium* transformation.[63] The use of plant cells as hosts for the insertion of foreign genes is an intriguing new area in plant culture research with many applications toward the production of pharmaceutically relevant compounds. In situations in which bacterial or yeast host systems are proven inappropriate, *Agrobacterium* transformation provides another system by which researchers can induce the formation of biologically active proteins.

Many physiologically active proteins normally made in humans are of great pharmaceutical utility as therapeutic agents. Unfortunately, these substances are often only produced in minute quantities and are very difficult to isolate. Prokaryotic bacterial cells can be induced to synthesize many of these substances (e.g., insulin) via the insertion of human cDNA (mRNA sequence DNA clones). However, in situations in which more advanced eukaryotic processes such as posttranscriptional mRNA splicing and protein glycosylation are required, different host cells must be used. Barta et al. were one of the first groups to attempt this procedure by fusing the complete gene for human growth hormone (hGH) into the T-DNA region of the Ti-plasmid from *A. tumefaciens*.[64] This recombinant *Agrobacterium* was then used to transform plant cells from tobacco and sunflower. Transcription of the hGH gene was detected in transformed cells by the isolation of foreign mRNAs via Northern blot analysis. The hGH protein itself was not produced owing to aberrant posttranscriptional modifications of the foreign mRNA. This early result demonstrated the need to insert cDNA clones of human genes rather than the complete sequence. In a more successful study,[65] cDNA from a mouse hybridoma that produced a human antibody was inserted into the T-DNA region of the Ti-plasmid (γ and κ chains on different plasmids), and the recombinant *Agrobacterium* strains were then used to transform tobacco leaf segments. Transformed cells were then manipulated to regenerate intact tobacco plants, which were then screened for recombinants that synthesized either the γ or κ immunoglobin chains from the antibody. After crossbreeding of the transgenic plants, F_1 progeny were produced that made both the γ and κ immunoglobulin chains simultaneously and were able to construct and accumulate fully functional antibodies at levels of up to 1.3% of the total leaf protein.

Working in another direction, Vanderkerckhove et al. transformed *Arabidopsis thaliana* and *Brassica napus* with a modified gene for the 2S albumin seed storage protein into which the cDNA for human Leu-enkephalin, a pentapeptide displaying opiate activity, was ligated.[66] Following regeneration of the transgenic plants, Leu-enkephalin was isolated from the seeds at a level of 200 nmol/g, equivalent to 0.1% of the total seed protein. Furthermore, the isolation of Leu-enkephalin was easily achieved by the inclusion of sequences for tryptic-digest sites on the borders of the Leu-enkephalin gene. The work of these researchers has provided results that demonstrate the utility of plant cells as hosts for the production of biologically active proteins from animal sources.

The production of foreign proteins by transgenic plant cells is not limited to those of animal or mammalian origin. A wide variety of genes have been introduced into plant cells via *Agrobacterium* transformation or other direct transfer techniques such as microinjection and protoplast fusion;[67] the transformed cell is then typically used to regenerate a transformed plant, which is the object of most of this type of research. Although the majority of these transformed plants are of agricultural (e.g., resistance to herbicides) or physiological (e.g., light regulated genes) interest and therefore only indirectly related to drug production, they are suggestive of future potential. One such example is the insertion of the gene for the proteinaceous toxin from *Bacillus thuringiensis* into tobacco plants via *Agrobacterium* transformation.[68] The *Bacillus* toxin is lethal to several species of lepidopteran larvae (e.g., tobacco hornworm) but is not toxic to other organisms, making it a "safe" insecticide. Regenerated transgenic tobacco plants in this report showed production of the toxin, in some cases, at levels of over 0.004%

of the total leaf protein. These concentrations exceeded those required for 100% mortality of the susceptible larvae.

In another study by Meyer et al. a gene for the enzyme dihydroflavonol-4-reductase from maize, which reduces dihydrokaempherol to the flavonoid pigment pelargonidin, was introduced into petunia cells that were previously unable to catalyze this reaction.[69] Transcription of the foreign genes was demonstrated in the regenerated transformed petunia plants by mRNA isolation and Northern blot analysis. Expression of the active enzyme was demonstrated by the presence of pelargonidin, a brick-red pigment, in the previously white flowers of regenerated plants. This report suggests that the ability to create side branches from the normal biosynthetic pathways by the insertion of unique plant-derived enzymes into heterologous plant cell hosts may be one of the simplest yet most effective applications of this recombinant DNA technology. Though the compound produced in this example is not of direct pharmacological interest, this technique could be used to produce biosynthetic enzymes necessary for the production of important secondary metabolites in the tissue cultures in which similar biosynthetic processes are already easily expressed.

In summation, there are many reasons to believe that the potential of plant tissue cultures to yield commercially significant quantities of secondary metabolites is very good. The use of nonclassical technologies, when combined with the classical tissue culture techniques, will undoubtedly play a role in the future production of pharmaceuticals.

References

1. Misawa, M., and Suzuki, T. 1982. Research on the production of useful plant metabolites in Japan. *Appl. Biochem. and Biotech.* 7:205–216.

2. Anderson, L.A., Phillipson, J.D., and Roberts, M.F. 1986. Aspects of alkaloid production by plant cell cultures. in *Secondary metabolism in plant cell culture.* P. Morris, A.H. Scragg, A. Stafford, and M.W. Fowler, eds. Cambridge: Cambridge University Press.

3. Yamada, Y. and Hashimoto, T. 1984. Secondary products in tissue culture. In *Applications of genetic engineering to crop improvement.* G.B. Collins, and J.G. Petolino, eds. Boston: Nijhoff/Junk, Pp. 561–604.

4. Murashige, T. and Skooge, F. 1962. A revised medium for rapid growth and bioassays with tobacco tissue cultures. *Physiol. Plant.* 15:473–481.

5. Gamborg, O.L., Miller, R., and Ojima, K. 1958. A medium for the growth of plant tissues. *Exp. Cell Res,* 50:151–158.

6. Knobloch, K.-H., Berlin, J. 1980. Influence of media composition on the formation of secondary metabolites in cell suspension cultures of *Catharanthus roseus* (L.) G. Don. *Z. Naturforsch.* 35c:551–556.

7. Lindsey, K. 1985. Manipulation, by nutrient limitation, of the biosynthetic activity of immobilized cells of *Capsicum frutescens* Mill. cv. annum. *Planta* 165:126–133.

8. De-Eknamkul, W., and Ellis, B.E. 1985. Effects of macronutrients on growth and rosmarinic acid formation in cell suspension cultures of *Anchusa officinalis. Plant Cell Rep.* 4:46–49.

9. Tanahashi, T., and Zenk, M.H. 1990. Elicitor induction and characterization of microsomal protopine-6-hydroxylase, the central enzyme in benzophenanthridine alkaloid biosynthesis. *Phytochemistry* 29:1113–1122.

10. Vogeli, U., Freeman, J.W., and Chappell, J. 1990. Purification and characterization of an inducible sesquiterpene cyclase in elicitor treated tobacco suspension cultures. *Plant Physiol.* 93:182–187.

11. Tyler, R.T., Eilert, U., Rijnders, C.O.M., Roewer, I.A., and Kurz, W.G.W. 1988. Semi-continuous production of sanguinarine and dihydrosanguinarine by *Papaver somniferum* L. cell suspension cultures treated with fungal homogenates. *Plant Cell Rep.* 7:410–413.

12. Kurz, W.G.W., Constabel, F., Eillert, U., and Tyler, R.T. 1987. In *Topics in pharmaceutical sciences*, D.D. Breimer and P. Speiser, eds. New York: Elsevier.

13. Cramer, C.L., Ryder, T.B., Bell, J.N., and Lamb, C.J. 1985. Rapid switching of plant gene expression induced by fungal elicitors. *Science* 227:1240–1242.

14. Heinstein, P.F. 1985. Future approaches to the formation of secondary natural products in plant cell suspension cultures. *J. Nat. Prod.* 48:1–9.

15. Cho, G.H., Kim, D.I., Pederson, H., and Chin, C.K. 1988. Etephon enhancement of secondary metabolite synthesis in plant cell cultures. *Biotech. Prog.* 4:184–188.

16. Mahady, G.B., O'Keefe, B., and Beecher, C.W.W. *P. expansum* preparation. Unpublished results.

17. Schumacher, H.-M., Gundlach, H., Fiedler, F. and Zenk, M.H. 1987. Elicitation of benzophenanthridine alkaloid synthesis in *Eschscholtzia* cell cultures. *Plant Cell Rep.* 6:410–413.

18. Apostol, I., Heinstein, P.F., and Low, P.S. 1989. Rapid stimulation of an oxidative burst during elicitation of cultured plant cells. *Plant Physiol.* 90:109–116.

19. Zambryski, P., Tempe, J., and Schell, J. 1989. Transfer and function of T-DNA genes from *Agrobacterium* Ti and Ri plasmids in plants. *Cell* 56:193–201.

20. Stachel, S.E., Timmerman, B., and Zambryski, P. 1986. Generation of single stranded T-DNA molecules during the initial stages of T-DNA transfer from *Agrobacterium tumefaciens* to plant cells. *Nature* 322:706–712.

21. Zambryski, P., Holsters, M., Kruger, K., Depicker, A., Schell, J., Van Montague, M., and Goodman, M. 1980. Tumor DNA structure in plant cells transformed by *A. tumefaciens. Science* 209:1385–1391.

22. DeBlock, M., Herrera-Estrella, L., Van Montague, M., Schell, J., and Zambryski, P. 1984. Expression of foreign genes in regenerated plants and their progeny. *EMBO J.* 3:1681–1689.

23. Caplan, A., Herrera-Estrella, L., Inze', D., Van Haute, E., Van Montague, M., Schell, J., and Zambryski, P. 1983. Introduction of genetic material into plant cells. *Science* 222:815–821.

24. O'Keefe, B., Schilling, A.B., and Beecher, C.Wm.W. 1991. Alkaloid production by hormone independent cell lines of *Catharanthus roseus* (L.)G. Don following transformation with *Agrobacterium* sp. Poster abstract. Annual Meeting Am. Soc. of Pharmacognosy, Chicago, Illinois.

25. Anderson, L.A., Keene, A.T., and Phillipson, J.D. 1982. Alkaloid production by leaf organ, root organ and cell suspension cultures of *Cinchona ledgeriana*. *Planta Med.* 46:25–27.

26. Staba, J.E., and Chung, A.C. 1981. Quinine and quinidine production by *Cinchona* leaf, root and unorganized callus. *Phytochemistry* 20:2495–2498.

27. Winjsma, R. 1986. Ph.D. thesis. Department of Pharmacognosy, University of Leiden, The Netherlands.

28. Robins, R.J., Payne, J., and Rhodes, M.J.C. 1986. Cell suspension cultures of *Cinchona ledgeriana*. 1. Growth and quinoline alkaloid production. *Planta Med.* 52:220–226.

29. Rhodes, M.J.C., Payne, J., and Robins, R.J. 1986. Cell suspension cultures of *Cinchona ledgeriana*. 2. The effect of a range of auxins and cytokinins on the production of quinoline alkaloids. *Planta Med.* 52:226–229.

30. Payne, J., Rhodes, M.J.C., and Robins, R.J. 1987. Quinoline alkaloid production by transformed cultures of *Cinchoma ledgeriana*. *Planta Med.* 53:367–372.

31. Hamill, J.D., Robins, R.J., and Rhodes, M.J.C. 1989. Alkaloid production by transformed root cultures of *Cinchona ledgeriana*. *Planta Med.* 55:354–357.

32. Flores, H.E., Hoy, M.W., and Pickard, J.J. 1987. Secondary metabolites from root cultures. *Trends Biotech.* 5:64–69.

33. Zenk, M.H., El-Shagi, H., Arens, H., Stockigt, J., Weiler, E.W., and Deus, B. 1977. Formation of the indole alkaloids serpentine and ajmalicine in cell suspension cultures of *Catharanthus roseus*. In *Plant tissue culture and its bio-technological application*. ed. W. Barz, E. Reinhard, and M.H. Zenk, eds. Berlin: Springer-Verlag. Pp. 27–43.

34. Kreuger, R.J., Carew, D.P., Lui, J.H.C., and Staba, E.J. 1982. Initiation, maintenance and alkaloid content of *Catharanthus roseus* leaf organ cultures. *Planta Med.* 45:56–57.

35. Knobloch, K.-H., and Berlin, J. 1980. Influence of medium composition on the formation of secondary compounds in cell suspension cultures of *Catharanthus roseus* (L.) G. Don. *Z. Natureforsch.* 35c:551–556.

36. Lee, S.-L., Cheng, K.-D., and Scott, A.I. 1981. Effects of bioregulators on indole alkaloid biosynthesis in *Catharanthus roseus* cell culture. *Phytochemistry* 20:1841–1843.

37. Arfmann, H.A., Kohl, W., and Wray, V. 1985. Effect of 5-azacytidine on the formation of secondary metabolites in *Catharanthus roseus* cell suspension cultures. *Z. Natureforsch.* 40c:21–25.

38. Eilert, U., DeLuca, V., Kurz, W.G.W., and Constabel, F. 1987. Alkaloid formation by habituated and tumorous cell suspension cultures of *Catharanthus roseus*. *Plant Cell Rep.* 6:271–274.

39. Parr, A.J., Peerless, A.C.J., Hamill, J.D., Walton, N.J., Robins, R.J., and Rhodes, M.J.C. 1988. Alkaloid production by transformed root cultures of *Catharanthus roseus*. *Plant Cell Rep.* 7:309–312.

40. Davioud, E., Kan, C., Hamon, J., Tempe, J., and Husson, H.-P. 1989. Production of indole alkaloids by *in vitro* root cultures from *Catharanthus tricophyllus*. *Phytochemistry* 28:2675–2680.

41. Yamada, Y., and Hashimoto, T. 1982. Production of tropane alkaloids in cultured cells of *Hyoscyamus niger*. *Plant Cell Rep.* 1:101–103.

42. Endo, T., and Yamada, Y. 1985. Alkaloid production in cultured roots of three species of *Duboisia*. *Phytochemistry* 24:1233–1236.

43. Christen. P., Roberts, M.F., Phillipson, J.D., and Evans, W.C. 1989. High production of tropane alkaloids by "hairy root" cultures of a *Datura candida* hybrid. *Plant Cell Rep.* 8:75–77.

44. Payne, J., Hamill, J.D., Robins, R.J., and Rhodes, M.J.C. 1987. Production of hyoscyamine by "hairy root" cultures of *Datura stramonium*. *Planta Med.* 53:474–478.

45. Jaziri, M., Legros, M., Homes, J., and Vanhaelen, M. 1988. Tropine alkaloid production by hairy root cultures of *Datura stramonium* and *Hyoscyamus niger*. *Phytochemistry* 27:419–420.

46. Kamada, H., Okamura, N., Satake, M., Harada, H., and Shimomura, K. 1986. Alkaloid production by hairy root cultures in *Atropa belladonna*. *Plant Cell Rep.* 5:239–242.

47. Mano, Y., Nabeshima, S., Matsui, C., and Ohkawa, H. 1986. Production of tropane alkaloids by hairy root cultures of *Scopalia japonica*. *Agric, Biol. Chem.* 50:2715–2722.

48. Wink, M., and Witte, L. 1987. Alkaloids in stem roots of *Nicotiana tabacum* and *Spartium junceum* transformed by *Agrobacterium rhizogenes*. *Z. Naturforsch.* 42c:69–72.

49. Hamill, J.D., Parr, A.J., Robins, R.J., and Rhodes, M.J.C. 1986. Secondary product formation by cultures of *Beta vulgaris* and *Nicotiana rustica* transformed with *Agrobacterium rhizogenes*. *Plant Cell Rep.* 5:111–114.

50. Parr, A.J., and Hamill, J.D. 1987. Relationship between *Agrobacterium rhizogenes* transformed hairy roots and intact uninfected *Nicotiana* plants. *Phytochemistry* 26:3214–3245.

51. Saito, K., Murakoshi, I., Inze, D., and VanMontague, M. 1989. Biotransformation of nicotine alkaloids by tobacco shooty teratomas induced by a Ti-plasmid mutant. *Plant Cell Rep.* 7:607–610.

52. Flores, H.E., Hoy, M.W., and Pickard, J.J. 1987. Secondary metabolites from root cultures. *Trends Biotech.* 5:64–69.

53. Flores, H., Pickard, J. and Signs, M. 1988. Elicitation of polyacetyline production in hairy root cultures of Asteraceae. *Suppl. Plant Physiol.* 86:108.

54. Croes, A.F., van den Berg, A.J.R., Bosveld, M., Breteler, H., and Wullems, G.J.

1989. Thiophene accumulation in relation in relation to morphology in roots of *Tagates patula*: Effects of auxin and transformation by *Agrobacterium*. *Planta* 179:43–50.

55. Yamazaki, T., and Flores, H. 1989. Production of steviol glucosides by hairy root cultures of *Stevia*. *Plant Physiol.* 89:10.

56. Yoshikawa, T., and Furuya, T. 1987. Saponin production by cultures of *Panax ginseng* transformed with *Agrobacterium rhizogenes*. *Plant Cell Rep.* 6:449–453.

57. Ushiyama, K., Oda, H., and Miyamoto, Y. 1986. Large scale tissue culture of *Panax ginseng* root. In *Proc. 6th International Congress of Plant Tissue and Cell Culture*. D. Somers, B.G. Gregenbach, D.D. Biesboer, W.P. Hackett, and C.E. Green, eds. Minneapolis: University of Minnesota.

58. Fujita, Y., Hara, Y., Suga, C., and Morimota, T. 1981. Production of shikonin derivatives by cell suspension cultures of *Lithospermum erythrorhizon*. *Plant Cell Rep.* 1:61–63.

59. Linsmaier, E.F., and Skoog, F. 1965. Organic growth factor requirements of tobacco tissue cultures. *Physiol. Plant.* 18:100–127.

60. Shimomura, K., Satake, M., and Kamada, H. 1986. Production of useful secondary metabolites by hairy roots transformed with Ri-plasmid. In *Proc. 6th International Congress of Plant Tissue and Cell Culture*. D. Somers, B.G. Gegenbach, D.D. Beisboer, W.P. Hackett, and C.E. Green. Minneapolis: University of Minnesota.

61. Klee, H., Horsch, R., and Rogers, S. 1987. *Agrobacterium*-mediated plant transformation and its further applications to plant biology. *Annu. Rev. Plant Physiol.* 38:467–486.

62. Chantal, D., Petit, A., and Tempe, J. 1988. T-DNA length variability in mannopine hairy root: More than 50 kilobasepairs of pRi T-DNA can integrate in plant cells. *Plant Cell Rep.* 7:92–95.

63. Wang, K., Hererra-Estrella, L., Van Montague, M., and Zambryski, P. 1984. Right 25bp terminus sequence of the nopaline T-DNA is essential for and determines direction of DNA transfer from *Agrobacterium* to the plant genome. *Cell* 38:455–462.

64. Barta, A., Sommergruber, K., Thompson, D., Hartmuth, K., Matzke, M.A., and Matzke, A.J.M. 1986. The expression of a nopaline synthase-human growth hormone chimaeric gene in transformed tobacco and sunflower callus tissue. *Plant Mol. Biol.* 6:347–357.

65. Haitt, A., Cafferkey, R., and Bowdish, K. 1989. Production of antibodies in transgenic plants. *Nature (London)* 342:76–78.

66. Vanderkerckhove, J., Van Damme, J., Van Lijsebettens, M., Botterman, J., De Block, M., Vandewiele, M., De Clercq, A., Leemans, J., Van Montague, M., and Krebbers, E. 1989. Enkephalins produced in transgenic plants using modified 2S seed storage proteins. *Bio/Technology* 7:929–932.

67. Weising, K., Schell, J., and Kahl, G. 1988. Foreign genes in plants: Transfer, structure, expression and applications. *Annu. Rev. Genet.* 22:421–477.

68. Vaeck, M., Reynaerts, A., Hofte, H., Jansens, S., De Beuckeleer, M., Dean, C.,

Zabeau, M., Van Montague, M., and Leemans, J. 1987. Transgenic plants protected from insect attack. *Nature (London)* 328:33–37.

69. Meyer, P., Heidmann, I., Forkmann, G., and Saedler, H. 1987. A new petunia flower colour generated by transformation of a mutant with a maize gene. *Nature (London)* 330:677–678.

13

Applications of Biotechnology in Drug Discovery and Evaluation

Cindy K. Angerhofer and John M. Pezzuto

13.1. Abbreviations

The following is a list of abbreviations used in this chapter.

CAT: chloramphenicol acetyltransferase

CCK: cholecystokinin

CPA: cytopathic effects

DFMO: difluoromethylornithine

EDTA: ethylenediaminetetraacetic acid

EGF: epidermal growth factor

ELISA: enzyme-linked immunosorbent assay

HIV: human immunodeficiency virus

HRE: hormone-responsive element

LTR: long terminal repeat

MDR1: the human gene responsible for multiple-drug resistance; MDR: the nonhuman counterpart of the MDR1 gene (definitions from Gottesman and Pastan[86])

ODC: ornithine decarboxylase

PAGE: polyacrylamide gel electrophoresis

PDBu: phorbol dibutyrate

PDGF: platelet-derived growth factor

PKC: protein kinase C

RT: reverse transcriptase

SDS: sodium dodecyl sulfate

SFU: syncytium-forming units

SSC: standard saline citrate

TBP-1: TAT-binding protein 1

TCDD: 2,3,7,8-tetrachlorodibenzo-*p*-dioxin

TGF: transforming growth factor

TPA: 12-*O*-tetradecanoylphorbol-13-acetate

TPK: tyrosine-specific protein kinase

XTT: 2,3-bis[2-methoxy-4-nitro-5-sulfophenyl]-5-[(phenylamino)carbonyl]-2*H*-tetrazolium hydroxide

13.2. Introduction

As should be expected, a large portion of this volume has been dedicated to products that are presently available through various biotechnological processes (e.g., human insulin, tissue plasminogen activator, interferons, and monoclonal antibodies). These substances have either demonstrated or theoretical utility in the treatment of human disease conditions and clearly will have increasing impact on the practice of contemporary pharmacy. From a pragmatic viewpoint, however, widespread clinical use of these products is generally limited by the necessity of parenteral administration. A direct method of circumventing this problem involves the development of novel drug-delivery systems (see Chapter 6, this volume). An alternative approach involves defining the structural and conformational characteristics of the active region of large-molecular-weight drugs and then producing relatively small molecular weight structural analogues that are anticipated to mediate similar biological responses (see Chapter 14, this volume).

Our efforts in the area of drug discovery have primarily involved the identification, isolation, and structure elucidation of secondary metabolites that are constituents of plants and other natural products. For individuals who have not previously been exposed to natural-product drug discovery programs, the rationale of such an endeavor may not be obvious. As an example, one may ask, Why should a plant produce a drug that is useful for the treatment of human cancer? An attempt to answer such a question is beyond the scope of this chapter, but it seems sufficient to point out that numerous essential drugs used in modern day medical practice are natural products. A few well-known examples are provided in Table 13.1.

In dealing with plants, given that conservative estimates suggest the existence of 250,000 species, a formidable task relates to the selection of starting materials. As has been described previously,[1] phytochemical screening or exploitation of chemotaxonomic relationships may be of some practical benefit, but these procedures are not directed toward the discovery of new therapeutically useful sub-

Table 13.1. *Plant-Derived Drugs of Known Structure Widely Used on a Global Basis*

Drug Name[a]	Chemical Class	Therapeutic Category
Atropine	Alkaloid	Anticholinergic
Codeine	Alkaloid	Antitussive/analgesic
Colchicine	Alkaloid	Antigout
Digitoxin	Cardiac glycoside	Cardiotonic
Digoxin	Cardiac glycoside	Cardiotonic
L-Dopa[b]	Amino acid	Parkinson's
Emetine[b]	Alkaloid	Antiamebic
Ephedrine[b]	Alkaloid	Bronchodilator
Vincristine	Alkaloid	Antitumor
Morphine	Alkaloid	Analgesic
Quinidine	Alkaloid	Antifibrillatory
Quinine	Alkaloid	Antimalarial
Reserpine	Alkaloid	Antihypertensive

[a]These drugs are all listed by the WHO as essential.

[b]These drugs can also be produced by chemical synthesis.

stances. Of course, serendipity may come into play, and these straightforward methods may lead to the discovery of extremely valuable therapeutic agents. For most drug discovery programs, however, it is not acceptable to await the occurrence of a serendipitous series of events that lead to success.

The approach generally regarded as most practical for novel drug discovery is referred to as bioassay-directed fractionation. In this approach, starting materials are selected (e.g., by random collection, information derived from ethnomedical systems of medicine, or literature surveillance[2]), and extracts are prepared that are suitable for biological evaluation. The materials are then tested in a bioassay system, and substances demonstrating a positive response are considered as active leads. After a number of active leads are identified, decisions are made to fractionate the most promising materials. Each fraction is monitored for the potential to mediate a positive response in the bioassay test system, and this process continues until a pure active substance is obtained. The resulting substance is then subjected to procedures of structure elucidation. Recent examples of this approach can be found in the literature.[3-7] Once an active isolate is obtained, more thorough biological evaluation procedures are often performed and, based on the accumulated data, the material is considered as a candidate for more advanced testing and development.

A question of paramount importance relates to the bioassay test system. In the area of antitumor drug discovery, a large number of in vitro test systems, and the issues that need to be considered when attempting to interrelate in vitro test results with in vivo efficacy studies, have recently been described.[8] In every case the "bottom line" should be whether the agent in question demonstrates therapeutic efficacy in humans. Therefore, the most certain method of achieving this goal

Table 13.2. Isolation of Biologically Active Substances Using Animal Models for Bioassay-Directed Fractionation

Sample	Test Material	No. of Animals[a]	Time Required (months)
Original extract	2 fractions	60	2
Column 1	10 fractions	260	2
Column 2	10 fractions	260	2
Column 3	10 fractions	260	2
Column 4	10 fractions	260	2
Column 5	10 fractions	260	2
Isolates	4	110	2
Totals		1,470	14

[a]Assumes five doses, five animals/group, positive and negative controls (no duplicate or confirmation tests).

would involve using human subjects from the outset to monitor the potential of a test substance to mediate a biological response. Under normal circumstances, to even suggest such a procedure is ludicrous.

An alternative test system could be an animal model that is ostensibly representative of a human disease condition. As summarized in Table 13.2, utilizing this approach, a "typical" isolation may require over one year to complete at the expense of well over 1,000 laboratory animals. It should also be recognized that this generalized example does not take into account potential differences between species or due to gender, or differences that may result from various modes of administration, dosage scheduling, and so forth. Thus, the numerical summary provided in Table 13.2 is probably grossly underestimated, and the use of a single animal model certainly does not assure success. Therefore, under normal circumstances, ethical considerations as well as fiscal constraints prohibit such procedures.

Factors such as these lead to the use of in vitro bioassay test systems. Traditional test systems dating back to the time of Ehrlich involve monitoring overt effects with intact organisms such as protozoa and bacteria. Recent efforts continue to monitor effects with intact organisms (protozoa, bacteria, mammalian cells, fungi, etc.) or the modulation of specific enzymatic reactions or metabolic pathways. There is certainly merit in these procedures.

More recently, advances in molecular biology have led to a greater understanding of the molecular basis of human disease states. As a correlate, in vitro systems that monitor a response that is either closely related to or identical with the molecular event yielding the disease condition can be devised. Alternatively, biotechnology can be used for the production of substances (such as enzymes) that can be used to monitor the activity of drug candidates. These systems can be employed to rapidly and inexpensively assess the biological potential of a large number of test materials, and several representative examples will be presented in this chapter. We will also describe some of the rationale involved in selecting novel biological test systems, as well as a few pragmatic considerations that inevitably need to be taken into account when devising or conducting such a research program.

13.3. Drug-Induced Modulation of Gene Expression

As will be illustrated by examples in this section, the deleterious effects of many pathological processes have been shown to originate at the level of gene expression. These advances are due to major technological advances over the past decade that have led to simplified methods for the manipulation and analysis of nucleic acids. Rather than limiting the investigation of new drugs to factors that are simply consequences of a disease state, it is now feasible to examine the direct effects of putative new drugs at the level of expression. While it should be noted that many clinically useful drugs act at the protein level (i.e., direct inhibition of an enzyme), it is still reasonable to assume that an agent capable of preventing or reversing a cellular event that occurs in the early stages of the disease state may be more efficacious than one that acts at a later stage; thus, evaluation of compounds that intervene at the level of gene expression is a logical approach for the discovery of new drug entities.

An early event in gene activation is the transcription of messenger RNA (mRNA) from a DNA template. The mRNA expressed in a cell closely reflects the activation status of that cell. Methods for isolating cellular RNA and transferring it to a solid support have now been firmly established, as have hybridization procedures with complementary DNA (cDNA) that enable accurate identification and quantitation of specific RNA transcripts.[9, 10] For purposes of drug discovery and evaluation, a routine assay to estimate the types and quantities of mRNAs expressed in cultured cell systems could be of great predictive value in assessing the therapeutic effects of potential new drugs. The analysis of mRNA expression subsequent to drug treatment has several advantages over the standard techniques of quantitating protein concentrations and enzyme activities, such as the following:

- Assays of enzymatic activity are highly dependent on incubation conditions (e.g., cofactors, pH, temperature), which can lead to variability in results.

- As was alluded to previously, since certain disease states presumably originate at the level of gene expression, evaluation of changes in transcription and translation of specific mRNAs are more direct measures of specific effectiveness when compared with assays of enzyme activities and protein concentrations (with the caveat that agents that interact only with the ultimate gene product or protein may go undetected).

- Unlike activity-based assays that must be developed and performed separately for each enzyme, a mRNA hybridization assay is more universally applicable, since its specificity changes solely on the basis of the cDNA probe that is employed. Cytosolic RNA can be prepared from cells that are treated with a given test substance and stored frozen for subsequent hybridization studies with different cDNA probes (an untreated control would be required in parallel). Therefore, one preparation or technique is sufficient for analysis with a wide range of cDNA probes, and this would greatly streamline the process of screening new compounds for different and possibly unrelated biological activities.

13.3.1. Applications of Hybridization Procedures in Drug Discovery Programs

13.3.1.1. RNA Isolation Methods and Cell Preparation for Hybridization

From a pragmatic viewpoint, one of the factors an investigator needs to be concerned with is the quantity of cells required to yield a sufficient quantity of mRNA for subsequent analysis. Another important consideration is the technical difficulty associated with the preparation. One of the earlier methods of total cell RNA isolation reported for hybridization protocols was that of Ross.[11] This method entails a phenol-chloroform extraction of RNA followed by CsCl gradient centrifugation. The technique has the advantage of giving a relatively pure RNA fraction, but it requires a large quantity of cells (i.e., tissue scale) and vanadate as an RNase inhibitor.

A less complicated method of preparing cellular RNA for hybridization involves guanidine-HCl precipitation of RNA as described by Chirgwin et al.[12] and later modified by Cheley and Anderson.[13] Guanidine-HCl (7.6 M in 0.1 M potassium acetate buffer, pH 5.0) is added to cells, and the suspension is homogenized, transferred to small sterile tubes, and treated with ethanol at $-20°C$ for 12 hours. The tubes are then centrifuged and the supernatant (containing DNA and protein) is removed and discarded. The RNA pellet is dissolved in formaldehyde and standard saline citrate and applied to nitrocellulose for hybridization. This method has been adapted for small multiple samples and is conservative of sample owing to the relatively few manipulations that are required. DNA and protein are removed from the sample during the course of the procedure, as is tRNA, since it is soluble in guanidine-HCl. This reagent also serves as an RNase inhibitor.

An alternative procedure is that recently reported by Pearse and Wu for the simultaneous preparation of RNA and DNA from a suspension of as few as 5×10^5 cells.[14] Briefly, Nonidet P-40 (NP-40) detergent is used to lyse the plasma membrane (without affecting the nuclear membrane), and the lysate is subjected to a low-speed centrifugation. Total cytoplasmic RNA can then be prepared from the supernatant, and DNA can be prepared from the nuclear pellet. These materials can be further processed and subjected to various procedures of electrophoresis and restriction enzyme digestions, or simply used for hybridization procedures.[14]

Although this procedure is straightforward and economical, the cytosol itself is all that is required for the analysis of RNA using hybridization procedures.[15] Therefore, similar to the method described previously, cells can be treated with NP-40 and centrifuged. The supernatant fraction, containing cytoplasmic mRNA, is denatured with formaldehyde in standard saline citrate buffer and directly applied to nitrocellulose filters for hybridization. The sensitivity of dot blot hybridization procedures achieved with this cytoplasmic lysate is about the same as that achieved with pure RNA, and the procedure can be performed with as few as 5×10^4 cells.

13.3.1.2. Immobilization of RNA Preparations on a Solid Support

Samples prepared as described previously can be serially diluted with $10-15\times$ standard saline citrate (SSC:0.15 M NaCl, 15 mM Na citrate, pH 7.0) buffer in a microtiter plate or applied directly to a sheet of nitrocellulose that has been prewet with $10-15\times$ SSC (approximately $1-10$ μg per dot). The high salt concentration is required for RNA to bind to the nitrocellulose membrane. The use of a commercially available filtration apparatus ensures rapid, uniform application of the samples in a reproducible pattern. If this type of apparatus is not available, samples can be applied directly to a prewet membrane placed on dry blotting filter paper.

To immobilize the RNA the sample-loaded membrane is baked in a vacuum oven at 80°C until dry (20–60 minutes). The dried membrane must then be equilibrated with hybridization buffer (at 42°C for at least 20 minutes) before addition of the specific probe. This "prehybridization" step ensures that the components of the hybridization buffer (bovine serum albumin, SDS, and low-molecular-weight DNA fragments) will block any active sites on the membrane so that nonspecific binding by the probe will be minimized.

13.3.1.3. Hybridization and Detection

One common procedure of detection involves the use of a ^{32}P-labeled cDNA probe. The probe is denatured and added to fresh hybridization buffer at a concentration of $0.1-50 \times 10^6$ cpm/ml, depending on the relative abundance of the sequence being analyzed. Hybridization should be carried out at 42°C for 12 to 18 hours. After hybridization the membrane is thoroughly washed [e.g., twice in $2\times$ SSC/0.1% SDS, 10 minutes at room temperature; twice in $0.1\times$ SSC/ 0.1% SDS, 10 minutes at room temperature, followed by 1 hour at a higher temperature (approximately 15°C below the T_m of the hybrid)]. Prehybridization and hybridization conditions should be optimized by adjusting temperature, ionic

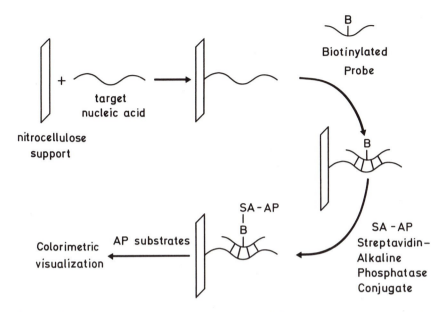

Figure 13.1. Schematic representation of the immobilization of target nucleic acid to nitrocellulose followed by hybridization with a biotinylated probe, binding of streptavidin conjugated with alkaline phosphatase, and colorimetric visualization.

strength, and the concentration of SDS and blocking agents to achieve the desired stringency of hybridization. Hybridized species are then detected by autoradiography. The washed, damp nitrocellulose membrane is wrapped in plastic and placed in a film cassette at $-70°C$ for 1 to 3 days, or until desired exposure is achieved. At this point, the membrane can be dried and stored, or treated with hot water to remove the first probe and rehybridized using another probe.

An alternative method of detection involves visualization using a streptavidin-biotin conjugate.[16] The probe is labeled via nick-translation, but using a biotin-derivatized deoxynucleotide triphosphate in place of a radiolabeled dNTP. These systems exploit the extraordinarily high binding affinity of the water-soluble vitamin, biotin, for the proteins avidin or streptavidin (a *Streptomyces*-derived biotin-binding protein with less nonspecific binding to nucleic acids).[17] After hybridization of the biotinylated probe to cellular mRNA, streptavidin coupled to an enzyme such as alkaline phosphatase is incubated with the blots. Each of the four identical subunits of streptavidin can bind biotin with extremely high affinity (about 10^{15} M^{-1}), and the conjugates are visualized by supplying a substrate for the enzyme that yields a colored product. A diagram of this reaction scheme is illustrated in Figure 13.1. There are significant advantages associated with this method of detection (e.g., no need to work with ^{32}P), but the quantity of probe required for similar sensitivity is greater than that required when using

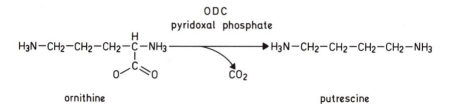

Figure 13.2. Conversion of ornithine to putrescine catalyzed by ornithine decarboxylase.

[32]P, and it is not possible to reevaluate the same preparation with secondary probes.

13.3.2. Quantitative Analysis of Ornithine Decarboxylase mRNA by RNA or DNA Hybridization: An Example

As an example of the potential usefulness of hybridization procedures, some studies conducted with ornithine decarboxylase (ODC) will be described. ODC catalyzes the reaction shown in Figure 13.2.

This is the first step in the synthesis of polyamines, and these products (such as spermidine and spermine) are known to be required for the proliferation of mammalian cells. Moreover, tumor cells characteristically have elevated levels of ODC and polyamines, as do cells stimulated by growth factors, hormones, and certain drugs.[18–21] One intensively studied compound known to elevate ODC activity is the tumor promoter TPA (12-O-tetradecanoylphorbol-13-acetate). As shown in Figure 13.3, experiments performed with cultured mouse epidermal 308 cells illustrate that one of the earliest cellular effects mediated by TPA is the induction of a rapid, transient increase in ODC activity. In addition to enzyme activity, protein and mRNA levels are elevated by treatment with TPA.[22–24] The increase in ODC mRNA is attributed to enhanced transcription rather than the stabilization of existing mRNA.[25]

Since the reaction catalyzed by ODC requires pyridoxal phosphate as a cofactor, any compound that interacts with pyridoxal phosphate is potentially capable of inhibiting the reaction. Obviously, this type of inhibition would not be specific for ODC activity. One important example of a specific ODC inhibitor is difluoromethylornithine (DFMO). DFMO has been found to prevent tumor promotion and progression in a number of in vitro test systems and with some animal tumor xenografts.[29, 30] DFMO has also been used in the treatment of parasitic diseases by inhibiting protozoal replication (e.g., *Trypanosoma brucei brucei*, *Plasmodium falciparum*, and *Pneumocystis carinii*). In addition, the inhibition of ODC by DFMO blocks retrovirus-induced transformation of murine erythroid precursors.[31] Therefore, compounds that inhibit the ODC metabolic pathway may be of therapeutic value.

One system that may be applicable for the discovery of such drugs has recently

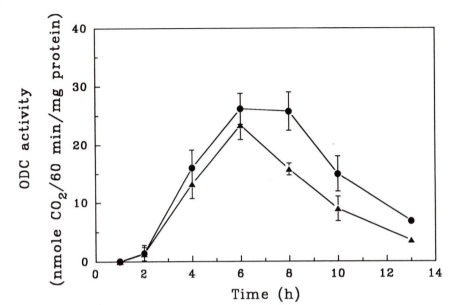

Figure 13.3. Mouse epidermal 308 cells[26, 27] were treated with PDBu or TPA (200 nM) to study the effect on ornithine decarboxylase activity.[28] Cells were plated in 24-well tissue culture plates and treated for the indicated time. Ornithine decarboxylase activity was measured in cell lysates (produced by two cycles of freeze-thawing) by providing L-[1-^{14}C]ornithine as a substrate in buffered medium replete with cofactors, and subsequently monitoring the evolution of $^{14}CO_2$. The $^{14}CO_2$ product of the enzymatic reaction was trapped on filter paper discs previously treated with 1 N NaOH and fitted tightly over each well of the tissue culture plate. Filter paper disks were then counted by liquid scintillation, and ODC activity was expressed as nmol CO_2/hour/mg protein. This assay is suitable for assessing the potential of test substances to inhibit phorbol ester-induced ornithine decarboxylase activity.

been described by Verma.[32] Primary cultures of newborn mouse epidermal cells were used to study the effect of retinoic acid on TPA-induced ODC mRNA. In brief, cells were treated with TPA, or TPA plus retinoic acid, in the presence of [³H]uridine. Therefore, all mRNA transcribed during the incubation period was radiolabeled. After incubation, the RNA was isolated and hybridized to a specific cDNA probe, pOD48 (unlabeled), which had been immobilized on a nitrocellulose filter.[33] ODC mRNA was then detected and quantitated by autoradiography. In this particular report, the assay permitted quantitation of retinoic acid-mediated inhibition of TPA-induced synthesis of ODC mRNA. However, the method is sufficiently general to be used for the identification of other agents that can inhibit TPA-induced ODC expression, although it would be more practical to use a continuous cell line such as T24 human epithelial[34] or mouse epidermal 308[26] cells.

Another possible method for detecting compounds that suppress ODC mRNA expression is standard dot-blot hybridization as described previously. Cell lines used for the assay could be plated with a very low serum concentration (so that basal ODC mRNA expression would be negligible) and treated with test compounds or TPA. Cytoplasmic extracts could be immobilized on nitrocellulose filters and probed with a ^{32}P- or biotin-labeled ODC cDNA, and visualized as described previously.

Gilmour and O'Brien have examined the TPA-induced accumulation of ODC mRNA in normal or transformed hamster fibroblasts.[35] Interestingly, by including cycloheximide in the incubation medium to block protein synthesis, a difference was found between the normal and transformed cells. ODC mRNA expression in the normal fibroblasts was dependent on protein synthesis, whereas the transformed cells continued to accumulate ODC mRNA irrespective of the presence of cycloheximide. This implies that the transformed cells have escaped the regulatory controls of the normal cells, although such results must be interpreted with caution, since other factors such as divergent pathways of gene regulation may affect the observations.[18, 36] Nonetheless, this differential regulation of normal and transformed cells could be exploited as a routine drug evaluation procedure. In essence, a test material could be evaluated for its potential to inhibit the expression of ODC mRNA in the presence and absence of cycloheximide. It may be speculated that an agent capable of inhibiting the expression of TPA-induced ODC mRNA in the presence of cycloheximide would be more selective against tumor cells, relative to normal cells. This could be a way of selecting compounds that would reverse tumor progression by restoring endogenous cellular control; agents functioning by this type of mechanism would be more selective for aberrant proliferation and demonstrate less nonspecific toxicity.

13.3.3. Nuclear Runoff Assays

As has been and will be noted elsewhere in this chapter, specificity is an extremely important aspect of drug action. In the case of transcription, it is not likely that any agent that does not specifically modulate expression (positively or negatively) will be of any interest. To determine whether an agent can regulate the transcription of a specific gene, a nuclear runoff assay can be employed.[37] This technique enables the level of transcription to be determined by allowing nascent mRNA transcripts (of one or more genes) to be elongated (or "run off") in the presence of radiolabeled nucleotides. The relative quantity of preinitiated transcripts associated with a particular gene at the time of nuclear DNA isolation is then estimated by hybridization of labeled mRNAs and cDNAs of interest, which have been bound to a support membrane using a Northern blotting protocol. Transcriptional regulation by phorbol esters, growth factors, and hormones have been investigated using this technology.[20, 34, 38] The basic procedure is as follows:

- Treat quiescent cells with the specific factors or compounds that are to be tested.

- Isolate nuclei from the cells at various time points after the addition of the test substance, and suspend the nuclei in a "runoff" buffer containing $[\alpha-^{32}P]UTP$ and an RNase inhibitor.

- Incubate the mixture for 15 to 30 minutes at a controlled temperature (30°–37°C).

- Terminate the reaction by the addition of RNase-free DNase I and *E. coli* tRNA; precipitate the RNA with ethanol.

- Extract the precipitate (^{32}P-labeled mRNA) with phenol and $CHCl_3$ (see previous discussion).

- Hybridize equivalent amounts of ^{32}P-labeled mRNA with various plasmid DNAs. (Plasmid DNAs complementary to mRNAs of interest can generally be obtained through individual investigators or commercial organizations. The material is linearized by restriction endonuclease digestion, denatured with 0.2 M NaOH, and spotted onto nitrocellulose.)

- Detect the presence of hybridized species (or the inhibition of transcription) by autoradiography (see previous discussion).

13.4. Oncogenes as Targets in Drug Therapy

The products of oncogene expression are generally thought to be causally associated with human cancers.[39, 40] For example, the product of oncogene expression is capable of inducing cell transformation in one or more cell lines, and a significant percentage of human tumors possess the abnormal transforming gene product of a cellular oncogene. Several specific examples of cellular oncogenes, their encoded protein products, and associated tumors follow:

- *src:* a tyrosine-specific protein kinase; gastrointestinal tumors[41]
- *erbB:* a cytoplasmic tyrosine-specific protein kinase domain of the EGF receptor; breast tumors[42]
- *ras:* a GTP-binding regulatory protein; multiple carcinomas[43, 44]
- *abl:* a tyrosine-specific protein kinase; chronic myelocytic leukemia[45]
- *myc:* nuclear DNA-binding protein; neuroblastoma, Burkitt's lymphoma[46]
- *neu:* a receptor-like, tyrosine-specific protein kinase; neuroblastoma[47]

Owing to correlations between oncogene expression and tumorigenesis, agents capable of limiting the expression or activity of oncogene products are of definite interest.[48] Such compounds could be identified through one of several possible screening strategies, several of which follow.

13.4.1. Specific Inhibition of Protein Expression (i.e., Translation) in Oncogene-Transformed Cultured Cells

The protein products induced by any cellular gene or oncogene can be quantitated by the technique of "Western" blotting. This method entails separating total cell proteins by polyacrylamide gel electrophoresis (PAGE),[49] transferring the separated proteins from the gel transversely to a solid support (such as nitrocellulose),[50] and identifying the protein(s) by incubating the support with radiolabeled or chromogen-labeled specific antibodies. Several oncogene-transformed cell lines are currently available, and the effect of potential inhibitors on the expression of oncogene product expression can readily be monitored using this procedure. Also, since DNA for many of the known oncogenes is commercially available, most cell lines of parochial interest can be transformed. However, to detect specific antagonists of oncogene protein induction, any changes in the level of expression of the oncogene protein product would need to be interpreted in terms of any change in nononcogenic proteins. To accomplish this, one or more constitutively expressed cellular proteins could be quantitated simultaneously in the transformed cells, thus enabling the development of a meaningful ratio of protein expression. Alternatively, to control for nonspecific inhibition of translation, bioassays could be conducted with an oncogenically transformed cell line concurrently with nontransformed cells of the same type.

13.4.2. Inhibition of Oncogene Transcription

As was described previously, the transcription of genomic DNA can be monitored by "Northern" hybridization technology. Therefore, using transformed cell lines, inhibitors of oncogene transcripts could be identified. Once again, an mRNA with a predictably constant rate of transcription (e.g., pHE-7 mRNA[51]) should be used to standardize the assay and to differentiate oncogene-specific antagonists from general transcription inhibitors.[51] Another possible method for attempting to differentiate nonspecific from specific suppression of transcription involves probing nontransformed cell lines (in addition to the transformed cells) with one or more cDNAs.

As examples of these types of experiments, effects on transcription have been studied with retinoic acid (with c-*myb* expression in human neuroblastoma cells)[52] and neplanocin A (with c-*myc* expression in human promyelocytic leukemia cells).[53]

13.4.3. Inhibition of the Activity Mediated by Oncogene Protein Products

This approach to screening for oncogene antagonists is based on the premise that the transforming abilities of oncogenic proteins are dependent on their functions, and that inhibition of function leads to the inhibition of oncogenic activity. What follows are two of many possible strategies for examining the inhibition of oncogene protein function.

13.4.3.1. Antagonism of the Interaction of the Oncogene Proteins with Cellular Receptors

A general description of these assays will be described in Section 13.7, Receptor Binding Assays.

13.4.3.2. Mechanism-Based Inhibition of Oncogene Action

One example of this approach is to search for inhibitors of tyrosine-specific protein kinase (TPK) activity. Since a major family of oncogenes codes for tyrosine-specific kinases,[54, 55] compounds that specifically suppress the activity of these kinases are potentially important drugs. If a particular protein kinase is available in purified form, the enzyme can be incubated with a known phosphate acceptor (substrate) in the presence of $[\gamma\text{-}^{32}P]ATP$ to measure its baseline activity. An agent included in the incubation medium may be a TPK antagonist if it decreases the incorporation of $^{32}PO_4$ into the substrate. Substrate phosphorylation can readily be quantitated by a number of procedures, including polyacrylamide gel electrophoresis (PAGE) followed by autoradiographic detection.[56] This technique has recently been employed to detect the inhibition of TPK activity by a series of flavonoid compounds.[57]

Given the central role of protein kinases in cell metabolism, it is likely that the therapeutic utility of any agents that function by TPK inhibition would be governed by the degree of their specificity. A protocol for the identification of specific protein substrates of TPKs has been outlined by Hunter.[39] This method makes use of two-dimensional polyacrylamide gel electrophoresis with an autoradiographic detection system. Oncogene-transformed cells in culture are incubated with $[\gamma\text{-}^{32}P]ATP$, allowing the substrates for all cellular kinases to become labeled with $^{32}PO_4$. Subjecting cells to electrophoresis by isoelectric point (in a urea-denatured polyacrylamide tube gel) followed by a second separation of these proteins through a sodium dodecyl sulfate PAGE slab gel yields a profile of total cellular protein;[58] each individual protein species is mapped to a unique two-dimensional location on the gel. While autoradiography at this point reveals all phosphoproteins produced during the radiolabeling period, an additional incubation of the final gel in alkali prior to autoradiography preferentially cleaves $^{32}PO_4$ from the phosphoserine residues. This step greatly reduces the number of phosphoprotein species on the autoradiograph; most of the remaining radiolabeled species are phosphorylated at tyrosine sites. By comparing analogous gels, with and without alkali treatment, specificity can be determined. This system seems appropriate for use as a screen for inhibitors of tyrosine-specific protein kinases, particularly since fast miniscale polyacrylamide gel electrophoresis equipment is now widely available.

13.4.4. Reversion of Oncogene-Transformed Cells to a Nontransformed Phenotype

Oncogenically transformed cells exhibit a characteristic phenotype that is distinct from their nontransformed counterparts.[41, 44, 59] Unconstrained by some of the regulatory controls of normal cells, transformed cells tend to proliferate more rapidly and the characteristic of contact inhibition may be lost (i.e., the cells continue to divide even when they are in direct physical contact with neighboring cells, resulting in an irregular monolayer punctuated with foci of stacked-up cells). Sometimes these cells lose their anchorage dependence altogether and begin to grow in suspension. Conscientious observation may allow some of these qualities to be exploited for a qualitative or semiquantitative assessment of oncogene expression. Some possibilities follow:

- Accelerated growth rates can be assessed by increased incorporation of [^{3}H]thymidine, [^{3}H]uridine, or [^{3}H]leucine into trichloroacetic acid–precipitable macromolecules (DNA, RNA, and protein, respectively).[60, 61] Alternatively, cell proliferation can be followed by measuring increases in total cell protein concentration.[61]

- The ability of transformed cells to form colonies in semisolid agar is well described and easily detectable.[60, 62]

- Morphological changes that accompany cell transformation can be monitored.

13.4.5. Use of Transgenic Animal Models

The introduction of selected DNA into fertilized mammalian ova (typically mouse) prior to cell division, and subsequent reimplantation of the transfected cells in utero, has led to the development of "transgenic" animals.[63] These animals can serve as dynamic models in which to study the expression of foreign genes or the altered expression of native genes. Neonatal mice possessing a transgene can be selected by gross phenotypic characteristics or by the direct identification of specific DNA or mRNA, respectively, on Southern or Northern blots, using a preparation derived from a small section of tail tissue.[64] Oncogenes such as H-ras[65] and c-myc[46] have been employed as transgenes, and tumor manifestation is observed in the transgenic animals. The commercial marketing of a patented, myc-transfected mouse by I. E. Du Pont[66] is notable and illustrates the importance of this technology in prospective studies of oncogene function. A reasonable bioassay to detect agents with the ability to suppress disease-associated gene expression could involve the administration of putative oncogene antagonists to transgenic animals, as well as to their nontransgenic counterparts, followed by analysis of phenotypic and genotypic effects. With judicious selection and administration of suspected oncogene antagonists, this type of assay could serve as a late-phase screen for specific inhibitors of oncogene expression that showed promise in other (less rigorous) bioassays.[8]

13.4.6. Positive Control Substances for Studies with Oncogenes

As a general policy, any experimental procedure that is conducted with a substance of unknown activity requires the simultaneous evaluation of a "positive control" substance. This convention assures that the assay was properly conducted and is capable of demonstrating the desired effect. However, if no drug is known that demonstrates the type of response for which the investigator is searching, the procedure becomes ambiguous. Fortunately, in absolute terms, this is rarely the case. For example, in studying the expression of ODC mRNA, retinoic acid would be a suitable assay control. In the case of oncogene expression, the answer is not as straightforward. One possibility involves the use of antisense mRNA transcripts that are complementary to the oncogene message. In some of the assays described previously, specific antisense mRNAs could antagonize the expression of specific proteins observed in Western blots, or inhibit the activity of oncogenic proteins that are monitored.[67–69]

13.5. Analysis of Expression Utilizing Luciferase or Chloramphenicol Acetyltransferase (CAT) Sequences as Reporter Genes

Inducible regulation of gene promoter sequences can be exquisitely investigated through the use of "reporter" genes such as those that code for chloramphenicol acetyltransferase (CAT) or luciferase. Expression vectors have been constructed possessing the 5'-regulatory region of the gene to be studied in a position that is upstream of the coding sequences for CAT or luciferase.[70, 71] The expression of the reporter gene is thus under the regulation of the specific promoter sequence of interest, and since these enzymes are not endogenous to mammalian cells, their biological activities directly correlate with induction of the promoters that precede them. The activity of CAT is measured by resistance to chloramphenicol, whereas the activity of luciferase is measured by the reactions shown in Figure 13.4.

Three examples will be given that show how this technology applies in the area of drug discovery.

13.5.1. Inducible Regulation of the Human-c-erbB2/neu Promoter

One example is the inducible regulation of the human c-*erb*B2/*neu* promoter. To study the regulation of this particular promotor, a 1,500-base pair fragment

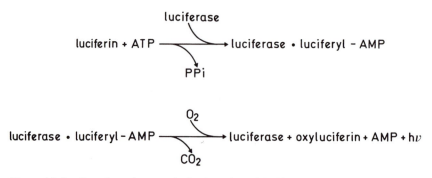

Figure 13.4. Reactions that permit the detection of luciferase activity.

derived from the gene was inserted immediately 5′ to the ATG initiator codon of an expression vector, along with the coding region of the luciferase gene. HeLa cells transfected with this expression vector were then treated with various substances to evaluate their potential for activating the c-*erb*B2/*neu* promoter.[72] As judged by luciferase activity, epidermal growth factor (EGF) activated gene transcription, as did TPA, dibutyryl cyclic AMP, and retinoic acid. Since overexpression of p185, the protein product of this gene, has been associated with cell transformation and tumorigenesis,[73] inhibition of p185 would be one mechanism by which an antitumor agent could function. This system could be used as a monitor for such an effect. Interestingly, the inducers of the c-*erb*B2/*neu* promoter listed previously were found to have synergistic effects on gene transcription, that is, the highest level of luciferase activity was observed when transfected cells were exposed simultaneously to all four compounds. These results are significant, since they imply that multiple mechanisms are involved in the process of inducing the expression of this gene product. Inhibitors of this activation would likely be of value in defining these various mechanisms.

13.5.2. Ligand-Receptor Interactions and Modulation of Hormone Responsive Elements

Once the receptors for steroid or thyroid hormones are occupied by their respective ligands, it is well known that they transduce signals by interacting with DNA and altering the transcription of specific genes.[74, 75] The sequence of DNA to which the receptor binds is known as a hormone responsive element (HRE), and this has been thoroughly investigated.[76–79] Advantage can be taken of this information to study the potential of test substances to interact with DNA at unique promoter regions, compete with ligand-receptor complexes for binding to a HRE, or antagonize ligand binding to receptor. Test substances found to be active in these processes would be expected to profoundly affect cellular metabolism. The technology involving the use of reporter genes is applicable to the discovery of such substances.

One straightforward system involves ligand–receptor interactions that can be monitored by the use of reporter genes. For example, a system could be devised wherein the expression of the luciferase gene could be under the control of a HRE that is activated by the estrogen–estrogen receptor complex. Therefore, luciferase activity would be elevated by treatment with estrogen, and the potential of test substances to inhibit the expression of estrogen-induced luciferase could be used as an assay for antiestrogenic activity. Using this type of methodology, it was demonstrated that the protein product of the v-*erb*A oncogene can bind to the thyroid HRE (in the absence of thyroid hormone) and thereby inhibit the normal hormone-activated transcription of DNA.[80] Since expression of the v-*erb*A oncogene product is associated with the development of erythroleukemia, this illustrates the influence that modifying the expression of a limited number of genes may have on the incidence of disease.

13.5.3. Regulation of tat Gene Expression

Specific control of *tat* gene expression will be given as a final example of this approach. The product of the human immunodeficiency virus (HIV) *tat* gene is a protein, designated Tat, that regulates expression of the remaining HIV genome. Tat interacts with viral RNA in a specific region of the long terminal repeat (LTR)[81] and causes transactivation of the gene by an unknown mechanism. Clearly, any factor that is able to interfere with this Tat–LTR interaction might be deleterious to the survival of the virus, so it is reasonable to investigate agents capable of mediating such an effect as potential antiviral agents.

Based on the premise that endogenous proteins capable of interacting with Tat would alter Tat-mediated gene transactivation, Nelbock et al. screened the proteins expressed from a λgt11 fusion cDNA library.[82] These proteins were adsorbed to nitrocellulose and probed with biotinylated Tat; those proteins capable of binding Tat were identified by introducing a streptavidin-alkaline phosphatase conjugate followed by a chromogenic substrate (see Figure 13.1). Ten Tat-hybridizing proteins were detected in a screen of 2×10^6 phages, and the first of these proteins to be further characterized was designated Tat-binding protein-1 (TBP-1). Chinese hamster ovary cells were transfected with the following expression vectors: (1) plasmid pSV-TBP-1, which expressed TBP-1 (under the control of the SV40 promoter); (2) plasmid pU3R-III, which expressed CAT (under the control of the HIV-1 LTR region); and (3) plasmid pSV-*tat*, which expressed the HIV-1 *tat* gene (under the control of the SV40 promoter). As should be anticipated, maximum CAT activity was obtained when (2) and (3) were cotransfected. This simply indicates transactivation of the HIV promoter by Tat. When (1) was cotransfected in addition to (2) and (3), a significant decrease in CAT activity was observed, and this decrease was dependent on the amount of plasmid (1) used to treat the cells. This result implies that the expression of a protein, TBP-1, which specifically interacts with Tat, can suppress Tat-mediated transactivation

Table 13.3. Relative Cytotoxic Potential of Several Agents Determined with KB (Normal) or KB-VI (Drug-Resistant) Cells in Culture[a]

Agent Tested	ED_{50} (μg/ml)	
	KB	KB-VI
Bruceantin	0.00007	0.62
Colchicine	0.0033	0.67
Homoharringtonine	0.00073	0.23
Taxol	0.00041	2.26
Vinblastine	0.0066	1.89
Vincristine	0.0064	9.61

[a]For a description of the procedure, see Swanson and Pezzuto.[61]

of the HIV promoter. It may therefore be proposed that this procedure is useful for identifying proteins (or other agents) that can bind to Tat and modulate its regulatory effects on HIV gene expression. This strategy could easily be generalized to include other DNA-binding proteins that have regulatory effects on gene expression.

13.6. Multiple Drug Resistance Phenotype as a Therapeutic Target

Multiple drug resistance has emerged as a dilemma in which the effective treatment of a variety of diseases including human cancer and malaria is seriously impaired.[83-87] Multidrug-resistant cells are characterized by resistance to the cytotoxic effects of several drugs that are structurally unrelated except for the fact that they are often hydrophobic natural products[86, 88] (see Table 13.3). A specific glycoprotein has frequently been found to be present on the plasma membrane of multidrug-resistant cells,[89-92] the "p-glycoprotein." This glycoprotein appears to be at least partially responsible for conferring drug-resistance to cells in which it is expressed, possibly by acting as an ATP-driven drug "pump" that actively prevents the intracellular accumulation of various chemotherapeutic agents.[93-96]

The suggestion that p-glycoprotein confers multidrug resistance is further supported by experiments in which drug-sensitive cells transfected with the *MDR1* gene (which codes for p-glycoprotein) demonstrate the response indicative of the MDR phenotype.[97-101] It has also been found that genomic[97, 102] or chromosomal[103] DNA from multidrug-resistant cells can be used to confer the MDR phenotype to drug-sensitive cell lines. More recently, plasmid vectors containing full-length cDNAs of the coding region for human *MDR1* have been constructed. Ueda et al. have inserted a 4.4-kilobase fragment of *MDR1* into the eukaryotic expression vector pGV16, which contains LTRs of the Moloney murine leukemia virus (to promote *MDR1* expression), and a G418-resistance gene (useful for selection procedures).[100] This expression plasmid, designated pGMDR, was transfected into drug-sensitive NIH 3T3 fibroblasts. Transformants were selected for resistance to G418, and these positives were then selected for colchicine resistance. All of the G418 and colchicine-resistant positives were found to contain the human *MDR1* gene, implying that the transfected gene was responsible for conferring colchicine-resistance. Southern blots reveal the overexpression of the *MDR1* gene in multidrug-resistant transformants; under hybridization conditions of high stringency, the *MDR1* DNA can be shown to originate from the transfecting cell line, not from amplification of endogenous DNA in the transformant.[102] As verified by the use of Western blots, cells having acquired the MDR phenotype overexpress p-glycoprotein.

On the basis of this fundamental information, it seems logical to anticipate that inhibition of the expression or activity of p-glycoprotein could prove useful in the treatment of certain disease conditions.[104] Several substances have already been identified as inhibitors of p-glycoprotein function (e.g., verapamil, trifluoperazine, cytochalasin B, quinidine, and chloroquine),[95, 105–109] but ancillary toxicity has thus far limited clinical utility.

Several specific strategies that are applicable for the discovery of drugs capable of interfering with *MDR1*-induced resistance follow.

13.6.1. Cell Survival Assays

Standard cytotoxicity assays (e.g., based on dye exclusion or metabolism or cell proliferation) can be used to determine the dose of known drugs required to inhibit cell growth by 50% (ED_{50}) in wild-type and multidrug-resistant cells in the absence or presence of potential antagonists of p-glycoprotein. For example, vinblastine-resistant KB-V1 (human epidermoid carcinoma) cells and the drug-sensitive parent line, KB, might be treated with vinblastine in the presence and absence of suspected reversing agents. A straightforward example of this procedure has been reported by Bellamy et al. wherein cell viability was assessed by metabolic conversion of a tetrazolium dye.[110] The data clearly revealed a reversion of doxorubicin-resistant myeloma cells to drug sensitivity that was related to the dose of verapamil added to the incubation medium. Unambiguous results can be obtained when the concentration of test compound (i.e., verapamil) that reverses drug resistance is not cytotoxic itself.

While a tetrazolium dye-based assay may be devised to yield results in a relatively short period of time, there may be some advantage to evaluating cytotoxicity in the presence of drug and potential MDR-reversing agent using an end point that requires cell proliferation. Most commonly, a standardized number of cells are inoculated into a series of culture dishes or tubes and treated with drugs to which the cells are resistant or potential reversing compounds according to the protocol outlined previously. After several days of culture, the cells are harvested and cell number or total protein is determined. This type of assay has the advantage of disclosing MDR antagonists that act as nonspecific metabolic inhibitors.

The availability of an expression vector such as pGMDR allows any cell line to be transformed to a multidrug-resistant type. In testing procedures using these transformed cells, results obtained with the wild-type cells serve as controls to differentiate between nonspecific cytotoxicity and selective MDR antagonism. An additional possibility, of course, is to generate drug-resistant cell lines by chronic treatment with escalating doses of a particular cytotoxic drug, retaining the parent cell line for use as a control.

13.6.2. Direct Determination of Drug Accumulation within MDR Cells

The most widely used procedure for studying the MDR phenomenon involves monitoring the intracellular retention of a drug. The procedure is greatly facilitated by the use of radiolabeled drugs, either commercially available or prepared through a custom synthesis. The protocol for measuring drug accumulation is basic and generally similar throughout the field: multidrug-resistant cells (or the corresponding wild-type line) are incubated in a medium containing radiolabeled drug, in the presence or absence of a putative MDR-reversing agent.[111] The cells are collected by centrifugation or filtration, and after additional washing procedures the isolate is subjected to liquid scintillation counting. This procedure reveals the amount of radiolabeled drug retained by the cells, and MDR antagonists found to enhance intracellular accumulation of the labeled drug are generally evaluated in a time- or dose-dependent manner.

Rather than measuring radioactivity, determination of fluorescence can be an effective method of monitoring the retention of drugs by wild-type and multidrug-resistant cells. If the primary drug used to select for the resistant cells is itself fluorescent (e.g., doxorubicin, adriamycin), it constitutes a logical choice for the assay.[112, 113] If this is not the case, however, the multidrug-resistant phenotype can be exploited. Many MDR cells have been observed to exhibit cross-resistance to fluorescent compounds such as rhodamine 123, ethidium bromide, and acridine orange.[114, 115] These dyes can be conveniently utilized to study drug resistance by (1) microscopic observation of intracellular fluorescence, (2) fluorescence-activated cell sorting (FACS analysis), or (3) extraction of the dye from cells and measurement of relative fluorescence directly with a spectrophotofluorometer[116] or by high-performance liquid chromatography.[110, 112]

13.6.3. Competitive Binding with p-Glycoprotein

Many drugs that reverse the multidrug-resistant phenotype have been found to bind to the p-glycoprotein, and this binding competes with the binding of drugs to which the cells are resistant, such as vinblastine.[95, 117, 118] Therefore, the evaluation of test substances to competitively inhibit the specific binding of vinblastine to multidrug-resistant cells (or cell membranes or semipurified p-glycoprotein) is a method of identifying potential MDR reversing agents. As an example, Cornwell et al. described the ability of calcium-channel blockers (which suppress the MDR phenotype) to inhibit the binding of an ^{125}I-labeled vinblastine photoaffinity analogue to membrane vesicles obtained from a multidrug-resistant human KB cell line.[95] Agents known to reverse multiple drug resistance have also been demonstrated to inhibit the binding of a photoaffinity label (an azido analogue of vinblastine) to the p-glycoprotein.[117] These procedures are therefore applicable for the general evaluation of reversing agents (also see Section 13.8, Receptor Binding Assays).

13.6.4. Evaluation of MDR Gene Expression

Antagonists of multiple drug resistance that inhibit transcription or translation of the *MDR1* gene could also be analyzed. The MDR phenotype has been studied by hybridizing total cellular mRNA with oligonucleotide or cDNA probes.[97, 102] Given the likelihood that transcriptional control plays a major role in distinguishing normal expression from aberrant expression of p-glycoprotein, Northern hybridization assays could prove to be powerful tools for the discovery of MDR-reversing agents. Alternatively, the quantity of p-glycoprotein could be directly measured. There are many examples of p-glycoprotein analysis using polyacrylamide gel electrophoresis (PAGE)[49] and Coomassie blue or silver staining, or autoradiography.[95, 119–121] However, the electrophoretic mobility of p-glycoprotein is variable,[120] and positive identification using this method may be difficult. Two-dimensional PAGE often affords a significantly greater separation of proteins than does a SDS-PAGE slab gel;[58] however, there have been reported difficulties in adequately resolving p-glycoprotein on 2-D gels, despite several different means of attempted visualization.[119, 122] For these reasons, Western blotting is one of the more reliable methods of detecting p-glycoprotein. In brief, this involves SDS-PAGE followed by electrophoretic transfer of the separated proteins to nitrocellulose. The MDR protein is then specifically localized with radiolabeled antibodies.[121, 123, 124] Once these analytical procedures have been standardized, screening assays for the detection of MDR-reversing agents (that act by influencing the translation or stability of p-glycoprotein) can be established.

13.6.5. Additional Factors Influencing the Success of MDR-Reversing Agents

Unlike aberrant oncogenic proteins, p-glycoprotein has been found to be expressed on the surface of a large number of "normal" human cell types, albeit at a much lower level than is observed in multidrug-resistant cells.[86, 125, 126] Although its function in "non-MDR" cells is still speculative,[97] the fact that p-glycoprotein is widely conserved and expressed in these tissues suggests that it has some fundamental metabolic role and that its transcription and translation are under some degree of normal cellular regulation. The observation that multidrug-resistant cells tend to revert to the nonresistant, wild-type phenotype when not continually exposed to a drug is further evidence that p-glycoprotein expression is under transcriptional control.[127] On the one hand, this suggests that novel endogenous or exogenous regulators of p-glycoprotein transcription are potentially important therapeutic agents. On the other, it is likely that the "MDR phenotype" of normal cells helps to reduce drug-induced toxicity. Therefore, an agent that reverses the MDR phenotype may be expected to enhance drug-induced toxicity by blocking the mechanism that would otherwise tend to protect normal cells.

Of additional interest are recent experiments in which various drug-sensitive cell lines expressing relatively low basal levels of p-glycoprotein were treated with differentiating agents such as sodium butyrate or retinoic acid.[128, 129] These agents enhanced the expression of p-glycoprotein as determined by Northern and Western blots; however, increased expression did not invariably correlate with the appearance of the multidrug-resistance phenotype. For example, while certain differentiating agents substantially increased the expression of p-glycoprotein in two colon carcinoma cell lines, SW-620 and HCT-15, increased resistance to vinblastine (and decreased intracellular accumulation) could only be demonstrated with HCT-15 cells. These findings serve as a caution against the assumption that p-glycoprotein expression is invariably accompanied by the multidrug-resistant phenotype, but also suggest that multiple biochemical avenues are available for therapeutic intervention.

Finally, although p-glycoprotein is certainly an important factor in multidrug resistance, it is not likely to be solely responsible for evoking the phenomenon in all drug-resistant cells and with all selecting agents.[130] Its expression in cells with normal drug sensitivity, coupled with the fact that many other proteins are known to be associated with the MDR phenotype,[119, 131–134] suggests that protocols intended to screen for antagonists of multidrug resistance should incorporate at least one general assay for MDR reversal that is not dependent on changes in the *MDR1* gene or its expression. A multifaceted scheme[110] for the discovery of novel compounds that reverse multiple drug resistance is likely to have a higher potential for predicting clinically important drugs as compared with a one-dimensional approach based on a single theory of drug resistance.

13.7. Methods for the Discovery of Antiviral Compounds

The search for new antiviral compounds has historically been based on examining the relationship between the live virus and its host cell. These assays have used a variety of end points to determine the pathological effects of the virus,[135] such as visual observation of cytopathic effects (CPE)[136] and viral plaque reduction.[137, 138] One plaque reduction assay that employs a non-substratum-dependent cell[139] has recently been utilized as a screen for antiviral compounds.[140, 141] Another recent improvement in the plaque reduction assay eliminates the agar overlay of the infected cultures by performing the experiments in microtiter plates;[142] this technique proved amenable to large-scale screening of new compounds for antiviral activity. Assays conducted with intact ("whole") virus and viable mammalian cells are obviously important, since they are capable of simultaneously monitoring numerous complex interactions between the virus and its host cell. The fundamental nature of viral invasion and subsequent manipulation of host cell metabolism for its own replication necessitates the involvement of cells at some point in the screening process; these assays can also be manipulated to differentiate cytotoxins from true antiviral agents, and to detect agents that may prevent infectivity as well as viral proliferation.[142]

However, a significant drawback of this type of antiviral assay is the requirement of using viable, infectious virus particles. Although suitable precautions can be implemented, it is logical to minimize the handling of an agent such as viable HIV. Some options for antiviral screening protocols have resulted from a better understanding of the etiology of AIDS and the elucidation of viral mechanisms in general. Advances in biotechnology have simultaneously provided the tools and techniques to permit the implementation of new antiviral screening systems. What follows are several specific approaches for the discovery of antiviral agents.

13.7.1. Inhibition of Cytopathic Effects Induced by HIV

As was described previously, the classical approach of evaluating antiviral potential involves the determination of viral-induced cytopathic effects. This section will be used solely to describe some screening procedures that are currently used to search for anti-HIV agents. However, the general approach should be applicable with other viruses as well.

13.7.1.1. Cells and Viruses

Viral stocks can be maintained in cultured cells [e.g., H9 (human T cells) infected with HIV-1 (RF)]. Cell cultures are diluted each week (e.g., to a concentration of 1×10^5 cells/ml), and at the time of dilution, uninfected cells are added. Centrifugation of infected cultures yields supernatant fractions that contain virus particles, and these fractions can be stored in liquid nitrogen. The virus stocks obtained in this manner can be evaluated by a syncytium assay (described below), and the number of infectious particles can be expressed in syncytium-forming units (SFU)/ml. Additional strains of HIV-1 (e.g., B, CC, MN) and various cell lines [e.g., CEM-SS (human acute lymphoblastic leukemia), MT-2 (human T cell leukemia cell), and MOLT-3 (human peripheral blood, ALL)] can be used for these test procedures.

13.7.1.2. Evaluation of the Potential of Test Substances to Inhibit Cytopathic Effects Mediated by HIV-1

For these evaluations, a semiautomated protocol can be used.[143–145] In brief, cells are pretreated with polybrene, followed by treatment with cell-free virus for a period of 1 hour at 37°C. The ratio of infectivity (i.e., number of infectious virus particles/number of target cells, based on syncytium analysis) varies with different cell types (e.g., 0.005 for MT-2 cells, 0.05 for CEM-SS cells). Cells are then distributed to microtiter wells (0.1 ml/well, containing 5×10^3 cells) to which various concentrations of test substances (0.1 ml/well) have previously been added. The plates are then permitted to incubate at 37°C in a humidified incubator containing 5% CO_2 in air for a period of 7 days. At this time, 0.05 ml of a solution containing 2,3-bis[2-methoxy-4-nitro-5-sulfophenyl]-5-[(phenylamino)carbonyl]-2H-tetrazolium hydroxide (XTT) and N-methylphenazonium methosulfate is added.[146] Following an additional incubation period of 4 hours, the plates are covered with adhesive plate sealers and absorbance is determined at 450 nm. Results are plotted as a function of drug concentration, and these plots are used to determine the concentration of drug required to decrease XTT formazan production to a level of 50% of that of untreated, uninfected control cells (EC_{50}).

In addition to the tests described previously, incubations are performed in which cells are treated with various concentrations of the test substances but not the virus. With the exception of treatment with the virus, all procedures are the same as previously. Plots are then constructed as a function of drug concentration, and the concentration of drug required to inhibit cell growth (XTT formazan production) by 50% is determined (IC_{50}).

For each of these studies, a positive control substance (e.g., AZT) is evaluated at the same time as the test substances. The information that is generally of greatest importance is the ratio of the EC_{50} and IC_{50} values (the in vitro "therapeutic index").

13.7.1.3. Evaluation of the Potential of Test Substances to Inhibit Syncytia Formation That Is Mediated by HIV-1

MOLT-3 cells (5×10^5 cells/ml) are added to various concentrations of test compound, and the mixture is treated with a suspension containing HIV-1 (2.5–5×10^8 virus particles). After an incubation period of 96 hours, the cells are triturated to produce an even suspension, and the number of syncytia are determined by counting. Each determination is performed in duplicate, and several fields are counted to assess the number of syncytia.[147, 148] Cell survival as a function of drug concentration is also assessed.

13.7.1.4. Evaluation of Test Substances to Inhibit p24 Core
Antigen Production

This procedure may be used to further assess the activity demonstrated by promising leads. Cells are first incubated with test substance (e.g., 30–45 minutes) and then treated with medium containing HIV of known titer (e.g., 1 hour). The cells are then washed, incubated (3–4 days), and centrifuged. Aliquots of the supernatant fractions are then treated with a 5% Triton X-100 solution at ambient temperature (30 minutes). The quantity of HIV p24 antigen can then be determined using a commercially available ELISA kit. This briefly entails dispensing aliquots of standards and test solutions into 96-well plates (coated with purified heat-treated human antibody to HIV) and incubating overnight (ambient temperature). The solutions contained in each of the wells are then removed and the plates are washed (six times). Aliquots of HIV detector antibody (murine monoclonal antibodies to HIV p24 antigen that have been coupled to biotin) are then added to each of the wells, and the plate is incubated (2 hours, 37°C). The wells are then aspirated, washed (six times), and treated with streptavidin-peroxidase reagent (15 minutes, 37°C). The wells are again aspirated, washed (six times), and treated with substrate solution (tetramethylbenzidine). After 30 minutes, the reaction is stopped by the addition of sulfuric acid, and absorbance is determined at 450 nm using a microwell plate reader. Finally, the concentration of HIV p24 antigen in the samples is determined by comparison with the standards (tested at five concentrations ranging from 0.025 to 0.4 ng/ml).

13.7.2. Inhibition of Retroviral Reverse Transcriptase Activity

Reverse transcriptase (RT) catalyzes the production of DNA from an RNA template. Since this activity is of pivotal importance in the replicative cycle of retroviruses, agents demonstrating selective inhibition may eventually become part of the armamentarium employed against retrovirus-induced human diseases.[149–151] Reverse transcriptase can be obtained from various retroviruses in pure form. Of particular importance, however, is the availability of recombinant RT from HIV-1; this material has enabled large-scale screening for inhibitors.[152] The preparation will be described in some detail, since the procedures are generally applicable for the provision of a variety of viral (and other) proteins. In nature, mature RT protein is produced by the proteolytic cleavage of a larger *gag–pol* polyprotein. Therefore, to produce a recombinant RT, the appropriate coding region must be modified to contain the signals that are necessary for the initiation and termination of transcription. This construct was prepared and ligated into the pUC plasmid pUC12N. The recombinant plasmid was then introduced into *E. coli* strain DH5,[153] and the transformed bacteria were used for the expression of large quantities of an active HIV-1 reverse transcriptase, having a molecular weight of 66,000.[154, 155]

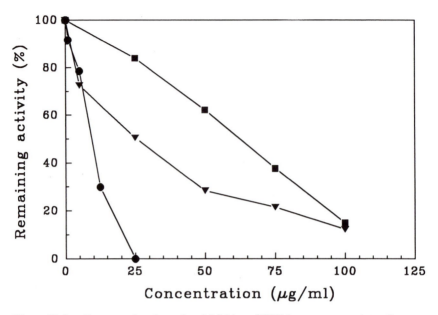

Figure 13.5. Concentration-dependent inhibition of HIV-1 reverse transcriptase by treatment with fagaronine (●), suramin (▲), or columbamine (■). The assays were performed as described by Tan et al.[152]

The catalytic activity of HIV-1 reverse transcriptase is determined by measuring the rate of transcription.[152] Transcription can be monitored under a variety of experimental conditions, but a common procedure involves incubating the enzyme in buffer containing a template ribonucleic acid (polyadenine [poly(A)]), a complementary oligonucleotide primer (oligodeoxythymidine), nucleotide [thymidine triphosphate (TTP)], and radiolabeled nucleotide (e.g., [^{3}H]TTP). This mixture is incubated at 37°C for a specified time before terminating the reaction by the addition of EDTA. Aliquots of the reaction mixture are spotted onto anion exchange filter paper, and these are dried and washed. The dried filters are then subjected to liquid scintillation counting to estimate the amount of [^{3}H]TTP incorporated. This assay can be used to screen for reverse transcriptase inhibitors; putative antagonists included in the reaction buffer should demonstrate a dose-dependent decrease in the incorporation of [^{3}H]TTP (see Figure 13.5).

13.7.3. Inhibition of Binding of Viral Coat Proteins to Target Cell Receptors

Enveloped viruses display surface glycoproteins that have high affinity for the plasma membrane proteins of their host cells. This "docking" of virus to cell appears to be essential for subsequent intracellular invasion; thus, antagonizing this interaction would seem to be an attractive antiviral strategy. For example, the CD4 receptor molecule of human helper T cells is known to be a site to which the HIV-1 envelope protein gp120 binds, and soluble forms of the CD4 receptor molecule have been shown to reduce binding of HIV-1 gp120 to cellular CD4.[156] Another example of this approach was reported by Sperber and Hayden;[157] monoclonal antibodies directed toward the cellular rhinovirus receptor were highly effective in blocking subsequent infection.

Bioassays to screen for inhibitors of viral binding to target cells can be designed using noninfective, purified viral envelope proteins, and cellular proteins, in either purified form or as a component of whole cell membranes. For example, whole cells, expressing the viral receptor molecule, can be seeded directly into microtiter plates and then treated with the viral component capable of binding to the receptor, in the presence or absence of a potential inhibitor. One method of determining the extent to which the two molecules interact involves the subsequent addition of a biotinylated antibody capable of reacting with the purified viral component. Assuming that the viral component is bound to the cell (not displaced by an inhibitor), this antibody would also remain bound to the complex. By the use of an avidin-conjugated dye or a chromogenic substrate procedure, the biotinylated antibody could easily be measured with a microtiter-adapted spectrophotometer. Since the resulting optical density is proportional to specific binding, an agent inhibiting the specific interaction of the viral component with the cell receptor would be expected to yield a corresponding decrease.

If both the viral and cellular proteins are pure, a similar procedure could be employed by selectively immobilizing one (either) of the components. One example of an immobilization procedure is the addition of a selected component to a red blood cell monolayer that had previously been treated with an antibody capable of binding to and retaining that component.

13.7.4. Inhibition of Viral Proteinases

All retroviruses consist of three major genes, *gag, pol,* and *env; gag* codes for structural proteins of the nucleocapsid, *pol* gives rise to replicative enzymes, and *env* encodes glycoproteins of the viral envelope. These virally encoded proteins are frequently translated as part of a large polypeptide precursor that requires proteolytic processing to yield mature protein species.[158] In addition to nucleases and reverse transcriptase, the *gag–pol* gene also codes for the protease(s) that is (are) responsible for this controlled, limited processing. To define specific proteolytic cleavage sites, the amino acid sequences of proteolytic fragments (generated from larger precursors) are directly determined.[159, 160] All the retroviral proteinases characterized thus far appear to be aspartic acid proteases. On the basis of this information, it is possible to establish bioassay screens for antagonists of viral proteases. Typically, a polypeptide substrate is prepared that contains a known cleavage site (nearly always Asp-Thr-Gly), and the quantity of a specific cleavage fragment (possibly radiolabeled), with respect to the concentration of a potential protease inhibitor, is monitored by HPLC or FPLC.

13.7.5. Inhibition of Other Virally Encoded Proteins

In addition to structural proteins and enzymes, viruses are known to code for proteins that are involved in the regulation of replication. Specific inhibition of the expression of these viral products is an obvious method of attempting to achieve antiviral activity. As was described in greater detail previously (Section 13.4), the *tat* gene of HIV-1 codes for a regulatory protein (Tat) that exerts positive control over the transcription of other viral genes, and one specific strategy is to discover agents that can interact with and inhibit Tat-mediated transactivation. A related strategy would be to identify a specific inhibitor of *tat* expression; such an agent would also be predicted to antagonize viral proliferation. As a screening procedure, Northern and Western blotting techniques could be employed to detect decreases in the transcription or translation of the viral genome, respectively, in the presence of putative antiviral agents. To control for compounds that are nonspecific inhibitors, the expression of one or more constitutive genes in the host genome could be simultaneously monitored (see Section 13.2).

13.8. Receptor Binding Assays

There are myriad molecules that mediate a biological response via interaction with a cellular receptor. A few interactions of this type have already been described in this chapter (e.g., steroid hormone binding to receptor, substances capable of binding to Tat or CD4, and vinblastine binding to p-glycoprotein), as have methods to take advantage of these interactions in drug discovery programs. However, there is virtually an unlimited number of possible ligand–receptor interactions that could be utilized in this capacity. Some examples of biologically relevant molecules that interact with receptors include prostaglandins, ion channel blockers, second messengers, and cholinergics. Additional examples that may directly relate to therapeutic intervention include estrogen-responsive breast cancers,[161] pain modulation via opiate receptors,[162] the regulation of fertility,[163] and the interaction of interleukin-2 (IL-2) with its receptor. Considering the latter, adult human T-leukemia cells constitutively express IL-2 receptors on their surface, and this apparently signals a constant state of activation.[164] Conversely, decreased interleukin-2 has been reported in various primary and acquired immunodeficiency diseases,[165] and as a result of treatment with certain immunosuppressive drugs (such as dexamethasone and cyclosporin A).[164] The importance of ligand–receptor interactions is therefore obvious, and it is certainly of interest to discover new agents that can interact with specific receptors.

Most assays in this area are performed on the basis of competitive binding that is dose dependent. Displacement of a radiolabeled ligand from a receptor by treatment with a test substance is considered a positive response.[166, 167] However, it is important to bear in mind that even specific displacement of a bound radioligand by a given compound indicates only that the compound competes for the same receptor site; it does not indicate whether the competitor is an agonist or an antagonist of the receptor. Therefore, assays that measure the consequences of receptor binding must also be performed to adequately assess the biological effects of displacing the natural ligand.

Two examples of this approach will be given, one with small molecular compounds (phorbol esters and staurosporine) and one with a proteinaceous substance (epidermal growth factor).

13.8.1. Antagonism of Phorbol Ester or Staurosporine Binding with Protein Kinase C

As was described previously, the expression of ornithine decarboxylase appears to be related to tumor promotion. The activity of protein kinase C (PKC) also correlates with this activity.[168–170] Since PKC has been shown to be a receptor that is activated by tumor-promoting phorbol esters,[171, 172] substances that antagonize protein kinase C activity may logically be considered as putative antitumor compounds.[173] A useful assay to detect molecules that compete with known ligands of PKC for binding has been described.[174, 175] Partially purified PKC is incubated with [^{3}H]phorbol dibutyrate [PDBu; which binds to the diacylglycerol (activating) site] or [^{3}H]staurosporine [which binds to the phosphorylating (catalytic) site], in the absence or presence of unlabeled test substance. After an incubation period the mixtures are filtered through glass fiber discs and washed with several volumes of buffer to remove unbound ligand. The binding of PDBu or staurosporine is then determined by subtracting nonspecific binding (radioactivity remaining on filters when excess unlabeled PDBu or staurosporine is used) from total binding.

By including test compounds during the incubation period, competitive antagonists of protein kinase C can be detected, as indicated by a decrease in the specific binding of known radioactive ligands (see Figure 13.6). This assay has been used to evaluate the activity of a diverse group of substances from plants and other natural products.[176] In our screening program a small but significant subset of compounds were found to inhibit the specific binding of PDBu or staurosporine to protein kinase C, and many of these positives also suppressed TPA-induced ornithine decarboxylase activity when tested with cultured mouse 308 cells (Mar and Pezzuto, unpublished observations).

13.8.2. Antagonism of Growth Factor Receptors

A large number of molecules are known that modulate cell growth and metabolism via specific receptor-mediated interactions.[177] For example, platelet-derived

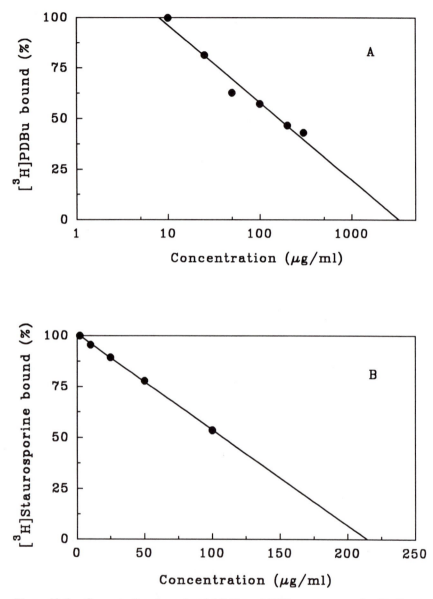

Figure 13.6. Concentration-dependent inhibition of PDBu or staurosporine binding to partially purified protein kinase C, by treatment with quercetin (**A**) or 18-*O*-β-glycyrrhetinic acid (**B**), respectively. See text for details.

growth factor (PDGF) is an endogenous mitogenic peptide that is transiently activated during normal cell proliferation.[178] PDGF has also been found to be constitutively secreted by human glioblastoma cells,[179] thus contributing to the process of tumor growth. In addition, the expression product of the simian sarcoma virus oncogene, *sis,* has been shown to be a PDGF-like molecule,[180] and PDGF can up-regulate the expression of the protooncogenes, c-*fos* and c-*myc*.[181] Clearly, it would be of interest to identify antagonists of PDGF.

Other examples of biological response modifiers with the potential of stimulating tumor growth are the peptide hormone, gastrin, and its relative, cholecystokinin (CCK).[182] Gastrin is normally a regulatory peptide of the gastrointestinal tract, but several malignant tumors are also known to secrete the substance.[183] In addition, gastrin has been reported to stimulate the proliferation of human colon cancer and stomach cancer cells.[43, 184] However, gastrin-induced tumor stimulation is suppressed by the gastrointestinal-inhibitory peptide, somatostatin, as well as by proglumide, a substance reported to antagonize gastrin–gastrin receptor interactions.[185, 186] These findings suggest that novel antagonists of regulatory peptide receptors may prove to be useful for the treatment of cancers and other diseases.

As a general observation, aberrant expression of growth factors by cells that simultaneously express the corresponding receptors can lead to autocrine regulation of tumor proliferation, ultimately relaxing the normal regulatory constraints on cell growth.[187, 188] Since these ligand–receptor interactions play such a central role in normal and aberrant metabolic regulation, bioassays to detect antagonism of specific binding at a given receptor could be instrumental in studying the etiology of disease and in designing effective therapeutic agents. Molecules that act at the epidermal growth factor receptor have been intensively studied, and this will be presented as an example of the relevant technology. Transforming growth factor α (TGF-α) is one endogenous polypeptide that stimulates the growth of a variety of mammalian cell lines.[189] It is structurally similar to epidermal growth factor (EGF) and competes for the same receptor; the EGF receptor apparently mediates the mitogenic effects of both these growth factors.[190] While EGF is expressed in several normal adult human tissues,[191] TGF-α has only been reported normally during fetal development and in human adult keratinocytes.[192] As implied by the name, TGF-α is capable of inducing the growth of nontransformed cells in semisolid agar.[62] Furthermore, the substance is secreted by many cultured human tumor cells[193] and tumors.[194] A truncated EGF/TGF-α receptor has actually been identified as the product of the *erb*B oncogene,[195] and a significant proportion of estrogen receptor-negative human breast cancers (generally indicating a poor prognosis) have been shown to overexpress EGF receptor protein.[161] All of these factors underscore the therapeutic potential of antagonists for these growth factors.

A typical protocol has been described by Rieman et al.[196] Briefly, membrane vesicles are prepared from cultured cells and incubated with ^{125}I-labeled EGF for

30 minutes at 37°C. The vesicles are then rapidly filtered through glass fiber discs and washed with isotonic saline. The radioactivity on the discs represents total ligand binding. Nonspecific binding is determined by performing the incubations in the presence of a 100-fold excess of unlabeled EGF, and this is subtracted from total binding to yield a value for specific binding. A test compound included in the incubation medium with ^{125}I-labeled EGF is considered a specific competitor if it decreases the amount of radioactivity on the filter in direct proportion to the dose of the compound used.

13.9. Cytochrome P-450: Emerging Technology with Potential Therapeutic Utility

We have thus far endeavored to correlate all of the enzyme reactions, theories, and biotechnology described in this chapter with some type of human disease state; we have also attempted to illustrate how these various domains may be integrated into drug discovery programs. In this section we will describe recent studies employing contemporary methods of biotechnology that have been performed with a predominant enzyme system involved in drug metabolism, the cytochrome P-450 mixed-function oxidase (P-450) system. The P-450 system is actually a "superfamily" of isozymes that has been extensively studied and characterized in terms of individual substrate specificities, amino acid sequences, and inducibility.[197, 198] The availability of cDNA probes for various isozymes has enabled subfamilies to be identified on the basis of DNA sequence homology; the probes have also proven useful in the assignment of gene loci and for in vivo studies examining expression.

Of particular relevance, certain human genetic polymorphisms of P-450 isozymes have been reported that correlate with the occurrence of cancers.[197–200] Since it is well known that drug metabolism catalyzed by this system may lead to metabolic activation (e.g., from a procarcinogen to an active carcinogen) rather than metabolic detoxification, it is reasonable that certain phenotypes may be associated with a higher incidence of tumorigenesis. It is apparent that complete inhibition of these enzymes would prevent the metabolic activation of carcinogens, and such a task could probably be accomplished on the basis of current knowledge. However, this certainly would not be an acceptable method of attempting to prevent chemically induced cancers. For example, many different forms of P-450 enzymes have been described that are constitutively expressed and catalyze biosynthetic and catabolic reactions of steroids and biogenic amines. Some of these constitutive P-450 isozymes, the catalytic activity of which obviously cannot be blocked with impunity, are also responsible for metabolizing a variety of xenobiotic compounds.[197, 201, 202]

Since general inhibition clearly is not reasonable, selective inhibition of P-450–catalyzed metabolism is a strategy that may be considered. Typically, liver microsomal preparations have been used for such studies,[203, 204] and whole cell systems would also be applicable.[205, 206] As one example of this approach, human liver microsomes were utilized to study the potential of test substances to inhibit the hydroxylation of (+)-bufuralol.[207] This reaction was studied based on the notion that inefficient catalysis of this hydroxylation reaction correlated with a low incidence of several cancers.[208] While this may ultimately be a valid approach, and agents capable of inhibiting this reaction may eventually be found to be useful, it must be borne in mind that some P-450 isozymes have overlapping substrate specificities;[197, 202, 209] furthermore, a given chemical has the potential to induce more than one P-450 family. These factors complicate the task of defining the in vivo consequences of modifying reactions catalyzed by individual isozymes.[198] Therefore, in terms of human health, the actual ramifications of such metabolic perturbations cannot be reliably predicted. Nonetheless, this is an extremely important and rapidly expanding field of research. As will be described below, a greater basic understanding of genetic control mechanisms is beginning to emerge, and at least some speculation as to health relevance is possible.

One major consequence of the metabolic activation of procarcinogenic substrates is the formation of covalent adducts with cellular DNA. Therefore, modulation of the quantity or quality of DNA adducts by a test substance may imply chemopreventive potential. It is possible to screen for compounds demonstrating this inhibitory activity by treating cultured hamster embryo cells with benzo[a]pyrene in the presence or absence of the test compound.[210] The profile of benzo[a]pyrene metabolites formed in the cells (with and without the test substance) indicates the overall effect of the substance on the activity of the cytochromes P-450 for which benzo[a]pyrene is a substrate. The test substance is then evaluated in cell culture for its ability to modify the formation of [^{3}H]benzo[a]pyrene metabolite adducts with cellular DNA.[211, 212] When a plant isothiocyanate (glucolimnanthin) was studied in this system, it was found to enhance the amount of water-soluble (detoxified) metabolites formed. However, this substance also caused an increase in the activation of benzo(a)pyrene to ultimate carcinogenic metabolites as evidenced by a significant elevation of benzo[a]pyrene–DNA adducts. This illustrates some of the complicating factors associated with interpreting the data obtained with this system; on the basis of binding, these data indicate that glucolimnanthin cannot be considered a putative chemopreventive compound. Other cruciferous isothiocyanates appear more promising.[213]

Rather than attempting to directly alter catalysis, an alternative approach would be to modulate the expression of P-450. Certain P-450 isozymes are proposed to correlate more directly with the incidence of cancer, particularly the 3-methylcholanthrene-inducible enzymes (e.g., P-450c and P-450d of the rat, and P_1-450 and P_3-450 of the mouse; also referred to as cytochromes P-448).[214, 215] Compared with other P-450 isozymes, a larger percentage of P-448 metabolism leads to substrate activation rather than detoxification, and this has been cited as direct evidence implicating P-448 in chemical carcinogenesis.[214] In support of this suggestion, genetic evidence has been obtained with laboratory animals indicating that selective inhibition of P_1-450 or P_3-450 expression may impart resistance to chemically induced carcinogenesis.[198, 216] Also, in various strains of inbred mice, the inducibility of various forms of cytochrome P-450 has been correlated with susceptibility.[201] Given the mounting evidence that implies that changes in the expression of at least several P-450 isozymes can influence susceptibility to cancer, it would seem prudent to identify novel compounds that are capable of modulating the expression or activity of selected cytochromes P-450 for potential use as cancer chemopreventive agents.

Regulation of the expression of 3-methylcholanthrene-inducible P-448 isozymes has been extensively studied. Hybridization of total cellular mRNA to specific cDNA probes for P-450 isozymes, as well as nuclear runoff assays, has been used to demonstrate the transcriptional activation of mouse P_1-450 and P_3-450 genes subsequent to treatment of animals with 3-methylcholanthrene.[203, 209] The induction is mediated via a cytosolic *Ah* receptor; the ligand (e.g., 3-methylcholanthrene) binds to the *Ah* receptor and the complex is translocated to the nucleus, where it interacts with DNA (at the *Ah* gene locus) and induces P-450 expression.[197, 216–218] High inducibility of cytochrome P-448 expression, as mediated through the *Ah* receptor, appears to predispose certain subjects to bronchogenic carcinoma. Thus, it should be of some value to identify inhibitors of this specific induction, and advantage can be taken of the *Ah* receptor. One substance known to bind to the receptor is 2,3,7,8-tetrachlorobidenzo-*p*-dioxin (TCDD), and this has been used in receptor binding assays for the characterization of inhibitors. For example, certain substrates for P-448 are able to displace [^{3}H]TCDD from the *Ah* receptor, whereas certain substrates for other forms of P-450 do not specifically displace the ligand.[219, 220] It is also of interest that certain molecules have been shown to compete very effectively with [^{3}H]TCDD for the *Ah* receptor without producing a biological response.[221] This tends to support the notion of finding competitive antagonists of the *Ah* receptor that are capable of preventing the undesired metabolic activation without leading to other complications. However, the provision of a substance that modifies only one specific isozyme will not be trivial. Also, while the available data regarding P-448 as an activating enzyme are certainly convincing, Paolini et al. point out that other isozymes of P-450 constitute the majority of the superfamily,[222] and many of these forms can also catalyze carcinogen activation. It is therefore dangerous to ignore the role that non-3-methylcholanthrene-inducible P-450 enzymes might play in the metabolism of procarcinogens under the assumption that only the P-448 isozymes are important.

The possibility of selectively controlling the expression of P-450 isozymes at the level of gene transcription has also been explored. Various regulatory sequences of the P_1-450 gene that are located 5′ to the enzyme-coding region were inserted into plasmid expression vectors that contained the chloramphenicol acetyltransferase (CAT) reporter gene.[218, 223] These plasmids were used to transfect mouse hepatoma cells, and CAT activity was then monitored after treatment with known inducers of the P_1-450 gene such as TCDD and cycloheximide. This type of assay could be used to investigate the potential of test substances to induce or inhibit the expression of P-450. In the cited studies, however, nucleotides were successively deleted from the 5′ region used to regulate CAT expression, and this permitted the identification of specific nucleotide sequences involved in gene induction. Thus, a P_1-450 promoter site was found to be located just 5′ of the initiation codon, a negative regulatory region was identified between nucleotides −823 and −389[223] or −990 and −650,[224] and a positive regulatory region was found between nucleotides −1642 and −823[223] or −1535 and −1265.[224] Now that these regulatory sequences have been characterized, they can be used to assay for specific binding proteins. For example, if DNA retards the progress of a protein through a nondenaturing gel, this suggests the formation of a specific protein–DNA complex.[80] Alternatively, "Southwestern" blots (binding of cellular proteins to specific DNA immobilized on nitrocellulose) can be used to identify proteins with a potential role in regulating gene transcription.[225]

13.10. Concluding Remarks

Drug discovery is a very challenging field of research. As biologists dealing with natural products, we are often presented with a substance that appears very similar to tar or some other type of waste material, and it is necessary to perform some type of assay procedure to determine if it is wise to invest substantial resources in isolation and corresponding follow-up work. In the majority of cases, if the decision is made to proceed, the agent being monitored can be purified and structurally characterized, even though it is generally present in the original extract at extraordinarily low concentrations (e.g., 0.001%, w/w). Given this level of sophistication, the burden of achievement often falls firmly on the shoulders of the bioassay. It is clear that if we were fortunate enough to obtain an extract containing a constituent as exciting as a useful remedy for the treatment of cancer or AIDS, the substance would simply be filed away with hundreds of other similar looking materials if the bioassay did not "speak out" and inform us of the presence of the active material.

As is illustrated by the examples given in this chapter, new avenues of drug discovery have recently opened owing to the advent of biotechnology. Whenever an investigator decides to enter into a new frontier of science, however, there are certain paradoxical situations that may ultimately need to be confronted. For example, let's assume that we recognize that no effective drug is available for a certain type of disease state and, on the basis of a solid scientific theory, we devise a new assay and search for a new chemical entity that mediates a positive response. This seems quite reasonable, but to what extent should we pursue this line of investigation without detecting a positive response? Would it be appropriate to analyze 1,000 compounds with completely negative results? Or perhaps 10,000 to 1 million compounds? In a sense, this is similar to searching for the Holy Grail, and intellectual foresight plays a key role in such matters. Unfortunately, the accuracy of such foresight can only be judged retrospectively, and there is often strong opposition to using limited financial resources for a risky, albeit innovative, approach.

It is of major advantage, however, that many recently developed bioassay systems are amenable to automation or semiautomation, and that test materials can thereby be assessed at impressive rates. Although screening capacity certainly varies owing to the complexity of the assay system, it is becoming increasingly common to evaluate 20,000 to 50,000 samples per year, and evaluation procedures capable of handling samples in the range of 1 million per year are under discussion. Of additional importance is the fact that only minute quantities of test materials are generally required to perform an evaluation. The minimum required quantity often is defined by the lower limit of accurate detection, since the precise mass of the substance being tested must be known. Once data can be accumulated with a large number of samples, assuming the test system is actually valid, the question tends to shift from Will an active substance be discovered in the test system? to What concentration of test substance should be considered as active? The issues of dose-response relationships and possible correlations with in vivo results have been discussed previously.[8] For in vitro test systems, once a large quantity of data has been accumulated, it is generally possible to retrospectively define a criterion of activity that places 1 to 5% of the test materials into the category of active, and follow-on studies can be performed with these leads.

Based on the mechanistic validity of contemporary bioassay systems and practical considerations such as those discussed previously, it is relatively certain that compounds will be uncovered during the next few years that will demonstrate novel structural features and biologic potential. It is possible that some of these substances will be found to mediate responses of therapeutic importance. Obviously, however, short-term in vitro assays are not invariably predictive of activity that may ultimately be observed with in vivo systems. Nonetheless, irrespective of therapeutic efficacy, these substances will serve as probes that will enable us to explore the molecular nature of various biological processes. The results of such studies should yield a clearer understanding of the molecular basis of various biological response mechanisms, which in turn may lead to the development of rational therapeutic strategies.

Typically, the response intensity observed with a short-term bioassay system correlates most accurately with an ultimately desirable in vivo response when dealing with structural analogues of a known active agent. The reasons for this are multiple and complex, relating to the myriad of physiological and biochemical pathways and interactions represented by in vivo systems. The possibility of devising a battery of bioassay test systems sufficiently representative of a mammalian species is impracticable. Conceivably, however, computer data bases could be created that comprise the entire anabolic and catabolic potential of a mammalian species, and queries could be made regarding the therapeutic effect of any given chemical compound without even needing to perform a single experiment. While it is interesting to muse over such possibilities, in the foreseeable future we will undoubtedly need to remain concerned with mammalian drug metabolism, disposition, pharmacokinetics, therapeutic indices, dose-limiting toxicity, metabolic detoxification and activation, and a host of other factors. For such reasons it is likely that laboratory animals will continue to play an important role in drug development programs. By making careful selections, however, the use of laboratory animals can be reduced to a minimum.

One important method of attempting to limit the development of a false positive lead is to construct a battery of bioassay test systems that are mutually supportive. As an example, in our program of attempting to discover agents capable of preventing cancer in humans, an initial evaluation is performed in the following test systems: (1) mammary organ culture to monitor reduced formation of carcinogen-induced nodulelike alveolar lesions,[226] (2) mouse epidermal cells to monitor inhibition of phorbol-ester–induced ornithine decarboxylase activity (this chapter), and (3) in vitro assays to monitor inhibition of binding of radiolabeled phorbol ester or staurosporine to partially purified protein kinase C (this chapter). Materials demonstrating activity in all three test systems are assigned highest priority, but the simplified receptor binding assay is used to direct fractionation. Resulting substances are thereby selected (predicted) to demonstrate the desirable activity in more biologically complicated (therapeutically relevant) test systems and are confirmed as active during the next level of testing.

Therefore, this approach of using batteries of bioassays tends to reduce the number of false positive leads. The other possibility, that of false negatives, is very important and must also be taken into account. It is likely that our concept of various disease states will rapidly evolve over the next few years; our concept of the most valid method of bioassay for drug discovery will likely evolve as well. As a result, assays indicative of a broader range of biological processes will probably be developed in the relatively near future, and it is also probable that new assays with greater predictive power will be devised. Thus, substances tested and found to be inactive with presently available bioassay systems will really need to be considered in the same category as untested materials once new evaluation procedures are available. Given the possibility of (re)evaluating large numbers of samples in a short period of time, this does not present a particularly cumbersome problem. It does become particularly important, however, to properly archive samples in a manner wherein chemical stability is assured.

Finally, returning to the discovery of drugs from plants, we are rapidly approaching a unique period in history wherein the entire flora of the earth could be monitored for potential to mediate a specific biological response in a period of less than one year, assuming the test materials were actually available. Unfortunately, it appears that deforestation and various other ecological disturbances will render numerous species extinct before they can be biologically evaluated and archived.

Acknowledgments

The authors are grateful to K. Lowell, W. Mar, S.M. Swanson, and G.T. Tan for contributing to some of the experimental studies described in this chapter. This experimental work was supported, in part, by Grant RO1 CA33047, awarded by the National Cancer Institute. J.M.P. is a recipient of a Research Career Development Award from the National Cancer Institute (1984–1989), and a Research Fellowship from the Alexander von Humboldt Foundation (1990–1991).

References

1. Farnsworth, N.R. and Pezzuto, J.M. 1983. Practical pharmacologic evaluation of plants. In *Pharmacological screening of plants and natural substances,* no. 13. Stockholm: International Foundation for Science. Pp. 138–160.

2. Cordell, G.A., Beecher, C.W.W., and Pezzuto, J.M. 1991. Can ethnopharmacology contribute to the development of new anticancer drugs? *J. Ethnopharmacol.* 32:117–133.

3. Duh, C.-Y., Pezzuto, J.M., Kinghorn, A.D., Leung, S.L., and Farnsworth, N.R., 1987. Plant anticancer agents. 44. Cytotoxic constituents from *Stizophyllum riparium. J. Nat. Prod.* 50:63–74.

4. Choi, Y.-H., Kim, J., Pezzuto, J.M., Kinghorn, A.D., Farnsworth, N.R., Lotter, H., and Wagner, H. 1986. Agrostistachin, a novel cytotoxic macrocyclic diterpene from *Agrostistachys hookeri. Tetrahedron Lett.* 27:5795–5798.

5. Jayasuriya, H., McChesney, J.D., Swanson, S.M., and Pezzuto, J.M. 1989. Antimicrobial and cytotoxic activity of rottlerin-type compounds from *Hypericum drummondii. J. Nat. Prod.* 52:325–331.

6. Kigodi, P.G.K., Blasko, G., Thebtaranonth, Y., Pezzuto, J.M., and Cordell, G.A. 1989. A new limonoid from *Ayadirachta indica.* Spectroscopic and biological investigation of nimbolide and 28-deoxynimbolide. *J. Nat. Prod.* 52:1118–1127.

7. Wong, S.-M., Oshima, Y., Pezzuto, J.M., Fong, H.H.S., and Farnsworth, N.R. 1986. Plant anticancer agents. 39. Triterpenes from *Iris missouriensis* (Iradaceae). *J. Pharm. Sci.* 75:317–320.

8. Suffness, S.M. and Pezzuto, J.M. 1991. Assays for Cytotoxicity and Antitumor Activity. In *Methods of plant biochemistry,* vol. 6, chap. 4, K. Hostettmann, ed. London: Academic Press. Pp. 71–133.

9. Kafatos, F.C., Jones, C.W., Efstratiadis, A. 1979. Determination of nucleic acid sequence homologies and relative concentrations by a dot hybridization procedure. *Nucleic Acids Res.* 7:1541–1552.

10. Thomas, P.S. 1980. Hybridization of denatured RNA and small DNA fragments transferred to nitrocellulose. *Proc. Natl. Acad. Sci. U.S.A.* 77:5201–5205.

11. Ross, J. 1976. A precursor of globin messenger RNA. *J. Mol. Biol.* 106:403–420.

12. Chirgwin, J., Przybyla, A., MacDonald, R., and Rutter, W.J. 1979. Isolation of biologically active ribonucleic acid from sources enriched in ribonuclease. *Biochemistry* 18:5294–5299.

13. Cheley, S., and Anderson, R. 1984. A reproducible microanalytical method for the detection of specific RNA sequences by dot-blot hybridization. *Anal. Biochem.* 137:15–19.

14. Pearse, M.J., and Wu, L. 1988. Preparation of both DNA and RNA for hybridization analysis from limiting quantities of lymphoid cells. *Immunol. Lett.* 18:219–224.

15. White, B.A., and Bancroft, F.C. 1982. Cytoplasmic dot hybridization. *J. Biol. Chem.* 257:8569–8572.

16. Bayer, E.A., and Wilchek, M. 1978. The avidin-biotin complex as a tool in molecular biology. *Trends Biochem. Sci.* 3:N257–N259.

17. Goding, J.W. 1986. The avidin-biotin system. In *Monoclonal antibodies: Principles and practice.* New York: Academic Press. Pp. 262–266.

18. Pegg, A.E. 1986. Recent advances in the biochemistry of polyamines in eukaryotes. *Biochem. J.* 234:249–262.

19. Pegg, A.E. 1988. Polyamine metabolism and its importance in neoplastic growth and as a target for chemotherapy. *Cancer Res.* 48:759–774.

20. Katz, A., and Kahana, C. 1987. Transcriptional activation of mammalian ornithine decarboxylase during stimulated growth. *Mol. Cell. Biol.* 7:2641–2643.

21. Feinstein, S.C., Dana, S.L., McConlogue, L., Shooter, E.M., and Coffino, P. 1985. Nerve growth factor rapidly induces ornithine decarboxylase in PC12 rat pheochromocytoma cells. *Proc. Natl. Acad. Sci. U.S.A.* 82:5761–5765.

22. Boutwell, R.K., O'Brien, T.G., Verma, A.K., Weekes, R.G., DeYoung, Y.M., Ashendel, C.L., and Astrup, E.G. 1979. The induction of ornithine decarboxylase activity and its control in mouse skin epidermis. *Adv. Enz. Reg.* 17:89–112.

23. Verma, A.K., and Boutwell, R.K. 1987. Inhibition of carcinogenesis by inhibitors of putrescine biosynthesis. In *Inhibition of polyamine metabolism. Biological significance and basis for new therapies,* ed. P.P. McCann, A.E. Pegg, and A. Sjoerdsma, eds. Orlando, Fl: Academic Press. Pp. 249–258.

24. Gilmour, S.K., Avdalovic, N., Madara, T., and O'Brien, T.G. 1985. Induction of ornithine decarboxylase by 12-*O*-tetradecanoylphorbol-13-acetate in hamster fibroblasts. *J. Biol. Chem.* 260:16439–16444.

25. Gilmour, S.K., Verma, A.K., Madara, T., O'Brien, T.G. 1987. Regulation of ornithine decarboxylase gene expression in mouse epidermis and epidermal tumors during two-stage tumorigenesis. *Cancer Res.* 47:1221–1225.

26. Yuspa, S.H., Morgan, D., Lichti, U., Spangler, E.F., Michael, D., Kilkeny, A., and Hennings, H. 1986. Cultivation and characterization of cells derived from mouse

skin papillomas induced by an initiation-promotion protocol. *Carcinogenesis* 7:949–958.

27. Kulesz-Martin, M.F., Koehler, B., Hennings, H., and Yuspa, S.H. 1980. Quantitative assay for carcinogen altered differentiation in mouse epidermal cells. *Carcinogenesis* 1:995–1006.

28. Hennings, H., Michael, D., Lichti, U., and Yuspa, S.H. 1987. Response of carcinogen-altered mouse epidermal cells to phorbol ester tumor promoters and calcium. *J. Invest. Dermatol.* 88:60–65.

29. Pegg, A.E., and McCann, P.P. 1982. Polyamine metabolism and function. *Am. J. Physiol.* 243:C212–C221.

30. Porter, C.W., and Sufrin, J.R. 1986. Interference with polyamine biosynthesis and/or function by analogs of polyamines or methionine as a potential anticancer chemotherapeutic strategy. *Anticancer Res.* 6:525–542.

31. Klinken, S.P., Castilla, M.J., and Thorgiersson, S.S. 1986. Effect of inhibitors of ornithine decarboxylase on retrovirus induced transformation of murine erythroid precursors *in vitro*. *Cancer Res.* 46:6246–6249.

32. Verma, A.K. 1988. Inhibition of tumor promoter 12-*O*-tetradecanoylphorbol-13-acetate-induced synthesis of epidermal ornithine decarboxylase messenger RNA and diacylglycerol-promoter mouse skin tumor formation by retinoic acid. *Cancer Res.* 48:2168–2173.

33. Maniatis, T., Fritsch, E.F., and Sambrook, J. 1982. *Molecular cloning: A laboratory manual*. Cold Spring Harbor, N.Y.: Cold Spring Harbor Laboratory.

34. Hsieh, J.T., and Verma, A.K. 1988. Involvement of protein kinase C in the transcriptional regulation of ornithine decarboxylase gene expression by 12-*O*-tetradecanoylphorbol-13-acetate in T24 human bladder carcinoma cells. *Arch. Biochem. Biophys.* 262:326–336.

35. Gilmour, S.K., and O'Brien, T.G. 1989. Regulation of ornithine decarboxylase gene expression in normal and transformed hamster embryo fibroblasts following stimulation by 12-*O*-tetradecanoylphorbol-13-acetate. *Carcinogenesis* 10:157–162.

36. Seely, J.E., Poso, H., and Pegg, A.E. 1982. Effect of androgens on turnover of ornithine decarboxylase in mouse kidney. *J. Biol. Chem.* 257:7549–7553.

37. Greenberg, M.E., and Ziff, E.B. 1984. Stimulation of 3T3 cells induces transcription of the c-*fos* protooncogene. *Nature (London)* 311:433–437.

38. Chinsky, J.M., Maa, M.-C., Ramamurthy, V., and Kellems, R.E. 1989. Adenosine deaminase gene expression. Tissue-dependent regulation of transcriptional elongation. *J. Biol. Chem.* 264:14561–14565.

39. Hunter, T. 1984. The proteins of oncogenes. *Sci. Am.* 251:70–79.

40. Weinberg, R.A. 1985. The action of oncogenes in the cytoplasm and nucleus. *Science* 230:770–776.

41. Uehara, Y., Murakami, Y., Suzukake-Tsuchiya, K., Moriya, Y., Sano, H., Shibata, K., and Omura, S. 1988. Effects of herbimycin derivatives on *src* oncogene function in relation to antitumor activity. *J. Antibiot.* 41:831–834.

42. King, C.R., Kraus, M.H., and Aaronson, S.A. 1985. Amplification of a novel v-*erb*B-related gene in a human mammary carcinoma. *Science* 229:974–978.

43. Tanaka, T., Slamon, D.J., Battifora, H., and Cline, M.J. 1986. Expression of p21 ras oncoproteins in human cancers. *Cancer Res.* 46:1465–1470.

44. Kung, H.-S., Smith, M.R., Bekisi, E., Manne, V., and Stacey, D.W. 1986. Reversal of transformed phenotype by monoclonal antibodies against Ha-*ras* p21 proteins. *Exp. Cell Res.* 162:363–371.

45. de Klein, A., van Kessel, A.G. Grosveld, G., Bartram, C.R., Hagemeijer, A., Bootsma, D., Spurr, N.K., Heisterkamp, N., Groffen, J., and Stephenson, J.R. 1982. A cellular oncogene is translocated to the Philadelphia chromosome in chronic myelocytic leukaemia. *Nature (London)* 300:765–767.

46. Leder, A., Pattengale, P.K., Kuo, A., Stewart, T.A., and Leder, P. 1986. Consequences of widespread deregulation of the c-*myc* gene in transgenic mice: Multiple neoplasms and normal development. *Cell* 45:485–495.

47. Maguire, H.C., Jr., and Greene, M.I. 1989. The *neu* (c-*erb* B-2) oncogene. *Sem. Oncol.* 16:148–155.

48. Huber, B.E. 1989. Therapeutic opportunities involving cellular oncogenes: Novel approaches fostered by biotechnology. *FASEB J.* 3:5–13.

49. Laemmli, U.K. 1970. Cleavage of structural proteins during the assembly of the head of bacteriophage T4. *Nature (London)*: 227:680–685.

50. Towbin, H., Staehelin, T., and Gordon, J. 1979. Electrophoretic transfer of proteins from polyacrylamide gels to nitrocellulose sheets: Procedure and some applications. *Proc. Natl. Acad. Sci. U.S.A.* 76:4350–4354.

51. Seshradri, T., and Campisi, J. 1990. Repression of c-*fos* transcription and an altered genetic program in senescent human fibroblasts. *Science* 247:205–259.

52. Thiele, C.J., Cohen, P.S., and Israel, M.A. 1988. Regulation of c-*myb* expression in human neuroblastoma cells during retinoic acid-induced differentiation. *Mol. Cell. Biol.* 8:1677–1683.

53. Linevsky, J., Cohen, M.B., Hartman, K.D., Knode, M.C., and Glazer, R.I. 1985. Effect of Neplanocin A on differentiation, nucleic acid methylation, and c-*myc* mRNA expression in human promyelocytic leukemia cells. *Mol. Pharmacol.* 28:45–50.

54. Hunter, T., and Cooper, J.A. 1985. Protein-tyrosine kinases. *Annu. Rev. Biochem.* 54:897–930.

55. Hunter, T. 1987. A thousand and one protein kinases. *Cell* 50:823–829.

56. Zioncheck, T.F., Harrison, M.L., and Gaehlen, R.L. 1986. Purification and characterization of a protein-tyrosine kinase from bovine thymus. *J. Biol. Chem.* 261:15637–15643.

57. Geahlen, R.L., Koonchanok, N.M., and McLaughlin, J.L. 1989. Inhibition of protein-tyrosine kinase activity by flavonoids and related compounds. *J. Nat. Prod.* 52:982–986.

58. O'Farrell, P.H. 1975. High resolution two-dimensional electrophoresis of proteins. *J. Biol. Chem.* 250:4007–4021.

59. Freshney, R.I. 1987. The transformed phenotype. In *Culture of animal cells.* New York: Alan R. Liss. Pp. 197–206.

60. Volm, M., Wayss, K., Kaufmann, M., and Mattern, J. 1979 Pretherapeutic detection of tumour resistance and the results of tumour chemotherapy. *Eur. J. Cancer* 15:983–993.

61. Swanson, S.M., and Pezzuto, J.M. 1990. Bioscreening for antitumor activity: Evaluation of cytotoxic potential and ability to inhibit macromolecule biosynthesis. In *Drug bioscreening, Drug evaluation techniques in pharmacology,* E.B. Thompson, ed. New York: VCH. Pp. 273–397.

62. Twardzik, D.R. 1985. Differential expression of transforming growth factor alpha during prenatal development of the mouse. *Cancer Res.* 45:5413–5416.

63. Cuthbertson, R.A., and Klintworth, G.K. 1988. Transgenic mice—A gold mine for furthering knowledge in pathobiology. *Lab. Invest.* 58:484–502.

64. Southern, E.M. 1975. Detection of specific sequences among DNA fragments separated by gel electrophoresis. *J. Mol. Biol.* 98:503–517.

65. Andres, A.-C., Schonenberger, C.-A., Groner, B., Henninghausen, L., LeMeur, M., and Gerlinger, P. 1987. Ha-*ras* oncogene expression directed by a milk protein gene promoter: Tissue specificity, hormonal regulation, and tumor induction in transgenic mice. *Proc. Natl. Acad. Sci. U.S.A.* 84:1299–1303.

66. Ezzell, C. 1988. First ever animal patent issued in United States. *Nature (London)* 332:668.

67. Holt, J.T., Gopal, T.V., Moulton, A.D., and Nienhuis, A.W. 1986. Inducible production of c-*fos* antisense RNA inhibits 3T3 cell proliferation. *Proc. Natl. Acad. Sci. U.S.A.* 83:4794–4798.

68. Nishikura, K., and Murray, J.M. 1987. Antisense RNA of proto-oncogene c-*fos* blocks renewed growth of quiescent 3T3 cells. *Mol. Cell. Biol.* 7:639–649.

69. Chambers, A.F., and Denhardt, D.T. 1990. Abatement of gene expression using antisense oligodeoxynucleotides. *Pharmaceut. Technol.* February:24–28.

70. Gorman, C.M., Moffat, L.F., and Howard, B.H. 1982. Recombinant genomes which express chloramphenicol acetyltransferase in mammalian cells. *Mol. Cell. Biol.* 2:1044–1051.

71. de Wet, J.R., Wood, K.V., DeLuca, M., Helinski, D.R., and Subramani, S. 1987. Firefly luciferase gene: Structure and expression in mammalian cells. *Mol. Cell. Biol.* 7:725–737.

72. Hudson, L.G., Ertl, A.P., Gill, G.N. 1990. Structure and inducible regulation of the human c-*erb* B2/*neu* promoter. *J. Biol. Chem.* 265:4389–4393.

73. Bargmann, C.I., and Weinberg, R.A. 1988. Increased tyrosine kinase activity associated with the protein encoded by the activated neu oncogene. *Proc. Natl. Acad. Sci. U.S.A.* 85:5394–5398.

74. Evans, R.M. 1988. The steroid and thyroid hormone receptor superfamily. *Science* 240:889–895.

75. Thompson, C.C., and Evans, R.M. 1989. Transactivation by thyroid hormone receptors: Functional parallels with steroid hormone receptors. *Proc. Natl. Acad. Sci. U.S.A.* 86:3494–3498.

76. Hard, T., Kellenbach, E., Boelens, R., Maler, B.A., Dahlman, K., Freedman, L.P., Carlstedt-Duke, J., Yamamoto, K.R., Gustafsson, J.-A., and Kaptein, R. 1990. Solution structure of the glucocorticoid receptor DNA-binding domain. *Science* 249:157–160.

77. Giguere, V., Ong, E.S., Segui, P., and Evans, R.M. 1987. Identification of a receptor for the morphogen retinoic acid. *Nature (London)* 330:624–629.

78. Evans, R.M. 1989. The v-*erb*A oncogene is a thyroid hormone receptor antagonist. *Int. J. Cancer.* 4: suppl. 26–28.

79. Umesono, K., and Evans, R.M. 1989. Determinants of target gene specificity for steroid/thyroid hormone receptors. *Cell* 57:1139–1146.

80. Damm, K., Thompson, C.C., and Evans, R.M. 1989. Protein encoded by v-*erb*A functions as a thyroid-hormone receptor antagonist. *Nature (London)* 339:593–597.

81. Muller, W.E.G., Okamoto, T., Reuter, P., Ugarkovic, D., and Schroder, H.C. 1990. Functional characterization of Tat protein from human immunodeficiency virus. *J. Biol. Chem.* 265:3803–3808.

82. Nelbock, P., Dillon, P.J., Perkins, A., and Rosen, C.A. 1990. A cDNA for a protein that interacts with the human immunodeficiency virus Tat transactivator. *Science* 248:1650–1653.

83. Krogstad, D.J., Schlesinger, P.H., Herwaldt, B.L. 1988. Antimalarial agents: Mechanism of chloroquine resistance. *Antimicrob Agents Chemother.* 32:799–801.

84. Fuqua, S.A.W., Moretti-Rojas, I.M., Schneider, S.L., and McGuire, W.L. 1987. P-Glycoprotein expression in human breast cancer cells. *Cancer Res.* 47:2103–2106.

85. Bradley, G., Juranka, P.F., and Ling, V. 1988. Mechanism of multidrug resistance. *Biochim Biophys. Acta* 948:87–128.

86. Gottesman, M.M., and Pastan, I. 1988. Resistance to multiple chemotherapeutic agents in human cancer cells. *Trends Pharm. Sci.* 9:54–58.

87. Kartner, N., and Ling, V. 1989. Multidrug resistance in cancer. *Sci. Am.* 260:44–51.

88. Thiebaut, F., Tsuruo, T., Hamada, H., Gottesman, M.M., Pastan, I., and Willingham, M.C. 1987. Cellular localization of the multidrug-resistance gene product P-glycoprotein in normal human tissues. *Proc. Natl. Acad. Sci. U.S.A.* 84:7735–7738.

89. Cornwell, M.M., Safa, A.R., Felsted, R.L., Gottesman, M.M., and Pastan, I. 1986. Membrane vesicles from multidrug-resistant human cancer cells contain a specific 150- to 170-kDa protein detected by photoaffinity labeling. *Proc. Natl. Acad. Sci. U.S.A.* 83:3847–3850.

90. Juliano, R.L., and Ling, V. 1976. A surface glycoprotein modulating drug permeability in Chinese hamster ovary cell mutants. *Biochem. Biophys. Acta* 455:152–162.

91. Kartner, N., Riordan, J.R., and Ling, V. 1983. Cell surface P-glycoprotein associated with multidrug resistance in mammalian cell lines. *Science* 221:1285–1288.

92. Ueda, K., Cornwell, M.M., Gottesman, M.M., Pastan, I., Roninson, I.B., Ling, V., and Riordan, J.R. 1986. The *mdr 1* gene, responsible for multi-drug resistance, codes for P-glycoprotein. *Biochem. Biophys. Res. Commun.* 141:956–962.

93. Fojo, A., Akiyama, S., Gottesman, M.M., and Pastan, I. 1985. Reduced drug accumulation in multiply drug-resistant human KB carcinoma cell lines. *Cancer Res.* 45:3002–3007.

94. Beck. W.T., Cirtain, M.C., and Lefko, J.L. 1983. Energy-dependent reduced drug binding as a mechanism of Vinca alkaloid resistance in human leukemic lymphoblasts. *Mol. Pharmacol.* 24:485–492.

95. Cornwell, M.M., Tsuruo, T., Gottesman, M.M., and Pastan, I. 1987. ATP-binding properties of P-glycoprotein from multidrug-resistant KB cells. *FASEB J.* 1:51–54.

96. Horio, M., Gottesman, M.M., and Pastan, I. 1988. ATP-dependent transport of vinblastine in vesicles from human multidrug-resistant cells. *Proc. Natl. Acad. Sci. U.S.A.* 85:3580–3584.

97. Deuchars, K.L., and Ling, V. 1989. P-Glycoprotein and multidrug resistance in cancer chemotherapy. *Sem. Oncol.* 16:156–165.

98. Deuchars, K.L., Du, R.-P., Naik, M., Evernden-Porelle, D., Kartner, N., van der Bliek, A.M., and Ling, V. 1987. Expression of hamster P-glycoprotein and multidrug resistance in DNA-mediated transformants of mouse LTA cells. *Mol. Cell. Biol.* 7:718–724.

99. Sugimoto, Y., and Tsuruo, T. 1987. DNA-mediated transfer and cloning of a human multidrug-resistant gene of adriamycin-resistant myelogenous leukemia K562. *Cancer Res.* 47:2620–2625.

100. Ueda, K., Cardarelli, C., Gottesman, M.M., and Pastan, I. 1987. Expression of a full-length cDNA for the human "MDR1" gene confers resistance to colchicine, doxorubicin, and vinblastine. *Proc. Natl. Acad. Sci. U.S.A.* 84:3004–3008.

101. Gros, P., Neriah, Y.B., Croop, J.M., and Housman, D.E. 1986. Isolation and expression of a complementary DNA that confers multidrug resistance. *Nature (London)* 323:728–731.

102. Shen, D.-W., Fojo, A., Roninson, I.B., Chin, J.E., Siffir, R., Pastan, I., and Gottesman, M.M. 1986. Multidrug resistance of DNA-mediated transformants is linked to transfer of the human *mdr1* gene. *Mol. Cell. Biol.* 6:4039–4044.

103. Gros, P., Fallows, D.A., Croop, J.M., and Housman, D.E. 1986. Chromosome-mediated gene transfer of multidrug resistance. *Mol. Cell. Biol.* 6:3785–3790.

104. Goldstein, L.J., Galski, H., Fojo, A., Willingham, M., Lai, S.-L., Gazdar, A., Pirkir, R., Green, A. Crist, W., Brodeur, G.M., Lieber, M., Cossman, J., Gottesman, M.M., and Pastan, I. 1989. Expression of a multidrug resistance gene in human cancers. *J. Natl. Cancer Inst.* 81:116–124.

105. Cano-Gauci, D.F., and Riordan, J.R. 1987. Action of calcium antagonists on multidrug resistant cells. Specific cytotoxicity independent of increased cancer drug accumulation. *Biochem. Pharmacol.* 36:2115–2123.

106. Tsuruo, T., Iida, H., Tsukagoshi, S., and Sakurai, Y. 1982. Increased accumulation of vincristine and adriamycin in drug-resistant P388 tumor cells following incubation with calcium antagonists and calmodulin inhibitors. *Cancer Res.* 42:4730–4703.

107. Tsuruo, T., and Iida, H. 1986. Effects of cytochalasins and colchicine on the accumulation and retention of daunomycin and vincristine in drug resistant tumor cells. *Biochem. Pharmacol.* 35:1087–1090.

108. Zamora, J.M., and Beck, W.T. 1986. Chloroquine enhancement of anticancer drug cytotoxicity in multiple drug resistant human leukemia cells. *Biochem. Pharmacol.* 35:4303–4310.

109. Krogstad, D.J., Gluzman, I.Y., Kyle, D.E., Oduola, A.M.J., Martin, S.K., Milhous, W.K., and Schlesinger, P.H. 1987. Efflux of chloroquine from *Plasmodium falciparum:* Mechanism of chloroquine resistance. *Science* 238:1283–1285.

110. Bellamy, W.T., Dalton, W.S., Kailey, J.M., Gleason, M.C., McCloskey, T.M., Dorr, R.T., and Alberts, D.S. 1988. Verapamil reversal of doxorubicin resistance in multidrug-resistant human myeloma cells and association with drug accumulation and DNA damage. *Cancer Res.* 48:6303–6308.

111. Inaba, M., Kobayashi, H., Sakurai, Y., and Johnson, R.K. 1979. Active efflux of daunorubicin and adriamycin in sensitive and resistant sublines of P388 leukemia. *Cancer Res.* 39:2200–2203.

112. Tapiero, H., Munck, J.-N., Fourcade, A., and Lampidis, T.J. 1984. Cross-resistance to rhodamine 123 in adriamycin- and daunorubicin-resistant Friend leukemia cell variants. *Cancer Res.* 44:5544–5549.

113. Munck, J.-N., Fourcade, A., Bennoum, M., and Tapiero, H. 1985. Relationship between the intracellular level and growth inhibition of a new anthracycline 4'-O-tetrahydropyranyl-adriamycin in Friend leukemia cell variants. *Leukemia Res.* 9:289–296.

114. Neyfakh, A.A. 1988. Use of fluorescent dyes as molecular probes for the study of multidrug resistance. *Exp. Cell. Res.* 174:168–176.

115. Lalande, M.E., Ling, V., and Miller, R.G. 1981. Hoechst 33342 dye uptake as a probe of membrane permeability changes in mammalian cells. *Proc. Natl. Acad. Sci. U.S.A.* 78:363–367.

116. Neyfakh, A.A., Dmitrevskaya, T.V., and Serpinskaya, A.S. 1988. The membrane transport system responsible for multidrug resistance is operating in nonresistant cells. *Exp. Cell. Res.* 178:513–517.

117. Akiyama, S.-I., Cornwell, M.M., Kuwano, M., Pastan, I., and Gottesman, M.M. 1988. Most drugs that reverse multidrug resistance also inhibit photoaffinity labeling of P-glycoprotein by a vinblastine analog. *Mol. Pharmacol.* 33:144–147.

118. Safa, A.R. 1988. Photoaffinity labeling of the multidrug-resistance-related P-glycoprotein with photoactive analogs of verapamil. *Proc. Natl. Acad. Sci. U.S.A.* 85:7187–7191.

119. Shen, D., Cardarelli, C., Hwang, J., Cornwell, M., Richert, N., Ishii, S., Pastan, I., and Gottesman, M.M. 1986. Multiple drug-resistant human KB carcinoma cells independently selected for high-level resistance to colchicine, adriamycin, or vinblastine show changes in expression of specific proteins. *J. Biol. Chem.* 261:7762–7770.

120. Greenberger, L.M., Williams, S.S., Georges, E., Ling, V., and Band-Horwitz, S. 1988. Electrophoretic analysis of P-glycoproteins produced by mouse J774.2 and Chinese Hamster Ovary multidrug-resistant cells. *J. Natl. Cancer Inst.* 80:506–510.

121. Kartner, N., Evernden-Porelle, D., Bradley, G., and Ling, V. 1985. Detection of P-glycoprotein in multidrug-resistant cell lines by monoclonal antibodies. *Nature* (*London*) 316:820–823.

122. Beck, W.T. 1983. Vinca alkaloid resistant phenotype in cultured human leukemia lymphoblasts. *Cancer Treat. Rep.* 67:875–881.

123. Danks, M.K., Metzger, D.W., Ashmun, R.A., and Beck, W.T. 1985. Monoclonal antibodies to glycoproteins of Vinca alkaloid-resistant human leukemic cells. *Cancer Res.* 45:3220–3224.

124. Volm, M., Efferth, T., and Lathan, B. 1987. Detection of murine S180 cells expressing a multidrug resistance phenotype using different *in vitro* test systems and a monoclonal antibody. *Arzneim-Forsch.* 37(2):862–867.

125. Fojo, A.T., Uead, K., Slamon, D.J., Poplack, D.J., Gottesman, M.M., and Pastan, I. 1987. Expression of a multidrug-resistance gene in human tumors and tissues. *Proc. Natl. Acad. Sci. U.S.A.* 84:265–269.

126. Hitchins, R.N., Harman, D.H., Davey, R.A., and Bell, D.R. 1988. Identification of a multidrug-resistance associated antigen (P-glycoprotein) in normal human tissues. *Eur. J. Cancer Clin. Oncol.* 24:449–454.

127. Biedler, J.L., Riehm, H., Peterson, R.H.F., and Spengler, B.A. 1975. Membrane-mediated drug resistance and phenotypic reversion to normal growth behavior of Chinese Hamster Cells. *J. Natl. Cancer Inst.* 55:671–677.

128. Bates, S.E., Mickley, L.A., Chen, Y.-N., Richert, N., Rudick, J., Biedler, J., and Fojo, A.T. 1989. Expression of a drug resistance gene in human neuroblastoma cell lines: Modulation by retinoic acid-induced differentiation. *Mol. Cell. Biol.* 9:4337–4344.

129. Mickley, L.A., Bates, S.E., Richert, N.D., Currier, S., Tanaka, S., Foss, F., Rosen, N., and Fojo, A.T. 1989. Modulation of the expression of a multidrug resistance gene (mdr-1/P-glycoprotein) by differentiating agents. *J. Biol. Chem.* 264:18031–18040.

130. Bell, D.R., Gerlack, J.H., Kartner, N., Buick, R.N., and Ling, V. 1985. Detection of p-glycoprotein in ovarian cancer: A molecular marker associated with multidrug resistance. *J. Clin. Oncol.* 3:311–315.

131. Beck, W.T. 1987. The cell biology of multiple drug resistance. *Biochem. Pharmacol.* 36:2879–2887.

132. Richert, N., Akiyama, S., Shen, D., Gottesman, M.M., and Pastan, I. 1985. Multiply drug-resistant human KB carcinoma cells have decreased amounts of a 75 kDa and a 72 kDa glycoprotein. *Proc. Natl. Acad. Sci. U.S.A.* 82:2330–2333.

133. van der Bliek, A.M., van der Velde-Koerts, T., Ling, V., and Borst, P. 1986. Overexpression and amplification of five genes in a multidrug-resistant Chinese Hamster Ovary cell line. *Mol. Cell. Biol.* 6:1671–1678.

134. Akiyama, S., Fojo, A., Hanover, J.A., Pastan, I., and Gottesman, M.M. 1985. Isolation and genetic characterization of human KB cell lines resistant to multiple drugs. *Som. Cell. Mol. Gen.* 11:117–126.

135. Hu, J.M., and Hsuing, G.D. 1989. Evaluation of new antiviral agents. 1. *In vitro* perspectives. *Antiviral Res.* 11:217–232.

136. Field, A.K., Davis, M.E., DeWitt, C.M., Perry, H.C., Schofield, T.L., Karkas, J.D., Germershausen, J., Wagner, A.F., Cantone, C.L., MacCoss, M., and Tolman, R.L. 1986. Efficacy of 2'-nor-cyclic GMP in treatment of experimental herpes virus infections. *Antiviral Res.* 6:329–341.

137. Hu, J.M., and Hsuing, G.D. 1988. Studies on two new antiviral agents against guinea pig lymphotropic herpesvirus infection *in vitro* (abstract). *Antiviral Res.* 9:83.

138. Collins, P., and Bauer, D.J. 1977. Relative potencies of anti-herpes compounds. *Ann. N.Y. Acad. Sci.* 284:49–59.

139. Harada, S., Koyanagi, Y., and Yamamoto, N. 1985. Infection of HTLV-III/LAV in HTLV-1-carrying cell MT-2 and MT-4 and application in a plaque assay. *Science* 229:563–566.

140. Nakashima, H., Matsui, T., Harada, S., Kobayashi, N., Matsuda, A., Ueda, T., and Yamamoto, N. 1986. Inhibition of replication and cytopathic effect of human T-cell lymphotropic virus type III/ lymphadenopathy-associated virus by 3'-deoxythymidine *in vitro*. *Antimicrob. Agents Chemother.* 30:933–937.

141. Montefiore, D.C., Robinson, W.E., Jr., Schuffman, S.S., and Mitchell, W.M. 1988. Evaluation of antiviral drugs and neutralizing antibodies to human immunodeficiency virus by a rapid and sensitive microtiter infection assay. *J. Clin. Microbiol.* 26:231–235.

142. Abou-Karam, M., and Shier, W.T. 1990. A simplified plaque reduction assay for antiviral agents from plants. Demonstration of frequent occurrence of antiviral activity in higher plants. *J. Nat. Prod.* 53:340–344.

143. Pauwels, R., Balzarini, J., Baba, M., Snoeck, R., Schols, D., Herdewijn, P., Desmyter, J., and DeClerq, E. 1988. Rapid and automated tetrazolium-based colorimetric assay for the detection of anti-HIV compounds. *J. Virol. Methods* 20:309–321.

144. Vince, R., Hua, M., Brownell, J., Daluge, S., Lee, F., Shannon, W.M., Lavelle, G.C., Qualls, J., Weislow, O.S., Kiser, R., Canonico, P.G., Schultz, R.H., Narayanan, V.L., Mayo, J.G., Shoemaker, R.H., and Boyd, M.R. 1988. Potent and selective activity of a new carbocyclic nucleoside analog (Carbovir: NSC 614846) against human immunodeficiency virus *in vitro*. *Biochem. Biophys. Res. Commun.* 156:1046–1053.

145. Weislow, O.S., Kiser, R., Fine, D.L., Bader, J., Shoemaker, R.H., and Boyd, M.R. 1989. New soluble-formazan assay for HIV-1 cytopathic effects: Application

to high-flux screening of synthetic and natural products for AIDS-antiviral activity. *J. Natl. Cancer Inst.* 81:577–586.

146. Paull, K.D., Shoemaker, R.H., Boyd, M.R., Parsons, J.L., Risbood, P.A., Barbera, W.A., Sharma, M.N., Baker, D.C., Hand, E., Scudiero, D.A., Monks, A., Alley, M.C., and Grote, M. 1988. The synthesis of XTT—A new tetrazolium reagent that is bioreducible to a water-soluble formazan. *J. Heterocyclic Chem.* 25:911–914.

147. Goodchild, J., Agrawal, S., Civeira, M.P., Sarin, P.S., Sun, D., and Zamecnik, P.C. 1988. Inhibition of human immunodeficiency virus replication by antisense oligodeoxynucleotides. *Proc. Natl. Acad. Sci. U.S.A.* 85:5507–5511.

148. Sarin, P.S., Agrawal, S., Civeira, M.P., Goodchild, J., Ikeuchi, T., and Zamecnik, P.C. 1988. Inhibition of acquired immunodeficiency syndrome virus by oligodeoxynucleoside methylphosphonates. *Proc. Natl. Acad. Sci. U.S.A.* 85:7448–7451.

149. Mitsuya, H., Popovic, M., Yarchoan, R., Matsushita, S., Gallo, R., and Broder, S. 1984. Suramin protection of T cells *in vitro* against infectivity and cytopathic effect of HTLV-III. *Science* 226:172–174.

150. DeClercq, E. 1986. Chemotherapeutic approaches to the treatment of the acquired immune deficiency syndrome (AIDS). *J. Med. Chem.* 29:1561–1569.

151. Spedding, G., Ratty, A., and Middleton, E., Jr. 1989. Inhibition of reverse transcriptases by flavonoids. *Antiviral Res.* 12:99–110.

152. Tan, G.T., Pezzuto, J.M., Kinghorn, A.D., and Hughes, S.H. 1991. Evaluation of natural products as inhibitors of human immunodeficiency virus type 1 (HIV-1) reverse transcriptase. *J. Nat. Prod.* 54:143–154.

153. Hizi, A., McGill, C., and Hughes, S.H. 1988. Expression of soluble, enzymatically active, human immunodeficiency virus reverse transcriptase in *Escherichia coli* and analysis of mutants. *Proc. Natl. Acad. Sci. U.S.A.* 85:1218–1222.

154. Clark, P.K., Ferris, A.L., Miller, D.A., Hizi, A., Kim, K.-W., Deringer-Boyer, S.M., Mellini, M.L., Clark, A.D., Jr., Arnold, G.F., Lebherz, W.B., III, Arnold, E., Muschik, G.M., and Hughes, S.H. 1990. HIV-1 reverse transcriptase purified from a recombinant strain of *Escherichia coli*. *AIDS Res. Hum. Retroviruses* 6:753–761.

155. Schinazi, R.F., Erikkson, B.F.H., and Hughes, S.H. 1989. Comparison of inhibitory activities of various antiretroviral agents against particle-derived and recombinant human immunodeficiency virus type 1 reverse transcriptases. *Antimicrob. Agents Chemother.* 33:115–117.

156. Lifson, J.D., and Engleman, E.G. 1989. Role of CD4 in normal immunity and HIV infection. *Immunol. Rev.* 109:93–117.

157. Sperber, S.J., and Hayden, F.G. 1989. Protective effect of rhinovirus receptor blocking antibody in human fibroblast cells. *Antiviral Res.* 12:231–238.

158. Krausslich, H.-G., and Wimmer, E. 1988. Viral proteinases. *Annu. Rev. Biochem.* 57:701–754.

159. Orr, D.C., Long, A.C., Kay, J., Dunn, B.M., and Cameron, J.M. 1989. Hydrolysis of a series of synthetic peptide substrates by the human rhinovirus 14 3C proteinase, cloned and expressed in *Escherichia coli*. *J. Gen. Virol.* 70:2931–2942.

160. Long, A.C., Orr, D.C., Cameron, J.M., Dunn, B.M., and Kay, J. 1989. A consensus sequence for substrate hydrolysis by rhinovirus 3C proteinase. *FEBS Lett.* 235:75–78.

161. Sainsbury, J.R.C., Sherbet, G.V., Farndon, J.R., and Harris, A.L. 1985. Epidermal-growth-factor receptors and oestrogen receptors in human breast cancer. *Lancet* 16:364–368.

162. Schusdziarra, V., Rewes, B., Lenz, N., Maier, V., and Pfeiffer, E.F. 1983. Carbohydrates modulate opiate receptor mediated mechanisms during postprandial endocrine function. *Regul. Peptides* 7:243–252.

163. Ulmann, A., Teutsch, G., and Philibert, D. 1990. RU 486. *Sci. Am.* 262:42–48.

164. Waldmann, T.A. 1986. The structure, function, and expression of interleukin-2 receptors on normal and malignant lymphocytes. *Science* 232:727–732.

165. Welte, K., and Mertelsmann, R. 1985. Human interleukin-2: Biochemistry, physiology, and possible pathogenic role in immunodeficiency syndromes. *Cancer Invest.* 3:35–49.

166. Sutherland, E.W. 1972. Studies on the mechanism of hormone action. *Science* 177:401–408.

167. Darnell, L., Lodish, H., and Baltimore, D. 1986. Cell-to-cell signalling: Hormones and receptors. *Molecular cell biology.* New York: Scientific American Books. Pp. 667–713.

168. Nishizuka, Y. 1984. The role of protein kinase C in cell surface signal transduction and tumour promotion. *Nature (London)* 308:693–697.

169. Nishizuka, Y. 1988. The molecular heterogeneity of protein kinase C and its implications for cellular regulation. *Nature (London)* 334:661–665.

170. Dreher, M.L., and Hanley, M.R. 1988. Multiple modes of protein kinase C regulation and their significance in signalling. *Trends Pharm. Sci.* 9:114–115.

171. Parker, P.J., Coussens, L., Totty, N., Rhee, L., Young, S., Chen, E., Stabel, S., Waterfield, M.D., and Ullrich, A. 1986. The complete primary structure of protein kinase C—the major phorbol ester receptor. *Science* 233:853–859.

172. Castagna, M., Takai, Y., Kaibuchi, K., Sano, K., Kikkawa, U., and Nishizuka, Y. 1982. Direct activation of calcium-activated, phospholipid-dependent protein kinase by tumor-promoting phorbol esters. *J. Biol. Chem.* 257:7847–7851.

173. Nakadate, T., Jeng, A.Y., and Blumberg, P.M. 1988. Comparison of protein kinase C functional assays to clarify mechanisms of inhibitor action. *Biochem. Pharmacol.* 37:1541–1545.

174. Leach, K.L., and Blumberg, P.M. 1985. Modulation of protein kinase C activity and [^{3}H]phorbol 12,13-dibutyrate binding by various tumor promoters in mouse brain cytosol. *Cancer Res.* 45:1958–1963.

175. de Vries, D.J., Herald, C.L., Pettit, G.R., and Blumberg, P.M. 1988. Demonstration of subnanomolar affinity of bryostatin 1 for the phorbol ester receptor in rat brain. *Biochem. Pharmacol.* 37:4069–4073.

176. Beutler, J.A., Alvarado, A.B., McCloud, T.G., and Cragg, G.M. 1989. Distribution of phorbol ester bioactivity in the Euphorbiaceae. *Phytother. Res.* 3:188–192.

177. Foon, K.A. 1989. Biological response modifiers: The new immunotherapy. *Cancer Res.* 49:1621–1639.

178. Zullo, J.N., and Faller, D.V. 1988. P21 V-*ras* inhibits induction of c-*myc* and c-*fos* expression by platelet-derived growth factor. *Mol. Cell. Biol.* 8:5080–5085.

179. Pantazis, P., Pelicci, P.G., Dalla-Favera, R., and Antoniades, H.N. 1985. Synthesis and secretion of proteins resembling platelet-derived growth factor by human glioblastoma and fibrosarcoma cells in culture. *Proc. Natl. Acad. Sci. U.S.A.* 82:2404–2408.

180. Schecter, A.L., Stern, D.F., Vaidyanathan, L., Decker, S.J., Drebin, J.A., Greene, M.I., and Weinberg, R.A. 1984. The *neu* oncogene: An *erb*-B-related gene encoding a 185,000-M$_r$ tumour antigen. *Nature (London)* 312:513–516.

181. Muller, R., Bravo, R., Burckhardt, J., and Curren, T. 1984. Induction of c-*fos* gene and protein by growth factors precedes activation of c-*myc*. *Nature (London)* 312:716–720.

182. Chang, R.S.L., and Lotti, V.J. 1986. Biochemical and pharmacological characterization of an extremely potent and selective non-peptide cholecystokinin antagonist. *Proc. Natl. Acad. Sci. U.S.A.* 83:4923–4926.

183. Gregory, H. 1975. Isolation and structure of urogastrone and its relationship to epidermal growth factor. *Nature (London)* 257:325–327.

184. Lamers, C.B.H.W., and Jansen, J.B.M.J. 1988. Role of gastrin and cholecystokinin in tumors of the gastrointestinal tract. *Eur. J. Cancer Clin. Oncol.* 24:267–273.

185. Beauchamp, R.D., Townsend, C.M., Jr., Singh, P., Glass, E.J., and Thompson, J.C. 1985. Proglumide, a gastrin receptor antagonist, inhibits growth of colon cancer and enhances survival in mice. *Ann. Surg.* 202:303–308.

186. Magous, R., and Bali, J.P. 1983. Evidence that proglumide and benzotript antagonize secretogogue stimulation of isolated gastric parietal cells. *Reg. Peptides* 7:233–241.

187. Wong, R.S., and Passar, E. Jr. 1989. Growth factors, oncogenes and the autocrine hypothesis. *Surg. Gynecol. Obstet.* 168:468–473.

188. Horowitz, J.M., Friend, S.H., Weinberg, R.A., Whyte, P., Buchkovich, K., and Harlow, E. 1988. Anti-oncogenes and the negative regulation of cell growth. *Cold Spring Harbor Symp. Quant. Biol.* 53:843–847.

189. Defeo-Jones, D., Tai, J.Y., Wegrzyn, R.J., Vuocolo, G.A., Baker, A.E., Payne, L.S., Garsky, V.M., Oliff, A., and Riemen, W.M. 1988. Structure-function analysis of synthetic and recombinant derivatives of transforming growth factor alpha. *Mol. Cell. Biol.* 8:2999–3007.

190. Todaro, G.J., DeLarco, J.E., Fryling, C., Johnson, P.A., and Sporn, M.B. 1981. Transforming growth factors (TGFs): Properties and possible mechanisms of action. *J. Supramol. Struct. Cell. Biochem.* 15:287–301.

191. Hirata, Y., and Orth, D.N. 1979. Epidermal growth factor (urogastrone) in human fluids: Size heterogeneity. *J. Clin. Endocrinol. Metab.* 48:673–679.

192. Coffey, R.J., Jr., Derynck, R., Wilcox, J.N., Bringman, T.S., Goustin, A.S., Moses, H.L., and Pittelkow, M.R. 1987. Production and auto-induction of transforming growth factor-alpha in human keratinocytes. *Nature (London)* 328:817–820.

193. Derynck, R., Goeddel, D.V., Ullrich, A., Gutterman, J.U., Williams, R.D., Bringman, T.S., and Berger, W.H. 1987. Synthesis of messenger RNAs for transforming growth factors alpha and beta and the epidermal growth factor receptor by human tumors. *Cancer Res.* 47:707–712.

194. Sherwin, S.A., Twardzik, D.R., Bohn, W.H., Cockley, K.D., and Todaro, G.J. 1983. High-molecular weight-transforming growth factor activity in the urine of patients with disseminated cancer. *Cancer Res.* 43:403–407.

195. Downward, J., Yarden, Y., Mayes, E., Scrace, G., Totty, N., Stockwell, P., Ullrich, A., Schlesinger, J., and Waterfield, M.D. 1984. Close similarity of epidermal growth factor receptor and v-*erb*B oncogene protein sequences. *Nature (London)* 307:521–527.

196. Riemen, M.W., Wegrzyn, R.J., Baker, A.E., Hurni, W.M., Bennett, C.D., Oliff, A., and Stein, R.B. 1987. Isolation of multiple biologically and chemically diverse species of epidermal growth factor. *Peptides* 8:877–885.

197. Nebert, D.W., and Gonzalez, F.J. 1987. P450 genes: Structure, evolution and regulation. *Annu. Rev. Biochem.* 56:945–993.

198. Guengerich, F.P. 1988. Roles of cytochrome P-450 enzymes in chemical carcinogenesis and cancer chemotherapy. *Cancer Res.* 48:2946–2954.

199. Gonzalez, F.J., Jaiswal, A.K., and Nebert, D.W. 1986. P-450 genes: Evolution, regulation, and relationship to human cancer and pharmacogenetics. *Cold Spring Harbor Symp. Quant. Biol.* 51:879–890.

200. Guengerich, F.P. 1989. Polymorphism of cytochrome P-450 in humans. *Trends Pharm. Sci.* 10:107–109.

201. Wolf, C.R. 1986. Cytochrome P-450s: Polymorphic multigene families involved in carcinogen activation. *Trends Genet.* 2:209–214.

202. Nebert, D.W., Eisen, H.J., Negishi, M., Lang, M.A., and Hjelmland, L.M. 1981. Genetic mechanisms controlling the induction of polysubstrate monooxygenase (P-450) activities. *Annu. Rev. Pharmacol. Toxicol.* 21:431–462.

203. Waterman, M.R., and Estabrook, R.W. 1983. The induction of microsomal electron transport enzymes. *Mol. Cell. Biochem.* 53/54:267–278.

204. Renton, K.W., Keyler, D.E., and Mannering, G.J. 1979. Suppression of the inductive effects of phenobarbital and 3-methylcholanthrene on ascorbic acid synthetic and hepatic cytochrome P-450-linked monooxygenase systems by the interferon inducers, poly rI · rC and tilorone. *Biochem. Biophys. Res. Commun.* 88:1017–1023.

205. Frey, A.B., Rosenfeld, M.G., Dolan, W.J., Adesnik, M., and Kreibich, G. 1984. Induction of cytochrome P-450 isozymes in rat hepatoma-derived cell cultures. *J. Cell Physiol.* 120:169–180.

206. Nowak, D., Schmidt-Preuss, U., Jorres, R., Liebke, F., and Rudiger, H.W. 1988.

Formation of DNA adducts and water-soluble metabolites of benzo[a]pyrene in human monocytes is genetically controlled. *Int. J. Cancer* 41:169–173.

207. Relling, M.V., Evans, W.E., Fonne-Pfister, R., and Meyer, U.A. 1989. Anticancer drugs as inhibitors of two polymorphic cytochrome P-450 enzymes, debrisoquin and mephenytoin hydroxylase, in human liver microsomes. *Cancer Res.* 49:68–71.

208. Nebert, D.W. 1981. Possible clinical importance of genetic differences in drug metabolism. *Br. Med. J.* 283:537.

209. Whitlock, J.P., Jr. 1986. The regulation of cytochrome P-450 gene expression. *Annu. Rev. Pharmacol. Toxicol.* 26:333–369.

210. Baird, W.M., Zennie, T.M., Ferin, M., Chae, Y.-H., Hatchell, J., and Cassady, J.M. 1988. Glucolimnanthin, a plant glucosinolate, increases the metabolism and DNA binding of benzo[a]pyrene in hamster embryo cell cultures. *Carcinogenesis* 9:657–660.

211. Pruess-Schwartz, D., and Baird, W.M. 1986. Benzo[a]pyrene: DNA adduct formation in early-passage Wistar rat embryo cell cultures: Evidence for multiple pathways of activation of benzo[a]pyrene. *Cancer Res.* 46:545–552.

212. Hesse, S., Cumpelik, O., Mezger, M., Kiefer, F., and Wiebel, F.J. 1990. Glutathione conjugation protects some, but not all, cell lines against DNA binding of benzo[a]pyrene metabolites. *Carcinogenesis* 11:485–487.

213. Chung, F.-L., Wang, M., and Hecht, S.S. 1985. Effects of dietary indoles and isothiocyantes on *N*-nitrosodimethylamine and 4-(methylnitrosoamino)-1-(3-pyridyl)-1 butanone alpha-hydroxylation and DNA methylation in rat liver. *Carcinogenesis* 6:539–543.

214. Ioannides, C., and Parke, D.V. 1987. The cytochromes P-448—A unique family of enzymes involved in chemical toxicity and carcinogenesis. *Biochem. Pharmacol.* 36:4197–4207.

215. Nebert, D.W., Nelson, D.R., Adesnik, M., Coon, M.J., Estabrook, R.W., Gonzalez, F.J., Guengerich, F.P., Gunsalus, I.C., Johnson, E.F., Kemper, B., Levin, W., Phillips, I.R., Sato, R., and Waterman, M.R. 1989. The P450 superfamily: Updated listing of all genes and recommended nomenclature for the chromosomal loci. *DNA* 8:1–13.

216. Nebert, D.W. 1979. Genetic differences in the induction of monooxygenase activities by polycyclic aromatic compounds. *Pharmacol. Ther.* 6:395–417.

217. Gonzalez, F.J., Tukey, R.H., and Nebert, D.W. 1984. Structural gene products of the *Ah* locus. Transcriptional regulation of cytochrome P_1-450 and P_3-450 mRNA levels by 3-methylcholanthrene. *Mol. Pharmacol.* 26:117–121.

218. Nebert, D.W., and Jones, J.E. 1989. Regulation of the mammalian cytochrome P_1-450 (CYPA1) gene. *Int. J. Biochem.* 21:243–252.

219. Poland, A., Glover, E., and Kende, A.S. 1976. Stereospecific, high affinity binding of 2,3,7,8-tetrachlorodibenzo-*p*-dioxin by hepatic cytosol. *J. Biol. Chem.* 251:4936–4946.

220. Guenthner, T.M., and Nebert, D.W. 1977. Cytosolic receptor for aryl hydrocarbon

hydroxylase induction by polycyclic aromatic compounds. *J. Biol. Chem.* 252:8981–8989.

221. Bigelow, S.W., and Nebert, D.W. 1982. The *Ah* regulatory gene product. Survey of nineteen polycyclic aromatic compounds' and fifteen benzo[a]pyrene metabolites' capacity to bind to the cytosolic receptor. *Toxicol. Lett.* 10:109–118.

222. Paolini, M., Bauer, C., Biagi, G.L., and Cantelli-Forti, G. 1989. Do cytochromes P-448 and P-450 have different functions? *Biochem. Pharmacol.* 38:2223–2225.

223. Gonzalez, F.J., and Nebert, D.W. 1985. Autoregulation plus upstream positive and negative control regions associated with transcriptional activation of the mouse P_1450 gene. *Nucleic Acids Res.* 13:7269–7288.

224. Jones, P.B.C., Galeazzi, D.R., Fisher, J.M., and Whitlock, J.P., Jr. 1985. Control of cytochrome P_1-450 gene expression by dioxin. *Science* 227:1499–1502.

225. Miskimins, W.K., Roberts, M.P., McClelland, A., and Ruddle, F.H. 1985. Use of a protein-blotting procedure and a specific DNA probe to identify nuclear proteins that recognize the promoter region of the transferrin receptor gene. *Proc. Natl. Acad. Sci. U.S.A.* 82:6741–6744.

226. Mehta, R.G., Cernz, V.L., Madigan, M., and Moon, R.C. 1984. Modification of mouse mammary gland organ culture technique. *J. Tissue Culture Methods* 8:27–30.

14

Peptide Turn Mimetics

Michael E. Johnson and Michael Kahn

14.1. Introduction

Despite the central position that peptides and proteins occupy in the regulation of virtually all biological processes, understanding their structure–function relationship at the molecular level remains in its infancy. Extensive efforts to modify peptides and proteins have been devoted to examining this interplay, and to developing therapeutic agents based on modified peptides and proteins.[1-3] Peptides and proteins have a number of disadvantages as therapeutic agents, including rapid metabolic degradation, lack of specificity, and difficulty in administration. The design of low-molecular-weight peptide and protein mimetics should have application as synthetic catalysts, pharmaceutical agents, and novel immunogens for the generation of antibodies.

Evidence obtained from comparison of the endogenous opioid peptides and the opiate analgesics suggests that the pharmacophoric interactions responsible for the biological activity of these peptides occurs between the peptide side chains and their target receptors. This information, coupled with the lessons of the modified peptide analogues, such as carba-, retro-, and depsipeptides,[4] suggests that the peptide backbone itself may not be essential, and that its role is to orient critical residues for interaction with complementary groups within the receptor. Based on this assessment, the next logical step is the design of new molecular frameworks to assume the role of the peptide backbone, thus providing unique tools for studying the relationship between structure and bioactivity.

14.2. Turns as Recognition Sites

Turns are common elements of secondary structure characterized by their reorientation of the peptide backbone.[5] The surface localization of turns in proteins, and the predominance of potentially reactive functional groups in their side chains, have led to the suggestion that turns function as recognition sites that trigger complex immunologic, metabolic, genomic, and endocrinologic regulatory mechanisms. Additionally, turns have been implicated as nucleation sites of protein folding,[6] and as markers for the co- and posttranslational events of phosphorylation,[7] glycosylation,[8] enzymatic processing,[9] and degradation.[10]

The surface localization of turns in proteins and the predominance of residues containing potentially reactive functional groups in their side chains (e.g., Asn, Ser, Arg, Asp, Lys, and Glu) have led to the hypothesis that β-turns play critical roles in a wide array of molecular recognition events. The role of turns has been unambiguously established in several peptide hormone systems. Although the importance of turns in antigenic recognition, phosphorylation, glycosylation, hydroxylation, and proteolytic processing remains to be established, the frequency with which turns coincide with these recognition processes suggests that conformational recognition plays a significant role. Evidence for the importance of turns in biological recognition is as follows.

14.2.1. Naturally Occurring Peptides with Turn-Containing Bioactive Conformations

Somatostatin is a tetradecapeptide with a cyclic structure formed by a cystine bridge between Cys residues 3 and 14 within the sequence, Ala-Gly-Cys-Lys-Asn-Phe-Phe-Trp-Lys-Thr-Phe-Thr-Ser-Cys. The cyclic nature of this potent natural peptide effector made it a promising target for correlating conformation and activity. Based on the assumption that the proposed bioactive conformation contained a β-turn at Phe^7-Trp^8-Lys^9-Thr^{10}, Veber et al. designed an extremely active cyclic hexapeptide analogue that contains a type II' β-turn at the D-Trp-Lys segment.[11] Thus, somatostatin provides an extremely well established example of a bioactive conformation that contains a β-turn (**1**). The results from somatostatin and analogues, including a retroenantiomeric analogue,[12] also illustrate the necessity of particular side-chain orientations for activity and demonstrate the relative unimportance of backbone conformation.

Examples of critical turn conformations involved in the recognition of bioactive peptides by their receptors have also been documented in the case of α-melanocyte–stimulating hormone,[13] bradykinin,[14] luteinizing hormone-releasing hormone,[15] and enkephalin.[16]

14.2.2. Immunological Recognition of Antigenic Sites

The mechanism by which the open surface of proteins generate highly specific antigenic responses when transferred to another species is a fascinating and critical question of modern biochemistry. One viewpoint is that certain parts of proteins are inherently antigenic and that this property is intrinsic to the nature of the protein molecule and independent of the host to be immunized. Alterna-

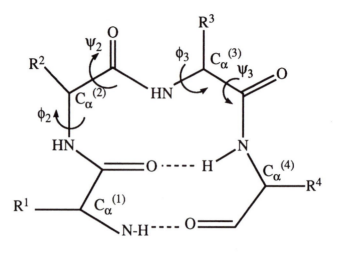

1

tively, the concept that virtually any accessible part of a protein is potentially an antigenic site, and that the choice of sites that elicit an immune response in a particular case depends largely on the bias of the immune system of a specific host, has been put forward. Whichever hypothesis one accepts, the antigenic determinants represent the accessible patches on the surface of a native protein that interact with antibody binding sites. The continuous epitopes comprise five to eight residues, and their uniform size in vastly different proteins highlights the uniform size of the complementary antibody binding sites. Although turns are often coincident with antigenic sites in proteins, it is difficult to determine whether this coincidence is due to specific recognition of turns, or to the frequency of turns among other surface features monitored by the immune system. For example, in sperm whale myoglobin all five continuous sites and two discontinuous sites coincide with turns.[17] The use of hybrid mimetics to inhibit antibodies against native proteins should shed new light on these questions. Moreover, the stabilized conformational epitopes of hybrid mimetics should allow for the generation of specific high-avidity antipeptide antibodies and should circumvent the problem of peptide hyperflexibility.[18]

14.2.3. Immunological Recognition of Viruses

There is increasing evidence that immune recognition of viruses focuses, in part, on immunogenic surface regions frequently coincident with turns. Analysis of the amino acid sequences predicts conformations with a preponderance of β-structure alternating with surface exposed β-turns for hepatitis-B surface antigen polypeptide (P25),[19] Rauscher murine leukemia virus polypeptide (p10),[20] adenovirus "spike" protein,[21] and herpes simplex virus type 1 glycoprotein D.[22] Interestingly, some of these turns are sites of posttranslational modification. Dreesman et al. chose a hydrophilic surface region (117–137) of hepatitis-B virus containing three predicted turns (118–121, 129–132, and 134–137) and synthesized a corresponding peptide with a disulfide bond between residues 124 and 137.[19] After immunization in mice, cross-reactive antibodies specific for hepatitis-B surface antigen developed. Thus, the frequent occurrence in turns with or without prolyl residues, and the propensity of these turns for surface distribution demonstrate their importance as immunogenic sites.

14.2.4. Posttranslational Modifications

14.2.4.1. Folding

Many protein folding mechanisms have been proposed. Some of these have implicated turns as nucleating sites that are formed early in the folding process and direct further folding events.[23] Alternatively, they have been depicted as essentially passive structures that arise in consequence of formative interactions in other parts of the chain.[24] These two extremes may not represent truly distinct alternatives.

Helices and β-sheets tend to run back and forth across protein molecules linked by intervening turns. Chain segments in a folding protein, which are destined to be helical, will exist in multiple conformational states, in which the linear sequence is known to be an important determinant of this equilibrium. However, it is probable that segments distant in sequence but near in space will shift this equilibrium. Regardless of the ultimate conformation adopted by folding segments, the interconnections between them will necessarily be turns. For this reason, turns may be less sensitive to the conformational constraints of spatial neighbors, and their positions better predicted from local sequence alone.

14.2.4.2. Phosphorylation

Protein kinase–dependent phosphorylation of proteins is one of the mechanisms leading to posttranslational protein modification. Intracellular protein kinases are known to transfer a phosphate from ATP to specific seryl, threonyl, and tyrosyl residues. Small et al. demonstrated that 24 of 30 seryl and threonyl phosphorylation sites in 14 proteins are coincident with predicted turn conformations.[7] Cochet and co-workers, using the empirical prediction method of Chou and Fasman, analyzed the available amino acid sequences in protein substrates for protein tyrosine kinases for their secondary structural features.[25] They found, in general, that the preferred conformation at this site was a β-turn with tyrosine in the $i+2$ position. Recently, Tinker et al.[26] synthesized three tyrosine containing peptides: Ala-Pro-Tyr-Gly-NHCH$_3$, Leu-Pro-Tyr-Ala-NHCH$_3$, and Pro-Gly-Ala-Tyr-NH$_2$, of which the first two peptides, but not the third, would be expected (and were shown by IR and CD) to adopt a β-turn with the tyrosine in the $i+2$ position. Peptides 1 and 2 (but not 3), were substrates (V_{max} = 232, 98 mmol/hour/mg and K_m = 1.76, 0.19 mM) for Moloney leukemia virus tyrosine kinase. This preference for a β-turn indicated that this conformation may serve as the recognition site for tyrosine phosphorylation.

14.2.4.3. Glycosylation

Two modes of glycosyl linkages exist in proteins: (1) N-glycosides linked to an asparagine residue of the protein, and (2) O-glycosides linked to a serine, threonine, hydroxylysine, or hydroxyproline residue. It has been suggested that glycosyltransferases recognize turn regions in either case. Glycoproteins containing carbohydrate linked N-glycosidically to an Asn residue invariably possess the sequence Asn-Xxx-[Thr-Ser]. Aubert et al. realized that asparagine, serine, and threonine occur commonly in turns, and by Chou–Fasman analysis demonstrated that 19 of 28 Asn glycosylation sites occur within turn regions in 14 glycoproteins analyzed.[27] Bause and Legler further extended this by examining the effectiveness of peptides with a proline in the $i+1$ position (thus favoring a turn) to act as substrates.[8]

However, the conformation of N-glycosylation sites is not invariably a turn. Wilson et al. (28) demonstrated by X-ray analysis of the Asn-linked oligosaccharide sites in influenza hemagglutinin, that for the six sites analyzed, two were coincident with α-helices, two with β-sheet regions, and two within turns.[28] The Asn residues in influenza hemagglutinin sites were located at the $i+2$ position in contrast to the $i+1$ position found in the Fc portion of IgG. In summary, the evidence that turn conformational recognition is sometimes required is suggestive but still inconclusive. The situation with O-glycosylation is similar, in that turns have been implicated based on secondary structure prediction, but experimental evidence is scant.[27, 28]

14.2.5. Recognition of Proteolytic Processing Sites

Activation of precursor molecules by proteolytic processing, although superficially appearing to be a waste of energy and biosynthetic capacity, is a ubiquitous mechanism in both lower and higher organisms. It appears to be essential in both the replication of picornaviruses[29] and the activation of either plasma proteins or zymogens.[30] In the case of secretory proteins, conversion of the biosynthetic precursor into the biologically active form(s) requires several posttranslational modifications, including proteolytic cleavage by presumed "selective" proteases. The mechanisms involved in controlling this processing are proving complex to unravel. The problem is complicated by the observation that proteolytic processing may be tissue specific,[31] although certain common features have emerged.[32]

Leader (signal) sequences have some common structural features. All have a hydrophobic core region ranging in length from 9 to 20 amino acids. A hydrophilic region containing one or two basic residues precedes the hydrophobic core, and the core region is often terminated by a proline or glycine residue. The cleavage site is on the C-terminal side of the hydrophobic core. Inouye and Halegoua proposed that turns promoted by glycyl or prolyl residues are essential for the proper processing of the signal peptide;[33] however, little direct evidence is available.

Examination of the amino acid sequences of peptide prohormones, as derived from the corresponding cDNA clones, has shown that the sites of cleavage of prohormones are characteristically marked by a pair (or quadruplet, occasionally singlet or triplet) of basic residues (Lys or Arg). Removal of these basic residues generates the free peptide and involves the consecutive action of two separate enzymes. The enzymes responsible for these processing steps are hypothesized to be a group of trypsin-like enzymes and a carboxypeptidase B–like enzyme. The origin of the selectivity of the relevant endoproteases does not appear to coincide with any "consensus" primary sequence, thus implicating a common secondary structure motif. Several types of cleavages occur (i.e., before, after, or within the pair of basic amino acids, thus implicating a common secondary structure motif). Several experiments using peptide analogues[34] or denaturing conditions[35] have implicated a role for a secondary structural motif, as well as the basic amino acid grouping, in recognition by endoproteases. Rholam and co-workers analyzed the amino acid sequences situated around the putative proteolytic cleavage sites in 20 different biosynthetic precursors of peptide hormones.[9] Chou–Fasman secondary structure analysis indicated that processing sequences that are cleaved in vivo are in all cases located within regions with high β-turn formation probability or else immediately adjacent to these structures. The β-turn-forming region at the cleavage locus is flanked on both sides by amino acid sequences with a high probability for forming highly ordered structures (i.e., β-sheet or α-helix). These features are not found around dibasic pairs that are not cleaved in vivo. Based on this analysis, it was hypothesized that β-turns including (or flanking) basic amino acid doublets that are flanked by highly ordered secondary structure units (β-sheet or α-helix) may constitute a minimal recognition motif for endoproteases.

14.3. Peptide Mimetic and Hybrid Mimetic Approach

14.3.1. Rationale

Understanding the relationship between the conformation and the activity of bioactive peptides and proteins is one of the critical goals of contemporary biochemistry. Peptides are highly flexible and can often adopt multiple low-energy conformations, thus presenting different arrays of pharmacophoric residues. Structural restrictions can reduce the difficulty of the investigation of this marriage of structure and function, yielding valuable information about the conformational requirements of peptide–peptide interaction. Additionally, peptides have limitations for biological investigations, in that the ease of proteolytic degradation and the lack of specificity often encountered can cloud and obscure the effect to be monitored.

The logical foundation for the design of peptide mimetics rests upon studies of peptide analogues, analysis of conformational properties of peptides and modified peptides, and on analogies to the relationship between opiate peptides and opiate analgesics. A wide range of modified peptide analogues have been studied. The major conclusion to be drawn from these modifications is that the peptide backbone can be extensively altered, provided that the conformation adopted by the side chains of the analogue approximates that in the native peptide.[36] Additionally, examples of irreversibly cyclized analogues suggest that equilibration to flexible open forms of cyclic peptides is not an essential step in binding to target receptors.[37]

14.3.2. Peptide-β-Turn Mimetic Design

Recent reviews suggest that research into the design and synthesis of peptide turn mimetics has barely scratched the surface.[1-3] The potential to examine hypothesized turn recognition elements in proteolytic processing, posttranslational enzymatic modification and degradation, folding, and antigenic recognition through the use of mimetic hybrids should produce extensive opportunities to examine structure-function relationships, and should allow for the design of new molecular entities that possess increased biological activity, greatly enhanced stability, and potential therapeutic efficacy.

From the schematic structure of a peptide β-turn (1) it can be seen that in forming the turn, the peptide backbone forms a pseudocyclic 10-membered ring structure, with the configuration depending on the ϕ and ψ peptide backbone torsional angles. As was noted previously, the principal function of the backbone appears to be to position the side chain functional groups, R^1 to R^4 as required for biological activity. Thus, the substitution of a ring structure that has similar configuration, and that can be functionalized at the appropriate locations, should provide a peptide β-turn mimetic with potential advantages of enhanced stability, specificity, and reduced antigenicity. An early approach to the design of a nonpeptide β-turn mimetic was that of Friedinger et al.; they reported the design of a lactam-based replacement for the Gly^6-Leu^7 residues in LHRH (2a) to produce a potent agonist with approximately nine times the in vitro biological activity of the native LHRH peptide.[38] A similar approach was used to design an analogue of substance P (2b).[39] The analogue was not particularly effective in binding to the substance-P receptor but was observed to be a strong eledoisin agonist. Since eledoisin and substance P are both tachykinins, this result suggested the existence

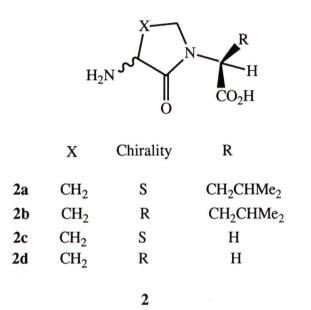

	X	Chirality	R
2a	CH$_2$	S	CH$_2$CHMe$_2$
2b	CH$_2$	R	CH$_2$CHMe$_2$
2c	CH$_2$	S	H
2d	CH$_2$	R	H

2

of multiple tachykinin receptors. Methionine enkephalinamide analogues were also constructed (**2c** and **2d**), with the lactam replacing the Gly^2-Gly^3 residues,[40] but were found to have low activity, suggesting that type II or II' 1→4 β-turns are probably not significant in the bioactive conformation of methionine enkephalin. The same lactam was used to design a gastrin analogue in which the Tyr^{12} and Gly^{13} residues were replaced with **2c** or **2d** and which had potent agonist activity.[41]

This lactam template was also used by Valle et al. to design a potent agonist (10^4 times higher activity) of the bioactive peptide, Pro-Leu-Gly-NH$_2$, with **2d** replacing the Leu^2-Gly^3 residues.[42] Conformational comparison of two isomers, only one of which had activity, suggested that the active conformation could contain a type II, but not a type II' β-turn.

Bicyclic β-turn templates, of the form **3**, have also been used to construct potent gramicidin-S analogues[43] and a moderately active LHRH agonist.[44] Another bicyclic structure (**4**) was designed to replace residues Gly^2-Gly^3-Phe^4-Leu^5 of leucine enkephalin; the resulting analogue was reported to have approximately one-third the analgesic activity of morphine.[45]

The jaspamide mimetic (**5**) based on a 9-member monocyclic ring, was designed to mimic a β-turn found in the jaspamide crystal structure and contains the two functional groups—the phenol and 2-bromoindole—believed to be important in jaspamide; preliminary biological results are encouraging.[46, 47] An 11-member bicyclic ring structure has also been used as a template for the design of thrombin inhibitors based on a β-turn believed to be present in the fibrinopeptide A bound to thrombin.[48] First generation substrate mimetics have K_m and k_{cat} values

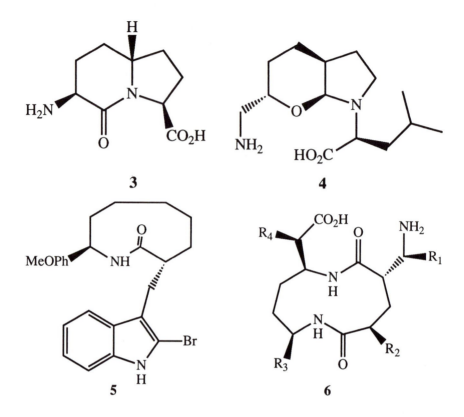

in the range of those for the native substrate, indicating that the mimetic-based species closely mimics the bound conformation of the endogenous substrate.[48] An HIV protease inhibitor with an $IC_{50} \sim 10^{-8}$ M has been developed from an analogous 11-member monocyclic ring system (**6**).[49] Very recent work has also shown that a β-turn mimetic of the β-turn region, Gln^{40}-Thr^{45} of the complementarity-determining 2 region of the CD4 protein exhibits significant activity as an inhibitor of the human immunodeficiency virus gp120 protein binding to CD4.[50]

In summary, the turn conformation is an important conformational element in peptide and protein structure. It also plays a key role in the biological activity of several peptides and proteins of therapeutic importance. The design of peptide turn mimetics, while still very much a developing area, holds promise of providing a number of important new therapeutic agents within the next few years.

References

1. Veber, D.F., and Friedinger, R.M. 1985. The design of metabolically-stable peptide analogs. *Trends Neurosci.* 8:392–396.

2. Friedinger, R.M. 1989. Non-peptide ligands for peptide receptors. *Trends Pharmacol. Sci.* 10:270–274.

3. Ball, J.B., and Alewood, P.F. 1990. Conformational Constraints: nonpeptide β-turn mimics. *J. Mol. Recogn.* 3:55–64.

4. Spatola, A.F., Gierasch, L.M., and Rockwell, A.L. 1983. Conformational comparison of cyclic peptide and pseudopeptide structures with intramolecular hydrogen-bonding. *Biopolymers* 22:147–151.

5. Rose, G.D., Gierasch, L.M., and Smith, J.A. 1985. Turns in peptides and proteins. *Adv. Prot. Chem.* 37:1–109.

6. Kim, P.S., and Baldwin, R.L. 1982. Specific intermediates in the folding reactions of small proteins and the mechanism of protein folding. *Annu. Rev. Biochem.* 51:459–489.

7. Small, D., Chou, P.Y., and Fasman, G.D. 1978. Occurrence of phosphorylated residues in predicted β-turns—implications for β-turn participation in control mechanisms. *Biochem. Biophys. Res. Commun.* 79:341–346.

8. Bause, E., and Legler, G. 1983. The role of the hydroxy amino acid in the triplet sequence Asn-Xaa-Thr(Ser) for the N-glycosylation step during glycoprotein biosynthesis. *Biochem. J.* 195:639–640.

9. Rholam, M., Nicholas, P., and Cohen, P. 1986. Precursors for peptide hormones share common secondary structure forming features at the proteolytic processing sites. *FEBS Lett.* 207:1–6.

10. Fontana, A., Fassina, G., Vita, C., Dalzoppo, D., Zamai, M., and Zambonin, M. 1986. Correlation between sites of limited proteolysis and segmental mobility in thermolysin. *Biochemistry* 25:1847–1851.

11. Veber, D.F., Freidinger, R.M., Perlow, D.S., Paleveda, W.J., Jr., Holly, F.W., Strachan, R.G.R., Nutt, F., Arison, B.H., Hommick, C., Randall, W.C., Glitzer, M.S., Saperstein, R., and Hirschmann, R. 1981. A potent cyclic hexapeptide analog of somatostatin. *Nature (London)* 292:55–58.

12. Freidinger, R.M., Colton, C.D., Perlow, D.S., Whitten, W.L., Paleveda, W.J., Veber, D.F., Arison, B.H., and Saperstein, R. 1983. Modified retro enantiomers are potent somatostatin analogs. In *Peptides, structure and function*. V. J. Hruby and D.H. Rich, eds. Rockford, Ill.: Pierce Chem. Co. Pp. 349–352.

13. Sawyer, T.K., Cody, W.L., Knittle, J.J., Hruby, V.J., Hadley, M.E., Hirsch, M.D., and O'Donohue, T.L. 1983. Design of conformationally-restricted cyclic α-melanotropins: comparison of melanocyte-stimulating and behavioral activities. In *Peptides, structure and function*. V.J. Hruby and D.H. Rich, eds. Rockford, Ill.: Pierce Chem. Co. Pp. 323–331.

14. Chipens, G.I., Mutulis, F.K., Katayev, B.S., Klusha, V.E., Misina, I.P., and Myshlyiakova, N.V. 1981. Cyclic analog of bradykinin processing selective and prolonged biological activity. *Int. J. Protein Res.* 18:302–311.

15. Struthers, R.S., Hagler, A.T., and Rivier, J. 1984. Design of peptide analogs—Theoretical simulation of conformation, energetics, and dynamics. *ACS Symp. Ser.* 251:239–267.

16. DiMaio, J., and Schiller, P.W. 1980. A cyclic enkephalin analog with high *in vitro* opiate activity. *Proc. Natl. Acad. Sci. U.S.A.* 77:7162–7166; DiMaio, J., Nguyen, T.M.D., Lemieux, C., and Schiller, P.W. 1982. Synthesis and pharmacological characterization *in vitro* of cyclic enkephalin analogs—effect of conformational constraints on opiate receptor selectivity. *J. Med. Chem.* 25:1432–1438.

17. Berzofsky, J.A., Buckenmeyer, G.K., Hicks, G., Gurd, F.R.N., Feldmann, R.J., and Minna, J. 1982. Topographic antigenic determinants recognized by monoclonal antibodies to sperm whale myoglobin. *J. Biol. Chem.* 257:3189–3198.

18. Shinnick, T.M., Sutcliff, J.G., Green, N., and Lerner, R.A. 1983. Synthetic peptide immunogens as vaccines. *Annu. Rev. Microbiol.* 37:425–446.

19. Dreesman, G.R., Sanchez, Y., Ionescu-Matiu, I., Sparrow, J.T., Six, H.R., Peterson, D.L., Hollinger, F.B., and Melnick, J.L. 1982. Antibody to hepatitis-B surface-antigen after a single inoculation of uncoupled synthetic HBSAG peptides. *Nature (London)* 295:158–160.

20. Henderson, L.E., Copeland, T.D., Sowder, R.C., Smythers, G.W., and Oroszlan, S. 1981. Primary structure of the low-molecular weight nucleic acid-binding proteins of murine leukemia viruses. *J. Biol. Chem.* 256:8400–8406.

21. Gingeras, T.R., Sciaky, D., Gelinas, R.E., Bing-Dong, J., Yen, C.E., Kelly, M.M., Bullock, P.A., Parsons, B.L., O'Neill, K.E., and Roberts, R.J. 1983. Nucleotide sequences from the adenovirus-2 genome. *J. Biol. Chem.* 257:13475–13491.

22. Watson, R.J., Weiss, H.J., Salstrom, J.S., and Enquist, L.W. 1982. Herpes-simplex virus type-1 glycoprotein-D gene-nucleotide sequence and expression in *Escherichia coli*. *Science* 218:381–384.

23. Zimmerman, S.S., and Scheraga, H.A. 1977. Local interactions in bends of proteins. *Proc. Natl. Acad. Sci. U.S.A.* 74: 4126–4129.

24. Karplus, M., and Weaver, D.L. 1976. Protein folding dynamics. *Nature (London)* 260:404–406.

25. Cochet, C., Gill, G.N., Meisenhelder, J., Copper, J.A., and Hunter, T. 1984. C-kinase phosphorylates the epidermal growth-factor receptor and reduces its epidermal growth factor-stimulated tyrosine protein-kinase activity. *J. Biol. Chem.* 259:2553–2558.

26. Tinker, D.A., Krebs, E., Fettham, I.C., Attah-Polu, S.K., and Ananthanarayanan, V.S. 1988. Synthetic beta-turn peptides as substrates for a tyrosine protein kinase. *J. Biol. Chem.* 263:5024–5026.

27. Aubert, J.P., Bisette, G., and Loucheux-Lefebvre, M.H. 1976. Carbohydrate-peptide linkage in glycoproteins. *Arch. Biochem. Biophys.* 175:410–418.

28. Wilson, I.A., Ladner, R.C., Skechel, J.J., and Wiley, D.C. 1983. The structure and role of the carbohydrate moieties of influenza-virus hemagglutinin. *Biochem. Soc. Trans.* 11:145–147.

29. Nicklin, M.J.H., Toyoda, H., Murray, M.G., and Wimmer, E. 1986. Processing in the replication of polio and related viruses. *Biotechnology* 4:33–36.

30. Barrett, A.J. 1979. *Proteinases in mammalian cells and tissues*. North Holland, Amsterdam.

31. Seizinger, B.R., Grimm, C., Hollt, V., and Herz, A. 1984. Evidence for a selective processing of proenkephalin-B into different opioid peptide forms in particular regions of rat brain and pituitary. *J. Neurochem.* 42:447–457.

32. Docherty, K., and Steiner, D.F. 1982. Post-transitional proteolysis in polypeptide hormone biosynthesis. *Annu. Rev. Physiol.* 44:625–638.

33. Inouye, M., and Halegoua, S. 1980. Secretion and membrane localization of proteins in *Escherichia coli. CRC Crit. Rev. Biochem.* 7:339–371.

34. Kojima-Mizuno, K.M., and Matsuo, H. 1985. A putative prohormone processing protease in bovine adrenal-medulla specifically cleaving in between Lys-Arg sequences. *Biochem. Biophys. Res. Commun.* 128:884–891.

35. Miura, S., Amaya, Y., and Mori, M. 1986. A metalloprotease involved in the processing of mitochondrial precursor proteins. *Biochem. Biophys. Res. Commun.* 134:1151–1159.

36. Garsky, V.M., Clark, D.E., and Grant, N.H. 1976. Synthesis of a non-reducible cyclic analog of somatostatin having only growth-hormone release inhibiting activity. *Biochem. Biophys. Res. Commun.* 73:911–916.

37. Freidinger, R.M., and Veber, D.F. 1979. Peptides and their retro enantiomers are topologically nonidentical. *J. Am. Chem. Soc.* 101:6129–6131.

38. Freidinger, R.M., Veber, D.F., Perlow, D.S., Brooks, J.R., and Saperstein, R. 1980. Bioactive conformation of luteinizing hormone releasing hormone: Evidence from a conformationally constrained analog. *Science* 210:656–658.

39. Freidinger, R.M., Brady, S.F., Paleveda, W.J., Perlow, D.S., Colton, C.D., Whitter, W.L., Saperstein, R., Brady, E.J., Cascieri, M.A., and Veber, D.F. 1987. Synthesis of new peptides based on models of receptor-bound conformation. In *Clinical pharmacology in psychiatry,* S.G. Dahl and L.F. Gram, eds. Berlin: Springer-Verlag. Pp. 12–19.

40. Freidinger, R.M. 1981. Computer graphics and chemical synthesis in the study of conformation of biologically active peptides. In *Peptides: synthesis-structure-function.* D.H. Rich and E. Gross, eds. Rockford, Ill.: Pierce Chemical Co. Pp. 673–683.

41. Douglas, A.J., Mullholland, G., Walker, B., Guthrie, D.J.S., Elmore, D.T., and Murphy, R.F. 1988. The preparation of a C-terminal gastrin peptide containing a synthetic β-bend mimic. *Biochem. Soc. Trans.* 16:175–176.

42. Valle, G., Crisma, M., Toniolo, C., Yu, K.L., and Johnson, R.L. 1989. Crystal state structural analysis of two γ-lactam-restricted analogues of Pro-Leu-Gly-NH$_2$. *Int. J. Pept. Protein Res.* 33:181–190.

43. Sato, K., and Nagai, U. 1986. Synthesis and antibiotic activity of a gramicidin S analogue containing bicyclic β-turn dipeptides. *J. Chem. Soc. Perkin Trans* 1:1231–1234.

44. Nagai, U., Nakamura, R., Sato, K., and Ying, S.Y. 1987. Synthesis of an LHRH analogue with restricted conformation by incorporation of a bicycle β-turn dipeptide unit. In *Peptide chemistry 1986,* T. Miyazawa, ed. Osaka: Protein Research Foundation. Pp. 295–298.

45. Krstenansky, J.L., Baranowski, R.L., and Currie, B.L. 1982. A new approach to conformationally restricted peptide analogs: Rigid β-bends. 1. Enkephalin as an example. *Biochem. Biophys. Res. Commun.* 109:1368–1374.

46. Kahn, M. and Su, T. 1988. Nonpeptide mimetics of jaspamide. In *Peptides: Chemistry and biology: Proceedings of the 10th american peptide symposium*. G. R. Marshall, ed. Leiden: ESCOM Science Publishers B. V. Pp. 109–111.

47. Kahn, M., Nakanishi, H., Su, T., Lee, J.Y.-H., and Johnson, M.E. 1991. The design and synthesis of nonpeptide mimetics of jaspamide. *Int. J. Peptide Protein Chem.*, 38:324–334.

48. Nakanishi, H., Chrusciel, R.A., Shen, R., Bertenshaw, S., Johnson, M.E., Rydel, T.J., Tulinsky, A., and Kahn, M. 1992. Peptide mimetics of the thrombin-bound structure of fibrinopeptide A. *Proc. Natl. Acad. Sci. U.S.A.* 89:1705–1709.

49. Kahn, M., Nakanishi, H., Chrusciel, R.A., Fitzpatrick, D., and Johnson, M.E. 1991. Examination of HIV-1 Protease Secondary Structure Specificity Using Conformationally Constrained Inhibitors. *J. Med. Chem.* 34:3395-3399.

50. Chen, S., Chrusciel, R.A., Nakanishi, H., Raktabutr, A., Johnson, M.E., Sato, A., Weiner, D., Hoxie, J., Saragovi, H.U., Greene, M.I. and Kahn, M. 1992. Design and synthesis of a CD4β-turn mimitic that inhibits human immunodeficiency virus envelope glycoprotein gp120 binding and infection of human lymphocytes. *Proc. Natl. Acad. Sci. U.S.A.* 89:5872-5876.

SECTION III

Biotechnology and the Practice of Pharmacy

15

Biotechnology Products: An Overview

Diana Brixner

15.1. Introduction

In essence, biotechnology defines the collection of industrial processes that involves the use of biological systems.[1] Compared with traditional pharmaceutical drug development, the history of biotechnology is relatively short. A chronological review was recently reported by the Ernst & Young High Technology Group.[2] Currently there are approximately 1,100 biotechnology companies nationwide (public, private, and subsidiary) employing an estimated 60,000 people.[1] The total U.S. drug market generates $50 billion annually in product sales, and the U.S. biotechnology market is predicted to generate over $1 billion in 1991 in the area of therapeutics and diagnostics alone.[3] Also, the sales growth rate for the biotechnology industry is estimated to continue at approximately 25% per year through the next decade, while the drug industry on the whole will maintain an average growth rate of 3%. This rate of growth in biotechnology product sales would result in an $11 billion industry by the year 2000. These predictions arise from the tremendous increased growth in biotechnology research. Over the past 4 years the number of biotechnology medicines in development has grown by 63%.[4] The United States holds three-fourths to four-fifths of the biotech world, compared with one-third of the total drug market, demonstrating a strong dominance in the biotechnology area by the United States.[3] Clearly, that biotechnology has made a significant contribution to health care over the last decade and promises to continue this trend into the future. Health care professionals need to be prepared to handle the impact of a technology with such an extensive potential for growth.

15.2. The Process of Approval for Biotechnology Drugs

In the context of pharmacy practice, biotechnology especially focuses on the application of recombinant DNA and monoclonal antibody technology toward

pharmaceutical products. Through the use of biotechnology a new class of therapeutic agents, generally referred to as biological response modifiers (BRMs), has
been designed. Since monoclonal antibodies are now available that can be directed
against a variety of specific targets, an entirely new dimension of clinical practice
can be realized. For example, these antibodies are used extensively for in vitro
diagnostic kits, the most commonly recognized being pregnancy[5] and ovulation
tests.[6] Current research is directed toward the expanded use of monoclonal
antibodies for in vivo imaging[7] and therapy,[8] and recombinant DNA technology
is being used to produce various different proteins, critical in the treatment of
disease, on a large enough scale for clinical use.

Once a potential product has been developed at the research and development
level, the next step is extensive animal testing. These studies are intended to show
efficacy of the product in animals. Toxicity studies are also done to determine a
LD_{50} (lethal dose for 50% of the population investigated). If the results in these
tests are satisfactory, and the product is believed to be beneficial to humans, one
submits an investigational new drug (IND) application to the Food and Drug
Administration (FDA). This document contains the results of animal studies. The
IND becomes effective if the FDA does not disapprove the application in 30
days.

At this point, human clinical trials can begin. There are four phases of clinical
trials. In Phase I trials the primary concern is to determine a safe dose. The
maximum tolerable dose (MTD) is established by gradually increasing the dose
in increments defined by the protocol, until unacceptable side effects are obtained.
The MTD is then set at one dose level below that point. Phase I studies are
generally done on healthy normal subjects. However, many of the biotechnology
drugs are for the treatment of debilitating diseases, which leads to a combination
of Phase I and Phase II trials. In Phase II trials efficacy studies are begun. Patients
are generally recruited from institutions within the local area. The patients are
studied to determine the efficacy and pharmacokinetics of the drug. Side effects
of the drug are also documented. In Phase III trials the study becomes much more
extensive. These trials are often referred to as multicenter clinical trials because
clinics and hospitals from all over the country become involved. Parameters
indicating efficacy, safety, pharmacokinetics, and side effects are all monitored
and documented. This process takes around 3 years. Once the Phase III trials are
complete all the information is gathered and submitted to the FDA as a new drug
application (NDA), also referred to as a product license application (PLA). The
review process generally takes around 2 to 3 years. Phase IV clinical trials are
ongoing after product approval. Two primary types of Phase IV clinical trials are
those that compare different drugs for a particular treatment and those that
investigate expanded indications for an already approved drug.

Of all the companies involved in biotech research, 58 have brought 132
products into human clinical trials and beyond.[4] The change in the number of
companies and products over the last several years is shown in Table 15.1. The

Table 15.1. Biotechnology Companies and Products in Development over the Last Few Years

	July 1988	October 1988	Annual 1989	Annual 1990	Annual 1991
Companies	48	54	45	51	58
Products in development	81	97	80	104	132
Marketed products	9	10	11	13	16

large increase of 52 products between 1989 and 1991 is indicative of a rapidly expanding industry. Monoclonal antibodies are the largest category of biotechnology drugs in development (58 products), although only one has reached the marketplace. Vaccines are the second largest category (18 products) followed by the interferons (16 products). Cancer or cancer-related conditions are the main target of over half of the biotech products in development.[1]

Currently, there are 17 products on the market resulting from biotechnology. Figure 15.1 shows the number of products per year that have been approved by the FDA since 1982. Table 15.2 lists the approved products in order of their approval and Table 15.3 lists products currently under review at the FDA.

15.3. A Review of Biotechnology Products on the Market

15.3.1. Insulin

The first recombinant DNA product to gain FDA approval was human insulin (Humulin, Eli Lilly) for the treatment of insulin-dependent diabetes. This oc-

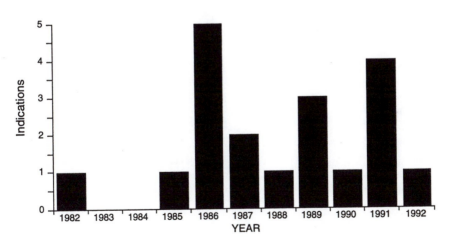

Figure 15.1. Approved biotechnology medicines by year.

Table 15.2.   Approved Biotechnology Drugs and Vaccines Tested by Approval Date

Product Name	Company	Approval Date
Humulin (human insulin)	Eli Lilly	October 1982
Protropin (somatrem)	Genentech	October 1985
Digibind (digoxin immune Fab)	Burroughs Wellcome	April 1986
Orthoclone OKT3 (muromonab CD3)	Ortho Biotech	June 1986
Intron A (interferon-α-2b)	Schering-Plough	June 1986
Roferon-A (interferon-α-2a)	Hoffmann-La Roche	June 1986
Recombivax HB (hepatitis-B vaccine)	Merck	July 1986
Humatrope (somatropin)	Eli Lilly	March 1987
Activase (alteplase)	Genentech	November 1987
HibTiter (haemophilus-B conjugate vaccine)	Praxis Biologics	December 1988
Epogen (epoietin-α)	Amgen	June 1989
Engerix-B (hepatitis-B vaccine)	SmithKline Beecham	September 1989
Alferon N (interferon-α-n3)	Interferon Sciences	October 1989
Actimmune (interferon-γ-1b)	Genentech	December 1990
Neupogen (filgrastim)	Amgen	February 1991
Procrit (epoietin-α)	Ortho Biotech	February 1991
Leukine (sargramostim)	Immunex	March 1991
Prokine (sargramostim)	Hoechst-Roussel	March 1991
Proleukin (aldesleukin)	Cetus	June 1992

Information, in part, from G. J. Mossinghoff, *Biotechnology Medicines*, Pharmaceutical Manufacturer's Assn., 1991 Annual Survey.

Table 15.3.   Biotechnology Drugs in Development

Product Name	Company	U.S. Developmental Status
GMCSF	Sandoz Schering-Plough Genetics Institute (licensor)	Application submitted
Marogen Sterile Powder (epoietin-β)	Genetics Institute Chugai-Upjohn	Application submitted
KoGENate (factor VIII)	Cutter Biological	Application submitted
Mono-1X (factor 1X)	Rhone Poulenc Rorer	Application submitted
Norditropin (somatropin)	Novo Nordisk	Application submitted
Saizen (somatropin)	Serono Laboratories	Application submitted
BioTropin	Genentech	Application submitted
Alferon Gel (interferon-α-2b)	Busch Biotech	Application submitted
Betaseron (interferon-β)	Berlex Laboratories	Phase III
Centoxin (HA-1A MAb)	Centocor	Application submitted
E5 (MAb)	Pfizer Xoma	Application submitted
XomaZyme-CD5 Plus (MAb)	Xoma	Application submitted
Orthozyme CD5 (muromonab CD5- RTA)	Ortho Biotech	Application submitted
DNase (rh DNase)	Genentech	Phase III

curred in 1982. The production of insulin has been by two different methods, both utilizing recombinant DNA technology.

Insulin is produced in the body by the enzymatic degradation of proinsulin. This leaves behind the A and B chain of insulin connected by disulfide bridges (Figure 15.2). Before 1986 human insulin was prepared by the production of the genetically engineered A and B chains in separate fermentations. These individual pieces were then isolated, purified, and chemically joined together. Since 1986 recombinant DNA technology has been used to produce proinsulin. The connecting peptide is then enzymatically cleaved from the human proinsulin to provide human insulin. Multicenter clinical trials have shown no pharmacokinetic or clinical differences between Humulin made by the two manufacturing processes.[9]

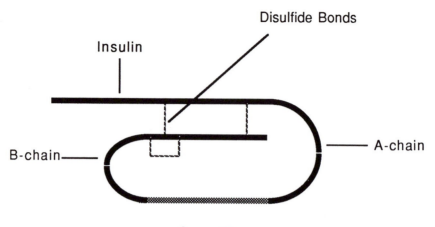

Figure 15.2. The structure of proinsulin. Insulin (solid black) is created by removal of the C-peptide, which leaves the A- and B-chains of insulin connected by disulfide bonds.

15.3.2. Growth Hormone

In 1985 the treatment of growth hormone deficiency in children was greatly improved by the use of recombinant DNA technology. For over two decades prior the only source of the hormone was the pituitary glands of human cadavers. Genentech developed an analogue of human growth hormone called somatrem (Protropin) and produced it on a large scale via recombinant DNA technology. This compound differs from the natural product, somatropin, by the presence of an extra methionine at the end of the molecule. The drug was granted approval by the FDA in 1985 for use in growth hormone deficiency. In 1987 Eli Lilly received approval for the genetically engineered somatropin (Humatrope) with the identical amino acid sequence as growth hormone of pituitary origin.[10] The effects of recombinant human growth hormone on body composition and metabolism in adults with growth hormone deficiency have recently been reported.[11] Mean lean body mass increased by 5.5 kg and fat mass decreased by 5.7 kg in the group treated with growth hormone. No significant changes were noted in the control group.

Currently, three other versions of somatropin are under review for approval by the FDA. These include Norditropin by Novo Nordisk, Saizen by Sereno Laboratories, and BioTropin by BioTechnology General. Norditropin and Bio-Tropin will also be used to treat growth hormone deficiency in children, and Saizen will be indicated for the long-term treatment of growth failure due to inadequate secretion of normal endogenous growth hormone. Genentech has submitted an application to the FDA for the use of Protropin in chronic renal failure. Phase III clinical trials are in progress for the use of Protropin in Turner's syndrome, a condition characterized by retarded growth and sexual development.

15.3.3. Antibodies and Monoclonal Antibodies

The first antibody-derived product to receive approval by the FDA was digoxin immune Fab (Digibind) by Burroughs Wellcome Corp. (Research Triangle Park, N.C.) in April 1986. The product is indicated for the treatment of digitalis toxicity. Digibind employs a unique application of biotechnology toward an antibody-based product. An appropriate antigen is created by conjugating the digoxin molecule to human serum albumin. This conjugate is then injected into sheep to produce digoxin-specific antibodies. The antiserum of the sheep that responded with high titers of antibody, with high affinity and specificity for digoxin, was collected.[12] These antibodies are digested using the enzyme papain, to produce digoxin-specific Fab fragments. The Fab fragments are purified by affinity chromatography.[13] More recently the hybridoma technology has been used to produce monoclonal murine antibodies.[14] Spleen cells from mice immunized with digoxin coupled to serum albumin were fused with a myeloma cell line to yield high-affinity monoclonal antibodies to digoxin. However, in a clinical setting it is often more desirable to use a less specific antidigoxin antibody. A greater cross-reactivity allows for interaction with digitoxin as well as digoxin.[15] After injection into a digitalis toxic patient these Fab fragments bind to digoxin and prevent binding at the site of action. The Fab fragment–digoxin complex then accumulates in the blood and is excreted by the kidneys. In a multicenter clinical trial 21 out of 26 patients with life-threatening toxicity saw a rapid reversal of cardiac rhythm disturbances and hyperkalemia, and a full recovery ensued. There were no adverse effects from the treatment.[12]

Although this drug has received orphan status owing to the relatively low incidence of digitalis toxicity, it is evident that this technology could be applied to various types of drug overdoses and has applicability in suicide rescue.

The first therapeutic application of monoclonal antibody technology was in the development of the immunosuppressant drug muromonab-CD3 (OKT3, Ortho Biotech, Nutley, N.J.). This molecule is used in the treatment of renal transplant rejection. OKT3 is a murine antibody that is directed to a 20,000 molecular weight glycoprotein on the surface of human T cells that is essential for T cell function. T cells play a major role in acute renal transplant rejection. The 20,000 MW molecule is called CD3 for the cell differentiation cluster 3.[16] This factor has been shown, in vitro, to be associated with the antigen recognition structure of T cells and is essential for signal transduction. Although OKT3 blocks all known T cell functions, B cells are still available to assist in fighting infections.

In a multicenter clinical trial OKT3 was compared with the conventional high-dose steroid therapy for treatment in acute rejection of cadaveric renal transplants in 123 patients.[17] Sixty-three patients who received OKT3 also had a concomitant lowering of the dosage of other immunosuppressant drugs. OKT3 reversed 94% of the rejections, whereas steroid therapy demonstrated a 75% reversal. Host antibodies to the murine monoclonal antibody were noted shortly after OKT3 treatment was stopped. This event needs to be considered when contemplating the use of OKT3 for subsequent episodes of rejection.

Because of the large success in the use of this agent for kidney transplants an application is currently under review at the FDA for the use of OKT3 in heart and liver transplants.

In addition to the application for the expanded use of OKT3, four other PLAs have been submitted to the FDA for monoclonal antibody-based products. Two were submitted by Xoma Corp. (Berkeley, Calif.). Xomazyme-CD5 Plus is indicated for the treatment of graft-versus-host disease in bone marrow transplants. This disease can arise when a patient requiring a transplant receives bone marrow from a healthy donor and the patient initiates an immune response.[18] Xomazyme-CD5 is a monoclonal antibody (MAb) directed against the CD5 (cell differentiation cluster 5) antigen of T cells involved in this response. Ortho Biotech has submitted an application for Orthozyme CD5, also for the treatment of graft-versus-host disease. The other MAb agents that have been submitted to the FDA are for the treatment of septic shock. Pfizer and Xoma have submitted an application for a monoclonal antibody (E5) directed against the endotoxins released in septicemia. Centocor (Malvern, Pa.) has also submitted an application for their product, Centoxin (HA-1A MAb). Centoxin is a human monoclonal IgM antibody that has been shown to reduce the lethal effects of gram-negative bacteremia (GNB) and sepis. In Phase III clinical trials mortality was reduced (57%) in patients with GNB and shock compared with the placebo group (33%). These beneficial effects were present as early as seven days after treatment. All patients treated tolerated HA-1A well, and there were no detectable human antibodies generated in response to HA-1A.[19] This result is certainly a reflection of the usefulness of a human monoclonal antibody.

Although monoclonal antibodies are the leading category of biotechnology medicines in development, only one such product has been approved to date (OKT3). The preparation of Digibind used clinically is accurately referred to as a polyclonal antibody. However, with FDA approval currently anticipated for four monoclonal antibody products and many other monoclonal-based agents in clinical trials, these products will have a tremendous influence on medicine over the next decade.

15.3.4. Lymphokines

Along with the introduction of OKT3 in June 1986, interferon-α was introduced into the marketplace by two major drug companies. The interferons are the first in the class of cytokines and lymphokines to receive FDA approval. In general, these substances are naturally occurring proteins that serve as messengers between cells to help coordinate an immune response against a foreign substance. Since one of the problems underlying cancer growth is a breakdown of communication between cells, these cytokines may help to establish a better immune response against a tumor cell. The development of recombinant DNA technology has allowed for the production of individual cytokines in a highly pure form and in large quantity.

The interferons can be further broken down into three basic types: α, β, and γ. They are distinguished by their antigenic specificity, source, and molecular content.[20] Within the α class there are known to be at least 16 different subtypes, each differing slightly in amino acid sequence.[21] The two primary types are α-2a and α-2b. These two interferons differ by only a single amino acid at the twenty-fourth position; the amino acid is lysine and arginine in α-2a and -2b, respectively.

The first indication to gain approval was for the treatment of hairy cell leukemia (HCL). Interferon-α-2a (Roferon-A) is manufactured by Hoffmann–LaRoche, and interferon-α-2b (Intron A) by Schering-Plough.

The results of a multiinstitutional study of 193 patients with hairy cell leukemia receiving interferon-α-2b were reported in 1988.[22] This study observed an 80% response rate, with the response rate being a combination of complete responses, pathological partial responses, and hematological partial responses. The range of all patients was 69 to 92% response. The 91 responding patients were divided into two groups, one group of 45 assigned to observation alone and one group of 46 to receive six additional months of interferon therapy. After 6 months, 6 relapses occurred in patients receiving continued therapy and 14 occurred in patients undergoing only observation. Although the results are encouraging, clinical trials comparing interferons with conventional HCL therapy need to be evaluated. A randomized trial comparing interferon and Pentostatin (2'-deoxyco-formycin) is currently underway.

Intron A and Alferon N (interferon-α-n3) were approved for use in the treatment of genital warts in June 1988 and October 1989, respectively. Roferon A and Intron A were granted approval for the treatment of AIDS-related Kaposi's sarcoma in November 1988. Schering-Plough is seeking to further extend the use of Intron-A and has recently received approval for use in the treatment of chronic hepatitis B, and non-A and non-B hepatitis. An application has been submitted for the use of Intron A in superficial bladder cancer and basal cell carcinoma.

In 1990 the first γ-interferon was approved by the FDA. Interferon-γ-1b (Actimmune) is manufactured by Genentech and is indicated for reducing the frequency and severity of infections associated with chronic granulomatous disease. In a controlled trial, gamma interferon was shown to reduce the frequency of serious infections in patients with this disease. Of 63 patients assigned to interferon, 14 had serious infections, as compared with 30 of the 65 patients with chronic granulomatous disease assigned to placebo.[23] This product is currently in Phase III clinical trials for use in small cell lung cancer and atopic dermatitis.

The interferons are only one type of lymphokine being developed via biotechnology. Like the monoclonal antibodies, there are a large numbers of lymphokine-related products in development. When all lymphokines in development are added together—interleukins (13 products), interferons (16 products), and colony-stimulating factors (8 products)—the total is 37, second only to monoclonal antibodies for the number of products of a particular class in development.

Several colony-stimulating factors, developed by various companies, are beginning to reach the marketplace. The two major types include granulocyte–monocyte colony-stimulating factor (GM-CSF) and granulocyte colony-stimulating factor (G-CSF). Although both GM-CSF and G-CSF will be used in related clinical settings, there are some biological differences. G-CSF specifically stimulates neutrophil granulocytes, whereas GM-CSF stimulates all granulocytes. Also, GM-CSF is a potent activator of monocytes and macrophages as well. This includes the induction of other cytokines, which is why some side effects are seen with GM-CSF that are not seen with G-CSF. Whereas GM-CSF is a potent inhibitor of neutrophil granulocyte migration, G-CSF enhances neutrophil migration.[24] These differences lead to a distinct clinical profile for each drug. Studies have shown GM-CSF to have a significant effect on chemotherapy-induced myelosuppression.[25, 26] One group consisted of 19 patients with either breast cancer or melanoma treated with high-dose chemotherapy and autologous bone marrow support. Leukocyte counts obtained 14 days after transplantation were $3,120 \pm 1,744$ in those given 32 μg/kg of body weight per day. Control groups had levels of 863 ± 645 per microliter.[26] G-CSF has shown a specific effect on patients with neutropenia-associated morbidity due to chemotherapy for transitional cell carcinoma of the urothelium.[27] Treatment with G-CSF before chemotherapy resulted in a dose-dependent increase in the absolute neutrophil count.

In a recent trial, G-CSF was used to reduce chemotherapy-related neutropenia in patients with small cell lung cancer. The results indicated a significantly lower incidence of fever for G-CSF-treated patients (4,090 vs. 7,790 placebo). In addition, the average duration of suboptimal neutrophil count was 6 days for the placebo group as opposed to 1 day for the G-CSF group. Finally, the amount of adjunctive therapy and length of hospitalization due to infections was reduced by approximately 50%.[28]

Two applications have been approved by the FDA for GM-CSF (sargramostim). The approved indication is to accelerate myeloid recovery in patients undergoing autologous bone marrow transplants. Immunex (Seattle, WA) will be marketing the drug under the name Leukine and Hoechst-Roussel (Somerville, N.J.) will market the product as Prokine. Both Hoechst-Roussel and Immunex have submitted an application for indication in the treatment of neutropenia secondary to chemotherapy. Sandoz (East Hanover, N.J.) and Schering-Plough have submitted a PLA together for the use of GM-CSF as an adjuvant to AIDS therapy, in allogenic and bone marrow transplants, and to treat low blood cell counts.

An application for approval of G-CSF (filgrastim) has been submitted by

Amgen for several indications. These include neutropenia resulting from AIDS, leukemia, and aplastic anemia. The FDA has already approved this product (Neupogen) for use in chemotherapy induced neutropenia.

The most recently approved biotechnology medicine is interleukin-2 (IL-2). This lymphokine binds to a specific, high-affinity cell surface receptor expressed on activated T cells. This interaction leads to a variety of effects including the induction of other cytokines such as tumor necrosis factor and interferon-γ.[29] The solid tumors most studied with IL-2 treatment are renal cell carcinoma and malignant melanoma. A 1988 report of a Phase II clinical trial demonstrated the efficacy of IL-2 in conjunction with an infusion of lymphokine-activated killer cells (LAK) against metastatic renal cancer.[30] The product name is Proleukin (aldesleukin) and is manufactured by Cetus Corporation (Emeryville, Calif.) for use in renal cell carcinoma only. Proleukin is currently indicated. Clinical trials are investigating use in other cancers as well.

15.3.5. Vaccines

In the summer of 1986 the hepatitis-B vaccine Recombivax HB (Merck, Rahway, N.J.) received FDA approval. This was the introductory product for the use of recombinant DNA technology in vaccine development.

Biotechnology has offered new strategies for engineering vaccines. Monoclonal antibodies have been a powerful tool in defining the specific nature of the antigenic portion of an infectious agent. Crystallization of the surface components that make up the antigenic determinant has assisted in the structural elucidation of these proteins. Once the portion of the molecule responsible for an immune response is identified, the appropriate genes can be cloned and the protein expressed. Many viral genes have been cloned and expressed in various types of cells. Although amounts of expressed protein have been high, the level of immunity has generally been quite low. An exception to this has been the surface antigen of the hepatitis-B virus (HBsAg).[31]

Before 1986 hepatitis-B vaccine was derived from the human plasma of screened, healthy, chronic HBsAg carriers. The particles were obtained by plasmaphoresis. Merck developed and sold this product as Heptavax-B. In 1986 a different version was obtained via recombinant DNA technology. The product is prepared using genetically engineered cells of the yeast strain *Saccharomyces cerevisiae*. These cells contain the gene for the hepatitis-B surface antigen (HBsAg). The desired protein is expressed, extracted from the cells, purified, and then adsorbed onto aluminum hydroxide.[32] Even though yeast-derived and plasma-derived HBsAg differ slightly in terms of physical and chemical properties, in terms of quality, the antibodies produced in response to the vaccines are indistinguishable.[33]

Merck had an exclusive market on recombinant hepatitis-B vaccine in the U.S. until September 1989, when SmithKline Beecham received FDA approval to market its recombinant version of the vaccine, Engerix-B. The primary difference between the two appears to be in dosing regimens.[34] Engerix-B follows the dosing regimen of plasma-derived Heptavax by Merck. Recombivax HB dosing is half that of Heptavax in adults and one-quarter the dose in children. In most cases the dose seems to provide adequate antibody titers.

There are no other hepatitis-B vaccine products under review by the FDA. Amgen has recently entered Phase III clinical trials for a hepatitis-B vaccine. Other vaccines that are beginning to enter early clinical trials target AIDS (Immune Response Corporation) and malaria (SmithKline Beecham).

In December 1988 another vaccine was granted approval for marketing by the FDA. This was a haemophilus influenza type B vaccine (Hibtiter) by Praxis Biologicals (Rochester, N.Y.). The original version of the vaccine was prepared from the polyribosyl ribitol phosphate (PRP) capsular polysaccharide from the influenza bacteria. However, this vaccine had limited immunogenicity in children under 2 years of age.[35] The immunogenicity was improved through the application of the carrier-hapten immunologic principle. A carrier is a molecule that renders a hapten linked to it able to stimulate antibody production.[36] The carrier most widely studied, and part of the Hibtiter vaccine, is diphtheria toxoid protein. Unlike the PRP vaccine, this vaccine induces high levels of antibody production against haemophilus B. Two recently reported studies both showed significantly improved responses to the haemophilus B conjugate over the haemophilus B PRP vaccine alone.[37, 38] The later study was reported by investigators at the FDA who monitored the occurrence of influenza in two populations, each receiving either conjugated or nonconjugated vaccine. The disease occurrence rate was 83% (106 out of 127) for those receiving Hib vaccine alone and 41% (7 out of 17) for those receiving the conjugate.[38] This approach to improving the antibody titer will possibly be of great use in the future development of vaccines.

15.3.6. Thrombolytics and Factor VIII

Biotechnology entered the thrombolytic market in November 1987 with the introduction of recombinant tissue plasminogen activator (Activase, Genentech). The primary indication is for use in the treatment of mycardial infarction. This drug was slated to be the first billion-dollar product of biotechnology. However, a combination of clinical trial results failed to produce a dramatic increase in the cost–benefit ratio for tissue plasminogen activator (TPA) when compared with other thrombolytics. This has left investors waiting for sales expectations to be met. Likewise, the medical community is also awaiting clinical trial results that effectively compare the thrombolytics.

All available thrombolytics act through the conversion of plasminogen to plasmin, the enzyme that degrades a fibrin clot (Figure 15.3). However, fibrin located throughout the systemic circulation can also be degraded. When this occurs a "lytic state" develops that may lead to excessive bleeding. Attempts to

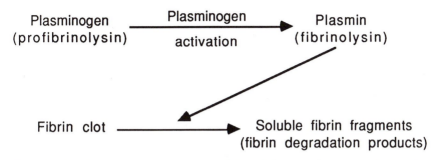

Figure 15.3. The mechanism of action of all thrombolytic agents is to break down the fibrin clot into soluble fibrin degradation products.

avoid the lytic state have centered around producing a clot-specific agent that will not generate plasmin activity outside the fibrin clot.[39]

A number of tissues secrete proteases that enzymatically activate plasminogen. The gene coding for tissue plasminogen activator has been cloned into vectors and expressed to produce recombinant TPA. This drug was initially thought to have two advantages: lack of antigenicity and clot specificity. Although clot specificity was noted in vitro, with purified components, this effect has not been nearly as evident in humans.[39] An explanation may be that a clot-specific agent will also dissolve hemostatic plugs, another contributor to bleeding complications.

Various clinical trials have compared TPA to other specific thrombolytics. The thrombolysis in myocardial infarction trial (TIMI) compared intravenous TPA with intravenous streptokinase (SK).[40] Unlike TPA, SK does not directly cleave plasminogen to produce plasmin. Instead, SK forms a 1:1 complex with plasminogen, which then converts free plasminogen to plasmin.[39] SK is derived from cultures of β-hemolytic streptococci, resulting in its low cost when compared with the recombinant-derived TPA.

Trial results demonstrated an overall advantage in reperfusion for TPA when treatment was begun at an average delay of 4.8 hours after chest pain. However, if therapy was begun during the first 4 hours, this difference was not statistically significant. Because the difference in efficacy is not evident during the crucial time interval for salvaging myocardium and saving lives,[41] the overall clinical advantage of TPA is yet to be determined.

A recent study, comparing TPA to streptokinase, involved the treatment of 12,381 heart-attack victims.[42] The report is referred to as the GISSI II study (Gruppo Italinao per lo Studio della Streptochinasi nell'Infarcto Miocardico). The patients on study all received either streptokinase or TPA, with or without heparin. Heparin was administered 12 hours after the thrombolytic. Mortality rates for either TPA or SK alone were 8.7 and 9.2%, and in combination with heparin rates were 9.2 and 7.9%. These results do not show a dramatic difference between the two products. The dosing of heparin was different from standard

U.S. protocol. In the United States heparin is administered along with the thrombolytic. This further decreases the applicability of the data for use in therapeutic decisions in the United States.

Another thrombolytic that has recently been introduced to the market is anisoylated plasminogen streptokinase activator complex, or APSAC (Eminase, SmithKline Beecham). This drug provides a time-released supply of thrombolytic activity by complex dissociation, allowing for administration via bolus injections over 5 minutes. TPA and SK require infusions from 1 to 6 hours. A study group has determined a mortality reduction from APSAC alone similar to that of either SK or TPA in combination with aspirin.[43] All three agents were directly compared in the Isis-3 study (International Study of Infarct Survival, Part 3). Results indicate that all three agents are equally effective in mortality reduction.[44] Again, the use of heparin in Isis-3 varied from standard U.S. practice. Results from a recent trial indicate that Eminase may offer a solution to the heparin controversy. The Duke University Clinical Cardiology Studies Group (DUCCS-1) gave two patient groups Eminase with or without heparin. Results demonstrated similar morbidity and mortality with a significantly lower incidence of bleeding in patients on Eminase alone.[45]

TPA is currently the only thrombolytic produced via recombinant DNA technology. This offers unique opportunities for drug delivery. The enclosure of cells containing the gene for TPA production, within a bioengineered membrane, would allow for a continuous internal infusion of the drug. Biofeedback stimulated by the TPA buildup in the blood stream could regulate the release of drug.

TPA has also received approval by the FDA for indication in the treatment of pulmonary embolism. Genetics Institute currently has its version of TPA undergoing Phase III clinical trials.

Another recombinant clotting protein that is being developed for use in bleeding disorders is Factor VIII. Hemophiliacs suffer internal bleeding because of a Factor VIII deficiency. Treatment requires infusions of the protein derived from human blood. In the early 1980s these transfusions could be contaminated with the human immunodeficiency virus (HIV). After 1984 manufacturers devised methods to kill the viral component. However, pharmaceutical products derived from human plasma were still highly impure ($< 1\%$ Factor VIII).[46]

The production of recombinant-derived Factor VIII allows a product that can truly be HIV free and of much greater purity. DNA clones encoding the amino acid sequence for human Factor VIII where isolated and used to produce the protein in cultured mammalian cells. The protein is purified by passing the fermentation over an affinity column containing the von Willebrand protein. This molecule complexes with Factor VIII as the cofactor circulates in plasma.[46] The recombinant clotting protein corrects the clotting time of plasma from hemophiliacs and has many similar biochemical and immunological characteristics to the serum-derived Factor VIII.

Currently, an application is under review by the FDA for the use of Factor VIII

(KoGenate; Cutter Biological, Berkeley, Calif.) in the treatment of hemophilia. Rhone-Poulenc Rorer (Fort Washington, Pa.) has submitted an application for Factor 1X produced using recombinant DNA technology. This product will also be used in the treatment of hemophilia.

15.3.7. Erythropoietins

The biotechnology product that truly has the potential to be biotech's first billion-dollar drug is recombinant-derived erythropoietin (Epogen; Amgen, Thousand Oaks, Calif.). Whereas sales of TPA are only expected to increase from $200 million to $300 million over the next 5 years, Epogen sales are projected to increase to $1.5 to $2.0 billion during the same amount of time.[47] Erythropoietin was granted FDA approval in 1989 for the indication of anemic kidney dialysis patients.

Erythropoietin is a glycoprotein hormone that is normally produced in the kidneys and is responsible for the regulation of red blood cell production, specifically. This hormone has been found to have specific receptor sites on erythroid progenitor cells in the bone marrow. Erythropoietin production in the kidneys is stimulated by renal hypoxia, a direct result of anemia. However, in the patient with chronic renal failure, blood flow to the kidney is continuously decreased, resulting in an incomplete erythropoietic response to anemia and a persistent hypoxic state. This deficiency of erythropoietin is thought to be the primary cause of anemia associated with chronic renal failure.[48]

Before the era of recombinant DNA technology, erythropoietin was available only from sheep plasma or human urine, where it was present in very small amounts. In 1983 the gene coding for the hormone was determined. This gene was then incorporated into the host chromosomes of Chinese hamster ovary cells and allowed to replicate and subsequently express erythropoietin on a production scale.[49] In 1986 human clinical trials were initiated using Epogen in the treatment of anemia due to end-stage renal disease. Phase I and II trials involved 35 patients in the United States and United Kingdom.[50, 51] All responded with an increase in red blood cell production, lack of transfusion requirements, normalization of hemoglobin and hematocrit, and improvement in general well-being. These results warranted a Phase III multicenter clinical trial involving 333 patients.[52] The positive results were further corroborated. Issues related to patient management that were identified in Phase I and II trials were further substantiated in the Phase III trial. Some of the more important issues include blood pressure management, iron-deficiency monitoring, and awareness of potential thrombotic events. The conclusion of the trial was that Epogen is an effective, safe, and well tolerated drug highly recommended in the treatment of anemic patients with end-stage renal disease. The decreased need for transfusions was well borne out in this trial. Using the group of patients (333) as a representation of all end-stage renal disease patients, the savings in blood nationally would be projected at 500,000 units a year.[52] This would help to avoid the inherent risks of blood transfusions.

Although Epogen is approved for the indication of chronic renal failure-induced anemia, the potential exists for treating other types of anemia as well. A particular area of interest is the use of Epogen in AIDS patients being treated with azidothymidine (Zidovudine, AZT). In a recent report of a study done on 63 patients (29 receiving Epogen, 34 receiving placebo), reductions in the number of transfusions required became evident for the treatment group in the second and third months of the trial.[53] This can have a significant impact on the treatment of bone marrow suppression and anemia resulting from the treatment of AIDS with zidovudine. In 1990, Epogen received FDA approval for use in Zidovudine-induced anemia in AIDS patients. Ortho Biotech recently received approval by the FDA to market their version of erythropoietin (Procrit). This product has indications for the treatment of anemia associated with chronic renal failure, AIDS, or zidovudine therapy.

Genetics Institute has submitted a product license application for its version of erythropoietin-β (Marogen) for anemia secondary to kidney disease.

15.3.8. Tumor Necrosis Factors and DNase

Another class of agents involving biotechnology are the tumor necrosis factors. These substances can serve as intercellular messengers to trigger an immune response against tumor cells. This leads to the destruction of blood vessels that supply the tumor and deprive the mass of nutrition.[1] Products based on this technology are still in the early stages of development. Biogen, and Genentech have compounds in Phase I and II clinical trials. All are directed toward cancer treatment.

A final product, which has moved rapidly through the drug development process, is recombinant human DNase. This enzyme has utility in the treatment of cystic fibrosis, a lung disease characterized by highly viscous, purulent secretions. Large amounts of DNA contribute to this condition, and bovine pancreatic DNase I has been shown to reduce the viscosity of these secretions. Genentech has used recombinant DNA technology to produce human DNase.[54] The clinical utility of this product has been demonstrated by the University of Washington Clinical Research Center. Patients receiving rhDNase via inhalation had improved lung function when compared to patients receiving placebo.[55] Genentech has recently submitted an application to the FDA for the use of DNase in cystic fibrosis.

15.4. The Pharmacist and Biotechnology

Clearly, the drugs described previously demonstrate the effect of biotechnology on the future of pharmacy practice. With a new class of therapeutic agents comes a tremendous need for the education of health care professionals and patients. It is clear that the pharmacist will play an integral role in this process. For example, as reported by Schweigert et al., a group of 60 pharmacists recommended changes in drug therapy in 62% of the cases, and 80% of the recommendations were implemented.[56]

Another important consideration in the use of biotechnology products is cost. Research and development costs are high, and this expense is reflected in the price of the product. The high cost of these products therefore creates a challenge for the pharmacist involved in formulary management.[57] Health benefits produced by these agents need to be weighed against costs. Many of these products offer new care or a marked improvement over traditional agents when used in appropriate clinical settings. One also needs to consider the economic impact of a shortened hospital stay, often a direct result of biotechnology therapy.[58] The clinical application of biotechnology products will provide fertile ground for clinical outcomes research and for the use of pharmacoeconomic principles. The pharmacist is the ideal person to coordinate such research. In addition, results will support a necessary increase in the pharmacy budget to affect a decrease in various other hospital budgets. Pharmaceutical companies that manufacture biotechnology products will need to form an alliance with pharmacists involved in administration at hospitals and managed care organizations to demonstrate the positive economic impact of their product to an institution. Without this partnership, many biotechnology products will not receive formulary approval.

The role of the pharmacist will also involve patient education on biotechnology products. According to a recent Gallup Poll survey,[59] the public considers the pharmacist knowledgeable and an important source of information. Seventy-nine percent of the respondents consider pharmacists to be very knowledgeable about medicines. This is similar to the percentage of respondents ranking medical doctors in this category (83%). Therefore, as prescriptions for biotechnology products increase in the future, it is clear that patients and other health professionals will rely on pharmacists for an explanation of how to use these products as well as a description of their mode of action.

References

1. Mossinghoff, G.J., 1989. *Biotechnology Medicines*. Pharmaceutical Manufacturer's Association, May.

2. Burrill, G.S., with the Ernst & Young High Technology Group. 1990. *Biotech 1991: A changing environment*. New York: Mary Ann Liebert.

3. Shamel, R.D. 1990. Biotechnology . . . What's in store for the 1990's? *Consul. Res. Corp. Newslett.* Spring.

4. Mossinghoff, G.J. 1991. Biotechnology Medicines Pharmaceutical Manufacturers Association Annual Survey.

5. McCombs, J. 1989. Home pregnancy tests. *Pharm. Times* September:35–37.

6. McCombs, J. 1989. Ovulation prediction kits. *Pharm. Times* August:39–41.

7. Serafini, A. 1990. Biocompatible MoAbs detect, stage disease. *Diagnos. Imaging* June:94–107.

8. Fritzberg, A.R., Berninger, R.W., Hadley, S.W., and Wester, D.W. 1988. Approaches to radiolabeling of antibodies for diagnosis and therapy of cancer. *Pharmaceut. Res.* 5:325–334.

9. Anderson, J.H., Jr. Personal communication. Clinical Investigation and Regulatory Affairs Division, Eli Lilly Corporation, Indianapolis.

10. Humatrope monograph. 1986. Eli Lilly Corporation, Indianapolis, Ind. January.

11. Salomon, F., Cuneo, R.C., Hesp, R., and Sonksen, P.H. 1989. The effects of treatment with recombinant human growth hormone on body composition and metabolism in adults with growth hormone deficiency. *N. Engl. J. Med.* 321:797–1803.

12. Smith, T.W., Butler, V.P., Jr., Haber, E., Fozzard, H., Marcus, F.I., Bremmer, W.F., Schulman, I.C., and Phillips, A. 1982. Treatment of life-threatening digitalis intoxication with digoxin-specific Fab antibody fragments. *N. Engl. J. Med.* 307:1357–1362.

13. Curd, J., Smith, T.W., Jaton, J.C., and Haber, E. 1971. The isolation of digoxin-specific antibody and its use in reversing the effects of digoxin. *Proc. Natl. Acad. Sci. U.S.A.* 68:2401–2406.

14. Hunger, M.M., Margolies, M.N., Ju, A., and Haber, E. 1982. High-affinity monoclonal antibodies to the cardiac glycoside, digoxin. *J. Immun.* 129:1165–1172.

15. Butler, V.P. 1986. Development of digibind. *Management of digitalis toxicity,* 33. Monograph, Burroughs Wellcome Co.

16. Muromonab-CD3. 1990. *Facts Comparisons* June 738f.

17. Ortho Multicenter Transplant Study Group. 1985. A randomized clinical trial of OKT3 monoclonal antibody for acute rejection of cadaveric renal transplants. *N. Engl. J. Med.* 313:337–342.

18. Roitt, I., Brostoff, J., and Male, D. 1989. *Immunology,* 18.17. St. Louis: C.V. Mosby, and New York: Gower Medical.

19. Ziegler, E., Fisher, C., Sprung, C., Straube, R., Sadoff, J., and the HA-1A Sepsis Study Group. 1991. Treatment of gram-negative bacteremia and septic shock with HA-1A human monoclonal antibody against endotoxin. *N. Engl. J. Med.* 324:429–436.

20. DeVita, V.T., Jr., Hellman, S., and Rosenberg, S.A., eds. 1985. *Cancer: Principles and practice of oncology.* 2nd ed. Philadelphia: J.B. Lippincott.

21. Kirkwood, J.M., and Ernstoff, M.S. 1984. Interferons in the treatment of human cancer. *J. Clin. Oncol.* 2:336–349.

22. Golomb, H.M., Fefer, A., Golde, D.W., Ozer, H., Portlock, C., Silber, R., Rappeport, J., Ratain, M.J., Thompson, J., Bonnem, E., Spiegel, R., Tensen, L., Burke, J.S., and Vardima, J.W. 1988. Report of a multi-institutional study of 193 patients with hairy cell leukemia treated with interferon-alfa 2b. *Sem. Oncol.* 15:7–9.

23. The International Chronic Granulomatous Disease Cooperative Study Group. 1991.

A controlled trial of interferon gamma to prevent infections in chronic granulomatous disease. *N. Engl. J. Med.* 324:509–516.

24. Gabrilove, J.L. 1989. Introduction and overview of hematopoietic growth factors. *Sem. Hematol.* 26:1–4.

25. Antman, K.S., Griffin, J.D., Elias, A., Socinski, M.A., Ryan, L., Cannistra, S.A., Oette, D., Whitley, M., Frei, E., III, and Schnipper, L.E. 1988. Effect of recombinant human granulocyte-macrophage colony-stimulating factor on chemotherapy-induced myelosuppression. *N. Engl. J. Med.* 319:593–598.

26. Brandt, S.J., Peters, W.P., Atware, S.K., Kurtzberg, J., Borowitz, M.J., Jones, R.B., Shpall, E.J., Bast, R.C., Gilbert, C.J., and Oette, D.H. 1988. Effect of recombinant human granulocyte-macrophge colony-stimulating factor on hematopoietic reconstitution after high-dose chemotherapy and autologous bone marrow transplantation. *N. Engl. J. Med.* 318:869–875.

27. Gabrilove, J.L., Jakubowski, A., Scher, H., Sternberg, C., Wong, G., Grous, J., Yagoda, A., Fain, K., Moore, M.A.S., Clarkson, B., Oettgen, H.F., Alton, K., Welte, K., and Souza, L. 1988. Effect of granulocyte colony-stimulating factor on neutropenia associated morbidity due to chemotherapy for transitional-cell carcinoma of the urothelium. *N. Engl. J. Med.* 318:1414–1422.

28. Crawford, J., et. al. 1991. Reduction by granulocyte colony-stimulating factor of fever and neutropenia induced by chemotherapy in patients with small cell lung cancer. *N. Engl. J. Med.* 325:164–170.

29. Parkinson, D.R. 1989. The role of interleukin-2 in the biotherapy of cancer. *Oncol. Nursing Forum* 16:16–20.

30. Fisher, R.I., Coltman, C.A., Doroshow, J.H., Rayner, A.A., Hawkins, M.J., Mier, J.W., Wien, P., McMannis, J.D., Weiss, G.R., Margolin, K.A., Gemlo, B., Hoth, D.F., Parkinson, D.R., and Paietta, E. 1988. Metastatic renal cancer treated with interleukin-2 and lymphokine-activated killer cells. *Ann. Intern. Med.* 108:518–523.

31. Fields, B.N., and Chanock, R.M. 1989. What biotechnology has to offer vaccine development. *Rev. Infect. Dis.* 11:S519–S523.

32. Hepatitis B vaccine. 1990. *Facts Comparisons* June 467a.

33. Ellis, R.W., and Gerety, R.J. 1989. Plasma-derived and yeast-derived hepatitis B vaccines. *Am. J. Infect. Control* 17:181–189.

34. Hepatitis B vaccine. 1990. *Facts Comparisons* June 468.

35. Stieb, D.M., Frayha, H.H., Oxman, A.D., Shannon, H.S., Hutchitson, B.G., and Crombie, F.S.S. 1990. Effectiveness of haemophilus influenzae type b vaccines. *Can. Med. Assoc. J.* 142:7.

36. Roitt, I., Brostoff, J., and Male, D. 1989. *Immunology,* 8.2. St. Louis: C.V. Mosby, and New York: Gower Medical.

37. Schneider, L.C., Insel, R.A., Howie, G., Madore, D.V., and Geha, R.S. 1990. Response to a haemophilus influenzae type b diphtheria CRM197 conjugate vaccine in children with a defect of antibody production to haemophilus influenzae type b polysaccharide. *J. Allergy Clin. Immunol.* 85:948–953.

38. Nelson, W.L., and Granoff, D.M. 1990. Protective efficacy of haemophilus influen- zae type b polysaccharide-diphtheria toxoid-conjugate vaccine. *ADJC* 144:292–295.

39. Gray, W.J., and Bell, W.R. 1990. Fibrinolytic agents in the treatment of thrombotic disorders. *Sem. Oncol.* 17:228–237.

40. Chesebro, J.H., et al. 1987. Thrombolysis in myocardial infarction (TIMI) trial, phase I: A comparison between intravenous tissue plasminogen activator and intravenous streptokinase. *Circulation* 76:142–154.

41. Marder, V.J. 1989. Comparison of thrombolytic agents: Selected hematologic, vascu- lar and clinical events. *Am. J. Cardiol.* 64:1A–7A.

42. Gruppo Italiano per lo Studio della Sopravvivenza nell'Infarto Miocardico, 1990. GISSI-2: A factorial randomised trial of alteplase versus streptokinase and heparin versus no heparin among 12,490 patients with acute mycocardial infarction. *Lancet* 336:65-71.

43. Chamberlain, D.A. chairman Aims Trial Study Group. 1988. Effect of intravenous APSAC on mortality after acute myocardial infarction: Preliminary report of a placebo- controlled clinical trial. *Lancet* 1:545–549.

44. ISIS-3 (Third International Study of Infarct Survival) Collaborative Group. 1992. ISIS-3: A randomised comparison of streptokinase vs tissue plasminogen activator vs anistreplase and of aspirin plus heparin vs aspirin alone among 41,299 cases of suspected acute myocardial infarction. *Lancet* 339:753-770.

45. Duke University Clinical Cardiology Study Group. 1992. Comparison of APSAC with and without adjuvant heparin therapy in the treatment of AMI. *New Engl. J. Med.* (submitted for publication).

46. Wood, W.I., Capon, D.J., Simonsen, C.C., Eaton, D.L., Gitschier, J., Keyt, B., Seeburg, P.H., Smith, D.H., Hollingshead, P., Wion, K.L., Delwart, E., Tudden- ham, E.G.D., Vehar, G.A., and Lawn, R.M. 1984. Expression of active human factor VIII from recombinant DNA clones. *Nature* (*London*) 312:330–337.

47. Kaweske, J.J. 1990. *Biotechnology in the market place; An investment in pharmacy's future.* 32nd Annual Pharmacy Congress, St. John's University, New York, N.Y., April.

48. Flaharty, K.K., Grimm, A.M., and Vlasses, P.H. 1989. Epoetin: Human recombinant erythropoietin. *Clin. Pharm.* 8:769–782.

49. Eschbach, J.W. 1989. Erythropoietin: Promise and fulfillment. *Renalife* Spring.

50. Eschbach, J.W., Egrie, J.C., Downing, M.R., Browne, J.K., and Adamson, J.W. 1987. Correction of the anemia of end-stage renal disease with recombinant human erythropoietin. *N. Engl. J. Med.* 316:73–78.

51. Winearls, G.C., Oliver, D.O., Pippard, M.J., Reid, C., Downing, M.R., and Cotes, P.M. 1986. Effect of human erythropoietin derived from recombinant DNA on the anaemia of patients maintained by chronic haemodialysis. *Lancet* 2:1175–1178.

52. Esbach, J.W., et al. 1989. Recombinant human erythropoietin in anemic patients with end stage renal disease. *Ann. Intern. Med.* 12:992–1000.

53. Fischl, M., et al. 1990. Recombinant human erythropoietin for patients with AIDS treated with zidovudine. *N. Engl. J. Med.* 322:1488–1493.

54. Shak, S., et. al. 1990. Recombinant human DNase I reduces the viscosity of cystic fibrosis sputum. *Proc. Natl. Acad. Sci. USA* 87:9188-9192.

55. Aitken, M.L., et al. 1992. Recombinant human DNase inhalation in normal subjects and patients with cystic fibrosis. A phase 1 study. *JAMA* 267:1947-1951.

56. Schweigert, B.F., Oppenheimer, P.R., and Smith, W.E. 1982. Hospital pharmacists as a source of drug information for physicians and nurses. *Am. J. Hosp. Pharm.* 39:74–77.

57. Herfindal, E.T. 1989. Formulary management of biotechnological drugs. *Am. J. Hosp. Pharm.* 46:2516–2520.

58. Grimm, A.M., Flaharty, K.K., Hopkins, L.E., Mauskopf, J., Besarab, A. and Vlasses, P.H. 1989. Economics of epoetin therapy. *Clin. Pharm.* 8:807–810.

59. Gallup, G., Jr. 1990. Public perceptions of pharmacists. NACDS/Gallup survey.

16

The Pharmacist Practitioner's Role in Biotechnology: Clinical Application of Biotechnology Products

Janet P. Engle, Donna M. Kraus, Louise S. Parent, and Mary Dean-Holland

16.1. Introduction

The profession of pharmacy has undergone many changes. Practice emphasis has shifted from a product focus to one that encompasses total patient care, regardless of the practice setting. Changes in the drug product have necessitated increased pharmacy participation in patient care. Biotechnology will dictate drug therapy in the future; therefore, it is critical for pharmacists to become familiar with these products as well as their impact on patient care. For instance, biotechnology products that are designed for use in the home require intense counseling and monitoring for maximal benefit.

Currently, there are more than 80 biotechnology products and vaccines being studied for human use in the United States.[1] The profession of pharmacy must meet the challenge that this technology presents for increased service to the public. Continued professional success will be determined by pharmacists' commitment to continuing their education and adapting to the new environment required to dispense and monitor drug therapy produced by biotechnology. Pharmacists must take the initiative to assume the authority for biotechnology drug usage. If pharmacists are not assertive in this role, other health care providers may claim control over these new drug products, as has already occurred in the area of radiopharmaceuticals.

This chapter is divided into two sections. The first discusses potential roles for the pharmacist in the clinical use of biotechnology agents. The second discusses several broad categories of drugs derived from biotechnology with the intent of familiarizing pharmacists with these agents and giving further examples of the role practitioners can play in their use.

16.2. The Pharmacist's Role in Biotechnology

Pharmacists can potentially be involved in all aspects of the biotechnology drug use process including product distribution, patient and physician education, clinical monitoring, administration, research, and social and ethical policy-making decisions.

16.2.1. Distribution

The vast majority of these products will not be marketed as oral preparations. Most of the currently produced biotechnology products are proteins and would be destroyed if administered orally. Hence, traditional nonoral routes of administration (im, sq, iv) as well as newer routes of administration such as nasal insufflation will be used.

Given the unique qualities of these products, storage requirements will likely be specialized. Larger refrigeration areas and separate storage sections may be required. Reconstitution of these products will require aseptic technique and special diluents. Owing to high overhead costs (products and special handling), not all pharmacies may be able to afford to carry these products.

16.2.2. Education

Today's health care system is characterized by early discharge from the hospital, with patients receiving treatment at home. Given the unique and complex properties of biotechnology drugs and products, these patients (or their care givers) must receive in-depth information of dilution, reconstitution, administration, storage, and a host of additional information including adverse effects and toxicities. Pharmacists are ideally suited to provide this information, particularly if the information is given in the patient's home, or in specialized administration clinics designed specifically for the patient receiving biotechnology drugs. Owing to the high cost of these agents, other health professionals involved in a patient's care also need information concerning delivery and monitoring techniques.

Although the pharmacist is often considered the drug information specialist, current pharmacy curricula, postgraduate training, and continuing education may be inadequate to serve the needs of the pharmacist involved in biotechnology drugs. The profession should seek to resolve the issues of curriculum changes, certification, required continuing education in biotechnology drugs, or, perhaps, relicensure, before these issues are resolved for us by consumer or governmental agencies. If the future of pharmacy is to include high-tech drugs and drug products, the pharmacists of today must prepare to face that future or be willing to see others have a more active role in high-tech health care.

With the advent of biotechnology drugs, it is likely that pharmacy practice acts will be updated to include regulations specific to these agents. Boards of pharmacy may choose to require continuing education in biotechnology for licensure renewal. Certification in the area of biotechnology may also become a reality, as it has in the area of nuclear pharmacy. The pharmacy board exam will also be updated to incorporate questions that test the practitioner's knowledge of biotechnology drugs.

16.2.3. Clinical Monitoring

The pharmacist must become involved with the dosing, kinetics, and therapeutic drug monitoring that will be necessary to adequately utilize these new agents. A potential role for pharmacists will be found in ambulatory clinics, where administration and long-term monitoring of these drugs will take place. This is already a reality in the cases of growth hormone and epoeitin-α.

Depending on the specific biotechnology agent, home monitoring systems will need to be developed. The pharmacist may have a significant impact in this area. This is especially true in the case where biotechnology agents are utilized for home diagnostics such as pregnancy tests. The pharmacist is in the ideal position to counsel the patient on the appropriate use of these agents.

16.2.4. Administration

Owing to the high cost of biotechnology drugs, the pharmacist must become involved in the administrative aspects of the drug use process. The practitioner must develop a formulary system that is capable of handling requests for these agents. A structure should be in place that ensures that formulary requests for these agents are adequately reviewed in terms of the scientific literature, comparison to currently available agents, indications, and cost. For each agent that is added to a formulary, drug use evaluation criteria should be established. It will be incumbent on the pharmacist to ensure that rational, safe, cost-effective therapy is provided to all patients.

Third-party reimbursement will be another important issue that must be faced by practitioners who utilize biotechnology drugs. Reimbursement for nondistributive services associated with the use of these agents, such as counseling and monitoring, should be considered by third-party payors. This would also include reimbursement for the pharmacist's decision *not* to dispense a biotechnology agent owing to lack of indication or availability of a less expensive alternative therapy. Drug utilization evaluation studies will also be critical in justifying the use of a biotechnology agent.

Another aspect of reimbursement will be payment for using a product for a non-FDA approved indication. As biotechnology products become more available and are used for a greater variety of disease states, third-party payors may withhold payments for uses that are not FDA approved.

Waste disposal may also become an issue, depending on the type of products developed. Patients will need counseling on how to dispose of these products if they are used in the ambulatory or home environment. Pharmacists and other health care providers will need education in this area as well.

16.2.5. Research

Owing to the nature and short plasma half-life of most biotechnology products, many will need to be conjugated with other substances (e.g., SOD-PEG) or made into unique dosage formulations (liposomal encapsulation) to decrease the product's clearance or to increase the product's effectiveness. The development of new biotechnology products utilizing these unique delivery systems requires the specialized training and viewpoint of a pharmacist. Unfortunately, today's graduate students are, more often than not, nonpharmacists. The lack of pharmacy-trained scientists has already affected pharmacognosy, the science of drug identification and development from natural sources. Schools and colleges of pharmacy must evaluate their basic science and pharmaceutics curricula in terms of attracting qualified graduate students with interests in biotechnological drug development. Ideally, these researchers can work more closely with clinical researchers when drugs are in the human-testing phase. The pharmacist as a clinical researcher has a definite role not only in determining dosing regimens and side effects but in defining incompatibilities with other parenterally administered drugs, interactions between biotechnology products and more traditional drugs in oral dosage forms, and the impact that biotechnology drugs have on major organ systems. The interferons, for example, inhibit the cytochrome P-450 enzyme system, increasing the therapeutic concentrations of some drugs at standard doses.[2] The clinical research pharmacist should also be involved with cost–benefit studies and drug utilization reviews, particularly since the biotechnology drugs are expensive and may have diverse indications.

16.2.6. Ethics

As with other new technologies, biotechnology drugs are, and will continue to be, a topic for ethical debate. Owing to the prohibitive cost of many of these agents, lower socioeconomic classes may have limited access to these drugs. There may be a temptation to use the drugs for inappropriate indications. Quality of life issues are also cogent, given the high cost and questionable benefit of these agents in certain situations. Perhaps the health care system needs to provide a mechanism whereby ethical issues surrounding these agents are addressed. As part of the health care team, pharmacists must become cognizant of these important matters and become involved in resolving the social and ethical issues surrounding biotechnology drugs.

16.3. The Products: A Clinical Focus

16.3.1. Superoxide Dismutase (SOD)

Superoxide dismutase is an intracellular enzyme that acts as a free radical scavenger by catalyzing the dismutation of superoxide anions to hydrogen peroxide and oxygen. Superoxide radicals are highly reactive substances that interact with cellular components to cause cell injury and death. Superoxide radicals, along with other activated oxygen species such as hydroxyl radicals and hydrogen peroxide, are considered important in the pathophysiology of oxygen toxicity, inflammation, radiation sickness, and cellular injury due to reperfusion following ischemic or hypoxic events.[3] Because cellular damage secondary to oxygen free radicals may occur from a variety of etiologies and in a wide range of tissues, the theoretical uses of superoxide dismutase are quite diverse. Currently, human recombinant SOD is being investigated for the treatment of reperfusion injury in acute myocardial infarction and renal transplantation, inflammatory diseases such as arthritis and colitis, and oxygen toxicity in premature neonates. Preliminary results in patients undergoing thrombolysis for acute myocardial infarction suggest that human recombinant SOD may decrease reperfusion arrhythmias.[4] Further studies are needed, however, to determine the beneficial effects, if any, on left ventricular function and long-term patient survival.

Intraarticular injections of bovine superoxide dismutase (orgotein) have been reported to be beneficial for osteoarthritis and periarticular inflammatory processes such as epicondylitis (tennis elbow) and other sports injuries. Bovine superoxide dismutase has also been used for the treatment of rheumatoid arthritis, radiation injury, and Crohn's disease.[5] In an effort to prolong the short serum half-life (0.5 hour) and to enhance tissue uptake, bovine SOD has been conjugated to polyethylene glycol, dextran, and albumin.[3] A polyethylene glycol-conjugated form of superoxide dismutase (PEG-SOD) is currently being investigated. Although a low incidence of allergic reactions has been reported with the bovine preparation (less than 1% of patients treated), anaphylactic reactions have occurred.[5] Allergic reactions should be minimal with human superoxide dismutase.

Beneficial effects of superoxide dismutase for particular types of cellular injury will depend on the extent of the role that free oxygen radicals play in the pathophysiology of that specific disease or injury. Possible future clinical uses of human superoxide dismutase, alone or in combination with other free radical scavengers, include adult respiratory distress syndrome, septicemia, prevention of cardiotoxicity and myelosuppression from chemotherapy, radiation protection, topical use for treatment of skin ulcers and thermal burns, and possibly the prevention of postischemic renal and CNS reperfusion injury.

16.3.2. Tumor Necrosis Factor (TNF)

Tumor necrosis factor, also known as cachectin, is a cytokine secreted predominantly by activated macrophages in response to cellular damage or invasion by bacteria or endotoxin. The physiological effects of TNF are multifold and complex. Tumor necrosis factor plays an important role in inflammation, immune modulation, tissue remodeling, and the regulation of tumor cell growth. The biological effects of TNF are listed in Table 16.1. Tumor necrosis factor's indirect tumor necrosis effects are most likely mediated by action on the tumor vascular endothelium. Tumor necrosis factor also suppresses lipoprotein lipase and with prolonged exposure may cause cachexia. In excessive amounts TNF mediates the

Table 16.1.   Biological Effects of TNF

Increases phagocytic activity of neutrophils
Promotes adherence of activated granulocytes to vascular endothelium
Increases procoagulant activity
Induces interleukin-1 synthesis
Possesses direct action on growth and differentiation of T and B cells
Stimulates fibroblasts
Involved in normal cell angiogenesis
Possesses direct and indirect cytotoxic effects on tumor cells

physiological derangements seen in septic shock.[6] The antitumor activity of TNF was first observed in the late nineteenth century when necrosis of tumors was noted in cancer patients with streptococcal infections. After decades of research, TNF was discovered and isolated, but it was not until 1984 that it was sequenced and cloned.

Tumor necrosis factor is currently in Phase I and Phase II trials for treatment of solid tumors. Owing to its short plasma half-life (10 to 20 minutes) and cell cycle-dependent effects, various dosing schedules and lengths of continuous infusions have been studied. The predominant side effects seen with TNF may vary with different dosing schedules. Fever, chills, rigor, headache, and fatigue are usually observed during short-term infusions, but hypotension is the dose-limiting effect seen with single-dose and 24-hour infusions. Thrombocytopenia and increases in bilirubin were the main dose-limiting toxicities seen with 5-day infusions.[7] Additional side effects of TNF are listed in Table 16.2. Although cachexia has been induced with TNF in animals, weight loss and cachexia have not as yet been observed in human trials. This may become an important side effect if prolonged treatment protocols are utilized.

Since TNF mediates the pathophysiologic effects of septic shock and cachexia, its role as an antitumor agent may be somewhat limited.[6, 8] During clinical trials, TNF-induced hypotension required treatment with aggressive fluid resuscitation and dopamine. Nonsteroidal antiinflammatory agents and meperidine were needed for the treatment of fever and rigors.[9, 10] Owing to its toxic side effects,

Table 16.2.   Adverse Effects of TNF

	Adverse Effects
CNS	Fever, chills, rigor, headache, fatigue
Cardiovascular	Hypotension, hypertension, peripheral cyanosis
Gastrointestinal	Nausea, vomiting, diarrhea, increases in liver enzymes, increases in bilirubin, lipolysis, hypertriglyceridemia
Hematologic	Leukopenia, leukocytosis, eosinophilia, monocytosis, thrombocytopenia
Other	Back pain, erythema, dyspnea, cachexia

TNF's future use as a single agent in the clinical setting is uncertain. In addition, consistent antitumor effects have not been noted with the use of TNF alone.[11, 12] Synergy of TNF with interferon has been noted, however, and clinical trials are underway investigating combination therapy of TNF with interferon as well as interleukin-2, chemotherapeutic agents, monoclonal antibodies, and radiation.[6]

In addition to combination therapy, recombinant TNF-like cytokines are also being investigated. Alteration of the N-terminal amino acid sequence of TNF may lead to TNF-like cytokines that may have greater anticancer effects but less systemic toxicity.[13] Induction of endogenous TNF in combination with lower doses of exogenously administered TNF may possibly decrease the frequency of severity of systemic toxicities.[13] Retroviral insertion of the TNF gene into melanoma-derived tumor-infiltrating lymphocytes may allow for higher concentrations of TNF in the local melanoma tumor region with possibly fewer systemic side effects.[14]

Monoclonal antibodies to TNF are also being studied for prevention and treatment of septic shock and cachexia and may have a greater therapeutic potential than TNF itself. In addition, owing to the increase in serum TNF concentrations that is seen in AIDS patients, TNF monoclonal antibodies may possibly play a role in combination regimens for the treatment of AIDS.[15]

16.3.3. Growth Hormone

Nonrecombinant human growth hormone obtained from cadaver pituitaries was removed from the market in mid-1985 owing to the associated development of Creutzfeldt–Jakob disease, a debilitating neurological slow-virus disease. The advent of recombinant DNA technology allowed the continued availability of growth hormone. Two recombinant growth hormone products are currently marketed in the United States, somatropin (Humatrope, by Lilly) and somatrem (Protropin, by Genentech). Both have amino acid sequencing identical to human growth hormone of pituitary origin; however, somatrem contains an additional methionine group. The metabolic effects of somatropin and somatrem are equal to pituitary human growth hormone and the two products are bioequivalent with respect to area under the plasma concentration time curve.[16] (For additional information regarding new growth hormone products under FDA review and the use of growth hormone in adults, the reader is referred to Chapter 15, this volume.)

Human growth hormone affects the growth and metabolism of multiple body systems (see Table 16.3) and is currently indicated as replacement therapy in children with growth failure secondary to growth hormone deficiency. Both currently available products are contraindicated in patients with closed epiphyses or active tumors. The two products are available as lyophilized powder for injection, which must be reconstituted before use. Somatrem's diluent contains benzyl alcohol and should therefore be avoided in newborns and in patients with

Table 16.3. Biological Effects of Growth Hormone

Stimulates linear, skeletal, skeletal muscle, and internal organ growth
Increases protein synthesis
Decreases body fat stores
Possesses initial insulin-like but long-term diabetogenic effects
Promotes sodium, potassium, and phosphorous retention
Helps maintain positive calcium balance

hypersensitivity to benzyl alcohol. Somatropin's diluent contains *m*-cresol and glycerin and must also be avoided in patients who are hypersensitive to these components. Both products may be reconstituted with sterile water for injection (SWI) to avoid the use of the supplied diluent.[17]

The optimal dose of human growth hormone for the treatment of growth hormone deficiency is yet to be identified. Weekly doses of somatropin, for example, have ranged from 0.5 to 0.7 IU/kg administered in divided doses three to seven times per week with resultant mean height velocities ranging from 5.8 to 13.9 cm/year.[18–20] Increases in height velocities are typically greater for younger children and support the practice of early treatment in growth hormone-deficient patients. Doses of human growth hormone should be individualized, and currently both products are recommended to be administered three times per week. Several studies, however, suggest that greater height velocity and bone density are observed with the same weekly dose administered in divided doses six to seven times per week rather than two to three times per week.[18, 21, 22] Further studies are required to determine if the final height attained is greater with a daily regimen versus a three-times-per-week administration schedule. In addition, optimal dosing regimens to achieve appropriate pubertal growth spurts for both sexes need to be identified.[23]

Adverse effects of human growth hormone are infrequent and minor; however, several case reports of symptomatic hypoglycemia and leukemia in children during prolonged use have been reported. The risk of leukemia has been estimated to be 1 in 2,400 with a 10-year course of growth hormone therapy.[24] Antibodies to growth hormone develop in approximately 30 to 40% of patients receiving somatrem and in 2% of patients receiving somatropin. Antibodies usually do not decrease the growth response, but failure to respond to somatrem secondary to antibody formation has been reported.[17]

Research efforts using human growth hormone for nondeficiency states were previously limited owing to the limited supply of growth hormone from cadavers. Now that a virtually unlimited supply of human growth hormone is available via recombinant DNA technology, human growth hormone is being studied for other indications such as chronic renal failure, Turner's syndrome, and extreme catabolic states (e.g., burns). Other areas of potential research include metabolic bone disease, obesity, hyperlipidemia, von Willebrand's disease, and aging.[25] In addition, the availability of human growth hormone and its possible unwarranted use in children with short statue who do not meet the classic diagnostic criteria for growth hormone deficiency will present both clinical and ethical dilemmas for the clinician.[26]

Because the use of growth hormone requires intramuscular or subcutaneous administration and prolonged courses of therapy, pharmacy practitioners can greatly affect the care of these patients. Opportunities for parent and patient education regarding the product, instruction in proper techniques for dosage preparation and administration, and long-term patient monitoring for this product are abundant.

16.3.4. Colony-Stimulating Factors

Colony-stimulating factors (CSFs) are a group of acidic, low-molecular-weight glycoproteins that are vital in the transformation of hematopoietic stem cells into peripheral blood cells (i.e., neutrophils and monocytes; see Chapter 15, this volume, for further information). Early in 1991 two CSFs received FDA approval. Filgrastim (Nupogen), a granulocyte colony-stimulating factor (G-CSF) was approved for decreasing the risk of post-cancer-chemotherapy neutropenia and infection; and sargramostim (Leukine, Prokine), a granulocyte–macrophage colony-stimulating factor (GM-CSF), was approved for enhancement of myeloid recovery after autologous bone marrow transplant in Hodgkin's and non-Hodgkin's lymphomas and acute lymphoblastic leukemia. GM-CSF has also been studied in the management of chemotherapy-related neutropenia, for the amelioration of zidovudine (AZT) related hematologic deficiencies in HIV-positive and AIDS patients, and as an enhancer of the cytotoxic effects of chemotherapeutic agents such as Ara-C in the management of myelodysplastic syndromes.[27–29]

Both agents have been generally well tolerated in clinical trials. The most common side effect of G-CSF is mild to moderate bone pain, which can usually be managed with oral analgesics.[30] Fever and bone pain have also been reported with GM-CSF.[31] Additional adverse reactions include myalgias, flushing, and phlebitis. These CSFs may transiently increase alkaline phosphatase levels as well as decrease cholesterol levels. One of the major concerns about CSFs is that they may stimulate the growth of malignant cells as well as normal cells or may induce malignancies due to the chronic stimulation of stem cells.[29, 32] In one study, progression to acute leukemia was noted in 3 of 18 patients receiving G-CSF. It is still unclear if this is an effect of the CSF or a natural progression of the underlying disease.[29] Patients should be counseled about this possibility and monitored closely after treatment with CSFs. Additionally, CSFs have been reported to enhance the replication of HIV, which may have negative implications for their use as adjuncts in HIV-positive subjects.[33] It must also be realized that as CSFs minimize the hematologic side effects of chemotherapeutic agents, allowing larger doses to be used, previously unknown side effects of the chemotherapeutic agents may become apparent.

Table 16.4. Administration of GM-CSF

1. Unreconstituted vials should be stored under refrigeration (2°–8°C). Do *not* freeze

2. Use 1 ml SWI (without preservatives) to reconstitute the powder. Inject the sterile water into the vial directed toward the sides and swirl gently to prevent foaming. Do *not* shake

3. The resulting isotonic solution should be clear and colorless. This is a single dose vial; do *not* reenter or reuse; dispose of any unused portion properly

4. Dilution of the solution should be done with 0.9% SCI. If the administered concentration of sargramostim (GM-CSF) is <10 µg/ml, albumin must be added to the sodium chloride solution before the addition of the GM-CSF to prevent its binding to the delivery system. A concentration of 0.1% human albumin should be used (1 mg albumin/1 ml of normal saline = 0.1% albumin)

5. The reconstituted and diluted solution should be stored at 2°–8°C. Do *not* freeze. The GM-CSF solution must be used within 6 hours after reconstitution, preferably as soon as possible, since it does *not* contain any bacterial preservatives. Discard any unused solution no longer than 6 hours after reconstitution.

6. Data on compatability and the stability of GM-CSF in the presence of other agents are unavailable. Do *not* add any medications other than albumin (at the concentration recommended above) to the 0.9% saline solution containing the GM-CSF.

Adapted from Colony Stimulating Factors. 1991. *Facts and Comparisons* May:84i–85.

G-CSF is given sc or iv at an initial dose of 5 µg/kg/day, starting no sooner than 24 hours after the conclusion of chemotherapy and continuing until the absolute neutrophil count (ANC) is $\geq$ 10,000/mm^3. G-CSF dosage may be increased in 5 µg/kg/day increments for subsequent chemotherapy cycles, depending on the patient's response.[34] Since many patients receive chemotherapy in the clinic setting, G-CSF will also be given on an outpatient basis, in the clinic or at home.

GM-CSF is also given iv, but unlike G-CSF, the product must first be reconstituted (see Table 16.4). For post-bone marrow transplant use, the recommended dose of GM-CSF is 250 µg/m^2/day for 21 days or until the ANC $\geq$ 20,000/mm^3.[34] In this situation, GM-CSF will be used on an inpatient basis, but as other uses become more common, some patients will receive GM-CSF treatment as outpatients, either in the clinic or at home. Pharmacists can play an important role in patient education regarding self-administration and in patient monitoring and dosage adjustment.

The estimated price tag for one chemotherapeutic cycle of G-CSF is $1,800 to $2,800, and that for one 21-day course of GM-CSF post-bone marrow transplantation is $4,200.[31] In this age of cost containment, these drugs will need to prove their worth before being routinely used. Since they do not appear to decrease overall mortality, cost justification will most likely be based on decreased morbidity and decreased use of other health care resources. Studies have documented shorter hospital stays and decreased use of antibiotics in patients treated with CSFs, which may offset some of the drug costs.[35] However, few cost justification studies have been done, and the bottom line is still unclear. Pharmacists should take an active part in determining the appropriate therapeutic role for these agents by becoming involved in quality assurance and drug utilization review programs within their institutions.

16.3.5. Peptides

Peptides are chains of amino acids that differ from proteins in that they are shorter in length. Because of their lower molecular weight, they travel around the body more readily than proteins. If a specific area of a protein is identified as having biological activity, that individual segment of amino acids can be synthesized by biotechnological methods and made available for use. Human insulin is an example of such a peptide available in the United States. Many other peptides are currently under investigation for disease states ranging from the common cold to congestive heart failure.

16.3.5.1. Human Insulin

Human insulin produced by biotechnological methods differs from the pork and beef insulins in that the peptide chain is identical to that secreted by humans instead of being different at one or two sites, respectively. "Synthetic" human insulin can be produced in three ways: full chemical synthesis, enzymatic conversion of pork insulin, or recombinant DNA technology. Full chemical synthesis is currently too expensive for routine use in the diabetic patient, but the human insulin produced via the two other methods is widely available[36] (see Chapter 15, this volume).

Human insulin is slightly less immunogenic than purified pork insulin and is the insulin of choice for patients who will be on intermittent insulin therapy, such as gestational diabetics. Others also recommend it for treatment of newly diagnosed insulin-dependent diabetics.[37] Since the cost of human insulin is only slightly higher than that of the purified pork insulins, changing to human insulin is also appropriate in patients who manifest difficulties with purified pork insulins. A decrease in antibody titers has been reported in patients switched from animal-derived insulins to human insulin.[37] However, allergic reactions to human insulin have been noted in patients without a history of previous insulin allergies.[36] Diabetics who have been receiving nonhuman insulin and are switched over to human insulin may notice a small change in their glycemic control, but unless significant insulin resistance was present insulin requirements are generally unchanged.[38] It has been reported that human insulin is absorbed more quickly from subcutaneous sites and is slightly shorter acting than the corresponding animal-derived insulins; patients may notice a more rapid drop in blood glucose coupled with a loss of glycemic control toward the end of the dosing interval.[39, 40] Occasionally, patients may need to increase the frequency of insulin dosing to compensate for this change.[37]

Since human insulin became widely available there have been a number of reported cases describing episodes of "hypoglycemic unawareness" in patients switched from animal-derived insulins to human insulin. In these reports, newly switched patients suffered hypoglycemia without developing the usual warning symptoms that usually accompany low blood glucose.[41–43] As unrecognized hypoglycemia may have severe consequences, this raised much concern over the widespread use of human insulin. Two studies that have addressed this issue have found small but significant differences in patient perception of, and physical response to, hypoglycemia induced by animal-derived versus human insulin.[44, 45] In the majority of patients the clinical significance of this will be minimal; however, patients and their families should be informed about this potential problem when a patient is changed to human insulin. Human insulin may also be beneficial for patients who are experiencing lipoatrophy as a result of using animal-source insulin; in some cases injection of human insulin into the lipoatrophied area (at a 45° angle) may lead to improvement or resolution of the atrophy. Human insulin does not confer much advantage over the highly purified pork insulins in retarding the development of lipohypertrophy, since this appears to be secondary to the lipogenic effect of insulin itself rather than the source of insulin.[37]

In conclusion, human insulin is an advance in the management of diabetes, albeit a small one for most patients.

16.3.5.2. Atrial Natriuretic Factor

Atrial natriuretic factor (ANF) is a 28-amino acid peptide that is released from the cardia atria on atrial distension. It causes natriuresis and diuresis with a much less pronounced affect on kaliuresis. Calcium, phosphorus, and magnesium losses may also be increased.[46, 47] Additionally, ANF causes a vasodilatory effect without triggering a rebound tachycardia or activation of the renin–angiotensin system.[47] In essential hypertension and congestive heart failure, levels of endogenous ANF have been found to be elevated. This has led to trials of intravenously administered ANF to examine the benefits of exogenous ANF in these disease states. Results have been promising; however, ANF has a half-life of approximately 3 minutes and must be administered intravenously. These limitations would restrict this agent's use to the inpatient or acute care setting. To widen ANF's therapeutic potential in the management of hypertension and CHF, attempts are being made to increase endogenous ANF levels through inhibition of its metabolism. Approaches include the development of ANF analogues that can preferentially bind to, and block the site of, ANF metabolism or compounds that inhibit the enzymes that degrade ANF.[47] An orally available ANF analogue that is resistant to enzyme breakdown would be ideal.

16.3.5.3. Peptide T

Peptide T is an octapeptide (Ala-Ser-Thr-Thr-Thr-Asn-Tyr-Thr) that appears to be very similar to a segment of the envelope glycoprotein (gp 120) of the human immunodeficiency virus (HIV).[48] In vitro studies have indicated that peptide T can block binding of the viral envelope to the CD4 receptor.[49] In several small studies, AIDS patients who intravenously and intranasally administered peptide T for 4 weeks demonstrated hematologic and clinical improvement with minimal toxicities.[49, 50] Additionally, there are several case reports of improvement in patients with psoriasis receiving peptide T.[51]

The availability of an intranasal formulation should significantly improve the ease of administration of this agent, allowing patients to administer their medication at home. However, intranasal doses need to be 5-fold the intravenous dose to achieve similar plasma levels.[50]

16.3.6. Vaccines

Biotechnology has many possible applications in the area of active immunization. A wide variety of disease states are the targets for the new vaccine development, including safer polio and pertussis vaccines and vaccines against cholera, influenza A, malignant melanoma, Epstein–Barr virus, and even a vaccine against pregnancy (see Chapter 15, this volume, for more information). Several biogenetically engineered vaccines are already FDA approved (hepatitis B, hemophilus influenzae type B), and vaccines against the AIDS virus (HIV) are currently undergoing intense research.

16.3.6.1. Hepatitis-B Vaccine

Recombivax HB, the first recombinantly engineered vaccine to receive FDA approval, provides several advantages over the previously available plasma-derived product (Heptavax). Production of Recombivax HB by yeast cells provides an unlimited source of vaccine, in comparison with the plasma-derived vaccine, Heptavax, which was dependent on the harvesting of antigen from the plasma of chronic hepatitis-B carriers. Also, concerns about transmission of other disease states, such as AIDS, that arose with the plasma-derived vaccine were eliminated with the recombinant product because it contains no human-derived elements.[52]

As a result of studies indicating that the recombinant vaccines (Recombivax HB and Engerix B) result in similar immunologic responses and have similar adverse-effect profiles, the recombinant vaccines have become the products of choice except in patients with significant yeast allergies.[52] (The plasma-derived vaccine is no longer produced in the United States.) The recommended dose for Recombivax HB is 10 μg/dose, and that for Engerix B is 20 μg/dose. A dosage of 40 μg is recommended for dialysis patients because they are generally poor responders to usual doses. Recombivax is available in a high-potency formulation (40 μg/ml) for these patients. Recombivax HB contains formalin and should not be used in patients who are formalin sensitive; the Engerix B product should be used instead. Dosing should be im, avoiding the gluteal muscle area, which has been associated with a decreased immune response to this vaccine.[53] Although sc dosing can be used in patients at significant risk for bleeding, this route of administration may be associated with a poor immune response. The standard dosing regimen for both recombinant vaccines is the same: 0, 1, and 6 months. If rapid onset immunity is desired for patients at high risk, data indicate that a four-dose regimen (0, 1, 2, and 12 months) may produce more rapid immunity. However, the administration of a fourth dose would significantly increase the cost of the series.[52, 54] Booster doses are not routinely recommended for healthy adults or children at this time, since the duration of protection is unknown. For patient groups with a history of poor response to the hepatitis-B vaccine, such as immunocompromised patients and patients on dialysis, monitoring of serum titers and administration of boosters as necessary based on titer results is warranted.[52]

Cost, one of the major obstacles to widespread compliance with hepatitis-B vaccination recommendations, was not overcome by development of the new vaccines. Both of the recombinant vaccines are approximately the same price as the previously recommended plasma-derived product.[52] Studies looking at intradermal administration of 1 to 2 μg of the hepatitis-B vaccine have shown that an adequate immune response develops in a significant portion of the subjects studied. The use of a lower intradermal dose would significantly decrease the cost of vaccination and has been shown to be well accepted by patients.[55, 56] However, the duration of protection from intradermal vaccination is not adequately delineated, and this route of administration has been associated with an increased incidence of local and systemic reactions. Additional research is needed before intradermal vaccination with hepatitis B can be recommended for routine use.

16.3.6.2. Hemophilus Influenzae Type B Vaccine

In 1985 the FDA approved a vaccine against hemophilus influenza type B (HIB), the number one cause of bacterial meningitis in children in the United States. It was indicated for use in children 24 months of age or older; in younger patients it produced limited immunogenicity.[57] However, it is estimated that 66 to 75% of all systemic HIB infections are seen in children less than 18 months of age with a peak incidence between 6 and 7 months of age.[58] Subsequently, the development of protein conjugate vaccines against HIB have led to enhanced immunogenicity in the 2- to 15-month-old age group[59, 60] (see Chapter 15, this volume). Additionally, these new vaccines appear to produce an improved immune response in patient groups with a history of poor antibody production with the older polysaccharide vaccine.[60, 61]

There are three protein conjugate HIB vaccines approved by the FDA and an additional agent is in testing (see Table 16.5). Only two (Pedvax HIB and HibTiter) are currently approved for use in children between 2 and 15 months of

Table 16.5. Haemophilus Influenzae Type B Protein Conjugate Vaccines

Vaccine Name (Abbreviation)	Trade Name	Carrier Protein	Manufacturer	FDA Licensed, Age for Use
PRP-D	ProHibit	Diphtheria toxoid	Connaught	Yes, $\geq$ 15 months
HbOC	HibTiter	Nontoxic diphtheria toxin variant CRM_{197}	Praxis (distributed by Lederle)	Yes, $\geq$ 2 months
PRP-OMP	Pedvax HIB	Outer membrane protein complex of *Neisseria meningitidis*	Merck, Sharp & Dohme	Yes, $\geq$ 2 months
PRP-T		Tetanus toxoid	Merieux (to be distributed by Connaught)	No

Adapted from Committee on Infectious Diseases. 1991. Haemophilus influenzae type B. *Report of the Committee on Infectious Diseases*, 22nd ed. Elk Grove Village, Ill.: American Academy of Pediatrics. 1991.

age. Primary dosing schedules for these two vaccines are not the same (see Table 16.6) and their interchangeability is unknown. Three injections are necessary when HibTiter is used, but the Pedvax HIB series only requires two injections. Given these differences, if the source of a previous vaccination is unclear, a 2- to 6-month-old infant should receive a primary series of *three* injections of protein conjugate vaccine (all doses given after 2 months of age can be counted toward this total). For previously unvaccinated infants 7 to 11 months of age, or infants 12 to 14 months of age, two doses 2 months apart or one dose, respectively, of either Pedvax HIB or HibTiter may be given. A booster at 15 months of age (at

Table 16.6. Vaccination Schedules for Haemophilus Influenzae Type B Protein Conjugate Vaccines

Age at First Dose (months)	HbOC (HibTiter)	PRP-OMP (Pedvax HIB)	PRP-D (ProHibit)
2–6 months	3 doses[a]	2 doses[a]	NR[b]
7–11 months	2 doses[a]	2 doses[a]	NR
12–14 months	1 dose	1 dose	NR
Booster for above primary series	1 dose at 15 months[c]	1 dose at 15 months[c]	NR
15–59 months	1 dose	1 dose	1 dose

[a]All doss should be given at least 2 months apart.

[b]NR, not recommended.

[c]Booster doses should be given at least 2 months after last dose.

Adapted from Committee on Infectious Diseases. 1991. Haemophilus Influenzae Type B. *Report of the Committee on Infectious Diseases*, 22nd ed. Elk Grove Village, Ill.: American Academy of Pediatrics and Hemophilus B Conjugate Vaccines. 1991. *Facts and Comparisons* April:459g–459i.

least 2 months after the last dose) is necessary for both Pedvax HIB and HibTiter. For unvaccinated children 15 months of age or older, one injection with any of the three available protein conjugate vaccines appears to provide adequate protection.[62, 63]

Adverse effects of the protein conjugate HIB vaccines are generally not severe. Local adverse effects, including erythema and swelling, and fever have been reported to occur in approximately 2 percent of children. Systemic side effects include irritability, drowsiness, gastrointestinal symptoms, respiratory symptoms, and rash. There does not appear to be any increase in the incidence of side effects with subsequent doses.[59, 60]

Simultaneous administration of protein conjugate HIB vaccines with other childhood immunizations (e.g., MMR, DTP, and OPV) does not appear to decrease the immunogenicity of either the HIB vaccine or the other vaccines. Adverse reactions may be increased by concurrent administration, but it appears to be an additive, rather than a synergistic, effect.[60, 64] However, the simultaneous administration of HIB with other injectable vaccine combinations should be done at different sites. With multiple doses of the protein conjugate HIB vaccines needed to achieve adequate protection in addition to the multiple other infant immunizations that are required, parents may feel overwhelmed, both medically and financially (if their health insurance does not cover well-child care). The pharmacist can play an important role in educating parents about the need for adequate vaccination and the risks of infection in unprotected children, especially those in high-risk situations, including day-care centers and in homes with older children in school.[58]

16.3.6.3. Human Immunodeficiency Virus Vaccine

A vaccine against the human immunodeficiency virus (HIV) is a much needed tool in the fight against the spread of AIDS. Two approaches can be taken toward vaccine development: a vaccine to prophylactically protect persons from initial infection, and the use of a vaccine product as an immunomodulator in HIV-positive subjects to delay the onset or decrease the severity of disease. There are, however, several obstacles to development of an effective vaccine of either type: increased risk of adverse reactions including disease transmission with the use of attenuated live-virus vaccines, decreased immune response in HIV-infected patients, and exacerbation of AIDS or ARC (AIDS-related complex) symptoms in HIV-positive patients due to immune system stimulation.[65] A biogenetically engineered vaccine that utilizes only selected portions of the retrovirus would eliminate the risk of disease transmission; however, the last two concerns may still pose problems.

The gene structure of HIV has been elucidated, and various proteins have been identified that can stimulate antibody production, including the group-specific antigen (*gag*) gene, the reverse transcriptase or polymerase (*pol*) gene and the envelope precursor (gp160) gene. The gp160 glycoprotein can be divided into two areas, an extracellular protein (gp120) and a transmembrane protein (gp41).[66] The envelope protein (gp120) is the site of virus attachment to the cell receptor CD_4 and facilitates fusion of the virus and host cell.[67] Therefore, this glycoprotein and its parent glycoprotein (gp160) have been the focus of much of the vaccine research regarding the AIDS virus. Phase I studies have been carried out with a recombinant gp160 vaccine in both HIV-positive and HIV-seronegative subjects. In the HIV-positive group, the vaccine appeared to enhance the natural immune response to HIV (as measured by antibody levels and CD_4 counts) in more than 50% of the treated group, with minimal toxicity.[68] In the HIV-seronegative subjects, antibody production and serum-neutralizing activity were documented in 91 and 21% of the vaccinated subjects, respectively.[69] These results appear very promising. However, one of the basic problems regarding an HIV vaccine is lack of knowledge regarding the quantity and quality of the immune response needed to prevent HIV infection. Much additional work needs to be done in this area.

16.3.7. Erythropoietin

Epoietin-α (Epogen, Procrit) is a glycoprotein that has the same amino acid sequence and biological activity as erythropoietin. This agent is utilized to treat anemia secondary to chronic dialysis, in patients with chronic renal failure who are not receiving dialysis, and in AIDS patients who are anemic secondary to zidovudine (Retrovir) therapy. Epoietin-α has also been used on an experimental basis to reverse anemia in cancer patients receiving chemotherapy.

Epoietin-α stimulates erythropoiesis in patients with anemia with or without regular dialysis. This results in a rise in red blood cell count, hematocrit, and hemoglobin, usually within a 6-week period. Clinically, patients receiving this drug appear to note improvements in sleep habits, sexual function, and an overall increase in quality of life. Generally, this agent is well tolerated, with hypertension being noted as a common side effect.

In patients with chronic renal failure, blood pressure should be carefully monitored while receiving epoietin-α. Initiation of antihypertensive medication and dietary restrictions or modification of current therapy and restrictions may be warranted. If the patient's blood pressure is not adequately controlled with these interventions, the dose of epoietin-α should be decreased. Patients with uncontrolled hypertension should not receive this agent. HIV-infected patients without a history of hypertension have not had significant increases in blood pressure. However, in HIV-infected patients with uncontrolled hypertension, epoietin-α therapy should not be initiated until the hypertension is controlled.

Iron depletion has also been noted in all patients receiving epoietin-α. This occurs as the rising red cell mass associated with an increasing hematocrit depletes the body's iron reserves. Transferrin saturation should be maintained at $> 20\%$.

Ferritin should also be monitored and should be at least 100 ng/ml for optimal results. It has been found that most patients receiving epoietin-α will need iron supplementation.

As with other biotechnology drugs, the availability of epoietin-α offers many challenges to the pharmacy practitioner. Health care professionals must be educated by the pharmacist as to the proper indications, pharmacology, side effects, and administration guidelines for the drug. Patients receiving epoietin-α therapy require careful monitoring throughout therapy. Pharmacists have an important role in that they may be responsible for monitoring the patient's blood pressure, transferrin saturation, and ferritin levels and assuring an optimal therapeutic outcome from the medication. Institution-specific protocols should be implemented that guide use and the monitoring of patients receiving this costly agent. Lastly, patients must be counseled as to the beneficial and harmful effects of the drug. Patients must be warned to strictly adhere to their medication, diet, and dialysis regimens even if they feel better. Reimbursement for the use of this agent by third-party payors must also be explored.

16.3.8. Monoclonal Antibodies

Monoclonal antibodies (MAbs) can be defined as antibodies with a single specificity for a target site on an antigen.[70] These agents can be utilized for diagnostic, therapeutic, or monitoring purposes.

Monoclonal antibodies utilized for diagnostic purposes are available to patients, health care professionals, and researchers. Pharmacists must be cognizant of products that are marketed directly to the consumer and must be capable of counseling and monitoring the patients who utilize these tests. One of the most common diagnostic uses for MAbs is in home pregnancy tests (see Table 16.7). The MAb in the test is specific for human chorionic gonadotropin (hCG). By the seventh day after conception, hCG produced by the placenta is excreted into the urine. By using a test that has a MAb specific for hCG, the patient can determine if she is pregnant. Pregnancy tests that are based on MAb technology tend to be easy to use, require less steps, and can be completed in less than 30 minutes. Other home pregnancy tests that utilize hemagglutination technology do not have these desirable features.

Another test available to the patient directly is ovulation prediction kits (see Table 16.8). In this test, MAbs are specific for luteinizing hormone (LH). Because a LH surge precedes ovulation by 20 to 48 hours and can be detected in the urine 8 to 12 hours after it occurs, the MAb specific for LH can help predict when ovulation is to occur.

Monoclonal antibody technology is being used to determine CA-125 blood levels in patients thought to have ovarian as well as cervical cancer.[71, 72] Centocor has marketed this blood test under the trademark CA 125. Blood tests that are currently investigational include CA 15-3, CA 19-9, and CA 72-4. These tests

Table 16.7. Pregnancy Tests

Test	Time (min)	Results	Manufacturer	Comments
Advance	30	Stick; blue color +	Advance Care	First A.M. urine
Answer Plus	3	Color bead change to blue-green +	Carter	First A.M. urine; can test day after period was to begin
Answer Quick and Simple	3	Pink-purple +	Carter	Can test day after period was to begin; does not require first A.M. urine
Clearblue	30	Stick, blue +	Whitehall	Can test day after period was to begin
Clearblue Easy	3	Stick, blue line +	Whitehall	Easy to use; can test day period is due; test stick is held in first A.M. urine stream
Fact Plus	5–8	+ appears on cube	Advance Care	Can test on first day of missed period; any urine
First Response	5	Pink +	Carter	Can test on first day of missed period; any urine

will aid clinicians in the management of breast, pancreatic, and gastric cancers. An assay that detects the presence of P-glycoprotein in identifying tumors with intrinsic or acquired resistance to chemotherapeutic drugs is also available for investigational use. Other diagnostic uses for MAbs include tests for chlamydia, hepatitis, and AIDS.

There are several imaging products based on MAb technology currently undergoing trials in the United States and Europe. The MAb is combined with a radiopaque dye or a radioactive isotope and administered. The combination will bind to specific tissues containing the antigenic structure against which the MAb is structured. An example is a MAb-based imaging agent that will be used in patients who have suffered a heart attack. The product will detect the location and extent of necrotic heart tissue.

Monoclonal antibody technology is also helpful in developing therapeutic agents. Muromonab-CD3 (Orthoclone OKT3) is the first MAb to be marketed for a therapeutic indication (see Chapter 15, this volume). The drug is indicated for the treatment of acute allograft rejection in renal transplant patients.

Pharmacists should be aware of the potential for adverse effects 30 minutes to 6 hours after administration of the first dose of muromonab-CD3. These adverse effects include fever, chills, dyspnea, and malaise. Therefore, the first dose of

Table 16.8.   Ovulation Prediction Test Kits

Test	Time (min)	Results	Manufacturer	Comments
Answer	30	Bead is significantly darker green or green-blue than day before +	Carter	Test urine same time each day; 6-day kit
Clearplan Easy	5	Color of line in large window is same or darker than line in small window +	Whitehall	Test urine same time each day; 5-day kit
First Response	10	Color in well is darker than reference color +	Carter	First A.M. urine; 5-day kit with 3-day refill
Q Test	35	Darker blue +	Becton Dickson	Test urine same time each day; 5-day kit

this medication must be given in an area where the patient can be closely monitored and cardiopulmonary resuscitation equipment is available.

In the 24 hours before initiating muromonab-CD3 therapy, the patient should receive a chest X-ray (CXR) to ensure that no fluid is present in the lungs. The patient's weight should be monitored, and no more than a 3% weight gain should be seen during the 7 days prior to the first injection. The patient's temperature should also be taken prior to administration of the first dose. If the temperature is greater than 37.8°C (100°F), it should be lowered with antipyretics before muromonab-CD3 is administered. The patient's white blood cell (WBC) count should also be measured before administration of the first dose and periodically thereafter. It is recommended that the patient also receive pretreatment with 1 mg/kg of methylprednisolone sodium succinate before the first dose as well as 100 mg of hydrocortisone 30 minutes postinjection. Antihistamines and acetaminophen may also be used concomitantly to decrease the adverse effects of the first dose.

Muromonab-CD3 is currently available in 5 ml ampules containing 5 mg of drug. The product should be refrigerated at 2° to 8°C. The product should not be shaken or frozen. To use, the solution should be drawn through a low protein binding 0.22 μm filter. The filter is then discarded and a needle attached for the iv bolus injection.

Other MAb therapeutic agents are being studied (see Chapter 15, this volume). E5 and HA-1A are MAbs being studied in patients with suspected gram-negative sepsis.[73–75] Both of these agents have been well tolerated, with flushing, hives, and other minor allergic reactions being reported in a very small percentage of patients receiving these agents. Further studies are needed to define the optimal dosing schedule in patients.

16.3.9. Antisense Oligodeoxynucleotides

Recent advances in molecular biology and synthetic chemistry have prompted the development of oligodeoxynucleotides as antisense inhibitors of gene expression. Gene expression may be negated at the level of messenger RNA (mRNA). Antisense nucleic acids are complementary to the sequence of mRNA. Through binding of the antisense to mRNA, transcription or translation of the gene can be selectively blocked. These agents may be helpful in the treatment of cancer and viral diseases.

Although oligodeoxynucleotides appear to be promising agents, there are several problems associated with the development of the oligomers. First, the technology is such that only gram quantities of these agents are being produced. Several hundred or several thousand grams will be needed for clinical trials. Second, the cost of manufacturing the needed quantities is prohibitive. One estimation is that a one-day supply of the drug would cost between $7,500 and $37,500.[76]

The concept of an antisense drug is attractive because it offers the opportunity to act pharmacologically in an early stage of gene expression. Although issues such as production, potency, cellular accessibility, and cost must be addressed before these agents are incorporated into the therapeutic armamentarium for cancer and AIDS and other viral illnesses, the research done to date on antisense modalities is promising. These agents may become a clinical reality in the future. Genta, Inc., has filed a new drug application (NDA) for an antisense compound that will be used to treat chronic myelogenous leukemia. The drug will be used to block protein translation in leukemia cells.

16.3.10. Interleukins

Interleukins are molecules that induce differentiation, replication, and activation of cells. They are produced predominantly by macrophage and T cells in response to antigenic or mitogenic stimulation. There are seven known interleukins. Although each may differ in their source and in the cells they target, they share the capability of transmitting signals between hematopoietic cells. These agents are primarily utilized in cancer immunotherapy. Interleukin-2 (IL-2) is currently being reviewed by the FDA for an indication in renal cell carcinoma. Interleukin-2 stimulates the body's immunologic effector cells to attack tumors rather than acting as a direct antitumor agent. Interleukin-2 is also being studied in combination with zidovudine for use in patients with AIDS, ARC, or HIV.

16.3.11. Thrombolytic Agents

MacFarlane defined thrombosis as "hemostasis at the wrong place".[77] Thrombi are combinations of fibrin, aggregated platelets, red blood cells, and leukocytes. Their exact makeup depends on how and where they were formed. Thrombus formation occurs when the walls and lining of the vasculature are not smooth, when blood flow is abnormal, or when blood coagulability increases. Although atheromas, or plaques, are not thrombi, some disease states (e.g., hypertension and hypercholesterolemia) tend to increase platelet aggregation, inciting thrombus formation at the plaque. Regardless of the cause, platelets and fibrin are deposited in an area that does not need the protective properties of a clot. The "protected" vessel then becomes occluded, generally resulting in an ischemic event.[78]

Thrombotic disorders are responsible for considerable mortality and morbidity: a coronary thrombosis is the precipitating event in an acute myocardial infarction (AMI); a deep vein thrombosis (DVT) may result in a life-threatening pulmonary embolism; cerebral vascular occlusions or strokes typically result from a thromboembolic process.[79]

Because an AMI is often caused by a coronary thrombosis, therapy is directed at dissolving the thrombi as soon as possible; permanent tissue damage and cell death will occur if ischemia persists for 30 to 45 minutes. The size of the infarcted area will extend with time, but reversal of ischemia minimizes necrotic spreading. A myocardial infarction usually affects the left ventricle, resulting in a loss of contractility in the necrotic tissue and impaired contractility in the ischemic areas immediately surrounding the infarct. In addition to reduced contractility, abnormal wall motion, reduced stroke volume and ejection fraction, and elevated end-diastolic volumes and left ventricular end-diastolic pressure occurs. This sequelae of events results in a common complication of an MI: congestive heart failure (CHF).[78]

The ideal therapeutic agent for clot lysis and reperfusion, then, should not only act quickly but also should improve left ventricular function, reduce mortality, have minimal side effects (e.g., bleeding), be nonantigenic, and relatively cost-effective. The thrombolytic agents streptokinase (SK), urokinase (UK), and recombinant tissue plasminogen activator (t-PA, alteplase) are all effective in lysing arterial and venous clots and reestablishing perfusion. SK and UK, the first generation agents, are not fibrin selective; their use may result in a lytic state and excessive bleeding. The development and release of alteplase (Activase, Genentech), a second generation thrombolytic agent, promised thrombus selectivity and therefore less serious bleeding; more effective lysis and reperfusion, resulting in greater protection of left ventricular function; and no antigenicity. Clot specificity was found in vitro, but not to any appreciable extent in humans (see Chapter 15, this volume). Owing to the cost differences between SK and t-PA ($100 and $2,400, respectively), most of the comparative research is based on these two agents.

Medical researchers have conducted numerous investigations comparing the effectiveness of t-PA and SK. A 1989 review of 13 studies found that t-PA does achieve early reperfusion (i.e., within 60 minutes) 50% more frequently than does streptokinase. However, angiography performed 24 hours after symptom onset revealed no difference between these two thrombolytics in achieving reperfusion.[80]

Table 16.9.   Comparison of Thrombolytic Agents

	Streptokinase (SK)	t-PA
Dose (AMI)	1.5 million units	100 mg
Selectivity	Low	High
Reperfusion (%)[a]	65	70
Bleeding incidence	+	+
Allergic reactions	+ +	0(?)
Hypotension	+ +	+
Approx. cost	$100	$2,400

[a]When administered within three hours of symptoms.

Adapted from *Applied Therapeutics*. 4th ed., p. 353. Koda-Kimble, M.A. Vancouver, Washington, Applied Therapeutics, Inc., 1988.

Differences in the ability of these two drugs to improve left ventricular function has also been the subject of research interest. Initially, t-PA improved cardiac function during convalescence,[81] but further investigation began comparing t-PA to SK. Magnani et al. that found t-PA provided better left ventricular improvement than did SK,[82] whereas White et al. found no differences between the two thrombolytics.[83] Given these findings, and the lack of significant differences in reperfusion, major differences between these two drugs' abilities to improve left ventricular function are unlikely.

Both SK and t-PA improve survival.[84–87] In addition, results by Magnani et al. and White et al. indicated a lower mortality rate in t-PA treated patients: 4.2 and 3.7% for t-PA versus 8.2 and 7.3% for SK respectively.[82, 83] However, the results of the GISSI-2 trial (discussed in Chapter 15, this volume) do not indicate a mortality-rate difference.

Although the possibility of an allergic reaction to SK exists, the benefit of nonantigenic t-PA seems questionable, given the cost differences and therapeutic equivalence to SK. (See Tables 16.9 and 16.10 for a comparison of thrombolytic agents.)

When using Alteplase, or any thrombolytic, patients must be monitored closely. Clinical monitoring would include observing for signs of overt or covert

Table 16.10.   Comparison of Thrombolytic Agents

	SK	UK	t-PA
Source	Bacterial	Human	Recombinant technology
Fibrin specificity	Lowest	Low	High
Thrombolytic potential	Good	Good	Improved
Side effects			
Allergic	Yes	No	No
Dose-related	Yes	No	Yes
Time-related	Yes	Yes	Not known
Hemorrhagic complications	Yes	Yes	Yes

bleeding, checking all intravenous sites, wounds, and puncture sites. A careful physical examination, looking for signs of internal bleeding, should also be conducted. Laboratory parameters to monitor include thrombin time (TT) (however, TT is sensitive to low levels of heparin and may be of little value if heparin is being used), fibrinogen levels, CBC, and clinical chemistry. There is a correlation between fibrinogen levels < 100 mg% and bleeding complications. Invasive monitoring is not recommended because, in this case, invasive overmonitoring *increases* complications.

Contraindications to thrombolytic therapy include a high probability of intracranial bleeding, active, severe, gastrointestinal bleed, a recent CVA (within approximately 2 months), recent cranial surgery or trauma (within 10 days), intracranial neoplasm, or severe, uncontrolled hypertension. In addition, these agents are contraindicated in a patient in which they would affect a beneficial hemostatic plug, such as after major surgery, significant trauma, childbirth, an organ biopsy, or in invasive procedures in which direct pressure cannot be applied.

Other contraindications include any condition in which the potential for embolization from the dissolving thrombus exists. Examples include cardiac mural thrombus (due to atrial fibrillation), vegetative valvular heart disease, bacterial endocarditis, septic thrombophlebitis, or cerebrovascular disease (where there is a potential for emboli to the CNS). A preexisting coagulation defect (e.g., platelet dysfunction and severe renal or hepatic disease) is also a contraindication for these agents.

If significant bleeding develops during thrombolytic therapy, the infusion should be discontinued and direct pressure instituted. If this fails to control the bleeding, exogenous fibrinogen-containing products such as cryoprecipitate or fresh frozen plasma can be used. Whole blood is rarely used.

Alteplase is officially indicated for the management of acute myocardial infarction in adults. Alteplase is also indicated for the treatment of acute massive pulmonary embolism (PE) in adults. Pulmonary embolism is the obstruction of blood flow to a lobe or multiple segments of the lungs, accompanied by unstable hemodynamics. It should be noted that the treatment of PE with alteplase is *not* adequate for the management of the underlying deep vein thrombosis. Reembolization may occur when the underlying thrombus is lysed. Therefore, standard management of PE or MI must be implemented concomitantly.

As is true for all thrombolytics, bleeding (either internal or external) is the most common complication of alteplase. Common sites include the gastrointestinal and genitourinary tract, and retroperitoneal and intracranial areas. The risk of intracranial bleeding is dose dependent: data indicate a 0.4% risk at a dose of 100 mg, and a 1.3% risk at a dose of 150 mg. Bleeding may also be observed at venous cutdowns, arterial puncture sites, or sites of recent surgical intervention. The concomitant use of heparin has been suggested to decrease the risk of rethrombosis, but this will contribute to bleeding. Additionally, any drug that

Table 16.11. Recombinant Tissue Plasminogen Activator Protocol

Laboratory Studies	PT on admission, then 24 hours later; PTT every 6 hours for 24 hours, then daily. LDH on admission and in 24 hours. Creatinine phosphokinase with isoenzymes on start of therapy and at 4, 8, 12, 16, and 24 hours later. Complete blood count on admission and 24 hours later. Type and screen blood on admission
ECG	Initial and every 30 minutes for 2 hours
Medications	Lidocaine bolus and drip per hospital protocol at the beginning of infusion
Heparin	Start heparin drip during first hour of initial infusion. Bolus with 5,000 units IVP followed by 1,000 units/hour adjusted per physician's orders

Adapted from M.A. Pelter, 1989. Thrombolytic therapy in acute myocardial infarction. *Crit. Care Nurs. Q.* 3:67.

alters platelet function (given before, during, or after alteplase) has the potential to increase bleeding risks. Careful monitoring, particularly at puncture sites, is essential.

Other adverse effects of alteplase include nausea, vomiting, hypotension, and fever. These are, however, frequent sequelae of a myocardial infarction and may not necessarily be due to alteplase therapy.

For the management of acute myocardial infarction, a total of 100 mg of alteplase is given, intravenously, over three hours. A total of 60 mg (34.8 million IU) is given the first hour, of which a 6 to 10 mg bolus is given in the first 1 to 2 minutes. The second and third doses are each 20 mg (11.6 million IU). Smaller patients (< 65 kg) may receive 1.25 mg/kg over three hours.

Alteplase should be reconstituted *only* with the accompanying SWI without preservatives. The resulting solution should be slowly swirled, not shaken. The solution should be transparent, clear to pale yellow, with a pH of 7.3. It should be used within 8 hours if stored at controlled room temperature. The solution can be administered as 1 mg/ml, or further diluted with an equal volume of 0.9% SCI or 5% DI to yield 0.5 mg/ml. Glass or polyvinyl chloride bags may be used. Other drugs should *not* be added to Altepase. (See Table 16.11 for a recombinant tissue plasminogen activator protocol.)

16.3.12. Interferons

The interferons are proteins and glycoproteins, either naturally occurring or synthesized by recombinant DNA biotechnology. Their therapeutic use as antiviral, antitumor, antineoplastic, and immune-modifying agents has been investigated by many researchers. At the present time, two have been approved for use: Roferon-A (interferon-α-2a, by Hoffman-LaRouche) and Intron A (interferon-α-2b, by Schering-Plough). These two α-interferons differ only by one amino acid at position 23. Roferon-A has been approved for use in hairy cell leukemia; Intron A has been approved for use in hairy cell leukemia and genital warts. Further uses for Roferon-A include AIDS-related Kaposi's sarcoma, chronic myelogenous leukemia, and renal cell carcinoma. Intron A is under study for use in Kaposi's sarcoma, malignant melanoma, multiple myeloma, superficial bladder cancer, upper respiratory infections, chronic myelogenous leukemia, basal cell carcinoma, ovarian cancer, viral hepatitis, and transitional-cell bladder cancer.[88] Interferon-α-2b is currently being studied for use in metastatic skin melanoma (MSM), which is usually unresponsive to chemotherapy.

The alpha interferons are biological-response modifiers, that is, they have immune-modifying and cytotoxic effects. The exact mechanism by which the α-interferons exert their activity is not clearly known at the present time. In 1957 interferons were first described by Isaac and Lindenmann.[89] On the basis of their experiments, interferons were defined as proteins or glycoproteins with nonspecific antiviral activity through the induction of cellular RNA and protein synthesis. At that time, quantities of the interferons sufficient for laboratory experimentation were difficult to obtain. At the present time, however, it is possible to isolate human interferon genes, clone them in bacteria, and produce human interferons in large amounts.[90] The naturally occurring interferons—α, β, and γ—are distinguished from recombinant interferons, and are named according to the cells from which they are produced; for example, α is induced by Sendai virus in leukocytes or lymphoblasts.

In general, treatment with the α-interferons is well tolerated. The most commonly noted adverse effects are influenza-like, described as fatigue and anorexia. The use and effectiveness of the α-interferons in the treatment of hairy cell leukemia are well documented. The interferons were introduced for the treatment of hairy cell leukemia in 1984. Hairy cell leukemia is a B cell lymphoproliferative disorder for which splenectomy has been the treatment of choice. In those patients who do not respond, chlorambucil, androgens, and other combination chemotherapies have been used, but with mixed results. Since their introduction, Intron A and Roferon-A have shown a high response rate, but not necessarily a high remission rate. They are particularly indicated for those patients in whom splenectomy is not desirable.[91, 92] Interferon-α-2b (Intron A) is also approved for use in the treatment of condylomata acuminata (genital warts). Intron A is injected intralesionally. Pharmacists should be aware that the manufacturer recommends that only the 10 million IU vial be used for this indication, since dilution of other available strengths results in a hypertonic solution.

Condylomata acuminata is gaining importance as a sexually transmitted disease in the United States. The overt "wart" is only one manifestation of the disease, and most infections are subclinical. Human papillomavirus (HPV) has been detected in up to 35% of some population groups without overt clinical disease using HPV antigen detection and DNA hybridization. Genital HPV infection has been correlated with genital neoplasms: HPV DNA has been demonstrated in cervical, vulvar, penile, and anal dysplasias and carcinomas. Current therapies include podophyllin, cryotherapy, electrical cautery, and laser therapy. These "destructive" therapies are associated with discomfort (at times significant), and recurrences are frequent.

Interferon alpha is active against HPV. Interferon alpha has been used experimentally in topical, intralesional, and systemic dosage forms. The topical use has the distinct advantage of reduced side effects, but results of this dosage form have been mixed. Systemic therapy, theoretically, treats all tissues infected with HPV, thereby eliminating clinical and subclinical infections; however, systemic side effects then become an issue.

To summarize, the alpha interferons do offer additional therapeutic modalities for the oncologist, but the risk-versus-benefit ratio needs further study.

16.4. Conclusion

Pharmacists are in an ideal position to expand their practices to include drugs derived from biotechnology. The pharmacist's role should not be limited solely to distribution. Activities in the areas of education, monitoring, drug utilization, and research will be critical to the safe, rational, and cost-effective use of these products.

References

1. Stewart, C.W., and Fleming, R.A. 1989. Biotechnology products: New opportunities and responsibilities for the pharmacist. *Am. J. Hosp. Pharm.* 46:S4–S8.

2. Miscellaneous antineoplastics. *Facts and comparisons* 1991. May:682–683n.

3. Petkau, A. 1986. Scientific basis for the clinical use of superoxide dismutase. *Cancer Treat. Rev.* 13:17–44.

4. Murohara, Y., Yoshiki, Y., Hattori, R., and Kawai, C. 1991. Effects of superoxide dismutase on reperfusion arrhythmias and left ventricular function in patients undergoing thrombolysis for anterior wall acute myocardial infarction. *Am. J. Cardiol.* 67:765–767.

5. Flohe, L. 1988. Superoxide dismutase for therapeutic use: Clinical experience, dead ends and hopes. *Mol. Cell. Biochem.* 84:123–31.

6. Old, L.J. 1988. Tumor necrosis factor. *Sci Am.* 258 (5):59–60, 69–75.

7. Sherman, M.L., Spriggs, D.R., Arthur, K.A., et al. 1988. Recombinant human tumor necrosis factor administered as a five day continuous infusion in cancer patients. Phase I toxicity and effects on lipid metabolism. *J. Clin. Oncol.* 6(2):344–350.

8. Rosenblum, M.G., and Donato, N.J. 1989. Tumor necrosis factor: A multifaceted peptide hormone. *Crit. Rev. Immunol.* 9(1):21–44.

9. Spriggs, D.R., Sherman, M.L., Michie, H., et al. 1988. Recombinant human tumor

necrosis factor administered as a 24 hour intravenous infusion. A phase I and pharmacologic study. *J. Nat. Can. Inst.* 80(13):1039–1044.

10. Creagan, E.T., Kovach, J.S., and Moertel, C.G. 1988. A phase I clinical trial of recombinant human tumor necrosis factor. *Cancer* 62:2467–2471.

11. Jones, A.L., and Selby, P. 1989. Tumour necrosis factor: clinical relevance. *Cancer Surv.* 8:817–836.

12. Balkwill, F.R., Naylor, M.S., and Malik, S. 1990. Tumour necrosis factor as an anticancer agent. *Eur. J. Cancer* 26:641–644.

13. Soma, G., and Mizuno, D. 1989. Exogenous and endogenous tumor necrosis factor therapy. *Cancer Surv.* 8:837–852.

14. Karp, J.E., and Broder, S. 1991. Oncology. *J. Am. Med. Assoc.* 265:3141–3143.

15. Odeh, M. 1990. The role of tumour necrosis factor-alpha in acquired immunodeficiency syndrome. *J. Intern. Med.* 228:549–556.

16. Wilton, P., Wodlund, L., and Guilbaud, O. 1987. Bioequivalence of genotropin and somatonorm. *Acta Paediatr. Scand. Suppl.* 377:188–121.

17. Growth hormone. *Facts and comparisons* 1989. October:116b–c.

18. Gunnarsson, R., and Wilton, P. 1987. Clinical experience with genotropin worldwide: An update March 1987. *Acta Paediatr. Scand. Suppl.* 337:147–152.

19. Hibi, I., Takano, K., and Shizume, K. 1987. Current clinical trials with authentic recombinant somatropin in Japan. *Acta Paediatr. Scand. Suppl.* 337:141–146.

20. Albertsson-Wikland, K. 1987. Clinical trial with authentic recombinant somatropin in Sweden and Finland. *Acta Paediatr. Scand. Suppl.* 331:28–34.

21. Albertsson-Wikland, K., Westphal, O., and Westgren, U. 1986. Daily subcutaneous administration of human growth hormone in growth hormone deficient children. *Acta Paediatr. Scand.* 75:89–97.

22. Zamboni, G., Antoniazzi, F., Radetti, G., Musumeci, C., and Tato, L. 1991. Effects of two different regimens of recombinant human growth hormone therapy on the bone mineral density of patients with growth hormone deficiency. *J. Pediatr.* 119:483–485.

23. Thompson, R.G., Conforti, P., and Holcombe, J. 1989. Biosynthetic human growth hormone: Current status and future questions. *J. Endocrinol. Invest.* 12 (suppl. 3):35–39.

24. Fisher, D.A., Job, J., Preece, M., et al. 1988. Leukaemia in patients treated with growth hormone. *Lancet* 1:1159–1160.

25. Williams, T.C., and Froman, L.A. 1986. Potential therapeutic indications for growth hormone and growth hormone-releasing hormone in conditions other than growth retardation. *Pharmacotherapy* 6(6):311–318.

26. Lantos, J., Siegler, M., and Cuttler, L. 1989. Ethical issues in growth hormone therapy. *J. Am. Med. Assoc.* 261:1020–1024.

27. Schulz, G., Frisch, J., Greifenberg, B., Nicolay, U., and Oster, W. 1991. New therapeutic modalities for the clinical use of rhGM-CSF in patients with malignancies. *Am. J. Clin. Oncol.* 14(S):S19–S26.

28. Scadden, D.T., Dering, H.A., Levine, J.D., et al. 1991. GM-CSF as an alternative to dose modification of the combination zidovudine and interferon-alpha in the treatment of AIDS-associated Kaposi's Sarcoma. *Am. J. Clin. Oncol.* 14(S):S40–S44.

29. Ganser, A., Seipelt, G., and Hoelzer, D. 1991. The role of GM-CSF, G-CSF, interleukin-3 and erythropoietin in myelodysplastic syndromes. *Am. J. Clin. Oncol.* 14(S):S34–S39.

30. Crawford, J., Ozer, H., Stoller, R., et al. 1991. Reduction by granulocyte colony-stimulating factor of fever and neutropenia induced by chemotherapy in patients with small-cell lung cancer. *N. Engl. J. Med.* 325(3):164–170.

31. Anonymous. 1991. Granulocyte colony-stimulating factors. *Med. Lett.* 33(847):61–63.

32. Ohno, R., Tomonaga, M., Kobayashi, T., et al. Effect of granulocyte colony-stimulating factor after intensive induction therapy in relapsed or refractory acute leukemia. *N. Engl. J. Med.* 323(13):871–877.

33. Groopman, J.E. 1991. Antiretroviral therapy and immunomodulators in patients with AIDS. *Am. J. Med.* 90(S4A):18S–21S.

34. Colony stimulating factors. *Facts and comparisons.* 1991. May:84i–85.

35. Neumanitis, J., Rabinowe, S.N., Singer, J.W., et al. (1991). Recombinant granulocyte-macrophage colony-stimulating factor after autologous bone marrow transplantation for lymphoid cancer. *N. Engl. J. Med.* 324(25):1773–1778.

36. Brogden, R.N., and Heel, R.C. 1987. Human insulin: A review of its biological activity, pharmacokinetics and therapeutic use. *Drugs* 34:350–371.

37. Davidson, J.A. 1989. Recombinant DNA human insulin: Clinical experience with the transfer of diabetic patients from animal-source insulins. *Res. Staff Physician.* 35:39–47.

38. Davidson, J.K. 1989. Transferring patients with insulin-dependent diabetes mellitus from animal-source insulins to recombinant DNA human insulin: Clinical experience. *Clin. Therapeut.* 11:319–330.

39. Botterman, P., Gyaram, H., Wahl, K., Ermler, R., and Lebender, A. 1981. Pharmacokinetics of biosynthetic human insulin and characteristics of its effect. *Diabetes Care* 4:168–169.

40. Marre, M., Tabbi-Anneni, A., Tabbi-Anneni, H., and Assan, R. 1982. Comparative study of NPH human insulin (recombinant DNA) and NPH bovine insulin in diabetic subjects. *Diabetes Care* 5(S2):63–66.

41. Burden, A.C. 1990. Increased hypoglycemia on insulin derived from yeast (letter). Lancet 335:485.

42. Cryer, P.E. 1990. Human insulin and hypoglycemia unawareness. *Diabetes Care* 13:536–538.

43. Pickup, J. 1989. Human insulin: Problems with hypoglycaemia in a few patients. *Br. Med. J.* 299:991–993.

44. Heine, R.J., and van der Veen, E.A. 1990. Human insulin and hypoglycaemia (letter). *Lancet* 335:62.

45. Jakober, B., Lingenfelser, T., Gluck, H., et al. 1990. Symptoms of hypoglycemia: A comparison between porcine and human insulin. *Klin. Wochenschr.* 68:447–453.

46. Richards, A.M. 1989. Atrial natriuretic factor administered to humans: 1984–1988. *J. Cardiovasc. Pharmacol.* 13(S6):S69–S74.

47. Jardine, A.J., Northbridge, D.B., and Connell, J.M.C. 1989. Harnessing the therapeutic potential of atrial natriuretic factor. *Klin. Wochenschr.* 67:902–906.

48. Ruff, M.R., Hallberg, P.L., Hill, J.M., and Pert, C.B. 1987. Peptide $T_{(4-8)}$ is core HIV envelope sequence required for CD_4 receptor attachment (letter). Lancet 2:751.

49. Wetterberg, L., Alexius, B., Saaf, J., Sonnerborg, A., Britton, S., and Pert, C. 1987. Peptide T in the treatment of AIDS (letter). Lancet 1:159.

50. Bridge, T.P., Heseltine, P.N.R., Parker, E.S., et al. 1989. Improvement in AIDS patients on peptide T (letter). Lancet 2:226–227.

51. Marcusson, J.A., Lazega, D., Pert, C.B., Ruff, M.R., Sundquist, K.G., and Wetterberg, L. 1989. Peptide T and psoriasis. *Acta Derm. Venereol. (Stockholm)*, suppl. 146:117–121.

52. Garrison, M.W., and Baker, D.E. 1991. Therapeutic advances in the prevention of hepatitis B: Yeast derived recombinant hepatitis B vaccines. *DICP Ann. Pharmacother.* 25:617–627.

53. Lindsay, K.L., Herbert, D.A., and Gitnick, G.L. 1985. Hepatitis B vaccine: Low post-vaccination immunity in hospital personnel given gluteal injections. *Hepatology* 5:1088–1090.

54. Weissman, J.Y., Tsuchiyose, M.M., Tong, M.J., Co, R., Chin, K., and Ettenger, R.B. 1988. Lack of response to recombinant hepatitis B vaccine in nonresponders to the plasma vaccine. *J. Am. Med. Assoc.* 260:1734–1738.

55. Parish, D.C., Muecke, H.W., Joiner, T.A., Pope, W.T., and Hadler, S.C. 1991. Immunogenicity of low-dose intradermal recombinant DNA hepatitis B vaccine. *S. Med. J.* 84:426–430.

56. Rivey, M.P., and Peterson, J. 1991. Intradermal hepatitis B vaccine. *DICP Ann. Pharmacother.* 25:628–634.

57. Shapiro, E.D., and Berg, A.T. 1990. Protective efficacy of Haemophilus influenzae type B polysaccharide vaccine. *Pediatrics.* 85(4pt2):S643–S647.

58. Wilfert, C.M. 1990. Epidemiology of Haemophilus influenzae type b infections. *Pediatrics* 85(4pt2):S631–S635.

59. Ahonkhai, V.I., Lukas, L.J., Jonas, L.C., et al. 1990. Haemophilus influenzae type b conjugate vaccine (meningococcal protein conjugate) PedvaxHIB: Clinical evaluation. *Pediatrics* 85(4pt2):S676–S681.

60. Madore, D.V., Johnson, C.L., Phipps, D.C., et al. 1990. Safety and immunologic response to Haemophilus influenzae type b oligosaccharide-CRM_{197} conjugate vaccine in 1–6 month old infants. *Pediatrics* 85(3):331–337.

61. Weinberg, G.A., and Granoff, D.M. 1990. Immunogenicity of haemophilus influenzae type b polysaccharide-protein conjugate vaccine in children with conditions

associated with impaired antibody responses to type b polysaccharide vaccine. *Pediatrics* 85 (4pt2):S654–S661.

62. Committee on Infectious Diseases. 1991. Haemophilus influenzae type B. In *Report of the Committee on Infectious Diseases*. 22nd ed. Elk Grove Village, American Academy of Pediatrics. Pp. 220–229.

63. Hemophilus B conjugate vaccines. *Facts and comparisons* 1991. April:459g–459i.

64. Dashefsky, B., Wald, E., Guerra, N., and Byers, C. 1990. Safety, tolerability and immunogenicity of concurrent administration on haemophilus influenzae type b conjugate vaccine (meningococcal protein conjugate) with either measles-mumps-rubella vaccine or diphtheria-tetanus-pertussis and oral polio vaccines in 14–23 month old infants. *Pediatrics* 85(4pt2):S682–S689.

65. LaMontagne, J.R. 1989. Immunization programs and human immunodeficiency virus. *Rev. Infect. Dis.* 11(S3):S639–S643.

66. Fischinger, P.J. 1988. Strategies for the development of vaccines to prevent AIDS. In *AIDS: etiology, diagnosis, treatment and prevention*, 2nd ed. J.B. Lippincott. DeVita, V.T., Hellman, S., and Rosenberg, S.A. eds. 1988 Pp. 87–92.

67. Schild, G.C., and Minor, P.D. 1990. Human immunodeficiency virus and AIDS: Challenges and progress. *Lancet* 335:1081–1084.

68. Redfield, R.R., Birx, D.L., Ketter, N., et al. 1991. A phase I evaluation of the safety and immunogenicity of vaccination with recombinant gp160 in patients with early human immunodeficiency virus infection. *N. Engl. J. Med.* 324(24):1677–1684.

69. Dolin, R., Graham, B.S., Greenberg, S.B., et al. 1991. The safety and immunogenicity of a human immunodeficiency virus type 1 (HIV-1) recombinant gp160 candidate vaccine in humans. *Ann. Intern. Med.* 114:119–127.

70. Tami, J.A., Parr, M.D., Brown, S.A., and Thompson, J.S. 1986. Monoclonal antibody technology. *Am. J. Hosp. Pharm.* 43:2816–2825.

71. Patsner, B. 1990. Preoperative serum CA-125 levels in early stage ovarian cancer. *Eur. J. Gynaecol. Oncol.* 11:319–321.

72. Goldberg, G.L., Sklar, A., O'Hanlan, K.A., Levine, P.A., and Runowicz, C.D. 1991. CA-125: A potential prognostic indicator in patients with cervical cancer? *Gynecol. Oncol.* 40:222–224.

73. Gorelick, K.J., Jacobs, R., Chmel, H., et al. 1989. Efficacy results of a randomized multicenter trial of E5 antiendotoxin monoclonal antibody in patients with suspected gram-negative sepsis (abstract). In *Program and abstracts of the 29th interscience conference on antimicrobial agents and chemotherapy*, 2A. Washington, D.C.: American Society for Microbiology.

74. Gorelick, K.J., Schein, R.M.H., Macintyre, N.R., et al. 1990. Multicenter trial of antiendotoxin antibody E5 in the treatment of gram-negative sepsis (GNS) (abstract). *Crit. Care Med.* 18(suppl.):S253.

75. Ziegler, E.J., Fisher, C.J., Jr., Sprung, C.L., et al. 1991. Treatment of gram-negative bacteremia and septic shock with HA-1A human monoclonal antibody against endotoxin. *N. Engl. J. Med.* 324:429–436.

76. Rothenberg, M., Johnson, G., Laughlin, C., et al. 1989. Oligodeoxynucleotides as anti-sense inhibitors of gene expression: Therapeutic implications. *J. Natl. Cancer. Inst.* 81:1539–1544.

77. MacFarlane, R.D. 1977. Hemostasis: Introduction. *Br. Med. Bull.* 33:183–194.

78. Price, S.A., and Wilson, L.M. eds. 1986. Coronary atherosclerotic disease. In *Pathophysiology.* 3rd ed. New York: McGraw-Hill.

79. Comeroto, A.J. 1988. *Thromboembolic therapy.* Philadelphia: Grune & Stratton.

80. Topol, E.J., and Califf, R.M. 1989. Tissue plasminogen activator: Why the backlash. *JACC* 7:1477–1480.

81. Armstrong, P.W., Baigrie, R.S., Daly, P.A., et al. 1989. Tissue plasminogen activator: Toronto (TPAT) placebo controlled randomized trial in acute myocardial infarction. *J. Am. Coll. Cardiol.* 13:1469–1476.

82. Magnani, B., for the PAIMS Investigators. 1989. Plasminogen Activator Italian Multicenter Study (PAIMS): Comparisons of intravenous recombinant single-chain human tissue type plasminogen activator (rt-PA) with intravenous streptokinase in acute myocardial infarction. *J. Am. Coll. Cardiol.* 13:19–26.

83. White, H.D., Rivers, J.T., Maslowski, A.H., et al. 1989. Effect of intravenous streptokinase as compared with that of tissue plasminogen activator on left ventricular function after first myocardial infarction. *N. Engl. J. Med.* 320:349–360.

84. Gruppo Italiano per lo Studio della Streptochinasi nell'Infarto Miocardico (GISSI). 1986. Effectiveness of intravenous thrombolytic treatment in acute myocardial infarction. *Lancet* 1:397–401.

85. ISIS-2 (Second International Study of Infarct Survival) Collaborative Group. 1988. Randomized trial of intravenous streptokinase, oral aspirin, both, or neither among 17, 187 cases of suspected myocardial infarction: ISIS-2. *Lancet* 2:349–360.

86. Wilcox, R.G., von der Lippe, G., Olsson, C.G., et al. 1988. Trial of tissue plasminogen activator for mortality reduction in acute myocardial infarction. *Lancet* 2:525–530.

87. ISIS-2 (Second International Study of Infarct Survival) Collaborative Group. 1988. Randomized trial of intravenous streptokinase, oral aspirin, both, or neither among 17, 187 cases of suspected acute myocardial infarction: ISIS-2. *J. Am. Coll. Cardiol.* 12 (suppl. A):3A–13A.

88. *Biotechnology products in development.* 1988. Washington, D.C.: Pharmaceutical Manufacturer's Association.

89. Isaac, A., and Lindenmann, J. 1957. Virus interference. I. The interferons. *Proc. R. Soc. London* 147:258–273.

90. Pestka, S. 1983. The human interferons-from protein purification and sequence to cloning and expression in bacteria: before, between, and beyond. *Arch. Biochem. Biophys.* 221:1–37.

91. Quesada, J.R., Reuben, J., Manning, J.T., Hersh, E.M., and Gutterman, J.U. 1984.

Alpha interferon for the induction of remission in hairy cell leukemia. *N. Engl. J. Med.* 310:15–18.

92. Romeril, K.R., Carter, J.M., Green, G.J., et al. 1989. Treatment of hairy cell leukaemia with recombinant alpha interferon. *New Zealand Med. J.* 102:186–188.

17

Human Trials of Biotechnology Products: A Perspective

W. Leigh Thompson

17.1. Introduction: The Purpose of Clinical Trials

Clinical trials estimate the safety and efficacy of drugs as they will be used (and misused) in the marketplace. The precision of this estimate required for a given drug should be determined by the target patient population's size, diversity, and severity of illness and by the availability, safety, and efficacy of alternative therapies.

Chemical, pharmacological, and toxicological characterization of the clinical trial material should be sufficient to make reasonable ethical judgments concerning its introduction into humans. Preclinical studies should identify potential human metabolites that may require special toxicology studies. Pharmacokinetics and dynamics should be assessed over a wide range of doses in several pertinent species.

17.2. Prerequisites

The potential for injury to somatic and germinal genomes should be assessed before the first humans are exposed. Before fecund women taking the drug become pregnant, an assessment should be made of risk to the fetus, the neonate, and the breast-fed infant. Preclinical studies also will provide the estimates of radiation dosimetry variables' values to prepare for the use of radiolabeled materials in humans.

Nucleic acids and peptides are labile; subtle changes in peptide tertiary structure—amidation, glycosylation, and so on—may alter biological effects. Materials used for nonclinical and clinical studies should be characterized carefully. Some biotechnology products derived by cells of higher mammals raise the specter of potential contamination by prions or viruses that could cause delayed disease. Such risks should be minimized and estimated realistically.

The normal secretion pattern of peptides may be intermittent, timed to other events, and restricted to special sites. This normal regimen may not be mimicked by exogenous injection of those peptides once daily into systemic circulation. Transport proteins may vary with species; large doses or rapid absorption may saturate binding sites and distort the natural peptide total concentrations and time-course of action. Although nonclinical studies can define the concentration and effect kinetics of various doses of natural and modified human peptides, these should focus primarily on the safety of the materials when injected in patterns that resemble the anticipated human trials.

Antibody response to peptides may restrict the duration of meaningful animal studies to a few weeks. Thus, the nonclinical pharmacology, toxicology, and kinetics of biotechnology products may be less predictive of prolonged human use than studies of small organic chemicals.

17.3. The First Humans

Scientific and ethical debates, among fully informed members of several groups, lead to a shared decision to give a new product to humans for the first time. The sponsor bears great responsibility for this decision; the sponsor may be a manufacturer, institute, laboratory, or individual scientist.

The sponsor should assemble a group of basic and clinical scientists and perhaps in addition attorneys or ethicists who may be employees but who are not directly engaged in the discovery and development of the product. This group should consider carefully all pertinent intramural and extramural data and be held accountable for their decision to initiate clinical trials. This group also should approve the protocol for initial human studies and the "full" disclosure of information to be given to potential human subjects as part of the consent procedure.

After the sponsor has committed to clinical trials, permission should be requested of regulators and institutional ethical committees. The ethical committee directly associated with the institution and physician who will be responsible for the patients should judge the circumstances of the proposed study and monitor the course of that study with appropriate attention to data from studies at other sites with the same and related drugs.

Regulators and ethical committees may be inconsistent in their assessment of the wisdom of starting (and continuing) clinical trials. In addition, some regulators have far more stringent standards for industrial sponsors than for academicians. Such variation is disturbing; from the subject's perspective the risks and benefits are not altered by the sponsor's status or institution's location.

Initial trials should be designed to maximize safety and minimize human exposure while acquiring critical information. The purposes of these trials are to

1. define drug absorption, distribution, metabolism, and excretion

2. confirm in humans critical elements of preclinical pharmacology

3. evaluate carefully all adverse events (including immunogenicity)

4. define dosage regimens for subsequent trials of efficacy

5. begin assessment of the immunogenicity of large molecules.

Usually trials begin with small single doses. The dose is increased gradually in the same and subsequent subjects. The time between such "rising doses" should allow appropriate safety assessments before each new dose. Risks may be reduced by using a small number of subjects each of whom will receive progressively larger doses. This phase is followed by repetitive "multiple" dosing. Risks may be lessened by slow progression through initial trials, but in patients one may need to escalate doses more quickly to achieve potential beneficial effects.

The risks of initial human studies should be balanced by the benefits. Healthy volunteers, who are usually young men, may accept the risks of relatively safe products and the discomforts of complex testing procedures. In many cases, however, patients who might have a direct benefit may be the most appropriate humans to receive the first doses. Infertile women should be included, but exposing a fetus to a drug with only a 10% chance of being marketed is problematic.

17.4. Safety and Efficacy

The first human studies may measure some surrogates of efficacy, such as changes in laboratory tests, but their primary purpose is to define safety, pharmacology, kinetics, and potential dosage regimens. Subsequent trials focus on tests of efficacy and safety in patients with the target illness. In the past these latter trials were divided, arbitrarily, into Phases II and III. Phase II included small trials with a few investigators and homogeneous patients who tended to have a single illness and take only the target medication. Phase III included larger trials with more diverse investigators using patients who were less homogeneous.

The separation of these phases leads to "serial" thinking, pauses to collect and analyze data, delays in planning and initiating trials, and a patchwork quilt of studies rather than a seamless plan. The average N.D.A. approved by the U.S. F.D.A. contains 80 clinical trials. It is possible to conduct most safety and efficacy studies in a single multiinvestigator protocol suitable for division into "two studies"; other strategies may also achieve the continuum that is most effective.

The rate of expansion of trials in patients is related to the risk, the indications of potential efficacy, and the ability of the sponsor to effect a rapid flow of data from investigative sites to a central monitoring group. This system should minimize the possibility of continuing patient exposure beyond the point at which inefficacy or excessive risk could be determined.

17.5. Trial Size

The trials of safety and efficacy may include a few hundred or thousands of patients. They may be performed by two or hundreds of investigators in one or many countries. Their duration should be guided by the expected use. Smaller trials should suffice if

1. the illness is serious
2. the new product is very effective
3. alternative treatments are unavailable or less safe or effective
4. there are few potential patients
5. the dosage regimen can be defined easily
6. the target diagnosis can be made unequivocally
7. small trials can encompass the diversity of potential consumers.

17.6. Trial Design

A large number of small studies may be conducted in different patient populations, in different countries, and with different doses. Although these may in aggregate establish some vision of the safety and efficacy of the drug, they are inefficient, add logistic complexity, and diminish the power of analyzing patients within the same protocol. Instead, large multiinvestigator, multinational, dose–response studies are more elegant and powerful, though they require innovative and resilient systems and statistical staff.

17.7. Dose-Response

The relationship of dose (size, frequency, time of day, route, formulation, and dietary status) to safety and efficacy should be assessed in all significant patient populations. Dosage regimens tend to become more constricted as trials proceed; therefore, initial trials should include as much diversity in doses as practical.

Traditionally, initial trials of efficacy were "dose ranging," in which investigators (or patients) varied the dose, within limits, according to response. This may be appropriate for drugs with an immediate unambiguous effect, such as titration of intravenous vasodilators to regulate arterial blood pressure. Most drugs, how-ever, have delayed or latent effects and dose ranging often results in determination of the maximum tolerated rather than the minimal dose for clinically acceptable efficacy.

If interim analyses may lead to early termination of the study and rejection of the null hypothesis because the test drug is deemed superior to the control(s), the α level (e.g., $P < 0.05$) used to assess the probability of a Type I error must be adjusted. This might occur as a result of either a larger than expected difference in the average effects or a smaller than predicted variability among patients. Stopping dose groups for either inefficacy or safety should not require adjustment of the α level used in subsequent hypothesis testing. It is best to resolve these questions of monitoring and interim analyses before the study is started.

17.8. Clinical Benefit

In addition to a focus on the pharmacology of the drug, there should be an assessment of true clinical benefits. This may include measurement of quality of life or economic effects of alternative therapies, and it may require long-term assessment of the outcome of treatment. Some historical or contemporaneous control groups with the illness might be valuable additions to the traditional biologically focused trial.

17.9. Special Populations

There are special concerns in the use of new products in women who may become pregnant or are pregnant or lactating, in the young, in the old, in the dying, in patients who are not certain to have the target illness, in patients with concomitant disorders, and in patients taking many other unofficial, nonprescription or pre-scription drugs and diets. To the extent that these groups will be represented in the user population they should be included, in appropriate numbers, within the trial population. Appropriate nonclinical studies can help estimate the risk in these special groups.

Clinical protocols should be designed to permit evaluation of variability among investigators and patient populations. This can be accomplished by using multiple investigators and diverse patient groups. Multinational protocols are especially powerful. Analyses of safety and efficacy should be made for subpopulations of patients as well as for the entire group. Patients who themselves are unusual and patients who experience unexpected effects of therapy should be studied intensively to detect abnormal patterns of drug absorption, distribution, metabo-lism, or excretion or drug–drug, drug–disease, or drug–diet interactions.

17.10. Controls

Concurrent placebo controls are stringent but may not be acceptable in serious illnesses. How long is it ethical to treat patients with duodenal ulcer, hypertension, or major depression with placebo therapy? There are different standards in North America, Europe, and Japan. This may complicate a global approach to trials; placebo may be required in the United States and forbidden in Europe.

The use of active controls usually requires the study of more patients than in trials with placebo controls to have sufficient power to avoid Type II errors. The active comparator must have *established* efficacy, and in many cases this is uncertain. Often a series of active drugs has been compared one to another with no placebo control. This represents a slippery slope that may lead to acceptance of inactive products that are only a little less active than their predecessor in the chain of comparators.

Active comparators must be reformulated to match the clinical trial materials. This requires purchase, donation, or synthesis (in concert with patent restrictions) of the materials, and testing to ensure that the comparator reformulation matches the marketed (and previously approved) product in kinetics and effects. If the same active comparator is not marketed (or acceptable for use) in all countries, global studies may be complicated by the use of different comparators in the "same" protocol. Also, the standards for care may vary in regard to dose, duration, diagnostic techniques, and nosology of disease. Each such variable complicates the interpretation of results and increases the duration and cost of trials.

International variation in standards is due in part to legitimate cultural differences in the population, illnesses, and health care. Other variation should be minimized by discussions with regulators before initiation of global pivotal trials.

17.11. Blinding

The patient, the investigator, and the sponsor should be blinded to the assignment of individual patients to treatment groups. The investigator and the sponsor must be able to unblind any patient immediately if such information is important to that patient's medical care. Such events do not arise often. All episodes of unblinding should be documented fully.

Most interim analyses may be performed blindly. If all treatment groups are equivalent, one gains no information by identifying them. Even if it becomes essential to identify the treatment groups, the individual patient assignments can be protected (unless there is an apparent and common effect that is linked by unblinding to some but not all of the treatment groups).

Although it is not a regulatory requirement that the sponsor remain blinded, it is a good practice as long as safety monitoring and interim efficacy analyses can still be performed efficiently. If the sponsor is not blinded, monitoring of the study can proceed on a continuous basis; blinding the sponsor should not impede the use of all data to make decisions to stop doses or other patient groups for whom the protocols are unsafe or ineffective.

17.12. Patient Diligence

An important variable is the patient's adherence to the treatment regimen. Patients who do not take the trial materials, when lumped in the total group, decrease the power of the comparisons. To avoid this, patients may be excluded if they are not appropriately diligent in following directions during a preliminary qualification period of placebo therapy. Patient diligence may also be assessed during the trial by measurement of drug, metabolites, or markers in the patient, by counting returned doses, by patient or observer diaries, and by automated devices that record the withdrawal of products from containers.

17.13. Safety Monitoring

All safety data should be gathered quickly and provided to a *single* group monitoring the safety of the trials worldwide. Delays between the patient visit and analysis of validated data should be minimized. Investigators must report *all* serious events immediately, and the sponsor should assist the investigator in making an immediate complete medical evaluation. The imputation of causality is very difficult for individual events, and rapid, thorough patient evaluation is critical.

The speed of data transmission is critical to the pursuit of multiinvestigator and multinational studies with a new drug, a new dosage regimen, or new indications. The investigative site should enter data during or immediately after the patient visit and send all data promptly to the sponsor. This can be accomplished best using personal computers at the site with intermittent, prompt telecommunication to the sponsor.

Remote data management permits the use of sophisticated editing during data entry and facilitates communication between investigators, sponsor, and regulators during the trial. The sponsor should analyze newly received safety data daily with frequent scheduled analyses of differences between treatment groups.

17.14. Adverse Events

Adverse events should be recorded in a single standard lexicon (COSTART or WHO) to facilitate the merging of data across studies and countries. It is best to train investigators in the selection of lexicon terms so that they may classify events directly in addition to providing their own free-text medical description of events.

17.15. Laboratory Tests

Laboratory tests may be performed by central laboratories located on each continent that report electronically to the sponsor. Results from all laboratories should be reported in comparable (SI) units. Reference ranges from many laboratories seem *not* to reflect the distribution of analyte values among patients in many clinical trials. Many reference ranges were derived from a small number of young, healthy male employees, with "aberrant" values censored from the data.

The meaning of "three times the upper limit of *normal*" is obscure under the best of cases, but if the limits are from another population such calculations become really obfuscatory. It is more powerful to develop customized reference ranges for the population being studied. Lilly has evolved such ranges, for groups defined by age, gender, and racial origin, derived from observations of more than 20,000 patients taking placebo. About half of these patients had two observations during placebo therapy, thereby establishing a population distribution of intrasubject variation. These ranges are available on request.

17.16. Rare Serious Events

A problem exists with serious events that occur rarely; in small trials they are difficult to detect. If an event will occur once per thousand patients treated with a drug, there is only a two-thirds chance that one such event will be observed in the first thousand patients treated in a similar fashion. If the first serious event, such as neutropenia or hepatitis, occurs in a patient given the test drug, should trials be stopped? Suppose that a related event occurs in a second patient?

Traditional hypothesis testing may not give much comfort in making these decisions. To reach an α level of $p < 0.05$ when no events have been observed in patients allocated to a comparator, there must usually be 5 or 8 or 12 events in the patients assigned to the test drug (depending on the tests chosen). Two lessons are apparent. First, studies should be controlled so that appropriate comparison groups are available. Second, good clinical judgment based on the best bedside patient evaluation, rather than on the comfort of Sir Ronald A. Fisher's p, will be the basis of these tough decisions on rare serious events.

17.17. Stopping Rules

When serious safety concerns are expected, such as frequent deaths in the patient population, stopping rules should be agreed on in advance. In the treatment of cancer, myocardial infarction, or AIDS, one should establish stopping rules for the largest tolerated difference in death rates among the treatment groups. During the trial, each event can be characterized and classified blindly to determine if it meets the criteria for its inclusion in the stopping rule. Then one unblinded member of the sponsor's team can total up these events to track their distribution among treatment groups.

Stopping rules may be two-sided but are asymmetric. An excess of deaths on the new treatment should lead to termination of the dose group or protocol. A significant reduction in death rate on the new treatment may raise ethical concerns about continuing the trial. The α levels for these two decisions may be quite different. If there is a possibility of stopping early because of extraordinary life-saving efficacy, this should be a conservative decision agreed on with regulators before the study is started.

17.18. Data Validity

Investigators should be chosen critically, and they and their study coordinators trained thoroughly and visited frequently. Electronic and human surveillance should be used to audit all data.

Inattention to detail is more common and serious than duplicity. Each significant member of an investigative site should be made to understand the high level of excellence demanded. The pressure of rapid enrollment of patients and completion of trials must be balanced with the overarching emphasis on validity: *Do the right thing, and do it right.* Continued participation in the trial and payments should be made dependent on compulsive collection, recording, and rapid transmission of valid data. Remedial courses should be considered for sites that make errors early in trials or when there is a change in study coordinators.

Each datum should be compared with previous values for that patient and with norms. Precision should be demanded and errors not tolerated. Original records should be audited on site visits. An auditing group should be employed that reports at a corporate level rather than to medical or regulatory staff.

17.19. Regulators

Regulators should strive to be fair and consistent, maintaining an even playing field for all sponsors. Differences among products, patients, and diseases make inflexible guidelines unrealistic, but regulators should achieve and advertise fairness. Regulators should utilize extramural consultants to ensure that they remain consistent with current scientific methods and opinions. Regulators should challenge sponsors to pursue future new knowledge without shackling them to an outdated past.

One problem arises when regulators change rules to make criteria more stringent. In such cases new treatments may be subjected to more rigorous, prolonged, and expensive trials than their predecessors. This becomes especially complex when the predecessors are used as active comparators for the new treatments.

Sponsors should seek the advice and consent of regulators to the design of trials at an early stage; regulators should honor such agreements despite changes in personnel. New scientific information, however, may require the alteration of agreements made many years before; trial celerity facilitates constancy of agreements.

Regulators should balance the public benefit of having new therapies with the risk and benefits to individual participants in trials. They should communicate their concerns candidly and quickly to sponsors and they should respond quickly to submission and questions.

Sponsors should treat regulators as scientific colleagues and describe *quickly* and *completely* the problems and safety concerns in each clinical trial. Frequent telecommunication and personal visits will do much to foster collegiality instead of an adversarial relationship.

17.20. Applications for Marketing

Frequent consultation with regulators before, during, and after the trial can evolve presentations of group and individual data that are more complete and powerful.

Data compression and display techniques are the responsibility of the sponsor, with the active assistance of regulators. Innovative data displays are useful for monitoring the study (by the sponsor, investigators, and regulators), for review after the study, and for transmission of the data to physicians, pharmacists, and nurses after marketing. There is no excuse for submission of paper copies of electronic data bases.

Regulators, insurers, sponsors, physicians, pharmacists, nurses, legislators, patients, and the public share an interest in the rapid approval of safe and effective new therapies. Regulators can help by encouraging frequent meetings with sponsors and consultants to discuss data presentations and by raising questions frequently, as they arise during the review process. Sponsors can help by performing clinical trials of the highest quality and presenting data in an effective fashion.

Health professionals assist by donating their expertise to ethical committees and advisory panels and by assisting in the recruitment of patients. It is important that health professionals educate legislators and the public in risk-benefit analyses, especially for biotechnology products.

Legislators can help by addressing public health risk–benefit assessments and providing oversight of regulators focused on decisions, not outcomes. The best decision may lead to a terrible outcome, but until regulators and sponsors become clairvoyant they should be judged on the quality of their decisions. The public through legislators can help by providing regulators with resources to effect promptly scholarly reviews of applications, which each year increase in number, size, and complexity.

17.21. Launch

Marketing *begins* a new, complex, and important clinical trial. The sponsor's research component should remain responsible for continuing to accumulate and evaluate intramural and extramural data, promptly responding to unusual events encountered in marketing, and quickly providing regulators and health professionals with relevant new information. New chemical, animal, and clinical studies must be performed quickly in response to signals hidden in the noise of marketed use.

Marketing and sales personnel have a great educational challenge. They must translate gigabyte data bases and megaword documents into a message that can be delivered in a few pages or a few graphics or a few minutes. Medical, pharmacy, and nursing schools have faculty that take years to transfer this amount of information. The sponsor's research organization must be responsible for the correctness and fair balance of all marketing presentations. The research scientists should publish and present their studies and answer questions from health professionals, consumers, and regulators.

Index*

Acetaminophen, 59
Acrylic resins, 126
Actimmune, 389
Activase [see Tissue plasminogen activator (TPA)]
Active targeting, 129
Acute lymphoblastic leukemia (ALL), 177
Acute myocardial infarction (AMI), 423, 425, 426 (see also Myocardial infarction)
Acute respiratory disease (ARD), 276
Acyclic nucleoside analogues, 254–55
Acyclovir, 254
Adeno-Hepatitis B vaccines, 282–85, 286
Adeno-HIV vaccines, 285
Adenoviruses
    background information, 275–76
    genomic organization, 277–78
    vaccines, 275, 276, 278–86
Ad5-herpes simplex virus (HSV), 285
Administration, biotechnology products
    GM-CSF and, **411**
    implantable delivery systems, 135, 137
    oral delivery, 124–27
    pharmacists role in biotechnology, 403, 404
    protein conjugate HIB vaccines, 417
    recombinant hepatitis-B vaccines, 415
    self-adhesive buccal delivery systems, **142**
    site-specific delivery of peptides, 128
Adrenocorticotropin (ACTH), 17
Affinity chromatography, 75–76, 102
Agrobacterium spp., 291–95, 304–306
AIDS (acquired immunodeficiency syndrome)

[see also Human immunodeficiency virus (HIV)]
    antiviral agents combined with lymphokines and monokines, 66
    as challenge to pharmaceutical sciences and biotechnology industry, 193
    colony-stimulating factors and, 410
    Epogen and AZT treatment, 396
    future prospects for development of therapies, 265–66
    GM-CSF and, 60, 390–91
    IFN-α and, 57
    interferon-β and, 389
    peptide T and, 414
    potential anti-AIDS agents, 253–64
    systemic or chronic release of TNF-α, 64
    targets for drug design, 252–53
    TNF monoclonal antibodies, 408
    vaccine development, 392
Ajmalicine, **297**
Alginate-polylysine-alginate system, 135, **136**
Alleles, immunoglobulin classes, 41
All possible sequences (ASOs), 172
Alpha satellite, 205
Alzheimers disease, 25–28
Ambulatory clinics, 404
Amgen, 390–91, 392
Amino acids
    chromatography and primary sequence of proteins, 76–77
    composition analysis, 103
    immunoglobulin molecules sequences, 40

*Page numbers printed in boldface type refer to tables or figures.

oxidation and loss of biological activity, 122
racemization, 123
structural information and molecular genetic techniques, 13
Amplification, genetic sequence, 175, **176**
Amyloid precursor protein(s) (APP), 25–28
Anagyrine, 300
Analytical chemistry, methodologies of, 71–72
Anaphylactic shock, 243
Anemia, 60, 395–96, 418
Angiotensin, 128
Animal models
   adenovirus-vectored vaccines, 281–85
   approval process for biotechnology products, 382
   bioassay-directed fractionation, 315
   impact of molecular genetic approaches on biological research, 29
   oncogenes and transgenic, 326–27
Anion exchange, 229
Anionic compounds, 258–62
Anisoylated plasminogen streptokinase activator complex (APSAC), 394
Anorexia, 58, 64
Antibiotics, **292**
Antibodies (*see also* Monoclonal antibodies)
   biotechnology products on market, 386–88
   defined, 39
   functions, 42–43
Antibody-dependent, cell-mediated cytotoxicity (ADCC), 43
Antigen-presenting cells (APC), 42
Antigens, 39, 367–68
Antimyosin MAb, 235
Antisense oligodeoxynucleotides, 422
Antisense oligonucleotides, 256–57
Aplastic anemia, 60
Apolipoproteins, 116–17
Ara-A (carbocyclic arabinosyladenine), 254–55
Arabidopsis thaliana, 305
Arginine vasopressin (AVP), 16
Arteriosclerosis, 117
Ascites, 228
Asparaginase, 128
α-terthienyl, **302**
Atmospheric pressure ionization (API), 109
Atrial natriuretic factor (ANF), 413
Atrial natriuretic peptide (ANP), 116
Atropa belladonna, 299–300
Atropine, **299**
Autoimmune diseases, 59, 242

Autosomal recessive diseases, 178–82
Azo sulfonic acid dyes, 259
AZT (3'-azido-2',3'-dideoxythymidine), 253–54, 396

Bacillus thuringiensis, 305–306
Bacteremia, 241–42
Bacterial meningitis, 415
Bacteriology, 6–7
Base pair mismatches, 157–58, 171–73
Becker muscular dystrophy (BMD), 173
Bioassay-directed fractionation, 314–15
Bioassays, 72, 347–50
Biological response modifiers (BRMs), 381–82
Bioluminescence, 89
Biotechnology products (*see also* Drugs, biotechnological)
   analytical demands of, 99–100
   as analytical tools, 98–99
   defined, 97
   future analytical needs, 107–10
   impact on role of pharmacist, 402
   methods characterizing, 101–107
   review of on market, 383–96
BI-RG–587, 263
Bleeding disorders, 394–95
Bone marrow transplantation, 185, 241, 388, 390, 411
Brassica napus, 305
Breast cancer, 175
Buccal administration, 141–42
Busch Biotech, 389
Butenenylbithiophene, **302**

Cachectin [*see* Tumor necrosis factor (TNF)]
Cachexia, 407
Caenorhabditis elegans, 201
Calcitonin gene-related peptide (CGRP), 17
Cancer (*see also* Oncogenes; specific types; Tumors)
   IFN-γ and, 57
   IL–2 and LAK cells, 59
   interferon-β and, 389
   monoclonal antibodies and scintigraphic detection of, 237–38
   monoclonal antibodies and therapy, 242–43, 419–20
   multiple drug resistance phenotype, 330
   research on molecular genetics of, 44–45
   systematic or chronic release of TNF-α, 64
   TNF-α and treatment of, 61–62, 64–65

CA 125, 419
Capillary electrophoresis, 82–84, 108–109
Capillary gel electrophoresis (CGE), 108–109
Capillary zone electrophoresis (CZE), 83
Capsaicin, 291
Carbocyclic nucleoside analogues, 254–55
Cardiovascular disease, 235–37, 242 (*see also*
    Congestive heart failure; Myocardial in-
    farction)
Catharanthine, **297**
Catharanthus roseus, 291, 296, 298
CD4 derivatives, 258
cDNA, cloning procedures, 10–12
Cell fusion, 47
Cell-mediated immunity (CMI), 39
Cell survival assays, 331–32
Centocor, 388
Centoxin, 388
Central Dogma of Biology, 3, 5
Cetus Corporation, 389
Chain terminators, **15**
Chemical ionization mass spectrometry
    (CIMS), 77
Chemiluminescence immunoassay, 88–89
Chemotherapy, 390, 391, 410, 411
Chiroptical methods, 106
Chloramphenicol actyltransferase (CAT), 327–
    30
Chromatography, 72–80, 229–30
Chromosomes
    jumping, 199–200
    location of DNA markers, 161–63
    organization at molecular level, **208**
    physical maps and structure and dynamics
        of, 217–18
    walking, 198, 199
Chromosome-specific library, 199
Chronic myelogenous leukemia (CML), 177
Cinchona ledgeriana, 295, 296
Circular dichroism (CD), 89–91, 103
Classes, immunoglobulin, 40–41
Classes, socioeconomic, 405
Classical complement cascade, 43
Clinical trials, 382, 435–45
Clonal selection theory, 44
Cloning
    "functional," 220
    introduction to recombinant DNA technol-
        ogy, 7–13
    monoclonal antibodies and background con-
        cepts, 48–49

"positional," 220
    yeast artificial chromosomes, 200–201
Cluster determinant (CD) system, 50
Codeine, 101
Colchicine, 331
Colony-stimulating factors (CSFs), 390, 410–
    12
Columbamine, **338**
Complement, humoral immune system, 43
Compositae, 303
Condylomata acuminata (*see* Genital warts)
Congestive heart failure, 413, 423
Conjugation, monoclonal antibodies, 230
Controlled precipitation, 230
Controls, experimental, 439–40
Coronary thrombosis, 423
Cosmid contig maps, 213–14
Cosmid libraries, 198–99, 207, 208
Cost, biotechnology drugs, 405, 411–12, 415,
    422, 423
Creutzfeldt–Jakob disease, 408
Crohns disease, 406
Cyclaridine (carbocyclic arabinosyladenine),
    254–55
Cycloheximide, 322
Cystic fibrosis (CF), 172, 177
Cytisine, **301**
Cytochrome P–450, 344–47
Cytogenetic map, 196
Cytokines
    defined, 39
    recombinant TNF-like, 408
    role in immune system, **54**
Cytotoxicity assays, 331–32

Data bases, public genetic, 163
Datura candida, 299
Datura stramonium, 299
DDI (2′,3′-dideoxyinosine), 253
Deep-vein thrombosis (DVT), 235–37, 423
Deletions, genetic, **169**, 173–76
Delivery (*see* Administration)
Detection, hybridization and, 318–22 (*see also*
        Direct detection; Indirect detection)
Dextran sulfate, 259–61
Diabetes mellitus, 59 (*see also* Insulin)
Diagnosis
    disease-related and recombinant DNA tech-
        nology, 170–85
    DNA sequence and physical maps, 219–20
    future prospects, 185–86

genetic diseases and recombinant DNA techniques, 156–70
monoclonal antibodies and, 420
pharmaceutical analyses using immunoassays, 100–101
Diagnostic imaging, 233–38, 420
Difluoromethylornithine (DFMO), 320
Digibind, 386–87
Digitalis toxicity, 387
Digoxin, 387
Direct detection, 155–56, 171–77
Disease (*see also* specific disease)
    disease-related diagnosis and recombinant DNA technology, 170–85
    disorders diagnosed by DNA analysis, **157**
    genetic mutation and diagnosis at DNA level, 155–56
    physical maps and DNA sequence, 218–19
    research on molecular biology of, 25–28
Disulfide exchange, 123
DNA (*see also* cDNA)
    Central Dogma of Biology, 3, 5
    chromosome location of markers, 161–63
    as contaminant of monoclonal antibody products, 231
    eukaryotic gene organization, **6**
    introduction to principles of cloning, 7–13
    mutations and disease in humans, 155
    sequence and physical maps, 217–19
    sequence variation in humans, 159
DNA hybridization assay, 99
Domains, globular, 40
Dose-response relationship, 438–39
Dot blot assay, 99
Double-focusing mass spectrometry (MS-MS), 92
Down's syndrome, 26
Doxorubicin, 243
Drug abuse testing, 101
Drugs, biotechnological (*see also* Biotechnology products)
    accumulation, 332
    AIDS and potential agents, 253–64
    AIDS and targets for design, 252–53
    approval dates of specific, **384**
    approval process for, 381–83
    bioassay systems and research concerns, 347–50
    development status of specific, **385**
    modulation of gene expression, 316–23
    monoclonal antibody conjugates, 243

monoclonal immunoconjugates, 241
natural-product discovery programs, 313–16
oncogenes as targets in therapy, 323–27
parenteral targeting, 127–30
resistance, 330–34
Duchenne muscular dystrophy (DMD), 173–75, 180, 182
Duplex, formation and stability, 157–58
Dystrophin, 173

Ecology, global, 349–50
Economics (*see* Cost; Industry; Marketing; Sales)
Edman degradation, 77, **78**
Education
    health care and biotechnology products, 403
    marketing and sales of biotechnology products, 445
    pharmaceutics curricula and biotechnological drug development, 405
E5, 421
18-*O*-β-glycyrrhetinic acid, **342**
Electrospray sample introduction-ionization (ESI), 109
Elicitation, plant tissue culture, 291, **292**
Eli Lilly Corp., 386
Emulsions (lipid microspheres), 133–34
Engerix-B, 391–92, 414–15
Enzyme immunoassays (EIA), 98
Enzyme-linked immunosorbent assay (ELISA), 48
Enzyme-mediated immunoassay technique (EMIT), 87, 101
Epidermal growth factor (EGF), 343–44
Epogen, 395–96
Erythroleukemia, 329
Erythropoietin (EPO), 106, 395–96, 418–19
Escherichia coli, 201
Ethics, medical, 405, 436
Ethylenediaminetetraacetic acid (EDTA), 126
Expression cassettes, 280–81

Fabaceae, 300
Fab fragments, 41, 386–87
Factor VIII, 394–95
Fagaronine, **338**
Fermentation, 97, 228–29
Fibrin, **393**
Fingerprinting methods, 203, 209–11
Fluorescence immunoassay (FIA), 85, 87–88, 332

Fluorescence polarization immunoassays (FPIA), 98
Folding, protein mechanisms, 369
Food and Drug Administration (*see also* Regulation)
  approval process for biotechnology drugs, 382
  development of influenza vaccines, 392
  guidelines for manufacture of MAb products, 231
  reimbursement for non-approved indications, 404
Formamide, 158
Formulation, monoclonal antibody processing, 230–31
Forward genetics, 194–95
Free radical analysis technique (FRAT), 101
Free-solution capillary electrophoresis (FSCE), 83, 108

γ-Globulins, 39
Gamma interferon (IFN-γ), 53–57 (*see also* Interferons)
Gas·chromatography-mass spectrometry (GC–MS), 91
Gastrin, 343
Gel electrophoresis, 80–82
Gel filtration chromatography, 229–30
Gel permeation chromatography (GPC), 74–75
Gene expression, 316–23, 333, 422
Genentech Corp., 386, 389, 394
Gene regulation, **20**
Gene therapy, retroviral gene delivery systems, 134
Genetic engineering, 47
Genetic linkage map, 196
Genetics Institute, 394, 396
Genital herpes, 389
Genital warts, 57, 389, 427–28
Genome Data Base, 220
Genomic library, 10
Ginsengosides, 303
Genta, Inc., 422
GISSI II study, 393
Glucosidase inhibitors, 257
Glycoproteins, 79–80, 370
Glycosides, 303
Glycosylation, 370
Government (*see* Food and Drug Administration; Regulation)
Graft-versus-host disease, 388

Gram-negative bacteremia (GNB), 388
Gram-negative sepsis, 421
Granulocyte colony-stimulating factor (G-CSF), 390, 410–11
Granulocyte macrophage-colony-stimulating factor (GM-CSF), 59–61, 390–91, 410–11
Granulomatous disease, 389
Growth factors, 341, 343–44
Growth hormone, 305, 385–86, 408–10

Hairy cell leukemia, 57, 389, 427
Haloperidol, 24
HA–1A, 388, 421
Hapten-carrier coupling, 86
Hard-sphere *exo*-anormeric (HSEA) effect calculations, 106–107
Health care system
  biotechnology and role of pharmacists, 396–97, 402–405
  education and biotechnology products, 403
  ethics and biotechnology drugs, 405
  impact of biotechnology, 381
Heavy chain, 40
Hemophilia, 394–95
Hemophilus influenza type B vaccine, 392, 415–17
Heparin, 393–94, 425
Hepatitis B vaccines
  clinical applications of, 414–15
  evaluation of adeno-hepatitis B, 282–85, 286
  recombinant on market, 391–92
HEPT (1-[(2-hydroxyethoxy)methyl]-6-(phenylthio)thymine), 262–63
Heptavax-B, 391–92, 414
Herpes simplex virus (HSV), 57
Heterogeneous immunoassays, 84–85
Heterokaryons, 44–45
HibTiter, 392, 415–17
High-performance liquid chromatography (HPLC)
  description of method, 72–80
  purification of oligonucleotides, 104
  purification of proteins, 102
  therapeutic drug monitoring (TDM), 100
Hodgkin's disease, 296
Hoechst-Roussel, Inc., 390
Homogeneous immunoassays, 85
Hormone responsive element (HRE), 328–29
Human antimouse antibodies (HAMA), 243–44

Human-*c-erb*B2/*neu* promoter, 327–28
Human chorionic gonadotropin (hCG), 98, 419
Human Gene Mapping Workshops, 156
Human growth hormone (hGH), 305
Human immunodeficiency virus (HIV) (*see also* AIDS)
    adeno-HIV vaccines, 285
    antiviral assays, 335–37
    colony-stimulating factors and replication of, 410
    retroviral reverse transcriptase activity, 337–38
    tat gene expression, 329–30
    vaccine development, 417–18
    viral genes and life cycle, 251–52
Human papillomavirus (HPV), 427–28
Humans (*see also* Clinical trials)
    antigenetic diversity among adenoviruses, 275–76
    blood transfusion contamination, 394
    construction of physical map of chromosome, 16, 205–15
    general features of genome organization, 205–207
    genomes and physical mapping problem, 201–203
    synthetic insulin, 412
Humatrope, 386, 408
Humoral immune system, 39, 41–42
Humulin, 385
Huntingtons disease (HD), 182–83, **184**
Hybridization
    chromosome location of DNA markers, 161
    drug discovery programs, 317–20
    nucleic acids and diagnosis of genetic diseases, 156–59
Hybridoma, 45, **46**
Hydrophobic interaction chromatography (HIC), 74
Hyoscyamine, 299–300
Hyoscyamus muticus, 300
Hyperimmunization, 41
Hypertension, 413, 418
Hypoglycemia, 409, 413
Hypotension, 58, 407
Hypoxanthineguanine phosphoribosyltransferase (HGPRT), 45

Imaging (*see* Diagnostic imaging)
Immunex, 390

Immunization, monoclonal antibodies, 47 (*see also* Vaccines)
Immunoanalytical chemistry, 84–89
Immunoassays, 98, 100–101 (*see also* Specific methods)
Immunochromatography, 98
Immunoconjugates, 48
Immunogenicity, monoclonal antibodies, 243–44
Immunogens, humoral immune response, 41
Immunoglobulins
    antibody functions, 43
    cell fusion, 47
    characterization and storage, 49
    defined, 39
    selection and screening, 48
    structure, 39–41
Immunoreaction, 86–89
Immunoscintigraphy (*see* Diagnostic imaging)
Implantable delivery systems, 135, 137
Indirect detection, 156, 177–85
Indium antimyosin, 235
Indole alkaloids, 296–99
Industry, biotechnology, 97, 117, 381 (*see also* Cost; Marketing; Sales)
Infections
    adenoviruses and, 276
    monoclonal antibody imaging of sites, 237
    TNF-α and production of symptoms, 64
    therapeutic applications of monoclonal antibodies, 241–42
Inflammation, TNF-α and, 64
Influenza (*see* Hemophilus influenza type B vaccine)
Infrared laser desorption (IRLD), 110
Inhibition, oncogenes, 325–26
Initial trials, 436–37
In situ hybridization histochemistry
    description of method, 19, **21**, 23
    examples of data obtained from, **22**
    influence of development on neurobiology, 28–29
Insulin
    biotechnology products on market, 383, 385
    clinical applications, 412–13
    structure of proinsulin, **386**
Interferons (*see also* Gamma interferon)
    approval process for biotechnology drugs, 383
    biotechnology products on market, 388–89
    clinical applications, 426–28

TNF and synergy with, 408
Interferon Sciences Corp., 389
Interleukins, 57–61, 422
Intron A, 389, 426–27
Investigational new drug (IND) application, 382
Ion-exchange chromatography (IE), 75
Iontophoresis, 138
Isoelectric focusing (IEF), 81–82, 108
Isotopes, monoclonal antibodies, 233–34

Jaspamide, 373
Johnson & Johnson Corp., 392
Jumping libraries, 200

Kaposi's sarcoma, 57, 389
Keratoconjunctivitis, 276
Kidney
    dialysis and erythropoietins, 395
    renal cell carcinoma, 391, 422
    renal failure, 386, 418
    renal transplant rejection, 387–88

Laboratory tests, 442
Length variation, 158, 166–67
Leu-enkephalin, 305
Leukemia
    acute lymphoblastic leukemia (ALL), 177
    chronic myelogenous leukemia (CML), 177
    colony-stimulating factors and, 410
    erythroleukemia, 329
    growth hormone therapy, 409
    hairy cell leukemia, 57, 389, 427
    indole alkaloids and, 296
Leukine, 390
Leukocytes, 60
Leukopheresis, 59
Light chain, 40
Linkage analysis, 161–62
Linking libraries, 200
Lipid microspheres, 133–34
Lipoatrophy, 413
Lipohypertrophy, 413
Lipopolysaccharide (LPS), 61–62
Liposomes, 131–32
Liquid chromatography-mass spectrometry (LC–MS), 91–92
Liquid secondary ion mass spectrometry (LSIMS), 92
Lithospermum erythrorhizon, 303–304
Localization, monoclonal antibodies, 231–33

Low-angle laser light scattering (LALLS) detection, 74–75
Luciferase, 327–30
Lung cancer, 390
Lupus erythematosus, 59
Luteinizing hormone (LH), 419
Lymphatic uptake, 127
Lymphokine-activated killer (LAK) cells, 59
Lymphokines [*see also* Interleukins; Tumor necrosis factor (TNF)]
    biotechnology products on market, 388–91
    development of humoral immune response, 42, **54**
    therapeutic applications of, 65–66
Lyophilization, 231
Lytic state, 392–93

Macrophages, endogenous production, 62–63
Macrorestriction maps, 197
Major histocompatibility complex (MHC), 42
Malaria, 295, 330
Mapping, reverse genetics and, 195–201 (*see also* Physical maps)
Marketing, biotechnology products, 444–45
    (*see also* Sales)
Marogen, 396
Mass spectrometry, 91–92, 109–10
Maximum tolerable dose (MTD), 382
Media, specialized, 291, 298
Melanoma, 391, 427
Memory cells, 42
Meningitis, bacterial, 415
Meperidine, 407
Merck Corp., 391–92
Metastatic skin melanoma (MSM), 427
Methionine enkephalin (ENK), 16–17, 24, 66
Methotrexate, 243
Micellar electrokinetic capillary chromatography (MECC), 83
Microspheres, 132–33
Mimetics, design of peptide, 371–74
Monitoring, clinical, 404, 424–25, 441
Monoclonal antibodies
    biotechnology products on market, 386–88
    clinical considerations, 419–21
    commercial manufacture, 228–31
    diagnostic imaging, 233–38
    in vitro diagnostic kits, 382
    localization, 231–33
    therapeutic applications, 238–44
    TNF and, 408

Monokines, **54**, 65–66 [*see also* Interleukins; Tumor necrosis factor (TNF)]
Morphine, 101
mRNA
  cDNA cloning procedures, 10–11
  in situ hybridization histochemistry, 19, 23
Multi-colony-stimulating factors (CSF), 59–61
Multidimensional maps, **218**
Multiple sclerosis, 57
Muromonab-CD3, 387–88, 420–21
Muscular dystrophy, 180 [*see also* Duchenne muscular dystrophy (DMD)]
Myelosuppression, 390
Myocardial infarction
  MAb and diagnostic imaging, 235, **236**
  superoxide dismutase (SOD), 406
  thrombolytic agents and therapy, 423
  tissue plasminogen activator (TPA), 392, 393, 425, 426
Myoscint, 235

Nanoparticles, 133
Naphthalenedisulfonic acid, 259
Naphthalene dyes, 259
Nasal administration, 139–40
N-Carbomethoxycarbonyl-propyl-phenylalanyl benzyl esters (CPFs), 263
Neupogen, 391
Neurobiology, molecular genetic techniques, 16–17, 28–29
Neurons, **4**
Neurotransmission, 25
New drug application (NDA), 382
Nicotiana spp., 300–302
Nicotine, 300–302
Norditropin, 386
Nornicotine, **301**
Northern Blot hybridization, 17–18
Nuclear magnetic resonance (NMR) spectroscopy, 103–104
Nuclear runoff assays, 322–23
Nucleic acids, 156–59
Nucleosides, 253–54
Nucleotides, 14–16, 165–66

Octapeptide (Ala-Ser-Thr-Thr-Thr-Asn-Tyr-Thr), 413–14
OKT3, 241, 387–88, 420–21
Oligodeoxynucleotides, 422
Oligonucleotides

antisense concept and anti-AIDS agents, 256–57
isolation and purification processes, 104–105
probes and detection of base pair changes, 172–73
sequence and structure of as biotechnology products, 105
Oligosaccharides, 106–107
Oncogenes, 323–27
Optical Rotatory Dispersion (ORD), 89–91, 103
Oral administration, 124–27
Organ transplantation, 59, 241, 387–88 (*see also* Bone marrow transplantation; Renal transplant rejection)
Ornithine decarboxylase (ODC), 320–22
Ortho Biotech Corp., 396
Osmotic pumps, 135, 137
Osteoarthritis, 58, 406
Ovulation tests, 382, 419, **421**
Oxathiin carboxanilide, 263
Oxygen toxicity, 406
Oxytocin, 16, 128

Pain, IL-1 and, 58
Panax ginseng, 303
Papain, 41
Papillomavirus-induced infections (Condylomata acuminata), 57
"Paralog" chromatography, 102
Parasitic diseases, 320
Parenteral delivery systems, 127–37
Particulate carrier systems, 128, 131, 134–35
Passive targeting, 129
PAVAS (vinyl alcohol), 261
Pedvax HIB, 415–17
Penetration enhancers, 126–27, 138, 139–40
Pepsin, 41
Peptidase inhibitors, 126
Peptides
  chromatography and mapping, 77–79
  clinical applications, 412–14
  delivery systems, 124–42
  isolation and purification processes, 101–102
  mimetic and hybrid mimetic approach, 371–74
  nasal mucosa and absorption, 139–40
  neurotransmitters and flow of information within cells, **4**

particulate carrier systems, 134–35
sequence and structure as biotechnology
products, 103–104
sequences and structural information impact
of molecular genetic techniques, 13–14
site-specific delivery, 128
stability of as genetically engineered drugs,
117–24
turns as recognition sites, 366–71
Peptide T, 413–14
Pfizer Corp., 388
P-glycoprotein, 330–31, 332–34
Pharmaceutical industry, 97, 117, 381 (*see
also* Cost; Marketing; Sales)
Pharmacists, biotechnology and role of, 396–
97, 402–405
Pharmacognosy, 405
Phenylheptatriyne, **302**
Philadelphia chromosome, 176–77
Phonophoresis, 138
Phorbol ester, 341
Phosphorylation, 369–70
Photofluorescence analysis, 89
Physical maps
defined, 197–98
DNA sequence and, 217–19
evolution of technology and considerations
for construction of, 215–17
general features of human genome organization, 205–206
human chromosome 16 and construction of,
206–15
integration with other kinds of maps, 217
paradigms for construction of large-scale,
203–205
scale of problem for human genomes, 201–
203
Placebos, 439–40
Plant tissue culture
description of nonclassical techniques, 290–
95
ecological concerns, 349–50
indole alkaloids, 296–99
miscellaneous alkaloids, 300–302
natural-product drug discovery programs,
313–16
nonalkaloidal substances, 302–304
quinoline alkaloids, 295–96
transformations involving foreign genes,
304–306
tropane alkaloids, 299–300

Plaque formation, 25–28
Plaque reduction assay, 334–35
Plasma cells, humoral immune response, 42
Plasmacytoma, 44
Plasma desorption mass spectrometry (PDMS),
92
Plasmids, 11–12
Platelet-derived growth factor (PDGF), 341,
343
PMEA [9-(2-phosphonylmethoxyethyl)ade-
nine], 254
PMEDAP [9-(2-phosphonylmethoxyethyl)-2,6-
diaminopurine], 254
Point mutations, 176
Poly(4-styrenesulfonic acid) (PSS), 262
Poly(vinylsulfonic acid) (PVS), 262
Polyacetylenes, 303
Polyacrylamide gel electrophoresis (PAGE),
104
Polyethylene glycol-conjugated superoxide dis-
mutase (PEG-SOD), 406
Polymerase chain reaction (PCR)
biotechnology products as analytical tools,
98–99
direct detection of sickle cell mutation, 171
DMD deletions, 174–75
indirect diagnosis of autosomal recessive dis-
eases, **181**
sequence variation and, 163–70
Polymer coatings, 126
Polymers, 261–62
Polynucleotides
isolation and purification processes, 104–
105
sequence and structure of as biotechnology
products, 105
Polysaccharides
isolation and purification processes, 106
sequence and structure of as biotechnology
products, 106–107
Preclinical studies, 435–36
Pregnancy
clinical trials and, 435, 439
tests for, 382, 419, **420**
Primary antibody response, 42
Probes, DNA, 162–63
Procrit, 396
Prodrugs, 138–39
Product license application (PLA), 382
Prohormones, 17
Proinsulin, 385, **386**

Prokine, 390, 410
Proopiomelanocortin (POMC), 17, 18–19
Protease inhibitors, 255–56
Proteinases, viral, 339–40
Protein C, 116
Protein kinase C, 264, 341, **342**
Proteins
  delivery systems, 124–42
  isolation and purification processes, 101–102
  nasal mucosa and absorption, 139–40
  oncogenes as targets in drug therapy, 324–25
  particulate carrier systems, 134–35
  peptide mimetic and hybrid mimetic approach, 371–74
  sequence and structure of as biotechnology products, 103–104
  site-specific delivery, 128
  stability of as genetically engineered drugs, 117–24
  turns as recognition sites, 366–71
Proteolysis, 123, 370–71
Protropin, 386, 408
Pulmonary embolism, 394, 425
Purification, monoclonal antibodies, 229–30
PVAS (vinyl alcohol), 261

Quality control, 231
Quercetin, **342**
Quinoline alkaloids, 295–96
Quinolizidine alkaloids, 300

Racemization, 123
Radiation hybrid map, 196–97
Radiation injury, 406
Radioimmunoassay, 86–87
Radioisotopes, 239–40, 243
Receptor binding assays, 340–44
Recombivax HB, 391, 414
Rectal administration, 140–41
Red blood cells, 134
Regulation, government, 381–83, 443–44 (*see also* Food and Drug Administration)
R82150, 262
Reimbursement, third-party, 404
Renal cell carcinoma, 391, 422
Renal failure, 386, 418
Renal transplant rejection, 387–88
Replication, viral, 340
Respiratory disease, 276

Restriction enzymes, **9**
Restriction fragment length polymorphism (RFLP), 26, 159, **162**
Restriction mapping, 203
Retroviral gene delivery systems, 134
Retroviruses, 251–52
Reversed-phase HPLC, 73–74, 104
Reverse genetics, 195–201
Reverse transcriptase (RT), 11, 262–63, 337–38
Rheumatoid arthritis, 57, 58, 242, 406
Ricin toxin immunoconjugates, 241
RNA (*see also* mRNA)
  Central Dogma of Biology, 3, 5
  eukaryotic gene organization, **6**
  immobilization of preparations on solid support, 318
  isolation methods and cell preparation for hybridization, 317–18
Roferon-A, 389, 426–27
Rosmarinic acid, 291

Saccharomyces cerevisiae, 201, 391
Safety, clinical trials, 437, 441
Saizen, 386
Sales, biotechnology products, 117, 381, 395, 445 (*see also* Marketing)
Salicylates, 126, 127
Salt concentration, 158
Salvage pathways, 45
Sandoz, 390
Sanguinarine, **301**
Sarcoma, 60
Sargramostim, 390
Schering-Plough Corp., 389, 390
Schizophrenia, 25
Scleroderma, 57
Scopalia japonica, 300
Scopolamine, 299, 300
Secondary antibody response, 42
Self-adhesive buccal delivery systems, **142**
Sepsis, 241–42
Septicemia, 64
Septic shock, 61, 388, 407
Sequence tagged sites (STS), 215, **216**
Sequence variation, 159–61, 163–70
Serpentine (indole alkaloids), **297**
Shikonin, **302**, 303–304
Shooty teratomas, 300–302
Sickle cell disease, **166**, **167**, 171–73

Single photon emission computed tomography (SPECT), 233
Site-directed mutagenesis, 123–24
Size exclusion chromatography (SEC), 74–75
Slow-wave sleep, 64
SmithKline Beecham, 391–92
Sodium dodecyl sulfate-polyacrylamide gel electrophoresis (SDS-PAGE), 81, 82
Solanaceae, 299
Soluble carrier systems, 128, 131
Soluble CD4 derivatives, 258
Somatic mutations, 175–77, 185
Somatostatin, 367
Somatrem, 386, 408–409
Somatropin, 386, 408, 409
Somnolence, 58
Southern blot analysis
  compared to Northern blot, **18**
  defined, 7
  deletion in dystrophin gene, 173
  detection of DNA fragments sharing homology with DNA probe, **160**
  diagnosis of autosomal recessive diseases, 178, **179**
  sickle cell anemia, 171–72
Spartium junceum, 300
Splenocytes, 47
Staurosporine, 341, **342**
Stevia rebaudiana, 303
Steviobioside, 303
Stevioside, **302**
Stopping rules, 442–43
Storage, biotechnology drugs, 403
Stratum corneum, 137–38
Streptavidin, 319
Streptokinase (SK), 393–94, 423–24
Strong anion-exchange chromatography (SAX), 80
Substance-P peptide (SP), 23–24
Suicidal rescue, 387
Sulfated polysaccharides, 259–61
Sulfonic acid, 258–62
Superoxide dismutase (SOD), 116, 405–406
Suramin, 258, 259, **338**
Surfactants, 126
Syncytia, 336

Tachykinins, 23–25
tat gene, 329–30
T-cells, 58–59
T-dependent antigens, 42

Telomeres, 206
Temperature, DNA reassociation reaction, 157
Testing, drug stability, 124
Tetrazolium dye-based assay, 331
Therapeutic drug monitoring (TDM), 100
Thin-layer chromatography (TLC), 76
Thiophenes, 303
Third-party reimbursement, 404
Thrombolytic agents, 392–94, 422–26
TIBO (tetrahydroimidazo[4,5,1-*jk*][1.4]-benzodiazepin-2(1*H*)-thione), 262
Time-of-flight (TOF) mass spectrometers, 109
Tissue plasminogen activator (TPA), 116, 392–94, 423
TNF-α (cachectin), 61–65
TNF-β (lymphotoxin), 65
t-PA [*see* Tissue plasminogen activator (TPA)]
Transcription, oncogene, 324
Transdermal administration, 137–38
Transformation
  oncogene-transformed cells, 326
  plant tissue culture, 291–95, 298–99, 304–306
Transforming growth factor (TGF), 343
Transgenic animals, 29, 326–27 (*see also* Animal models)
Translation, posttranslational modifications, 369–70
Translocations, genetic, 168, 170, 176–77
Transplantation (*see* Bone marrow transplantation; Organ transplantation; Renal transplant rejection)
Triplet coding system, 5
Trisodium citrate, 126–27
Tropane alkaloids, 299–300
Tuberculosis, 64
Tumor necrosis factor (TNF), 61–65, 396, 406–408
Tumors (*see also* Cancer; Oncogenes; specific types)
  indirect diagnosis of somatic changes, 185
  molecular methods of diagnosis, 175–77
  monoclonal antibody imaging, **237**
Turners syndrome, 386
Turns, peptide, 366–74
12-Methoxydodecanoic acid, 264
Two-dimensional chromatography, 76
Tyrosine phosphorylation, 369–70
Tyrosine-specific protein kinase (TPK), 325–26

Ultrafiltration, 230
Urokinase (UK), 423
UV absorption spectra, 89–91

Vaccines (*see also* Immunization)
  adenoviruses, 275, 278–86
  approval dates of specific types, **384**
  biotechnology products on market, 391–92
  clinical considerations, 414–18
Variable domains, 40
Vasopressin, 128
Venereal warts, 57 (*see also* Genital warts)
Veratridine, 19
Vesicular stomatis virus (VSV), 285–86
Vinblastine (VLB), 296, **297**, 298
Vincristine (VCR), 296, **297**, 298
Vindoline, **297**
Vinyl alcohol, 261
Viral hepatitis, 57, 193

Viruses (*see also* Adenoviruses; Retroviruses)
  infections and monoclonal antibodies, 242
  interferons and treatment of diseases, 57
  methods for discovery of antiviral compounds, 334–40
  retroviral gene delivery systems, 134
  turns and immunological recognition of, 368–69
VNTR polymorphism, **163**

Waste disposal, biotechnology products, 404

Xoma Corp., 388
Xomaxyme-CD5, 388
X-Ray crystallography, 106

Yeast artificial chromosomes (YACs), 200–201, 220

Zidovudine (*see* AZT)